Plant Ecological Anatomy

Marius-Nicusor Grigore

Plant Ecological Anatomy

Springer

Marius-Nicusor Grigore
Doctoral School of Biology, IOSUD-UAIC;
Editura Universității "Alexandru Ioan Cuza" Iasi;
Faculty of Psychology and Educational Sciences
"Alexandru Ioan Cuza" University of Iasi
Iasi, Romania

ISBN 978-3-032-01331-6 ISBN 978-3-032-01332-3 (eBook)
https://doi.org/10.1007/978-3-032-01332-3

This Springer imprint is published by the registered company Springer Nature Switzerland AG
The registered company address is: Gewerbestrasse 11, 6330 Cham, Switzerland

Introduction

Although the field of plant morphology and anatomy is considered a "classi with consistent works and results delivered since the second half of the nin century, it is surprising, not only at first sight, that there was no monograp dedicated exclusively to ecological morphology and plant anatomy. Of cours are scattered data related to the anatomy of many plant species and, imp belonging to many ecological groups, but, as far as we know, in fact such sy work was missing worldwide.

The classic treatises on plant ecology and plant geography (with aspects of iology—although the term physiology at the end of the nineteenth century de a slightly different reality than the same phrase today) (see iconic names of War Schimper, Clements, Coulter, Barnes, and others) that we have cited and used i work, excellent in vision and execution, naturally included notions of plant ad tion at the morphological and anatomical level. These treatises—unmatched now—had the "chance" to bear the imprint of the times, when the holistic visi an approach invariably included elements of plant morphology and anatomy, i strict sense of the word.

Approaching some aspects of the ecological anatomy of plants may seem only at first sight; in reality, discussing ecological meanings of anatomical st tures is very demanding and not without challenges. From the nineteenth cen until now, when the volume of knowledge has grown enormously, many asp once considered "classic," universally valid and accepted, have now come to called into question. In the field of plant anatomy, the passage of time does always clarify things, and new, modern, interdisciplinary studies regarding a cert species often refer to classical studies, from the nineteenth century, or from the tu of this century and the next.

Another challenge, which applies to an elementary reality, is related to the ve definition and approach of some key concepts in the field of plant ecological ana omy. The most convenient example is that of "xerophyte," a term so often used in a botanical literature from ancient times. As in many other situations, the etymolog cal definition is the simplest and operational (etymologically—"plants of dr places," thus vegetating in arid, xeric environments), but, upon a particular analysis

the term hides many nuances, interpretations, pitfalls, and, therefore, unlimited possibilities of discussion. The theory of physiological drought (habitats affected by physiological drought) explains much of this approach. For this reason, mainly, the xerophytes are treated in the present work from a didactic point of view, but also with the awareness of the fact that the xerophytes represent, in fact, a (super) group of plants, comprising, in reality, several ecological separated categories ("classes") of plants: desert plants, halophytes (salt-plants), psammophytes (plants on sand), oxylophytes (plants vegetating on peatlands), psychrophytes (arctic, alpine plants), etc. Therefore, when we say "xerophytes," we have in mind a much more complex and delicate reality than at first sight.

It must be emphasized that the present work is not a proper work on plant ecology (not even in the sense of the definition of ecology, from the nineteenth century); therefore, in the contents of these pages only general elements of the ecological anatomy of plant organs are found, plus some nuances, as much as time and space allowed, for the moment. This would deserve a separate work, strictly focused on the discussion of different ecological groups of plants, from an adaptive point of view (morphological, anatomical and functional), a working idea that has been in the author's attention for over 15 years.

In the present book, we presented some aspects of the ecological anatomy of the vegetative organs of gymnosperms and angiosperms—dicotyledons and monocots; also, for reasons of space, the anatomy of the reproductive organs of gymnosperms and angiosperms is not treated here either purely descriptively or ecologically. Therefore, the pages of this book contain information on the ecological anatomy of the root, stem (aerial and rhizome), and leaf of some major ecological categories of plants: xerophytes (generally approached, plus some sub-categories of xerophytes: alpine plants—general considerations—, tropical alpine plants, steppe plants, etc.), hydrophytes (water plants), helophytes (marsh plants), halophytes (salt plants), epiphytes, lianas, mangroves.

We are aware that such work can never be exhaustive; each ecological group of plants may deserve a separate book, and perhaps even then the subject cannot be treated completely.

In the present work, the general anatomy of each vegetative organ is described first, then ecological anatomy of that organ, with emphasis on the last part. Regarding the plant species mentioned in this work, cited in classic papers on plant anatomy, probably some of them no longer correspond to the current plant taxonomy; it was not our intention to update them taxonomically (in some situations, that would certainly be impossible). Therefore, in these cases, the emphasis falls exclusively on the particularities of the structure of the respective species, which do not lose their anatomical and ecological significance, regardless of the taxonomic status of the species.

It was my intention to deal with this book in a "classic" manner; this is because the essence of this approach is deeply rooted in the work carried out by a plethora of plant anatomists and ecologists who activated mainly in the mid-nineteenth century to the first decades of twentieth century. Due to their work and power of fine understanding of some subtle interrelationships between anatomical adaptations

and environmental factors, some basics on the true nature of these adaptations have been documented and established ever since. These findings lasted throughout the time, and many modern studies on the anatomy of plants belonging to different ecological groups are documented with the help of these "classic" studies. For this reason, the present work is greatly indebted to this fruitful epoque of plant anatomy and, consequently, it uses many data and references dating back to nineteenth century-beginning of twentieth century.

Competing Interests: The author has no competing interests to declare that are relevant to the content of this manuscript.

[illegible]

[illegible]

Contents

List of Figures

Chapter 3

Chapter 4

Chapter 5

Chapter 6

Chapter 7

List of Tables

Primary Structure (Growth) *Versus* Secondary Structure (Growth): General Considerations

In plant anatomy, the terms "primary" and "secondary" are very often used, becoming classics in morphological and anatomical literature. However, sometimes the terms can generate confusion, as there are no such categorical (firm) differences between a "primary" and a "secondary" structure, within the same plant organ (Table 1). In addition, to even complicate things, some authors also talk about a "tertiary" structure, as for example, in the case of vegetative organs that present the phenomenon of successive (additional) cambia (Grigore 2008; Grigore and Toma 2021a, b—and all references included and reviewed).

Broadly speaking, primary growth (structure) is the initial growth of the aerial part or root and the generation of leaves and flowers. Many plants, especially annual monocots and dicots, show only primary growth (Crang et al. 2018).

Plants with primary structure tend to be short-lived (ephemeral, annual, or biennial plants) or have other mechanisms to generate permanent (longer-lasting) structures.

When using a no demanding language, the term "herbaceous plants" is generally used to describe dicotyledons that are limited to primary growth only.

Primary growth (structure) is produced by primary meristems, which are of two types—shoot and root apex. They are found at the tip of the stem and the root, respectively; therefore, primary growth results in longer stems and roots.

Primary growth allows plants to explore and occupy a larger aboveground volume, which is vital for their ability to compete for light, but also a larger underground volume of soil, from which they extract water and mineral elements (Crang et al. 2018; Grigore 2022).

Secondary growth (structure), on the other hand, increases the circumference of existing stems and roots and produces permanent structures.

Leaves and flowers, although sometimes containing sclerenchyma tissues (with secondary cell walls), show no true secondary growth.

Secondary growth is produced by *two secondary meristems*, *the vascular cambium* and *the phellogen*. In each growing season, the vascular cambium produces a

M.-N. Grigore, *Plant Ecological Anatomy*,
https://doi.org/10.1007/978-3-032-01332-3_1

Table 1 The terms "primary" and "secondary" used in anatomical language - comparative analysis (Crang et al. 2018)

Primary structure (growth)	*Secondary (growth) structure*
The initial growth of a plant organ Generated by primary meristems It results in an increase in the length of the organ Found in annual monocots and dicots (grasses)	Further growth of the root or stem Generated by secondary meristems It results in an increase in thickness (circumference) of the organ Found in woody plants and many herbaceous, perennial, and annual dicots
Primary meristem	*Secondary meristem*
Located at the tip of the root or stem Two types: shoot and root apical meristem	Located on the periphery of the root and stem Two types: vascular cambium (generates xylem and secondary phloem) and phellogen (generates cork and phelloderm—periderm)
Primary xylem and phloem	*Secondary xylem and phloem*
Generated by the shoot or root apical meristem	Generated by vascular cambium
Primary cell wall	*Secondary cell wall*
It is formed first Thin, cellulosic, rarely lignified Nonliving, but contains active enzymes; able to expand Found in all plant cells (parenchyma, collenchyma, sclerenchyma)	It forms after the primary wall Thick, multilayered, impregnated with lignin Nonliving and incapable of extension It is usually found in sclerenchyma cells

secondary xylem towards the interior and secondary phloem towards the exterior of the organ. The secondary xylem ("wood"), which is dead at maturity, accumulates in annual growth rings, while the phloem is replaced by a new tissue as the increasing diameter of the cylindrical xylem pushes outward and crushes the phloem of the previous (previous) year. Outside the phloem, the phellogen continuously generates the tissues that become the periderm or "bark" of a tree (although the term "bark" is problematic and inconsistent, as we shall discuss) (Grigore 2022).

It is very important to note that, since secondary growth can only occur in existing tissues, all woody plants start as seedlings and engage in primary growth during their first growing season.

Once the basic plan of the organ has been established (albeit with many variations), the vascular cambium and phellogen develop and give rise to secondary growth (structure).

Primary meristems generate the primary conducting bundles, which contain a xylem and primary phloem.

Vascular (conductive) bundles of herbaceous plants have no cambium, cannot develop further, and are therefore called closed vascular bundles (Fig. 1—just an example).

Open vascular bundles are those that have a vascular cambium (= secondary meristem). They pass from the primary to the secondary structure, with the

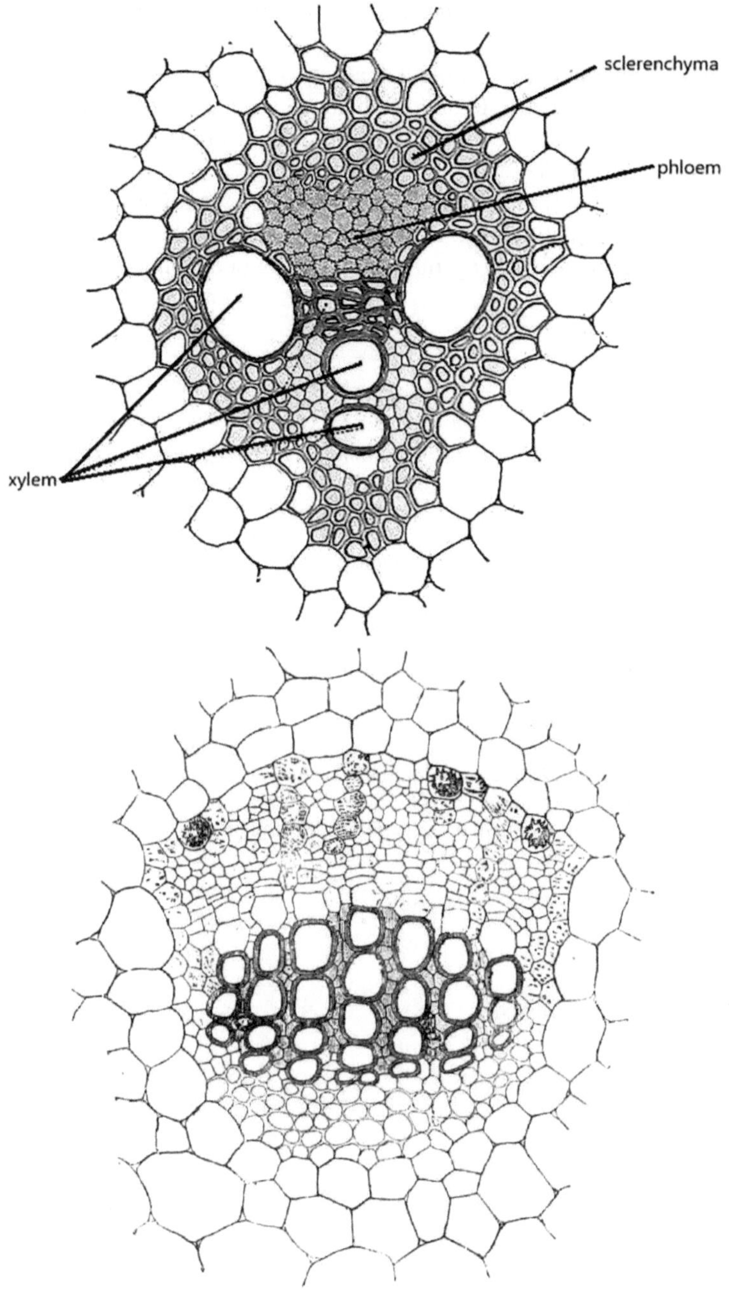

Fig. 1 Closed vascular bundle: (top)—in the stem of *Zea mays* and open vascular bundle—bottom—in the stem of *Malva sylvestris* (Bonnier and du Sablon 1905; Haberlandt 1914)

activation of the vascular cambium (Fig. 1), which then produces a secondary xylem and phloem.

Therefore, in classic works of plant anatomy, many statements such as a plant or an organ is "in the primary state of growth/shows primary structure," or in "secondary state of growth/shows secondary structure" or the organ (plant) is in "transition from primary to secondary structure" can be found.

In addition, by a similar semantic extension, one can distinguish between the primary body (monocotyledons, herbaceous dicotyledons, and first-year woody plants) and the secondary body (gymnosperms and woody dicotyledons) of the plant.

Bibliography

Bonnier G., du Sablon L., 1905 – *Cours de Botanique. Phanérogames*, Paris, Librairie Générale de l'enseignement

Crang R., Lyons-Sobaski S., Wise R., 2018 – *Plant anatomy. A concept-based approach to the structure of seed plants*, Springer Nature Switzerland AG

Grigore M.-N., 2008 – *Introducere în Halofitologie. Elemente de anatomie integrativă*, Editura PIM, Iaşi

Grigore M.-N., 2022 – *Morfologia plantelor. Aspecte ecologice*, Editura Universităţii „Ştefan cel Mare" Suceava

Grigore M.-N., Toma C., 2021a – Morphological and anatomical adaptations of halophytes. A review. In: Grigore M.-N. (Ed.), *Handbook of halophytes. From molecules to ecosystems towards biosaline agriculture*, Springer, pp. 1079–1222

Grigore M.-N., Toma C., 2021b – Integrative anatomy of halophytes from Mediterranean climate. In: Grigore M.-N. (Ed.), *Handbook of halophytes. From molecules to ecosystems towards biosaline agriculture*, Springer, pp. 1293–1328

Haberlandt G., 1914 – *Physiological plant anatomy*, Macmillan and Co, London

Root

Root

1 Anatomy (Structure) of the Root

A brief review of root morphology is useful, with reference, in particular, to the regions (zones, layers) of the root, or, more precisely, of a growing root tip (Fig. 1) (detailed discussion—Grigore 2022).

Zones of a root can be easily observed in a seedling immediately after germination. Thus, the following regions can be distinguished, starting from the tip of the root (however, this zonation may slightly vary among authors):

- Root cap (calyptra—according to some authors, but rarely used today) that covers and protect the vegetative tip.
- Zone of cell division (that correspond, morphologically, to a smooth region of a root).
- Zone of cell elongation and maturation, the last being the area where differentiation of tissues takes place, involving vascular tissues, the cortical parenchyma, and the rhizodermis.
- Above these zones, towards soil surface, there is a rough area of the root, which is limited towards the stem by the root-stem transition zone (some old botanists nominated the place where the root meets the stem as *collum*—Lindley 1861).

At the same time, an overview is also necessary regarding the integration of these zones (regions) of the root with the meristematic tissues that are involved in the generation of the primary structure of the root (Fig. 1), as well as the ways in which a tissue (meristematic—definitive) is correlated with the organ, and then finally within the plant, as a global system (Grigore 2022).

M.-N. Grigore, *Plant Ecological Anatomy*,
https://doi.org/10.1007/978-3-032-01332-3_2

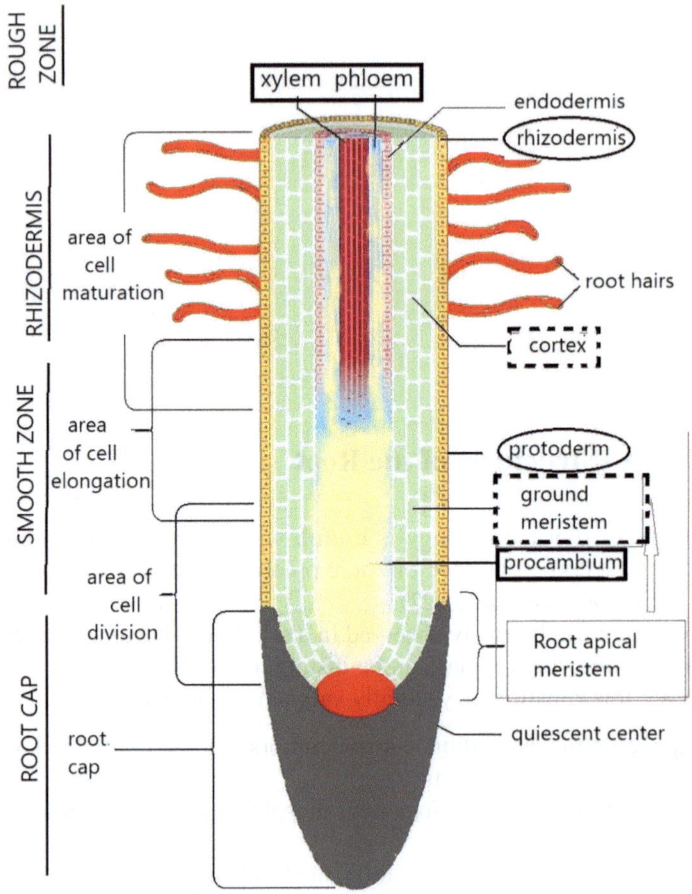

Fig. 1 Integrated diagram illustrating the areas (regions) of the root tip (uppercase, left) and the mersitematic tissues (right, bottom): root apical meristem, procambium, ground meristem and protoderm), from the activity of their cells resulting from the primary (definitive) structure of the root, visible completely next to the rhizodermis; as can be seen, the xylem and phloem result from the procambium (notice the rectangle in continuous line); the cortex derives from the ground meristem (notice the rectangle in dotted line); and the rhizodermis results from the protoderm (notice the black oval) (modified and adapted from Crang et al. 2018)

2 Primary Structure (Growth) of Root

The origin of the primary structure it is in the root apical meristem.

The definitive tissues of the primary structure are formed from the root meristems following repeated divisions and the process of cell differentiation. The transition from the meristematic to the definitive state occurs in the region of cell division ("smooth" zone or area) (Toma et al. 2003) (see also Fig. 1).

We will briefly describe the anatomical areas of a root with primary structure; the general discussion refers to dicotyledons, following the typical example of the buttercup species (*Ranunculus* sp.); several major structural scheme differences from monocotyledons are dotted in square brackets.

Considering a cross section through a young root, at the level of the rhizodermis, from the exterior towards the center of the organ, three anatomical areas can be delimited and described (Figs. 2, 3, and 4):

- *Rhizodermis.*
- *Cortex.*
- *Central cylinder* (stele).

Each of these areas consists of one or more definitive tissues, thus resulting from the activity of some primary meristem tissues.

1. *The rhizodermis* is located on the outside, fulfilling the function of water uptake and protection, being initially formed of two types of cells: atrichoblasts (*a*, gr., *devoid of*, *without*, and *trichos*, gr., *hai*, so cells that do not form root hairs) and

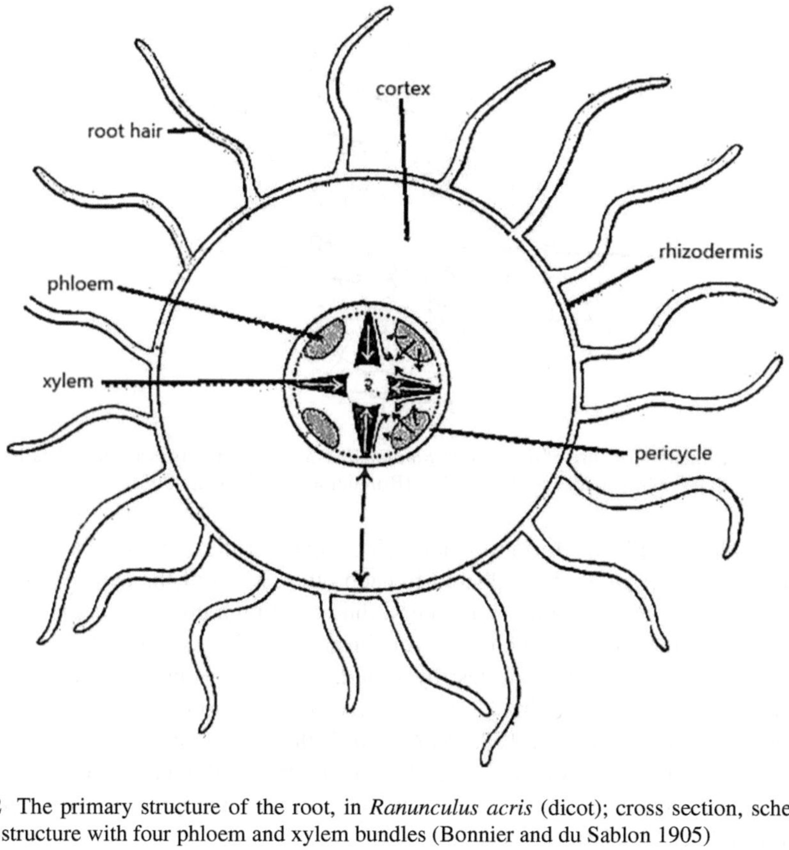

Fig. 2 The primary structure of the root, in *Ranunculus acris* (dicot); cross section, schematic view; structure with four phloem and xylem bundles (Bonnier and du Sablon 1905)

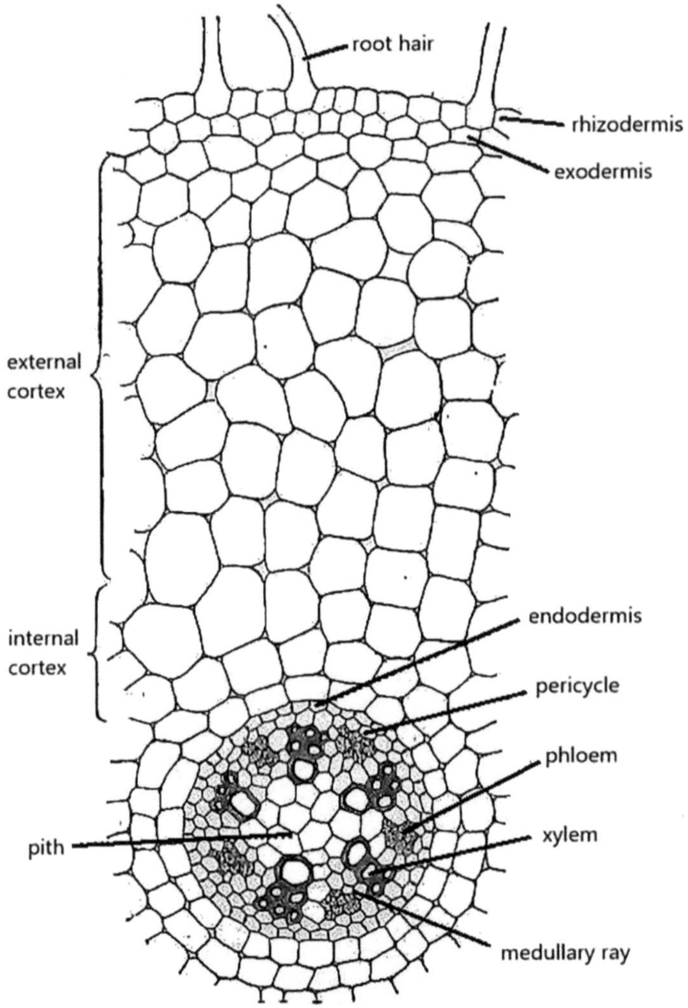

Fig. 3 The primary structure of the root, in *Ranunculus acris* (dicot); cross section, detailed view; structure with five phloem and xylem bundles (Bonnier and du Sablon 1905)

trichoblasts (cells that form root hairs). The rhizodermis consists of a single layer of cells, which are tightly adhered to each other. Most of the cells of the rhizodermis are transformed into root hairs, thus creating a greatly increased surface area for water uptake at the root level (Figs. 2 and 3). As the root hairs die, the cells of the rhizodermis are destroyed and, for this reason—in a section made, for example, above the rhizodermis—these cells are no longer visible (Morariu 1965; Zanoschi and Toma 1985; Sitte et al. 1998). The ecological implications of root hairs and of roots in general in different ecological groups of plants (especially xerophytes) have been reviewed in the monographic work on plant morphology (see Grigore 2022).

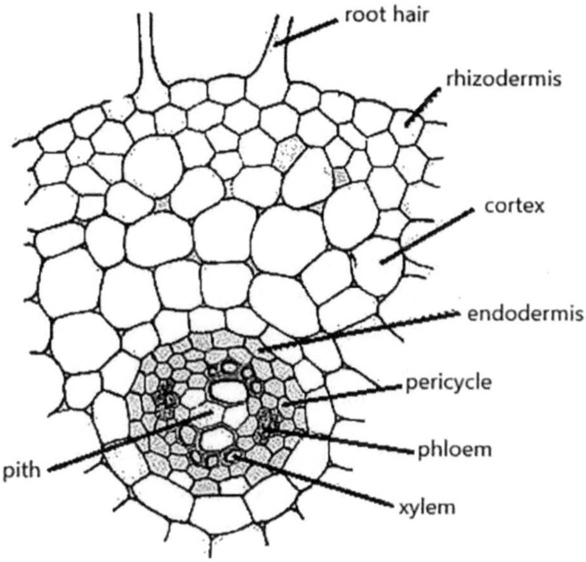

Fig. 4 The primary structure of a thin root, in *Ranunculus acris* (dicot); cross section, detailed view; structure with two phloem and xylem bundles (Bonnier and du Sablon 1905)

2. *The cortex* is located under the rhizodermis and consists of large, parenchymatous, polyhedral, or round cells, which form large intercellular spaces between them. The endodermis represents the innermost layer of the cortex, and its cells differ, by specific characters, from the other cells. Generally, in the same species, the cortex is thicker than the central cylinder, having a storage role, especially in the cortical parenchyma (Fig. 3). [*In monocots, the cortex has usually fewer layers than in dicots*].

From an anatomical point of view, the cortex can be divided into three distinct sub-areas in most plants (however, there are exceptions and variations):

- *The exodermis* is made up of closely packed cells with suberized walls (Figs. 3, 10, 11, 14, and 16). The exodermis has a protective role, especially after the rhizodermis has exfoliated. In some dicotyledons and many grasses, the exodermis, together with cells from the external cortical parenchyma, is sclerified and lignified, and thus the cortex also fulfills a mechanical role. The exodermis and, especially, the endodermis are very interesting tissues, being discussed, in the light of new research, in relation to control mechanisms related to entering of the soil solution in the root (Grigore 2008; Grigore and Toma 2021a), but also in relationship with transport mechanisms of sap through the root, to the xylem from the central cylinder (Sadava et al. 2011; Evert and Eichorn 2013; Crang et al. 2018). As in the case of the endodermis, as we will see, at the level of the exodermis, smaller, living cells, called passage cells, can remain from place to place; they have cellulosic walls and, therefore, are not thickened.
- *The cortical parenchyma* represents the thickest subzone of the cortex; it consists of cells that leave intercellular spaces between them or, in some ecological groups of plants (water plants, hydrophytes, and swamp plants, helophytes), aer-

enchyma (in different forms and degrees); the cells contain reserve substances (usually starch, but in some species, sucrose or inulin); the cortical parenchyma can display a great anatomical plasticity; for example, in the aged roots of some grasses, part of the cortical parenchyma disorganizes, resulting in wide air cavities (aerenchyma) (Figs. 12 and 14);

- *The endodermis* is the innermost layer of the cortex (therefore, it does not belong to the central cylinder) (Figs. 3, 4, 5, 6, 8, 10, 11, 12, 13, 15, and 16), being made up of relatively elongated cells, closely united to each other, of a parallelepipedal shape. The peculiarity is that the endodermis is made up of cells whose lateral walls are suberized in the form of a band around the cell. In cross section, these bands appear under the microscope as thickened dots, called Caspary 's thickenings (strip, girdles), after the botanist Robert Caspary (1818–1887), who discovered them. The endodermis acts as a "barrier," controlling the diffusion of solutions passing from the cortex into the central cylinder and vice versa; this is due to some cells with non-thickened walls, called passage cells, which have the role of facilitating the passage of water with minerals and assimilated substances in the xylem and phloem bundles (Moruzi and Andrei 1967; Doblas et al. 2017).

In the development of the endodermis, three phases are distinguished: primary, secondary, and tertiary. These phases are determined by the specific conformation of the cell wall (Toma 1977), although sometimes these developmental phases do not always follow each other. It should only be mentioned that endodermis differentiation, as well as its nutrient control role, is related to very complex histological and functional events (Miyashima and Nakajima 2011; Barberon 2017; Doblas et al. 2017).

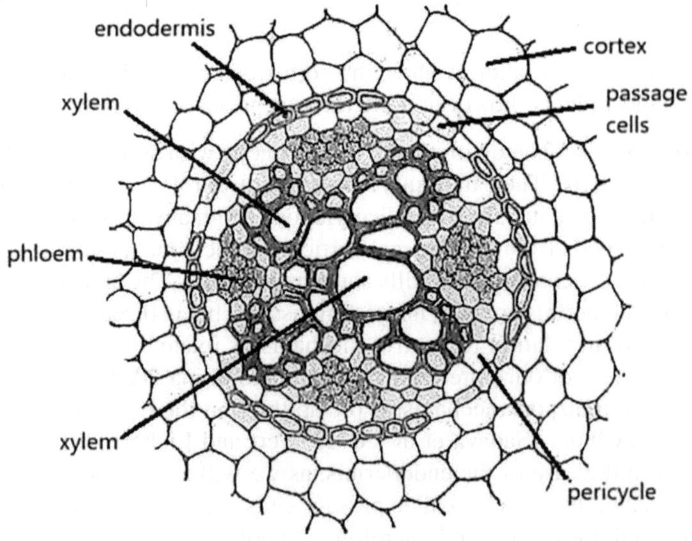

Fig. 5 The primary structure of the root, in *Ranunculus acris* (dicot), grown in dry soil; cross section, detailed view; structure with four phloem and xylem bundles—the pith is absent and the four xylem bundles result from the fusion of two xylem bundles—see Fig. 3, for comparison (Bonnier and du Sablon 1905); compare with Fig. 4, to reveal anatomical variations along the root

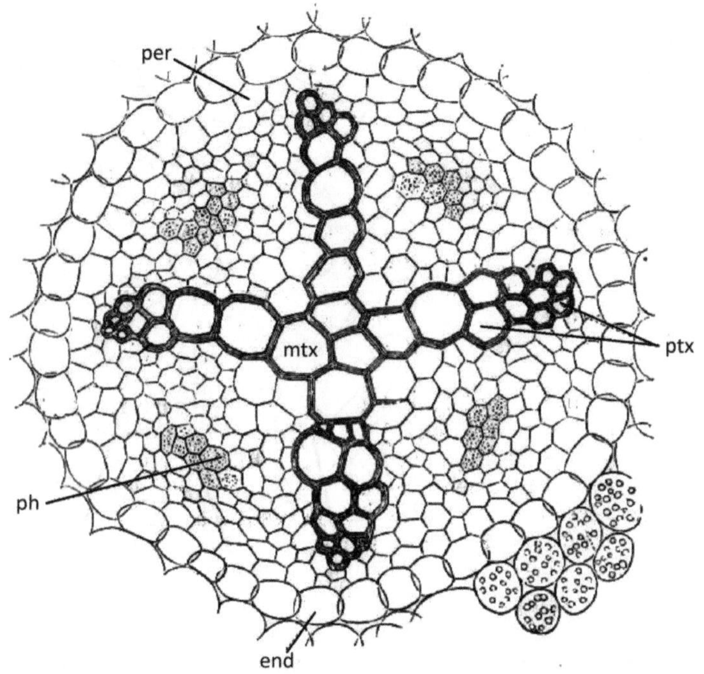

Fig. 6 The primary structure of a root of *Ranunculus acris* (dicot); cross section, detailed view; structure with four phloem and xylem bundles; end, endodermis; per, pericycle; ph, phloem; mtx, metaxylem; ptx, protoxylem (Gérardin and Guède 1895)

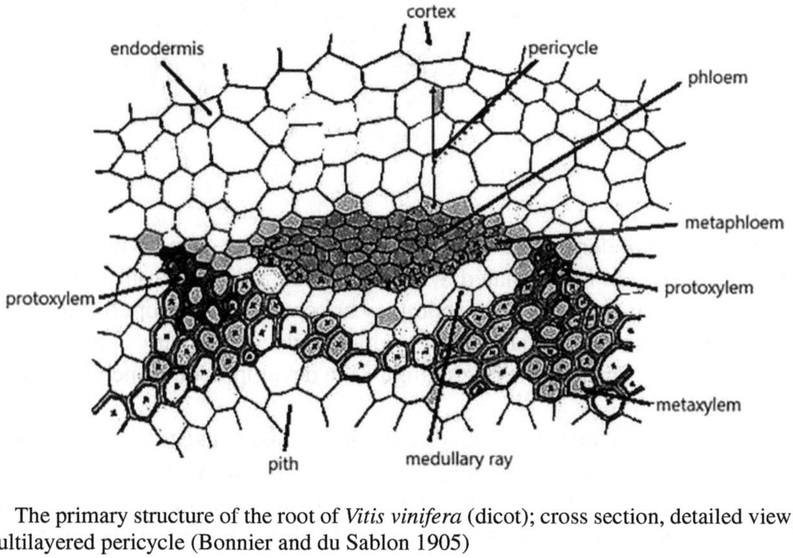

Fig. 7 The primary structure of the root of *Vitis vinifera* (dicot); cross section, detailed view; note the multilayered pericycle (Bonnier and du Sablon 1905)

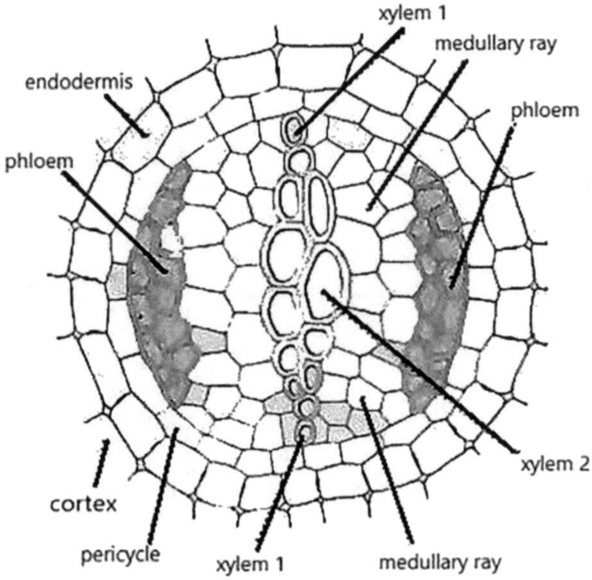

Fig. 8 The primary structure of the root of *Leontodon autumnalis* (dicot); cross section, detailed view; note that there is a vascular blade, formed by the fusion of two opposite xylem bundles (Bonnier and du Sablon 1905)

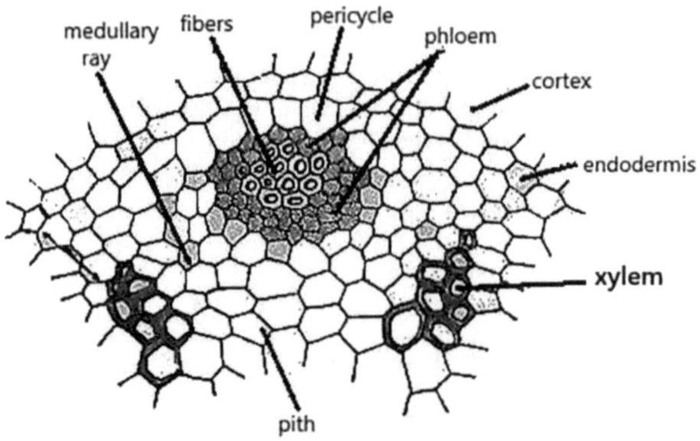

Fig. 9 The primary structure of the root of *Phaseolus* sp. (dicot); cross section, detailed view; note here the phloem fibers, similar to those found in the secondary phloem of the stem (Bonnier and du Sablon 1905)

The primary endodermis presents cells with Casparian strips, which represent, as we have seen, bands impregnated with lignin and suberin, from the radial cells walls; in cross section through the root, they appear as circular or elliptical thickenings, separated from the median lamella, which make up Caspary's points. This type of endodermis is formed at the beginning of the development of the young seedling (Toma 1977).

The secondary endodermis, found in dicotyledons, consists of cells with all the walls—uniformly and centripetally thickened—partially impregnated with a suberin-like substance (but lacking phellonic acid, as in the case of the primary endodermis), therefore having reduced permeability, although the contents of the cells remain alive for a long time.

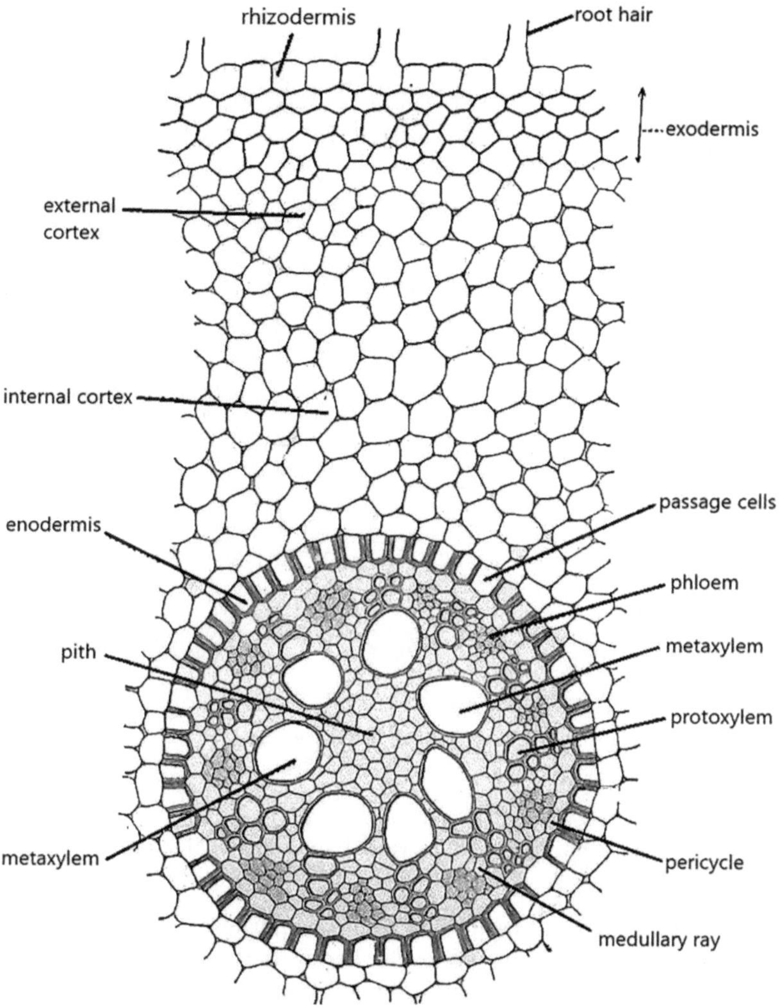

Fig. 10 The primary structure of the root, in *Iris germanica* (monocot); cross section, detailed view (Bonnier and du Sablon 1905)

Tertiary endodermis [*found especially in monocots*—Fig. 10] consists of cells with thickened and often lignified walls (cellulose and lignin are deposited, in many cases, over already formed suberin lamellae).

Thickening of cell walls can occur either uniformly and all around, or only the internal and lateral (radial) walls thicken; portion thickened part of the cell has—in cross section—the shape of the letters "C" or "U" and resembles a horseshoe (Toma 1977).

As we have already stated, in the secondary and tertiary endodermis, there still remain, from place to place (near the xylem poles of the conducting bundles in the

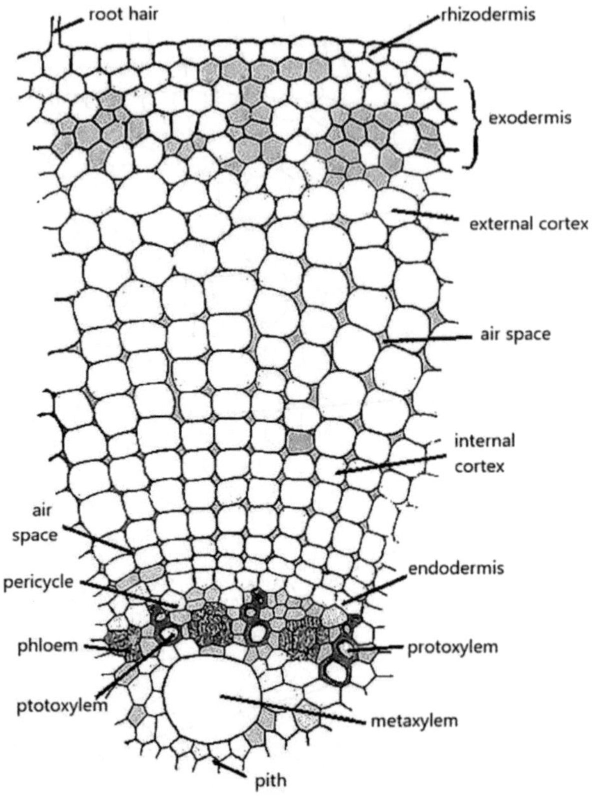

Fig. 11 The primary structure of the root of *Phragmites communis* (monocot); cross section, detailed view (Bonnier and du Sablon 1905)

central cylinder), smaller, living, thin-walled, cellulosic cells, called passage cells (Figs. 5, 10, and 15).

3. *The central cylinder* represents the axial (central) area of the root, delimited towards the periphery by the pericycle. Pericycle is missing in typical pteridophytes, being monolayered in most dicotyledons, and its cells alternating with those of the endodermis; it is multilayered in gymnosperms and many monocotyledons [*In old roots of many monocotyledons, the walls of the pericycle cells thicken and lignify, except for those near the xylem bundles*]. [*Considering that the root of monocotyledons remains its entire life with only the primary structure, the central cylinder, in monocotyledons, is thicker than in dicotyledons* (Figs. 10, 12, and 16)].

In the exemplary dicot species, *Ranunculus* sp., the pericycle consists of a single layer of cells. Inside the zone delimited by the pericycle, there is a fundamental tissue, in which the xylem (simple) and the phloem bundles (simple) are embedded and alternated with each other (Toma 1977; Moruzi and Andrei 1967). They are arranged on a sinuous circle, the phloem showing a centripetal development of its component elements.

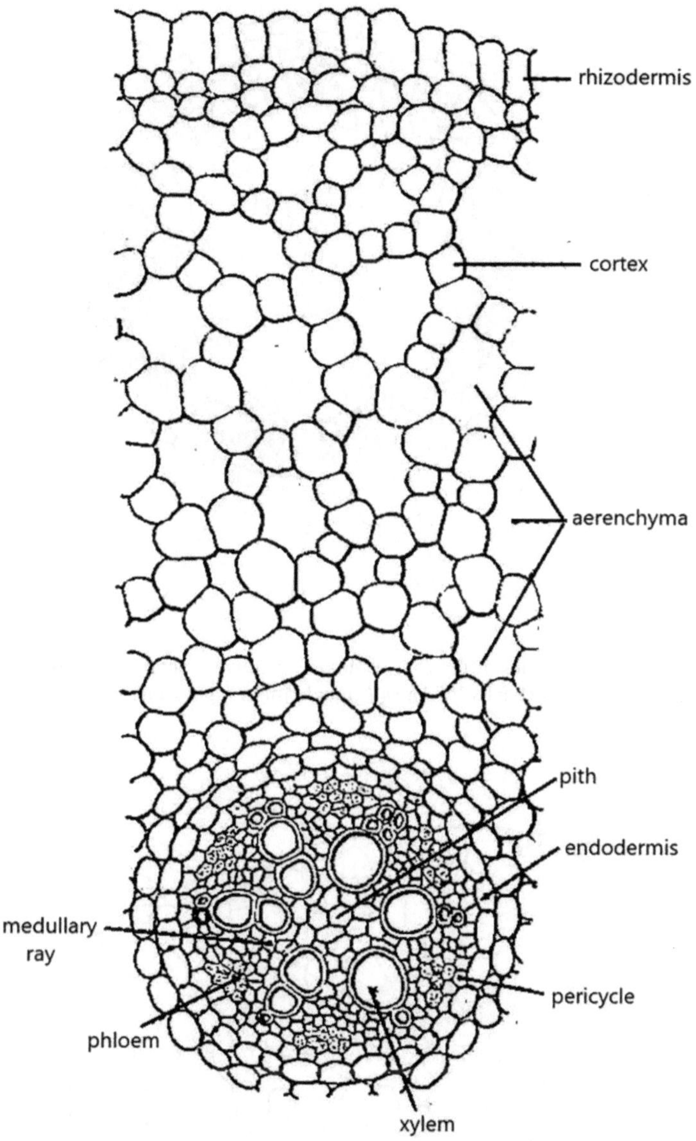

Fig. 12 The primary structure of the root of *Acorus calamus* (monocot), general view; cross section, detailed view (Belzung 1900); this is a marsh plant—note the development of aerenchyma

The number of xylem bundles (that is equal to that of phloem bundles—Figs. 5 and 6) is reduced in pteridophytes: in some, 1, monarch structure, such as in *Selaginella*, *Ophioglossum*, and *Trichomanes*; in others, 2, diarch structure, as in the majority of pteridophytes; and in gymnosperms and angiosperms, higher: 3, triarch structure in some gymnosperms and *Fabaceae* species, such as *Pisum*, *Medicago*,

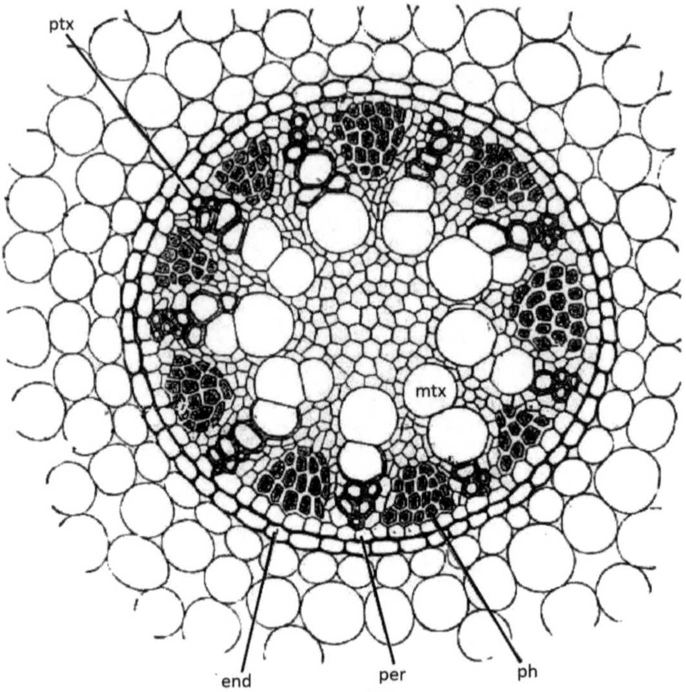

Fig. 13 The primary structure of the root of *Acorus calamus* (monocot), central cylinder view; cross section, detailed view (Sachs 1874); *end* endodermis, *ph* phloem, *per* pericycle, *mtx* metaxylem, *ptx* protoxylem

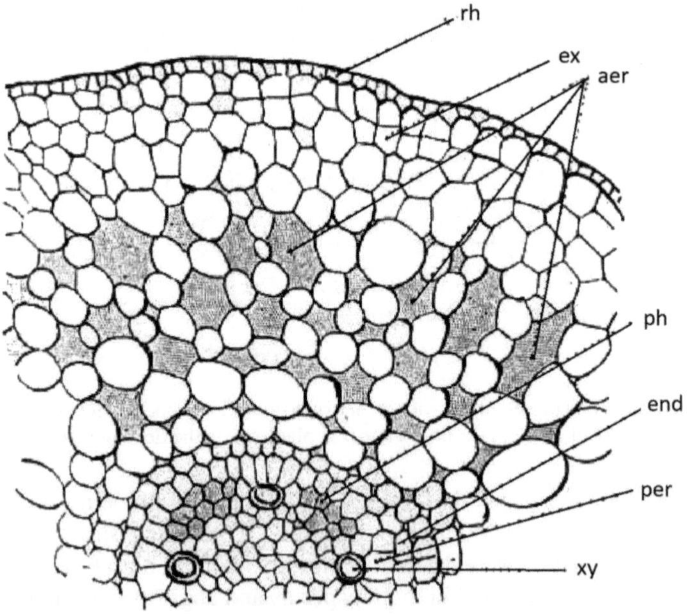

Fig. 14 The primary structure of the root of *Trapa natans* (monocot); cross section, detailed view (Sachs 1874); *end* endodermis, *ph* phloem, *per* pericycle, *ph* phloem, *xy* xylem, *aer* aerenchyma, *ex* exodermis, *rh* rhizodermis (Gérardin and Guède 1895)

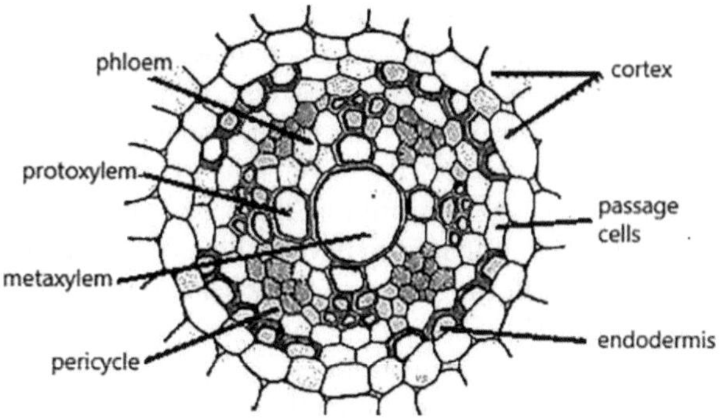

Fig. 15 The primary structure of the root of *Allium sativum* (monocot); cross section, detailed view; note that the metaxylem occupies the center of the organ (Bonnier and du Sablon 1905)

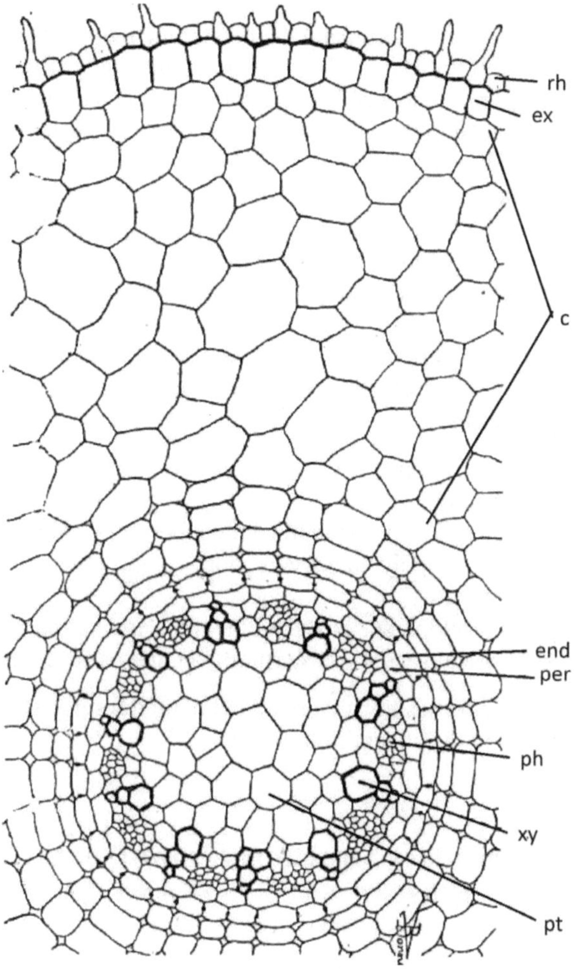

Fig. 16 The primary structure of the root of *Aloe* sp. (monocot); cross section, detailed view (Sachs 1874); *cp* cortical parenchyma, *end* endodermis, *ph* phloem, *per* pericycle, *ph* phloem, *xy* xylem, *pt* pith, *ex* exodermis, *rh* rhizodermis (Gérardin and Guède 1895)

Lens, and *Phaseolus*; *Ranunculaceae* and *Cucurbitaceae*; and 4, tetrarch structure, in many *Ranunculaceae*; several bundles—polyarch structure (as in many conifers, *Citrus*, *Quercus*, and *Fagus* species). In monocotyledons, as a rule, the number of xylem bundles is greater than seven, sometimes reaching ten (grasses) (Toma 1977).

The xylem vessels from the vicinity of the pericycle have a smaller lumen, a ring, and a spiral shape, constituting the protoxylem. As they move away from the pericycle, the vessels that form have an increasingly larger lumen. They consist of reticulated, scalariform, and sieve tracheae. These vessels make up the metaxylem (Figs. 6 and 10).

Phloem bundles are arranged, as already stated, alternately with the xylem ones (Figs. 7, 8, and 9) and are made up of sieve tubes, companion cells, and phloem parenchyma. Like xylem bundles, a *protophloem* and a *metaphloem* are distinguished within a phloem bundle. The protophloem is formed in the growth zone of the primary meristem, and the metaphloem appears later, but also from the primary meristem. Noticed in their development, the phloem bundles and xylem bundles develop in a centripetal sense (Moruzi and Andrei 1967).

It shall be reminded that the phloem and xylem arise from the activity of the procambium.

Sometimes the last and largest metaxylem vessels occupy the central part of the root (as in many dicotyledons), which explains the absence of pith in the root of dicotyledons. [*In monocotyledons, however, the xylem bundles remain individualized and, in this case, a pith remains in the center of the root* (Figs. 10, 12, 13, 16, and 17), *whose cells can sclerify and lignify in older roots*] (Toma 1977).

Summing up, the primary *structure of the root* shows certain common characters in all vascular plants, therefore both in pteridophytes and in gymnosperms and angiosperms:

- The rhizodermis is devoid of stomata and cuticle, having the ability to produce root hairs.
- A distinct, well-differentiated endodermis, being always present.
- Xylem and phloem vessel bundles are arranged alternately and accumbent on the pericycle.
- The development of the elements in the xylem and phloem bundle occurs in a centripetal sense, i.e., starting from the pericycle inwards.

The same basic structure is found, therefore, in the main root, in the lateral roots, but also in the adventitious roots.

3 Secondary Structure (Growth) of the Root (Growth in Girth)

The primary root structure is found and remains unchanged in pteridophyte roots, as well as annual herbaceous dicotyledons and monocotyledons, with some exceptions (*Dracaena*, *Yucca*, *Cordyline*, and *Aloe*) (Morariu 1965).

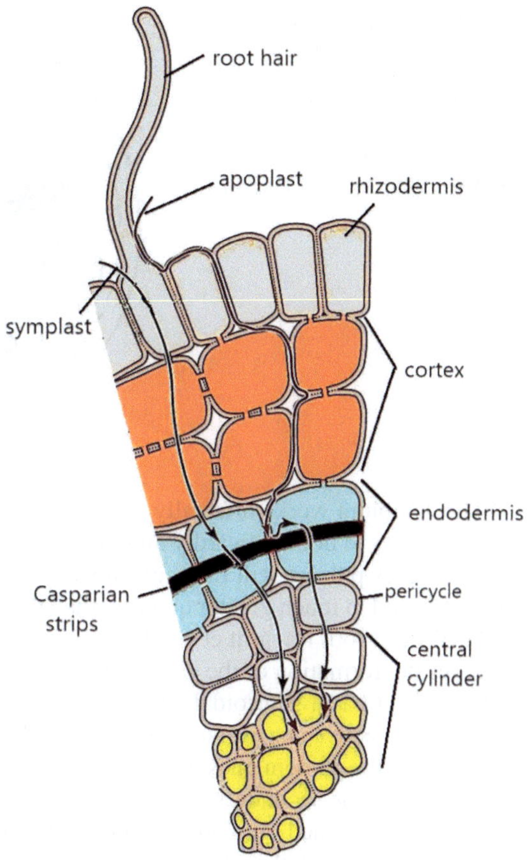

Fig. 17 Apoplast and symplast. Their integration with the endodermis and the other tissues of the root. The cell walls and intercellular spaces constitute the apoplast. Symplast refers to living cells, which are connected by plasmodesmata. To enter the symplast, water and solutes must pass through the plasmalemma. No such selective barrier limits movement through the apoplast. The Casparian strip in the endodermis is impregnated with the hydrophobic substance suberin and separates the apoplast in the cortex from the apoplast in the central cylinder (edited by: Sadava et al. 2011; Evert and Eichorn 2013; Crang et al. 2018; Taiz and Zeiger 2002)

However, the primary structure is transient in gymnosperms and most dicotyledons; further, growth in thickness is ensured by the functioning of two secondary (lateral) meristematic tissues (vascular cambium and phellogen), also known as secondary meristematic tissues. These secondary tissues, through repeated divisions, form new definitive, secondary tissues (Figs. 18, 19, 20, 21, 22, and 25), which significantly modify the structure of the root (Toma 1977).

The vascular cambium is formed deeper, mostly due to the remnants of the procambium, and occurs at the internal part of the phloem bundle and at the external part of the xylem bundle (Figs. 18 and 21). The first portion of the cambium occurs in the parenchyma inside the metaphloem and appears in cross section as an upturned arch with the concavity on the outside. The pericycle outside the protoxylem also enters division when the edge of the cambial arch approaches it and forms the second cambial portion in front of the xylem bundle (Fig. 21). The edges of the cambial zones appearing in the pericycle—and with those of the cambium from the parenchyma—join and form a continuous zone, which is sinuous at first, but then becomes circular.

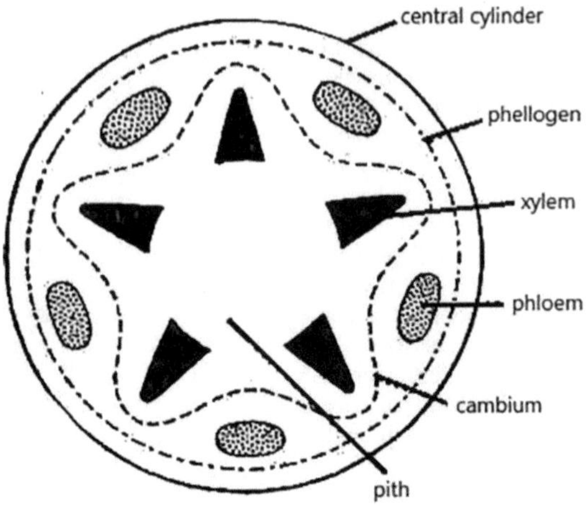

Fig. 18 General schema of a cross section through a root of the vine with five vascular bundles, to show the position of secondary meristems (Bonnier and du Sablon 1905)

The cambium works bifacially, that is, the cambial cells divide tangentially and give rise to cells both outwardly and inwardly; towards the outside, the elements of the secondary phloem will form, and towards the inside, the elements of the secondary xylem, all in much greater quantity than the phloem and primary xylem (Anghel and Chirilei 1960; Răvăruţ et al. 1967; Toma 1977; Morariu 1965).

After the formation of the cambium is finished, following the activity of the initial arches, it has a sinusoidal shape in cross section (Fig. 20), which is due to its undulations between the bundles, alternately—outwardly, in xylem ones—and inwardly, in the phloem ones. Later, through a more active division, followed by the production of the secondary xylem in greater quantity, in front of the phloem bundles, they are pushed out and the cambium acquires a circular shape. The cambial portions in front of the phloem bundles, which at first were concave outwards, change, becoming concave inwards; at this stage, the cambium acquires a continuous cylindrical sleeve shape.

Next, the vascular cambium operates from here on through a relatively simple division process. Its cells, which form a single layer, grow radially, in thickness, and divide tangentially (periclinally). As new cells are generated, they pass alternately from one side of the cambium to the other, leaving a few young and undifferentiated cambium-like layers. At a short distance from the cambium, cells differentiate, the internal ones turn into xylem ones and form the secondary xylem (xylem), which overlaps the primary one and forms a common tissue. The layers of cells, located outside the cambium, differentiate and become the secondary xylem. The primary phloem is pushed outwards and is visible only for a limited period, being then embedded in other tissues and crushed, its conductive function being taken over by the secondary xylem (Figs. 23 and 24) (Morariu 1965; Răvăruţ and Turenschi 1973; Toma 1977; Throm 1996).

In this way, the newly formed xylem and phloem tissues determine the thickening of the root, and the cambial zone gradually moves outwards (Morariu 1965).

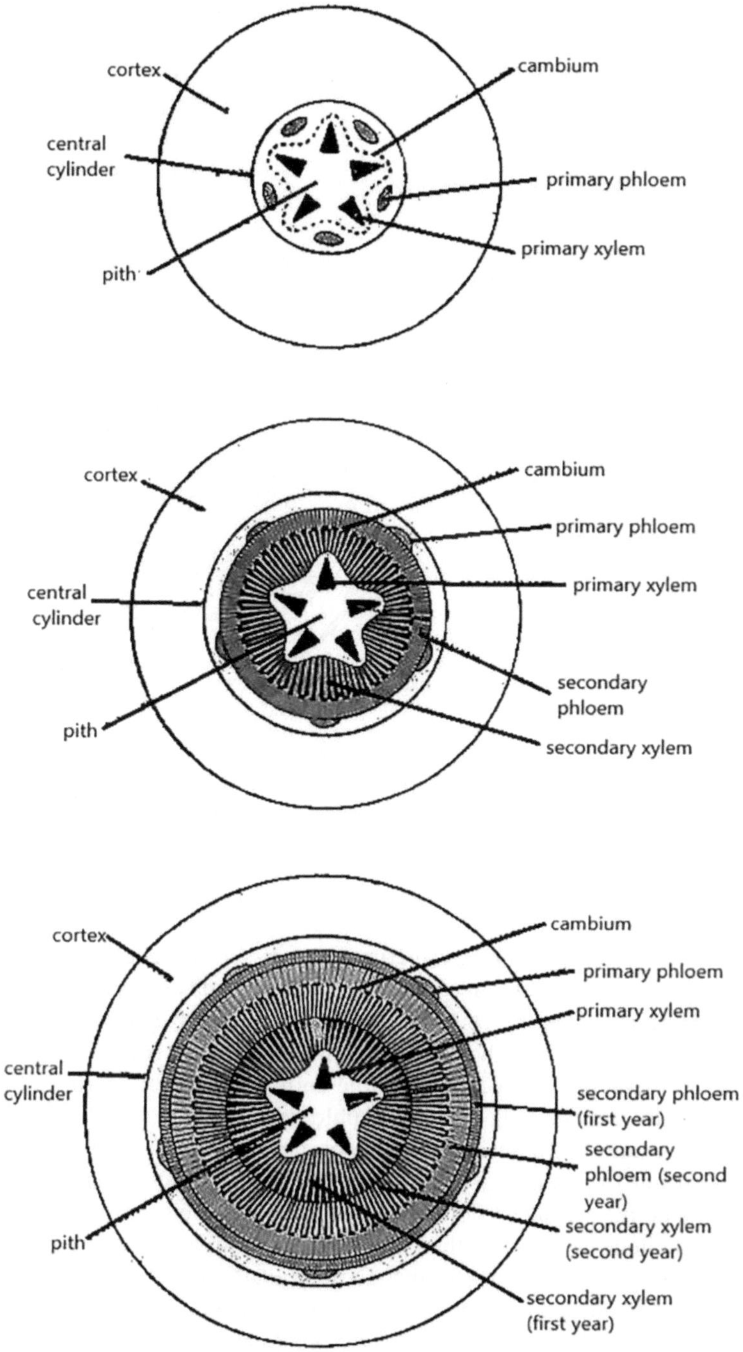

Fig. 19 Secondary root development in a dicot with five vascular bundles (general scheme); (**a**), beginning of cambium formation; (**b**), at the end of first year; (**c**), at the end of the second year (Bonnier and du Sablon 1905)

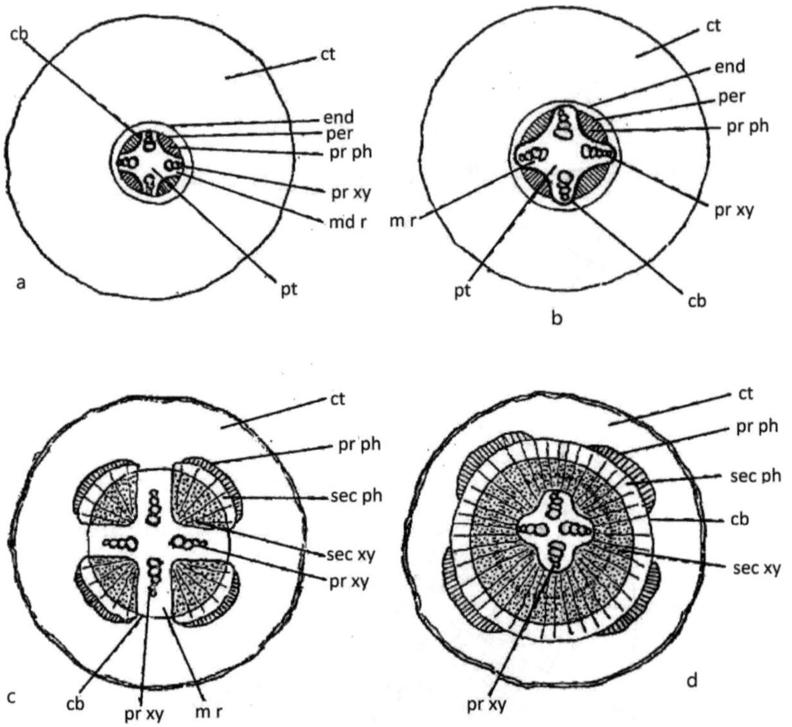

Fig. 20 Secondary root development in a dicot with four vascular bundles (general scheme); (**a**), beginning of arch cambium formation; (**b**), transformation of arches to continuous layer; (**c**), cambium generates a vascular cambium in front of each isle of primary phloem; (**d**), cambium produces a complete phloem and xylem ring; *cb* cambium, *ct* cortex, *end* endodermis, *pt* pith, *pr xy* primary xylem, *pr ph* primary phloem, *sec xy* secondary xylem, *sec ph* secondary phloem, *per* pericycle (Gérardin and Guède 1895)

Phellogen is another secondary meristematic tissue; it arises mainly from the cells of the pericycle or, sometimes, from the parenchyma cells of the primary cortex. The outermost layer of cells, of pericyclic origin, becomes the phellogen. Following the activity of the phellogen, two tissues develop: cork (a secondary protective tissue)— on the outside, towards the cortex—and phelloderm (a living tissue), on the inside, towards the secondary phloem.

It shall be recalled, as a purely didactic observation, the fact that primary definitive tissues result from a primary meristematic tissue, and secondary definitive tissues result, naturally, from a secondary meristematic tissue.

Cells of phelloderm maintain their cellulosic and thin walls, with a parenchymal appearance, and may even perform the storage functions that cortical parenchyma typically has. The ensemble of tissues generated by phellogen forms the periderm. All these three tissues (phellogen, cork, and phelloderm) make up the periderm or the secondary cortex (Toma 1977; Morariu 1965; Morlova et al. 1966; Moruzi and Andrei 1967).

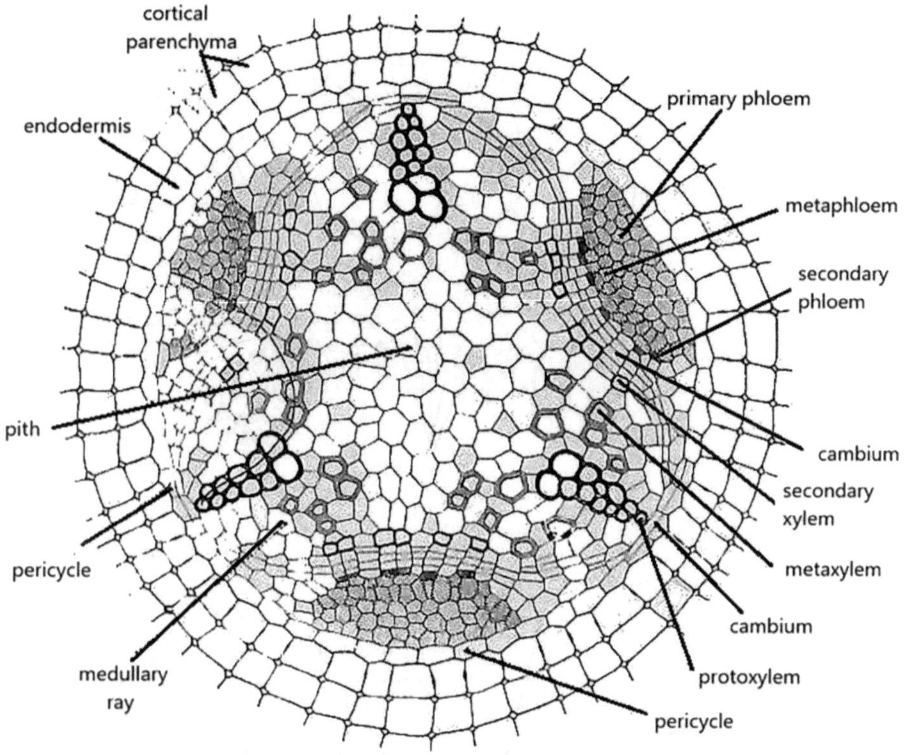

Fig. 21 Scheme showing the debut of transition from growth (primary structure) to secondary growth (structure) of a dicot root with three vascular bundles (Bonnier and du Sablon 1905)

In temperate climates, in many woody plants, but also in some perennial herbaceous plants, a periderm is formed every year; all of them and the tissues between them are called rhytidome (Grigore 2022).

The cambium and phellogen divide only during the growing season and then cease their activity in autumn. The cambium remains the same throughout the life of the roots, entering activity again every spring. The phellogen, on the other hand, can completely end its activity after a certain number of years, being replaced by another phellogen (therefore new), having a deeper position (Crang et al. 2018).

In the case of the root, in many plant species, the definitive secondary tissues are no longer grouped in bundles, but form concentric rings of cork, phelloderm, phloem, and xylem (Fig. 23), which are traversed by parenchyma rays (more often cellulosic), uniseriate or multiseriate.

In many plants from temperate zones (such as Romania), rings formed annually can be distinguished within the wood (secondary xylem). This is because, as already stated, the cambium works only during the growing season and stops working at the end of summer. The early xylem, which forms in the spring, comprises several vessels in the center of the root, usually having a much larger diameter than the vessels of the late xylem, which is oriented outwards.

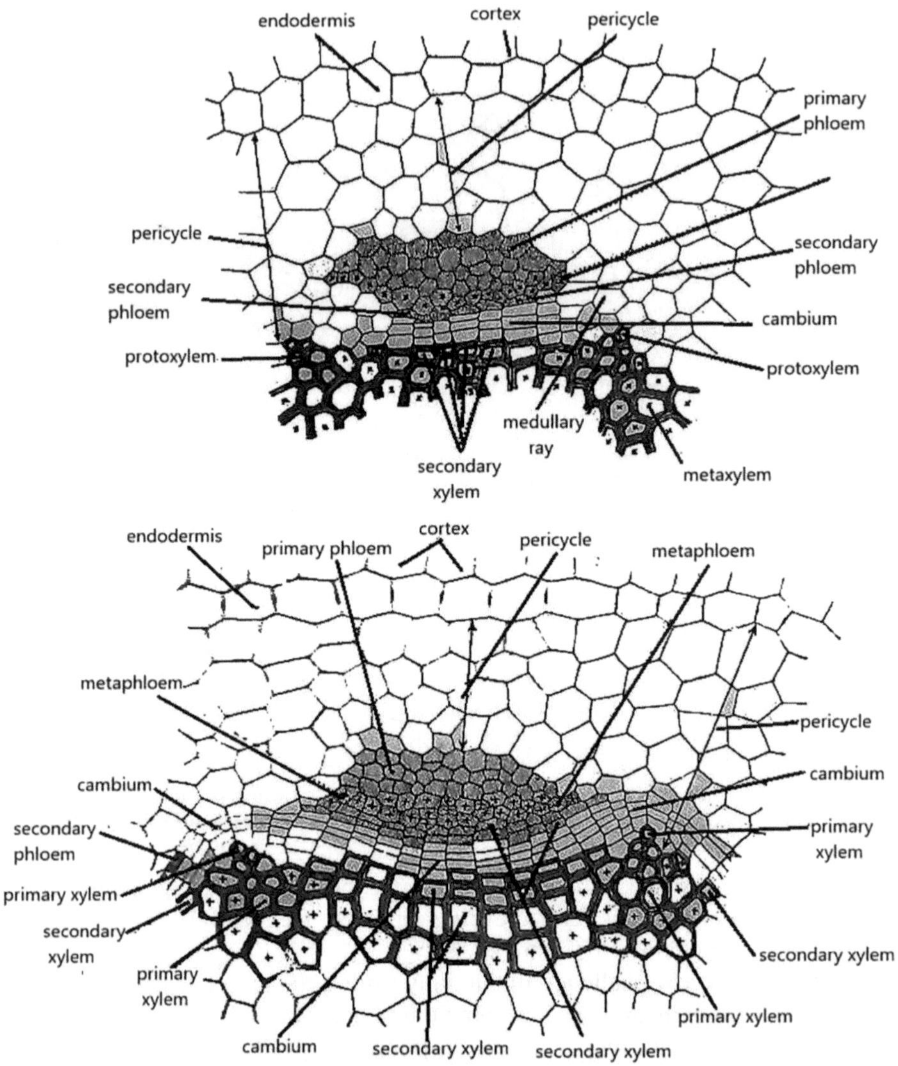

Fig. 22 Fragments of cross sections through the root of the vine, to see the development of the cambium: above, the start of cambium formation; bottom, a more advanced stage with the complete formation of cambium (Bonnier and du Sablon 1905)

In other words, in the secondary state (structure), the root will have approximately the same structure as the stem, the two axial organs being distinguished from each other only by what remains of their primary structure (Toma 1977).

In the root, in consequence of the centripetal differentiation of the primary xylem, the pith often disappears before the appearance of the cambium, and the secondary root may practically be said to be recognized by the absence of the pith; the whole center of the organ is occupied by xylem and it is not possible, as a rule,

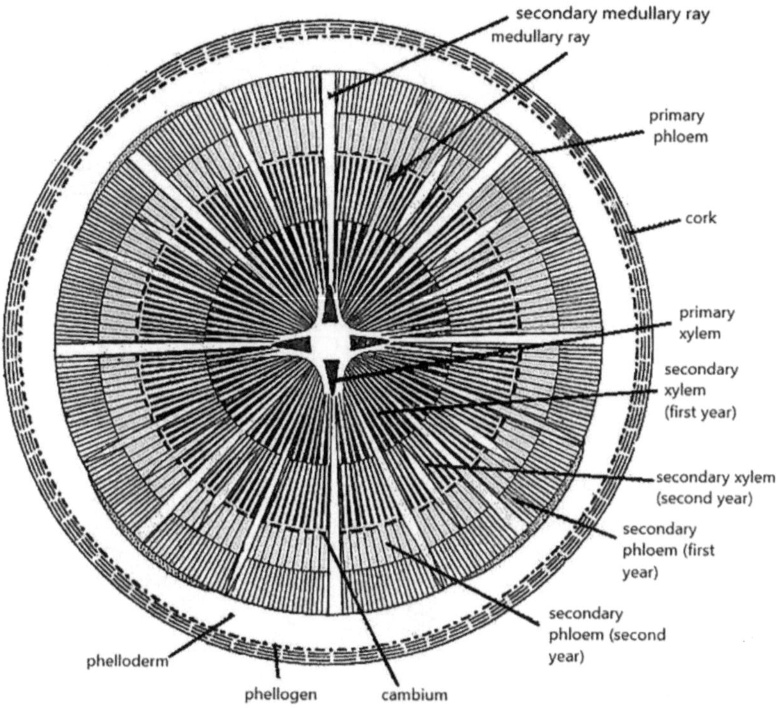

Fig. 23 General schema of a cross section through the two-year-old root of vine, with four vascular bundles (Bonnier and du Sablon 1905)

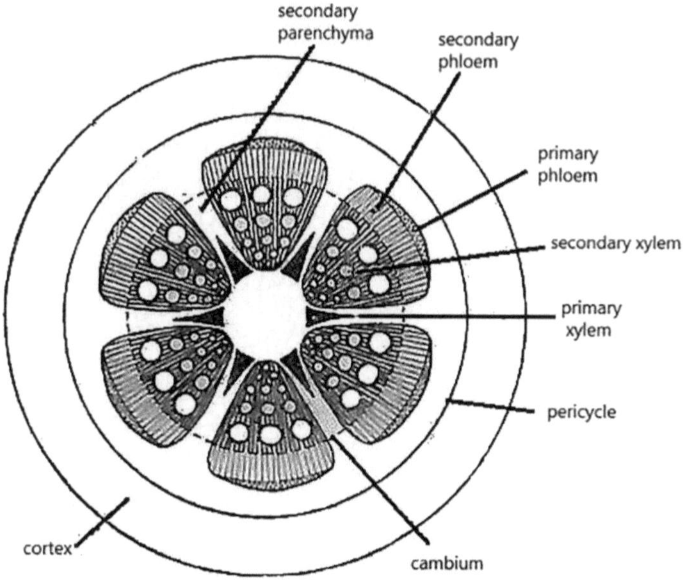

Fig. 24 General schema of a cross section through the root of *Aristolochia* sp., at the end of the first year (Bonnier and du Sablon 1905)

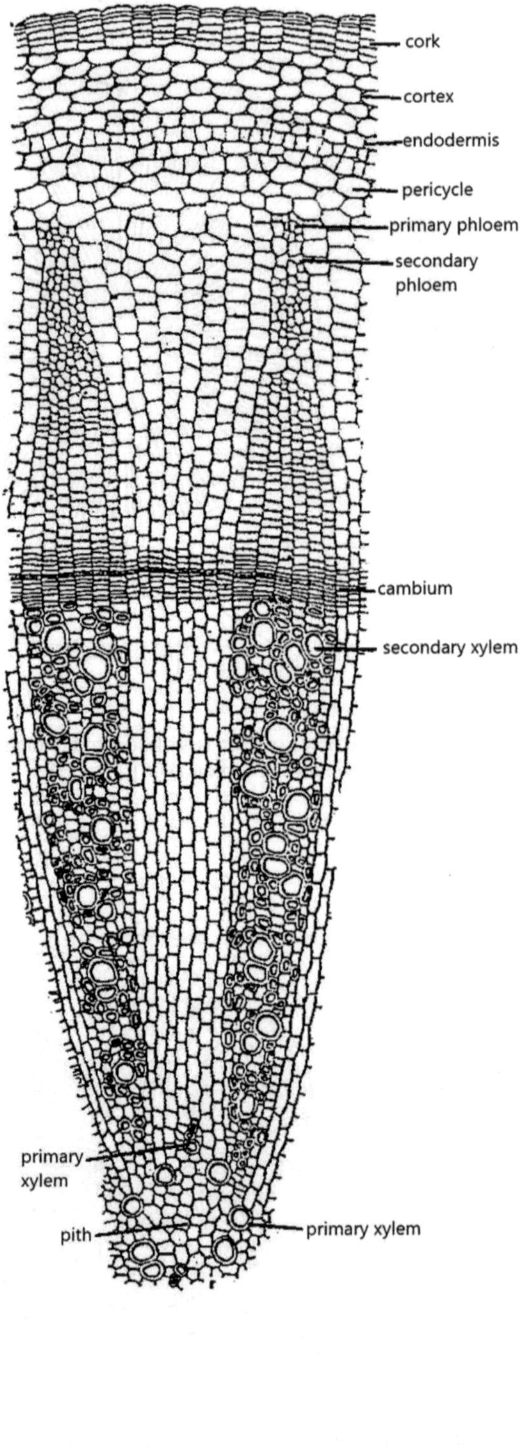

Fig. 25 Diagram of the cross section of the root with secondary structure of the *Lappa major* (dicotyledonous angiosperm) (Belzung 1900)

to distinguish the primary from the secondary xylem (Foster and Gifford 1959; Toma 1977).

A persistent pith is rarely distinguished, and, in this case, it is easy to recognize the primary xylem, arranged in the shape of a star (on a cross section), in roots with pentarch (five primary xylem bundles) or hexarch (six bundles of primary xylem) structure (Toma 1977).

4 "Abnormal" (Anomalous) Structures of the Root

The roots of some plant species show deviations from the "classical" structural organization described above for the primary and secondary root structures. These special arrangements are usually related to the function played by the root (tuberous roots), to the habitat in which the plant lives (e.g., water plants, epiphytes with aerial roots), or simply to structural characters typical for a certain botanical family. In all cases, regardless of whether the "abnormality" has an explicit ecological or functional significance, the general structural characters of the organ are preserved.

As long as these deviations of anatomical organization reflect, in fact, precisely the large structural, functional, and ecological diversity encountered in the plant world, probably the term structural "anomaly," used in classical, but also in modern anatomical literature (e.g., Crang et al. 2018), should be further reconsidered.

Among the "anomalies" of the root structure, we only mention:

- The changes induced by the tuberization of the root (which can come from the hypertrophy of the cortical parenchyma, as in *Ficaria* species, or from the development of a secondary tissue—either phloem parenchyma, as in carrots, or a xylem parenchyma, as in *Raphanus* species—radishes).
- Roots affected by the activity of successive (additional) cambia (in many species of halophytic *Chenopodiaceae/Amaranthaceae*);
- Intraxylary phloem—in *Atropa*, some species of *Gentiana*.

The phenomenon of successive (additional) cambia, which is found in the roots, but sometimes also in the stems of many species of salt-tolerant plants (halophytes) from the family *Chenopodiaceae* (*Amaranthaceae*), will be treated in detail; it is a structural "anomaly" for which certain hypotheses have been proposed that try to explain the ecological and functional significance of this phenomenon (Grigore and Toma 2006, 2010, 2021a, b, Grigore 2008).

The presence and activity of additional cambia are treated here as a fact encountered in many species of salt-tolerant plants that have been investigated by us, or by other authors. Moreover, we try now, as on another occasion (Grigore and Toma 2006), to attribute an ecological, adaptive significance to this phenomenon, which, it should be noted, is also widely found in non-halophilic species.

In the lines below, some elements related to the activity of the additional cambia within the stem have been also included, in order to have an overview of this phenomenon at the level of the axial vegetative organs (in this case, as a rule, the basal part of the stem is interested) (Grigore and Toma 2005, 2007b, 2008, 2010, 2017, 2021a, b; Grigore et al. 2013).

Coming back to additional cambia, it is a phenomenon known primarily due to the successive, concentric rings of *Beta vulgaris*. However, it also occurs in the roots and stems of other genera, such as *Amaranthus*, *Atriplex*, *Bougainvillea*, *Chenopodium*, *Cycas*, *Mirabilis*, *Phytolacca*, and *Welwitschia*, and species of *Petrosimonia*, *Halimione*, *Camphorosma*, *Suaeda*, and *Salicornia*, all representatives of the family *Chenopodiaceae,* include halophytic species that are among the most representative for the ecological group we are studying (for salt-tolerant plants of Romania, see Grigore 2012; Grigore and Cojocariu 2020).

The phenomenon of successive (additional, supernumerary, abnormal) cambia is considered by some authors (Hérail 1885; Metcalfe and Chalk 1950; Şerbănescu-Jitariu and Toma 1980; Fahn and Zimmermann 1982) to be a structural anomaly. It consists, in brief, of the following sequence of histological events: the typical general structure is generated by a "normal" cambium, which generally produces few vessels of xylem and secondary phloem. Later, on account of the pericycle, an additional (additional) cambium is born, and it will generate a ring of fundamental parenchyma, in which the vascular bundles are arranged on a circle, with the phloem to the outside and the xylem to the inside (Fig. 26). Each "abnormal" (successive) cambium then arises from the phloem parenchyma produced by the previous cambium (Esau and Cheadle 1969; Rajput and Rao 1999).

However, there is no uniform pattern governing the formation of successive cambia. The first additional cambium may arise in the primary phloem parenchyma, between the pericycle and the secondary phloem (Hayward 1938). The formation of rings, the activation of the formation of successive secondary cambia, represents a progressive, centrifugal process. In addition, the aforementioned author suggests that the way the vascular elements are distributed suggests that these secondary cambia do not appear as complete rings, but consist of discrete sectors derived from the pericycle.

It is obvious that this phenomenon occurs in the form of bands or strips of phloem and secondary xylem (growths, "vascular increments"), which are embedded in a "background" (fundamental tissue) of parenchymal or fiber-containing tissue.

It must be said that, from what we know, anatomical literature in general was limited to highlighting this phenomenon and characterizing it, most of the time, rather in a didactic, schematic, generalized manner. But the problem is much more complicated and complex, since a part of literature often uses different terms, new and difficult to equate and adopt in plant anatomy—the situation is also available for Romanian language. For example, the term "conjunctive tissue" is used to designate this fundamental tissue, parenchymatous or not, in which the productions of successive cambia are noticeable. It is a term first used by Schenck (1893) and Pfeiffer (1926); the latter uses it precisely to apply it to the genera in which this phenomenon occurs. Other authors, however, refer to this phenomenon, using terms such as "abnormal secondary structure" or "included phloem," although the secondary phloem is not completely included in the xylem, as Stevenson and Popham (1973) showed.

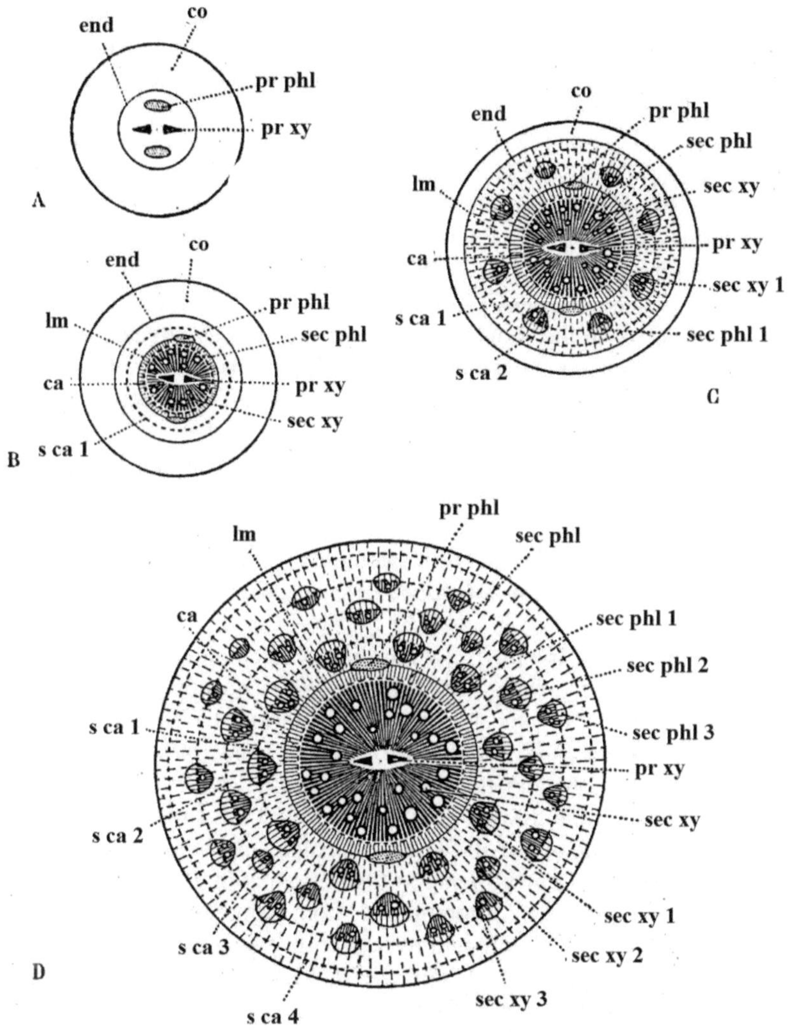

Fig. 26 Schema of functioning of the successive cambia in the root of *Beta* (*ca* normal cambium, *co* cortex, *end* endodermis, *pr xy* primary xylem, *pr phl* primary phloem, *sec phl* secondary phloem—1, 2, 3 indicate progressive stages, *sec xy* secondary xylem—1, 2, 3 indicate progressive stages, *s ca* 1, 2, 3—supernumerary cambia; *l m*, limit of the normal products) (Bonnier and du Sablon 1905)

Carlquist (2007) gives a good review of the phenomenon of successive cambia, discussing, among other things, the problems raised by the adequate understanding of this phenomenon.

A first problem is related to the use of a certain terminology, determined by the maintenance of some vague terms, which could have been perpetuated by those involved in the identification of xylem (xylotomists) and who, therefore, would have looked for simpler terms. It would not be excluded that the desire to use terms such as "included phloem" precisely represents the desire to consider this "background" tissue (conjunctive tissue) as "woody"; although often xylemic in structure, it should not be considered xylem in the narrow sense. Most specialists who have studied genera that present successive cambia use the term conjunctive tissue to denote the fundamental, fibrous, or parenchymal tissue, in which these vascular productions ("vascular increments") are embedded.

Second, although the authors described the histology of the species with sufficient precision, the understanding of the ontogeny of the successive cambia is still far from being understood. This and for a reason, let us call it rather, technical: the soft and hard tissues in the roots and stems of the species affected by successive cambia are actually mixed in the topography of the organs, so the soft ones and their developmental stages are often destroyed in sectioning and become practically uninterpretable.

A third cause of misinterpretations is related to the fact that xylotomists who highlighted this phenomenon often studied a single genus, sometimes even a single species. Although we generally accept a single plan to explain the operation of this phenomenon in different cases, its expression in the case of some "extreme" species (*Stayneria*, *Gnetum africanum*, *Mendonica*) can generate interpretation problems (Carlquist 2007).

This phenomenon is found in 34 families of dicotyledons (Carlquist 2001), within which the species of the order *Caryophyllales* occupy the first place; but it also appears in *Welwitschia*, species of *Gnetum* (Carlquist 1996), or in genera such as *Cycas*, *Encephalartos*, and *Macrozamia* (Greguss 1968). However, the number of species exhibiting this phenomenon is only a small part of the total number of species, but it still raises the following question: Why did this structural "pattern" need to evolve repeatedly, and above all, what is its value adaptive?

Finally, it is also worth mentioning in this introduction that the phenomenon of successive cambia must be differentiated from another structural "anomaly," namely, from the "interxylary phloem formed from a single cambium". It occurs in *Combretaceae* (van Vliet 1979), *Onagraceae* (Carlquist 1975), *Strychnos*, and other genera of the family *Loganiaceae* (Cockrell 1941). In this case, the islands or strips of interxylary phloem, formed by the single cambium, do not form corresponding pairs with the strips of vessels or groups of vessels; the vessels are located within the secondary xylem. In the case of additional cambia, the islands or bands of secondary phloem appear on the outside of those of secondary xylem. These bands (strands), considered as vascular productions (additions), are separated from each other by conjunctive tissue. As long as this conjunctive tissue is not secondary xylem, the term "interxylary" cannot be used in reference to plants with successive cambia (see Stevenson and Popham 1973; Carlquist 2007).

Carlquist (2007), discussing the ontogeny and consequences of the activity of successive cambia, provides a hypothetical, schematic overview of the functioning of successive cambia. On this occasion, we bring into discussion a histological reality little discussed in literature. It is the initiator of the whole sequence of histological events, which finally lead to the expression of successive cambia. The author promotes the idea of the existence of a master cambium, which produces external secondary cortex, rays, conjunctive tissue, *vascular cambium* (emphasis added), secondary xylem, and phloem, internally. In each new vascular product ("vascular increments"), the secondary xylem and phloem are formed by the vascular cambium. The vascular cambium functions indefinitely, so that the initiating cambium and a string of vascular cambia (each from the newly formed vascular complex) function indefinitely. The master cambium either remains active as long as the axial organ is actively growing (although it may become inactive after the formation of each vascular complex with its corresponding conjunctive tissue has been initiated), or, less commonly, it may be renewed from the secondary cortex. In certain taxa, such as *Gnetum africanum* or to certain tropical lianas (*Bauhinia*, family *Menispermaceae*), the master cambium is absent; new vascular cambia arise from one or two divisions in the cortical parenchyma, followed by a rapid tangential expansion of the vascular cambium. In fact, the term "master cambium" is preferred by Carlquist (2007) instead that of "lateral meristem."

As it is known, the first vascular cambium, in species with successive cambia, produces secondary xylem and phloem, in the same way as the vascular cambium produces in plants that have only one vascular cambium, which provides the secondary structure of the roots and stems. As we can see from the diagram in Fig. 27a (A), the master cambium arises following periclinal divisions, from the cortical cells, in the case of the stem, or from the pericycle, in the case of the root. From these divisions, a strip that has an unlimited circumference or a cylinder around the axis of the organ results. In stems, the parenchyma between the secondary phloem of the first cylinder of vascular tissues and the master cambium is the parenchyma from the primary cortex. As we already stated, the conjunctive tissue (hatched in black) is formed from the master cambium; it does not arise from the first adjacent cylinder of vascular tissues, but is formed inside each corresponding vascular complex (additions) (by vascular complex, we mean the secondary phloem and xylem, resulting from the activity of each additional cambium—"vascular increments").

Conjunctive tissue differs from cortical parenchyma in that its cells are arranged in radiating rows, and cortical cells are larger in diameter. It must be remembered that, after the formation of the master cambium, each vascular cambium continues to form secondary xylem and phloem, within the vascular complex that it "governs." Therefore, secondary xylem and phloem continue to add to the original vascular cylinder; therefore, a master cambium and a practically undefined number of vascular cambia coexist.

However, there is no uniformly expressed point of view on this concept of master cambium. Often, as Carlquist (2007) showed, there is a heterogeneous language, which contains terms that could be correlated with that of "master cambium". Thus, "anomalous cambium," "lateral meristem," "primary thickening meristem," "second cambium," and "additional cambium" are some examples of terms used to characterize this phenomenon.

Esau and Cheadle (1969) use the terms "old cambium" and "new cambium" in *Bougainvillea*; there is some inconsistency in this case, because sometimes we talk about the vascular cambium that produces phloem on the outside and xylem on the

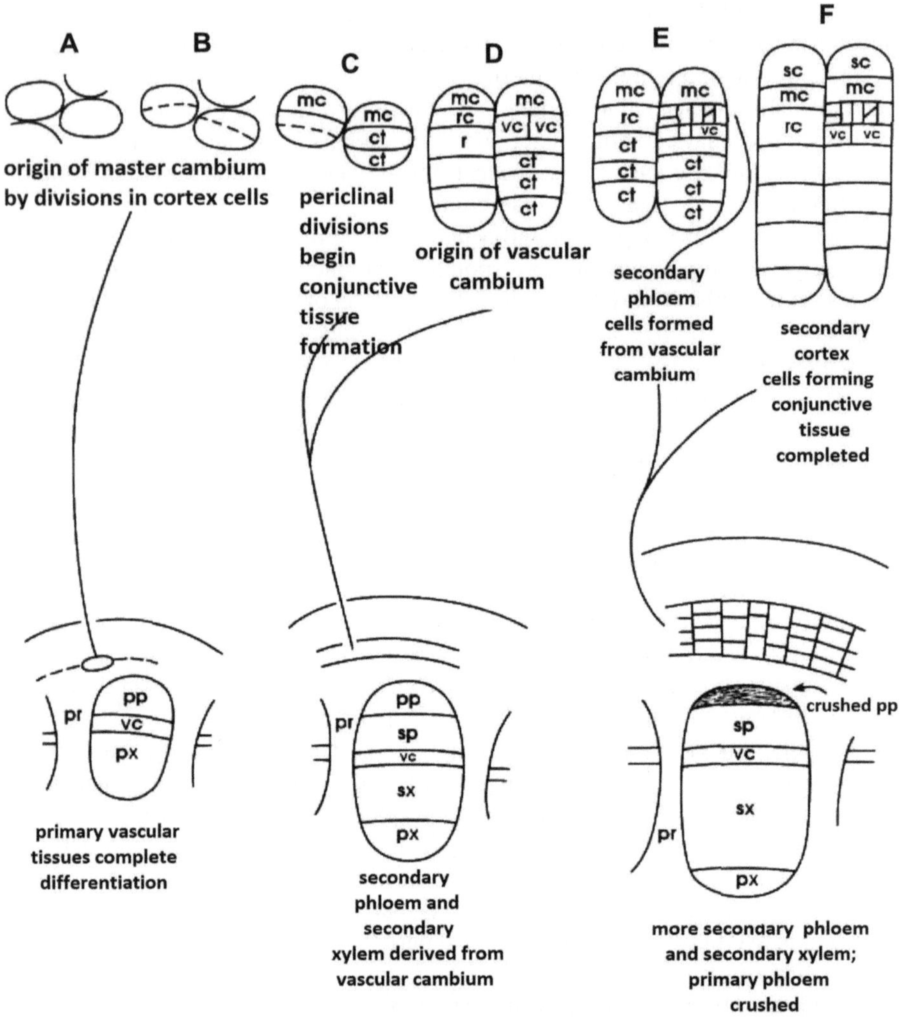

Fig. 27 a. Origin of the master cambium (in a hypothetical stem cross section) and stages in the production of cell types. (*c* cortex, *ct* conjunctive tissue, *mc* master cambium, *pd* periderm, *pp* primary phloem, *pr* primary ray, *r* ray, *rc* ray cambium, *sc* secondary cortex, *sp* secondary phloem, *sx* secondary xylem, *vc* vascular cambium) (Carlquist 2007)

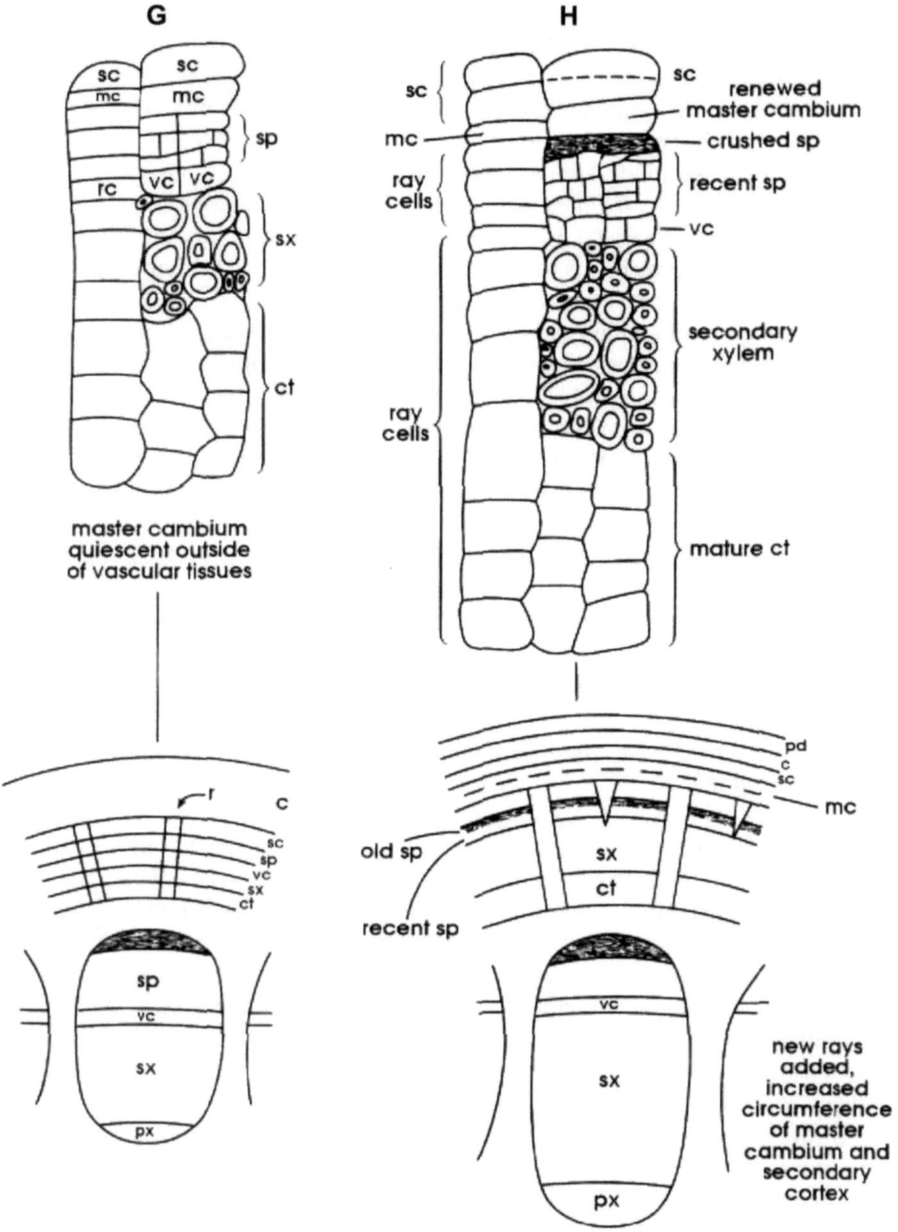

Fig. 2.27 (continued)

inside, and—in a more advanced stage—we talk about the fact that each of the cambium produces xylem and conjunctive tissue on the inside and phloem and conjunctive tissue on the outside.

The expression "master cambium," although used cautiously by Carlquist (2007), is considered by the same author to be the most appropriate to explain the formation of conjunctive tissue and vascular cambium inside, a situation observed in many taxa of the order *Caryophyllales*, but also in many other situations.

As far as we are concerned, we consider the introduction of this term particularly interesting, which corresponds to a histological reality. From the literature we have been able to go through so far, we can state that Carlquist (2007) is the first to distinguish two types of cambia (master cambium and vascular cambium) and consequently, two different types of specific activities, each with well-specified results. Most authors framed the problem based on a single type of cambium, even if, as explained above, they themselves were using inconsistently or imprecisely vague expressions. Carlquist (2004) outlines this differentiation, referring then to the "lateral meristem" to what will later become the "master cambium." By anatomically studying the roots and stems of some species of *Nyctaginaceae*, the author makes novel contributions regarding the lateral growth (in thickness) of these representatives. He states that a lateral meristem produces external secondary cortex and conjunctive tissue (both parenchymatous and fibrous), true rays (except *Bougainvillea* and *Heimerliodendron*), and a succession of vascular cambia. As each vascular cambium is formed, the lateral meristem on the outside of the vascular cambium tends to become inactive, returning to activity when the vascular cambium on its inner face becomes less active. It is, therefore, the cambium that will later be called the "master cambium."

Carlquist's point of view deserves to be discussed in more detail, as he provided a well-documented and quite critically exposed general plan. However, we will also present other points of view, which are not based on the two types of cambia.

It must be said that this master cambium is much easier to distinguish when it is actively dividing. The radial rows of conjunctive tissue cells, between the master cambium and the preceding vascular complex, allow the master cambium to be clearly distinguished; it is practically single-layered. In some species, the master cambium never ceases to function; in many species, the master cambium apparently functions for a long period of time. It may become inactive, but may regain activity; in many species, it does not change its position.

As for the vascular cambium, it arises from the master cambium, which, during the process, can maintain its identity (Fig. 27a-d). In this case, the way vascular cambium is formed is not different from that characteristic of most vascular plants: cells are produced, both inside and outside, and these transform into mature, visible cells.

The master cambium produced initial cells of conjunctive tissue prior to the formation of sieve tubes. Anticlinal divisions are involved in the formation of the vascular cambium.

Cells of master cambium, formed by periclinal divisions in cortical parenchyma cells, tend to have thick tangential walls, as do the cortical cells from which they originated. Xylem and phloem cells tend to be narrower than parenchyma cells. The intervention of the periclinal divisions of the vascular cambium is necessary to produce xylem and phloem. In fact, the number of rows of vascular cambial cells and their derived productions is greater than the number of initiating cambial cells and their derivatives. For example, in *Stegnosperma*, the ratio of vascular cambial cells to master cambial cells per surface unit is about 18:11; in *Heimerliodendron*, it is approximately 17:11. We believe that this aspect is important, because it represents an additional element for distinguishing the two types of cambia, vascular and master cambium. Since most authors considered only one type of cambium, nothing was mentioned in the literature about the process of transformation of the master cambium derivatives into cambial (vascular) initials.

Of note in all this complex and subtle sequence of cytohistological events is the fact that, when the vascular cambium is formed, the fusiform cambial initials are longer and more numerous than the master cambium cells, conjunctive tissue cells, or secondary cortex cells. These are considered by Carlquist (2007) as an obvious correlation with the fact that conjunctive tissue and secondary cortex are not produced by vascular cambium.

The secondary cortex, as stated before, arises from the master cambium. Generally, it is mono- or bi-layered. In certain cases, some authors have considered this cortex as conjunctive tissue (Esau and Cheadle 1969). We recall that, in the broadest sense, the term conjunctive tissue is reserved for the "background" tissue that connects one vascular complex ("vascular increment") to another.

As for the formation of vascular and master cambia, it must be said that in some tropical lianas (*Bauhinia*), vascular cambia form frequently, a fact noted by the appearance of vascular strips, isolated in the conjunctive tissue. Other species renew the master cambium irregularly, often only once in the life of the axial organ. In this direction, certain arguments can be made, according to Carlquist (2007):

1. *Beta* produces numerous vascular complexes ("rings") in a single season, each vascular complex having its own vascular cambium. This seems to be compatible with Artschwager (1926) that there is a single master cambium, which produces a small amount of secondary cortex. Otherwise, we should consider the numerous vascular complexes to be produced by two or more master cambia formed; this fact should be considered an uneconomical way to acquire the derived structures.
2. Cross sections made in many species of *Chenopodiaceae* show intact radial series of segments of vascular cambia, rays, etc. If this master cambia were continuously renewed, interruptions, discontinuities, and breaks should appear on these cross sections.
3. Many species of the order *Caryophyllales* have a distinct cylinder of cortical fibers, which delimits the outer cortex from the inner one. If the new master cambium were continuously produced externally, some of these fibers would be expected to be incorporated into the secondary tissues of the stem or root, which was not observed. The master cambium may be present in the layer of cells on the inner face of such fibers.

4. Although there are many differences in the schemes of the successive cambia, both among *the Caryophyllales* and between this order and others, there are no significant deviations from the basic plan, although in some genera (*Gnetum africanum* and some tropical lianas), the master cambia are absent.
5. The repeated renewals of the master cambia would raise problems in the formation of the interconnections between the vascular complexes; however, such interconnections are found (Kirchoff and Fahn 1984).

As mentioned before, the conjunctive tissue can be parenchymal, as in *Beta*. The fact that this example is the most "popular" has led some to consider this tissue to be parenchymal in nature. However, reality shows that fibers, sclereids, and idioblasts can be added to this parenchyma. Naturally, their proportion varies from species to species.

We also remember that, from the activity of the vascular cambium, secondary phloem results on the outside and secondary xylem on the inside. In species affected by successive cambia, the secondary phloem differentiates before the secondary xylem, in any of the vascular complexes (increments) (Artschwager 1926; Esau and Cheadle 1969); the phloem may contain fibers, or may consist mainly of parenchyma cells, as in roots of the *Beta* (Artschwager 1926) or *Mirabilis* (Mikesell and Popham 1976).

Hayward (1938), dealing with the structure of the organs in *Beta vulgaris*, refers to a secondary structure, related to the activity of these additional cambia. He believed that the point of origin of these exchanges varies with the level of the considered axis. Thus, at the level of the root and in the lower part of the hypocotyl, the first additional cambium arises in the area of the primary phloem parenchyma, between the pericycle and the secondary phloem. In the upper part of the hypocotyl, it arises from the pericycle; in the intermediate region, either from the pericycle or from the phloem parenchyma.

Here is how Hayward explains the formation of additional (additional) cambia: when the original (initial) cambium undergoes its first division, the outer cell of the two daughter cells becomes the initial cell of a new additional cambium, while the inner daughter cell divides further and produces xylem, phloem, and medullary rays (Fig. 28). This process is repeated until all additional cambia are formed.

An alternative explanation of the mechanism that produces the so-called secondary structure is also offered, an explanation that refers to the origin of secondary cambia because of the continuous activity of the pericycle. Early in ontogeny, it becomes a multilayered zone active in division, which keeps pace with the enlargement of the axis. It seems plausible that these secondary cambia occur in very rapid succession in the pericyclic parenchyma. During this interval, these cambia already formed are actively working, producing tissues of their respective rings, until the pericyclic tissue is no longer increased by radial and tangential divisions. In this situation, the pericycle perpetuates itself in the external area of the axis and produces phellogen, which in turn forms the cork.

Regardless of the hypothesis advanced as to the origin of these successive cambia, there is agreement as to the corresponding development of the tissues derived

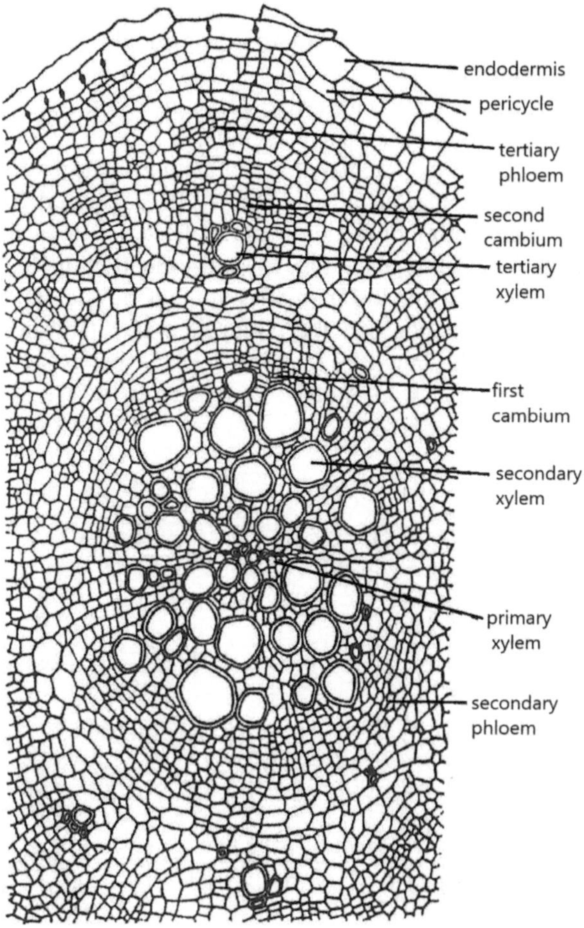

Fig. 28 Cross section through the central cylinder of the root of *Beta vulgaris*, highlighting—the primary, secondary structure and the initiation of the "tertiary" one (Hayward 1938)

from them. In the situation analyzed by Hayward, there are five to six relatively wide concentric rings, outside each of which there may be a few narrow ones; internal rings are not equal in width; this could suggest that, in the formation of the ring, the activation of the formation of successive cambia is progressive, centrifugal, and that some of them work simultaneously. Thanks to this mode of development, it is possible to determine the ontogeny of each ring in a centripetal direction: the outermost, placed immediately next to the periderm, is entirely meristematic, consisting of cambial cells (derived from the pericycle), parenchyma, and undifferentiated vascular elements. The first to differentiate are the sieve tubes and companion cells.

In most cases, the xylem elements are arranged in narrow radial strips and are separated by areas of parenchymal rays. In general, the formation of phloem cells precedes the differentiation of xylem cells.

The way the vascular elements are distributed indicates that these successive cambia do not appear as complete rings, but as discrete sectors derived from pericycle. The parenchymal cells intervening between the radiating rows of vascular elements are of pericyclic origin and should, according to this author, be regarded as pericyclic rays.

The secondary and tertiary structures in the case of the root and stem of *Beta* have the conformation specified in Figs. 28 and 29; the structure of the stem is here mentioned only to achieve a better picture on the way in which successive cambia act.

On this occasion, a short parenthesis is required. Since much has been said about the intervention of the pericycle in the series of events characterizing the phenomenon of successive cambia, we must mention the fact that Hérail (1885) discusses this aspect within the anomalies of the pericycle. So, the close relationship between this phenomenon and the pericycle was established long before. Unfortunately, however, the language used by Hérail, specific to that era, is not very explicit, because there were still no certain terms as we know and use them today. For

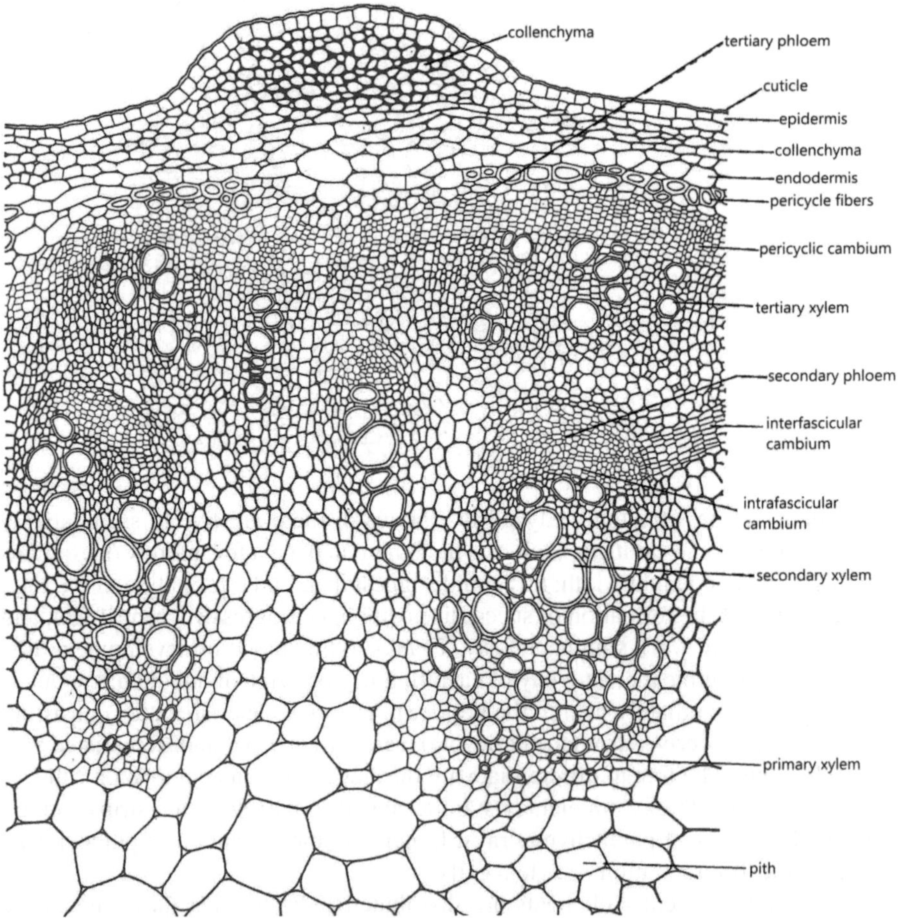

Fig. 29 Cross section through the stem of *Beta vulgaris* (Hayward 1938)

example, the word cambium is not used, although sometimes it can be intuited from the context when it is used. However, the similarity of this phenomenon in *Chenopodiaceae* with that happening in the case of the *Phytolaccaceae* family is recognized. In Hérail's text, there are mentions about a fragmentation (differentiation?) of the pericycle, with the differentiation *of the new meristem* (our emphasis), which forms vascular bundles. Between them, conjunctive tissue is located. Sometimes, he refers to *generative zones*, equated with *consecutive meristems*, sometimes with *successive meristems*; finally, a *primitive* meristem is also mentioned, "which passes outside the phloem bundles, due to the thickening of the pericycle. This thickening always starts on one side and goes into the interfascicular meristem."

The phenomenon has been observed in many genera belonging to the *Amaranthaceae* family: *Achyranthes*, *Acnida*, *Aerva*, *Alternanthera*, *Amaranthus*, *Bosea*, *Celosia*, *Chamissoa*, *Deeringia*, *Froelichia*, *Gomphrena*, *Hermbstaedtia*, *Iresine*, *Pfaffia*, *Psilotrichum*, and *Pupalia* (Metcalfe and Chalk 1972). Joshi (cited by Metcalfe and Chalk 1972) believed this kind of structural anomaly, in *Amaranthaceae* and *Chenopodiaceae*, may be present in the root or base of the stem, when absent from the rest of the stem. Conjunctive tissue is more abundant in the root than in the stem and decreases as the distance from the root increases. For example, in *Achyranthes*, in the root, successive cambia form wide arcs or complete rings, and in the stem, their activity is limited to the formation of small arcs. Joshi concluded that the formation of the succession of cambia is an ancestral character in these families, the primitive forms having several areas of vascular bundles embedded in the underlying parenchymal tissue, formed as a result of the activity of a corresponding number of secondary cambial rings. Evolution would have led either to the loss of this anomaly from the stem, or from both stem and root, or to the reduction of successive cambia to smaller and smaller segments.

Our own investigations have highlighted the phenomenon of successive cambia in many halophyte species of the family *Chenopodiaceae* (included, according to some modern classification systems, in *Amaranthaceae*) from different countries (Romania, Spain) and climates (temperate, Mediterranean): *Atriplex prostrata*, *A. littoralis*, *A. tatarica*, *Halimione verrucifera*, *Petrosimonia triandra*, *P. oppositifolia*, *Suaeda maritima*, *Salicornia europaea*, *S. ramosissima*, *Camphorosma annua*, *Sarcocornia fruticosa*, and *Arthrocnemum macrostachyum* (Grigore and Toma 2005, 2006, 2007a, 2008, 2010, 2017, 2021a, b; Grigore 2008; Grigore et al. 2014).

Thus, in the *Atriplex prostrata* investigated by us (Grigore and Toma 2005), the phenomenon affects the root, along its entire length. In the lower third (Fig. 30a), the primary structure is more often of the diarch type, and in the first ring, produced by the additional cambium, the vessels have an irregular arrangement and are of large diameter, and the walls are moderately thickened and lignified; between the vessels, the xylem parenchyma cells have cellulosic walls, and the libriform is represented by fibers with moderately thickened and lignified walls. Gradually, a second additional cambium is formed, which produces a relatively compact ring of xylem towards the inside and a phloem one towards the outside. In the xylem ring,

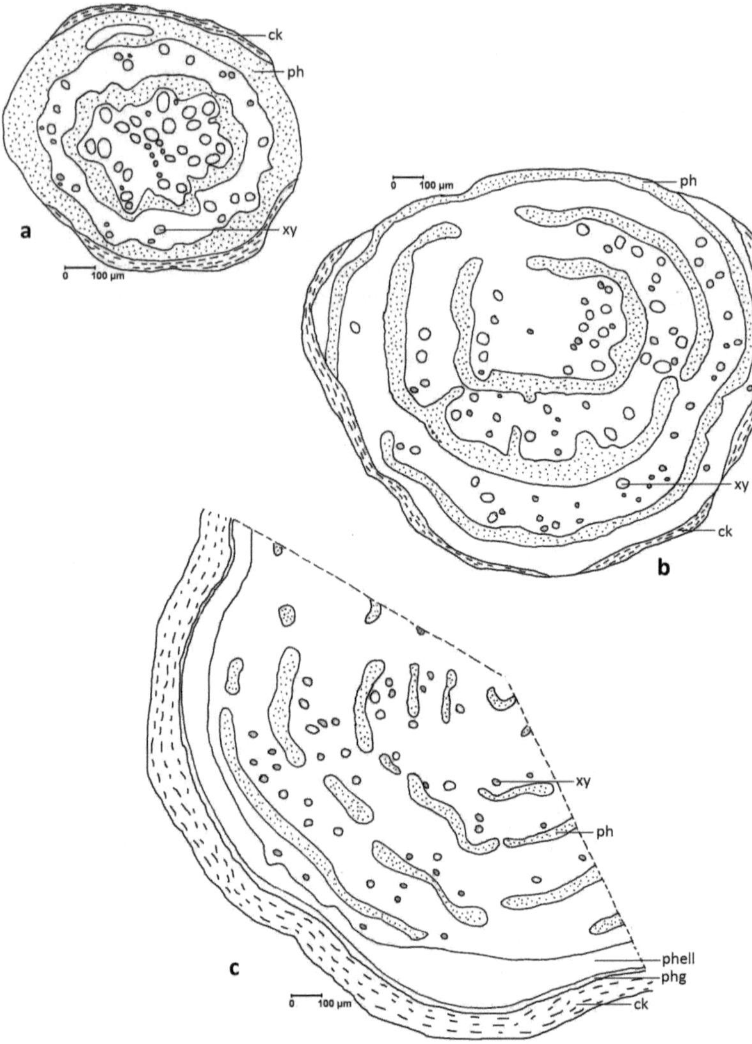

Fig. 30 Cross section through root from *Atriplex prostrata*. (**a**), lower third; (**b**), middle third; (**c**), upper third (*phg* phellogen, *phell* phelloderm, *xy* xylem, *ph* phloem, *ck* cork (Grigore and Toma 2005)

the libriform predominates, the parenchyma cells being very rare, as are the vessels, the latter being of different diameters and irregularly dispersed. The phloem ring comprises sieve tubes, companion cells, and phloem parenchyma cells.

In the middle third of the root (Fig. 30b), four xylem and phloem rings are visible, the general structure presenting a visible asymmetry; on one of the faces of the section, only two xylem and phloem rings are noticeable. All the xylem rings are intensely lignified, in which the libriform predominates, and the vessels are dispersed irregularly. The phloem rings are entirely cellulosic, comprising rings of

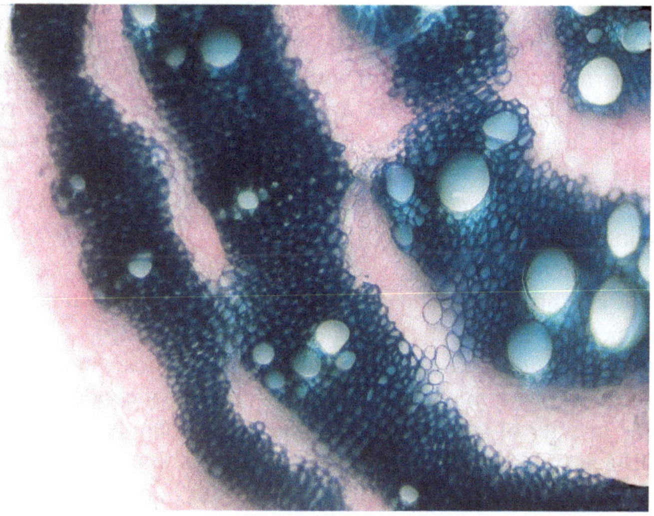

Fig. 31 Cross section through the root—middle third—in *Atriplex prostrata* (Grigore and Toma 2010)

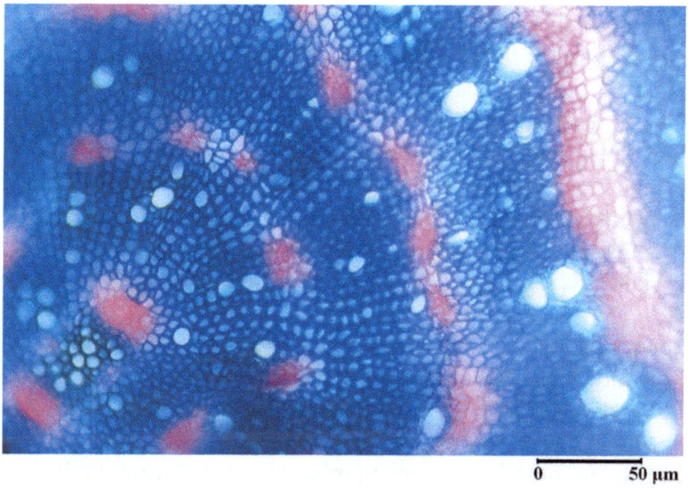

Fig. 32 Cross section through the root—basal third—in *Atriplex prostrata* (Grigore and Toma 2010)

conducting phloem (sieve tubes and companion cells) separated by rings of phloem parenchyma (Fig. 31). In places, the xylem rings are pierced by cellulosic parenchymatous rays, and in the phloem rings, thin bands of elements with lignified walls penetrate.

Towards the basal level of the root (Figs. 30c and 32), one can observe six to seven concentric rings of xylem, completely lignified, separated by an equal number of phloem rings, completely cellulosic. Xylem and phloem rings are more tortuous, of different thickness, fragmented in places or in contact with each other. In the central portion, xylem tissue predominates, and the phloem forms small islands completely surrounded by elements with thickened and lignified walls; a similar situation can also be observed towards the outside of the root, where the rings of cellulosic phloem are fragmented by radiating bands of lignified tissue.

The center of the organ is occupied by a compact mass of xylem, intensely lignified, having on one side and the other, so in opposite disposition, two bundles of phloem visible collenchymatized.

Interestingly, at the level of the stem, only the basal part is affected by the activity of successive cambia (Fig. 33), a situation, as we had already anticipated, common to other species that exhibit this phenomenon.

At *Atriplex littoralis* (Grigore and Toma 2007a), in the lower third of the root, the central cylinder shows a number of three to four rings of conductive tissues, resulting from the activity of successive cambia, each of them having thicker xylem on the internal face and, on the external face, thinner phloem.

In each ring, the xylem is entirely sclerified and lignified, with few irregularly dispersed vessels; the xylem fibers, which predominate in the xylem structure, have a strongly thickened and moderately lignified wall.

The phloem consists of sieve tubes, companion cells, and phloem parenchyma cells. The last ring of conducting tissues is, on the analyzed section, still forming, being unequal in thickness on the circumference of the root.

In the middle third of the root (Fig. 34), the general structure remains the same, with the same number of concentric conducting rings. The xylem vessels in the axial area have a smaller diameter, but with more intensively lignified walls. Libriform has fibers with extremely thickened walls, but only partially lignified.

Towards the upper third (Fig. 35), the root is thicker; in its structure, there are five (six) concentric rings of conducting tissues, but the phloem no longer forms continuous rings, but discontinuous ones, in the form of islands surrounded by xylem; the latter is strongly sclerified and lignified, in which the libriform predominates.

In the upper third of the stem, the structure remains normal, a fact also emphasized by Metcalfe and Chalk (1972), who indicate that the phenomenon affects species with relatively thick stems, but sometimes it was not observed in the upper parts of the stem and was not observed at all in species with thin stems. At this level, the central cylinder comprises a large number (14–16) of vascular bundles, of open collateral type and of different sizes; the largest bundles project deep into the medulla.

In the middle third, in the central cylinder, in addition to the mentioned bundles from the previous level, a thick libriform ring can be observed, in which numerous bundles (blades) of xylem vessels and islands of phloem elements are located on the outside. The phloem islands have visibly collenchymatized elements, and, on their exterior, cords (caps) of sclerenchymatous fibers with thickened but cellulosic walls can be observed.

The additional cambium, which gave rise to the aforementioned xylem and phloem, is continuous and multilayered.

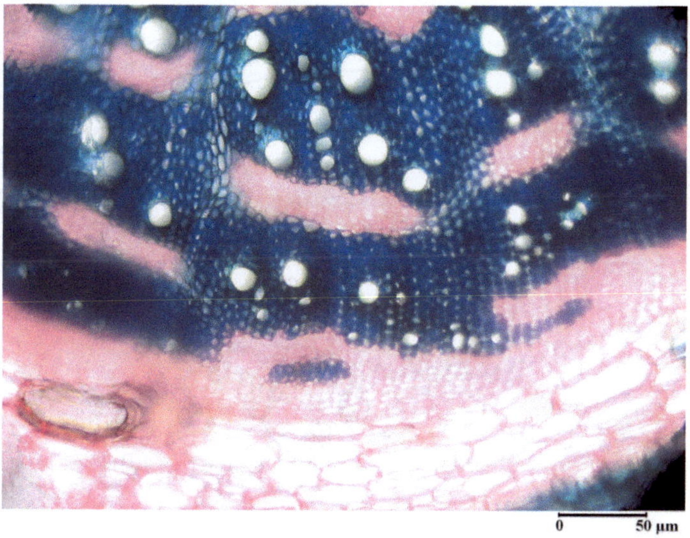

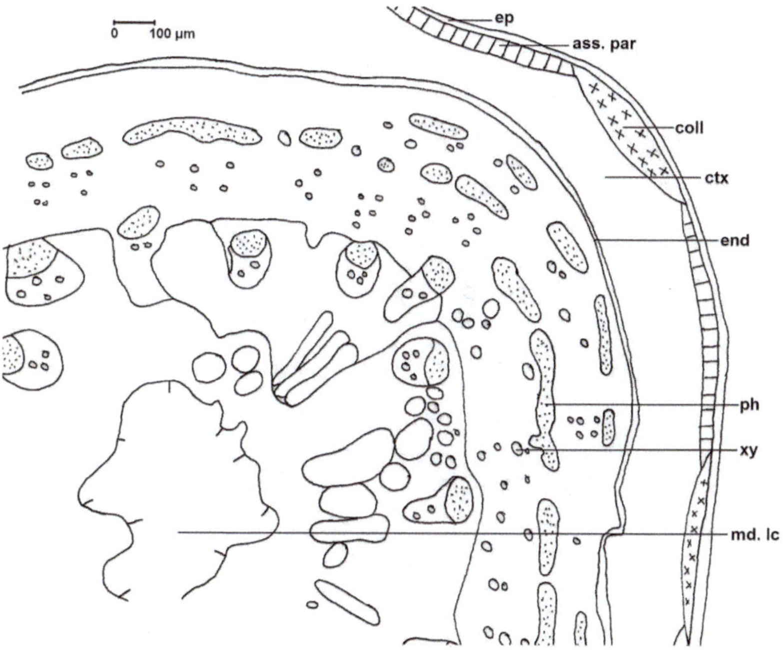

Fig. 33 Cross sections through the stem of *Atriplex prostrata*, basal third (top, photo taken on light microscope; bottom, ink diagram); *coll* collenchyma, *end* endodermis, *ep* epidermis, *ph* phloem, *xy* xylem, *md. lc.* medullary lacuna, *ass. par.* assimilating parenchyma, *ctx* cortex (Grigore and Toma 2010)

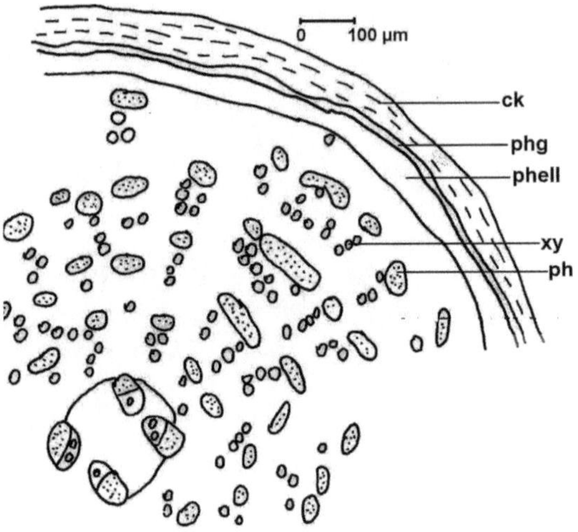

Fig. 34 Cross section through the root from *Atriplex littoralis*, middle third; *phell* phelloderm, *phg* phellogen, *ph* phloem, *xy* xylem, *ck* cork (Grigore and Toma 2010)

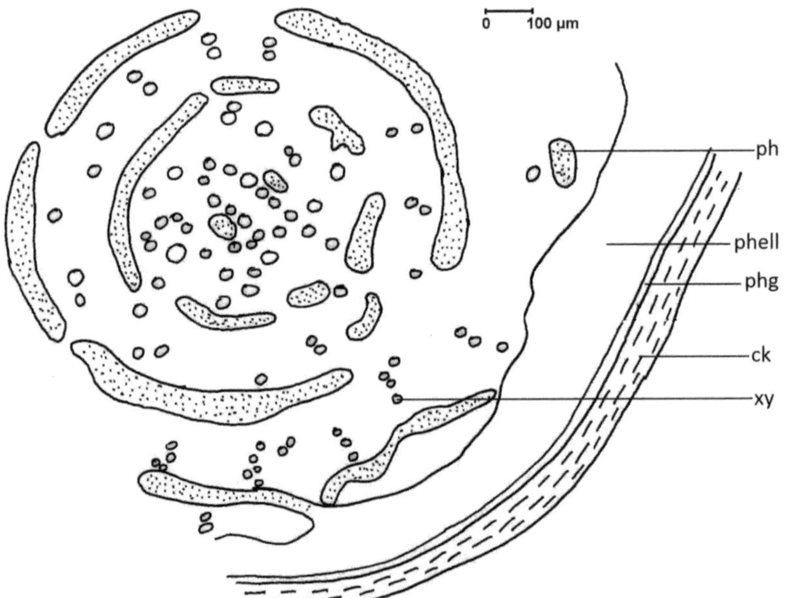

Fig. 35 Cross section through root from *Atriplex littoralis*, upper third; *phell* phelloderm, *phg* phellogen, *ph* phloem, *xy* xylem, *ck* cork (Grigore and Toma 2010)

Towards the base of the organ (Figs. 36 and 37), in the central cylinder, following the activity of successive cambia, three to four concentric rings of conducting tissues resulted, the phloem one being in the form of islands surrounded by xylem tissue. The fascicular structure observed in the previous levels is no longer so visible. The libriform is more intensely sclerified and lignified, the xylem vessels are

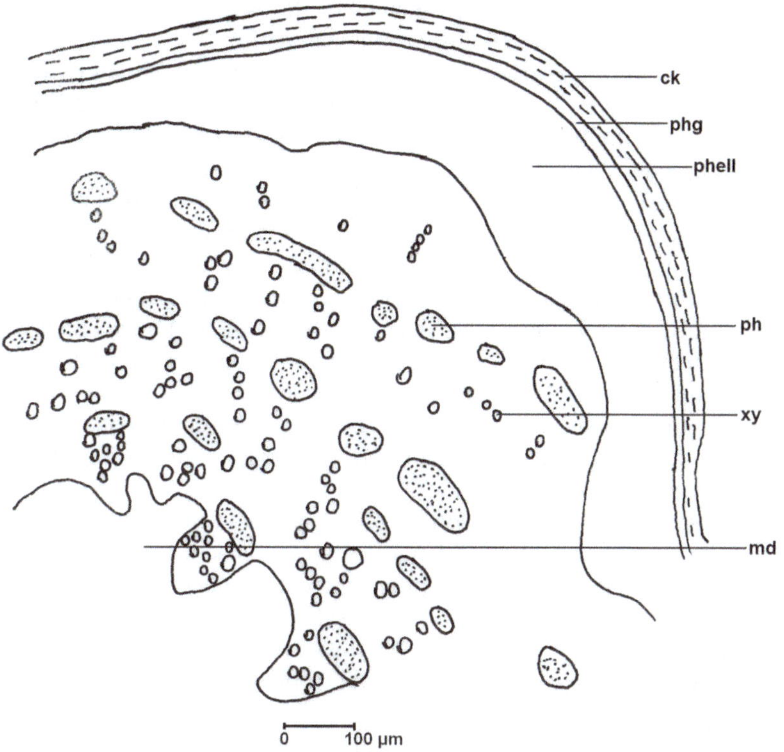

Fig. 36 Cross section through stem from *Atriplex littoralis*, lower (basal) third; *phell* phelloderm, *phg* phellogen, *ph* phloem, *xy* xylem, *ck* cork, *md* medulla (Grigore and Toma 2010)

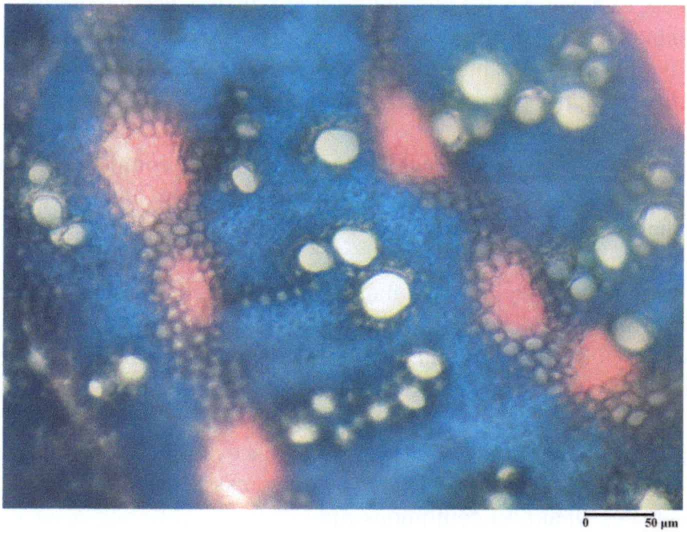

Fig. 37 Cross section through stem from *Atriplex littoralis*, lower third (Grigore and Toma 2010)

more irregularly dispersed, and the bundles of the original primary structure are completely embedded in the cellulose parenchymal pith.

At *Halimione verrucifera*, we noticed, regarding the root structure, that, in the central cylinder, it has a structure characteristic of the representatives of the *Chenopodiaceae* family, the structure resulting mostly from the activity of additional cambia (Figs. 38 and 39).

The axis of the root is occupied by a completely sclerified and lignified mass of xylem, in which three radial blades of primary xylem vessels can be observed, arranged in a single row; according to these primary xylem blades, it can be considered that the central cylinder with primary structure is of the triarch type.

This compact central xylem mass is surrounded by three phloem bundles, which alternate with the three radiating blades of primary xylem and are separated by elements of cellulosic parenchyma.

Next are two sinuous rings of xylem, resulting from the activity of additional cambia, separated by phloem zones, entirely cellulosic, resulting from the activity of the same meristems.

The xylem rings are strongly sclerified and lignified, in which libriform predominates, and the vessels have different diameters; more often, these vessels are grouped together and next to them, on the periphery, there are phloem islands; these islands are separated by radiating blades, formed by sclerified and lignified elements.

The second (external) xylem ring is still under organization, having few vessels, and on their exterior, the cambium produced few phloem elements.

Towards the middle third part of the organ, the general structure remains the same, with the difference that the second discontinuous (island-type) phloem ring was also formed, surrounded on the outside by the xylem of the third ring.

As in the previous level, the central part of the central cylinder has a slightly trilobed appearance of the xylem superimposed on the three phloem bundles.

At the base of the organ, we mention, in addition, the fact that the second discontinuous (island type) phloem ring was also formed, surrounded on the outside by the xylem of the third ring. As in the previous level, the central part of the stele has a slightly trilobed appearance of the xylem superimposed on the three phloem bundles.

In the stem, the central cylinder, in the upper third, comprises several (eight) vascular bundles of different sizes, with most of the xylem and phloem of primary origin; in the vicinity of the phloem, the xylem vessels are separated by few libriform elements.

At the periphery of the bundles, and between them, there is a sclerenchymatous tissue that forms a thick ring, with cells having extremely thickened and intensely lignified walls; in this way, the phloem of the conducting bundles appears to us in the form of islands surrounded by a sclerified and lignified tissue.

At the periphery of the sclerenchyma ring, the first additional cambium (of pericyclic origin) can be observed, which in some places formed several xylem vessels to the inside and several phloem elements to the outside; the latter, together with the cambium, creates an almost continuous ring.

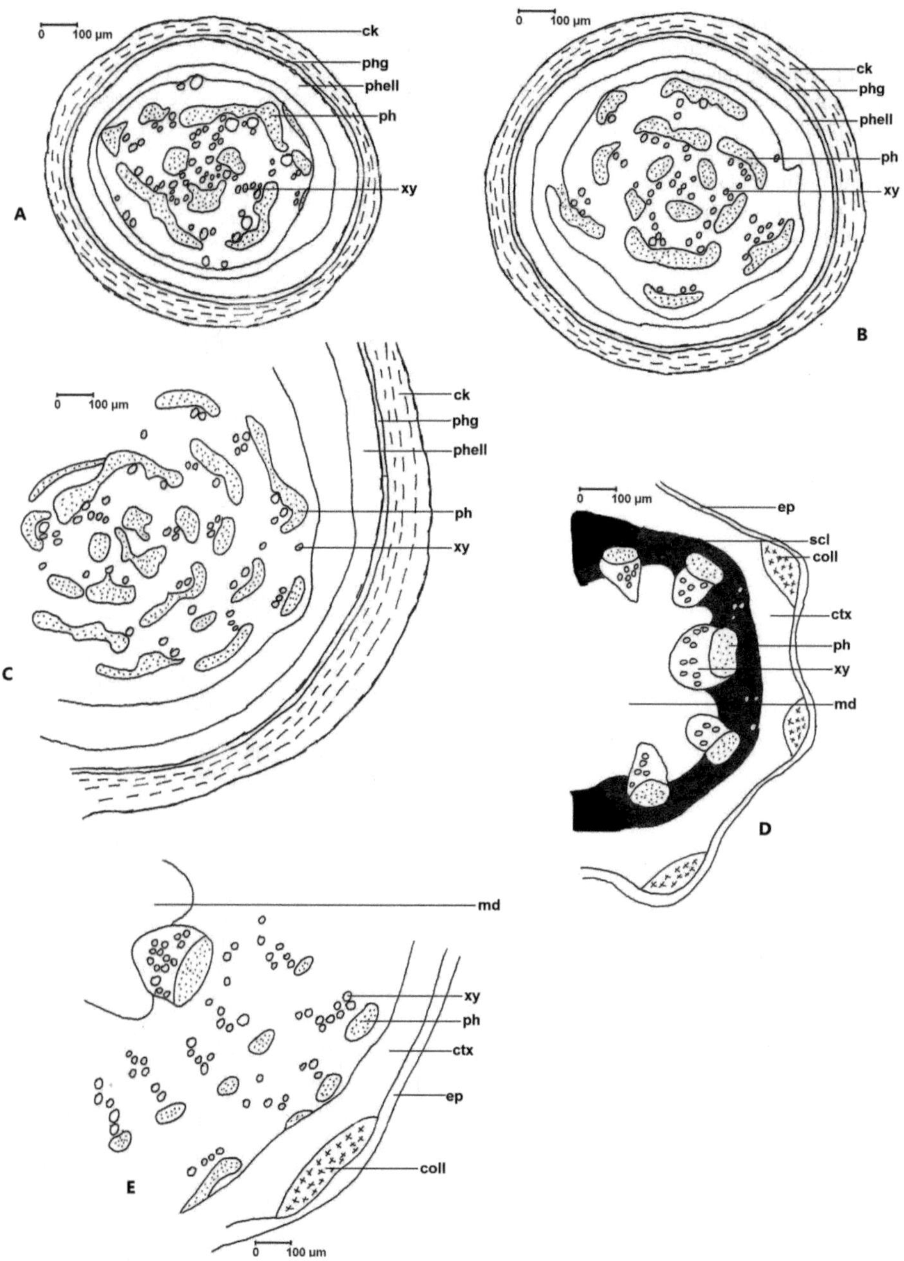

Fig. 38 *Halimione verrucifera*—Cross section through the root, (**a**), lower third; (**b**), middle third; (**c**), upper third, and stem, (**d**), upper third; (**e**), middle third; *coll* collenchyma, *ep* epidermis, *phell* phelloderm, *phg* phellogen, *ph* phloem; *xy* xylem, *md* medulla, *ck* cork, *ctx* cortex, *scl* sclerenchyma (Grigore and Toma 2010)

Fig. 39 *Halimione verrucifera*: (**a**–**c**)—Cross sections through the stem, lower third (diagrams); *phell* phelloderm, *phg* phellogen, *ph* phloem, *xy* xylem, *md* medulla, *ck* cork, *ctx* cortex (Grigore and Toma 2010)

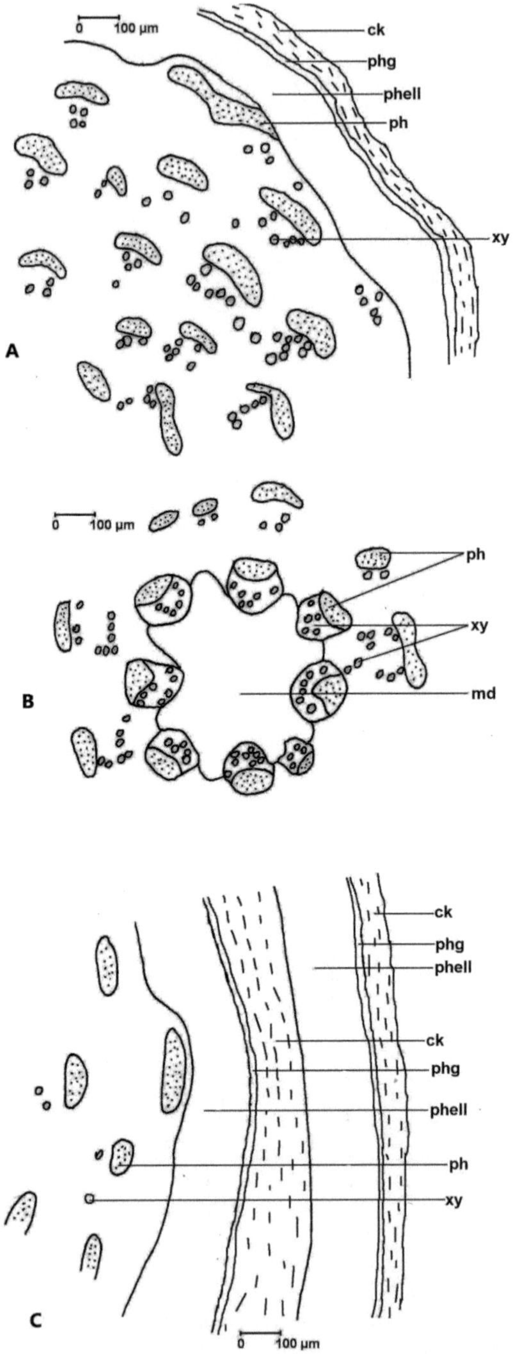

Compared to the previously analyzed level, the central cylinder in the middle region of the stem is thicker, comprising a thick, strongly sclerified, and lignified ring of fundamental parenchyma and a pith in which the eight conducting bundles of the primary, normal structure strongly protrude.

In this sclerified and lignified ring, xylem vessels with an irregular arrangement, but often forming discontinuous radiating rows, can be observed from place to place; each of these strings has an island of phloem tissue at its periphery.

The first row of xylem vessels, together with the corresponding phloem island, forms a thick ring of vascular "bundles" embedded in the libriform groundmass.

The second ring of bundles is much thinner, also having the phloem islands (caplets) surrounded by intensely lignified sclerenchyma tissue.

In some places, on the inner face of the primary cortex, a few sclerenchyma cells can be observed, often solitary, located on the periphery of the last ring (in state of forming) of additional cambium and late-formed phloem elements.

Finally, the central cylinder is much thicker in the lower third than at the previously analyzed levels, comprising a number of four to five rings of cellulosic phloem islands; they correspond, in fact, together with the xylem on the internal face, to the same number of rings of conducting bundles, of the collateral type, embedded in the fundamental libriform mass, extremely strongly sclerified and lignified.

In another halophytic species from the family *Chenopodiaceae* investigated by us, *Petrosimonia triandra* (Grigore and Toma 2007a, b), in the general structure of the root (Fig. 40a, b), in the lower third, six to seven concentric rings of conducting tissues can be distinguished, resulting from the activity of a corresponding number of additional cambial zones. A large amount of libriform, a smaller number of irregularly dispersed xylem vessels, and numerous thin tangential bands of libriform tissue are observed, which mark the place and number of conducting bundles, which arose from the activity of additional cambia.

All mechanical elements of sclerenchyma (libriform) have extremely thick walls and, for the most part, intensely lignified.

In the central part of the root, at the middle level, the primary structure, of the diarch type, can still be distinguished, followed by a relatively homogeneous structure, represented by several concentric zones of vascular bundles, between which are sclerified and lignified rays.

The xylem vessels in the primary structure have a small diameter, and the walls are intensely lignified; the xylem vessels in the bundles resulting from the activity of successive additional cambia have a much larger diameter and a very thick wall, but very poorly lignified, dispersed irregularly and with a lot of libriform between them.

The phloem of the conducting bundles forms thin tangential bands, interrupted here and there by narrow rows of mechanical cells with strongly thickened and moderately lignified walls, like those of the libriform fibers in the structure of xylem.

At the sectioned level, approximately four concentric rings (circles) of conductive tissue can be observed, resulting from the activity of four additional cambia.

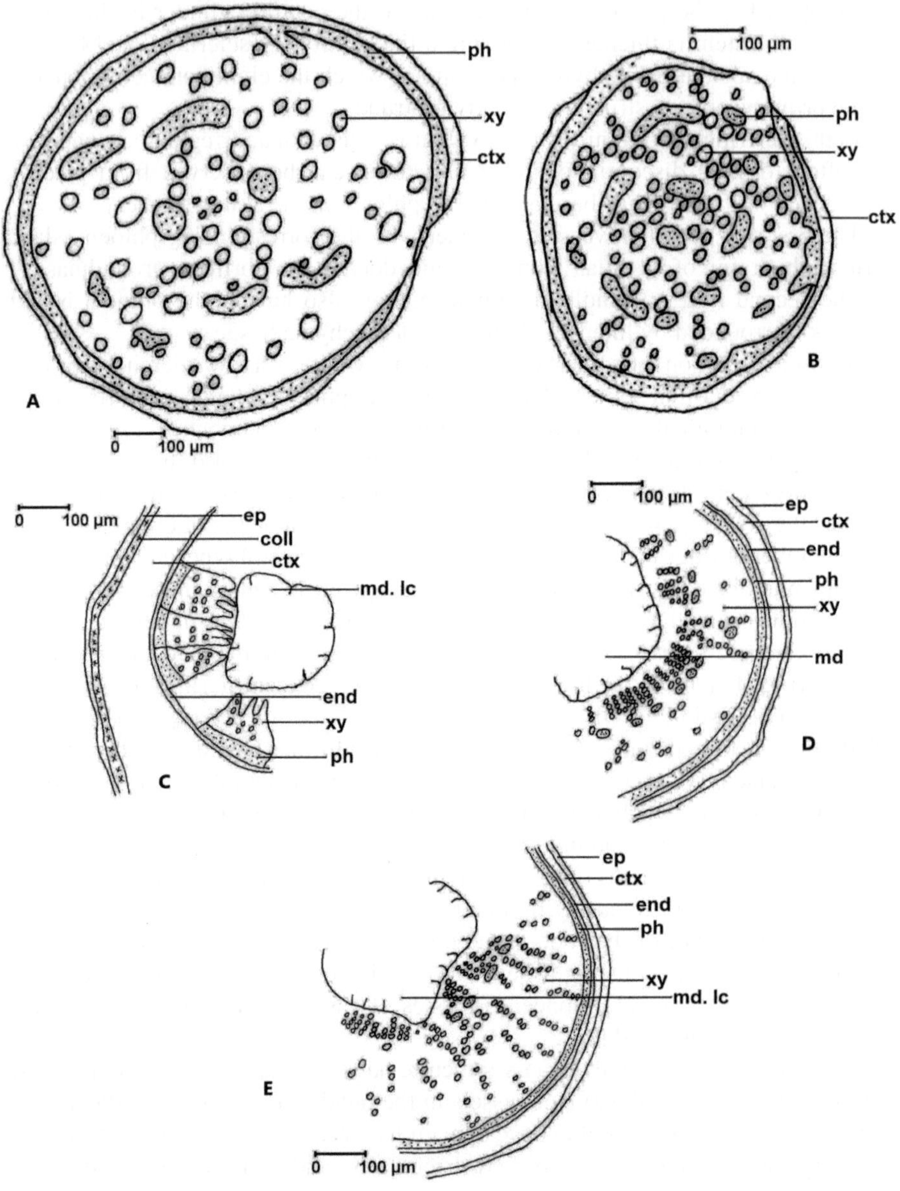

Fig. 40 *Petrosimonia triandra*: Cross sections through the root, lower third (**a**), middle third (**b**), and stem, upper third (**c**), middle (**d**), lower (**e**); *ph* phloem, *xy* xylem, *md* medulla, *ck* cork, *ctx* cortex, *end* endodermis, *md. lc* medullary lacuna, *coll* collenchyma; *ep* epidermis (Grigore and Toma 2010)

As for the stem (Fig. 40c, d, e), at the level of the upper third, the central cylinder includes several (five to seven) vascular bundles, of different sizes, separated by medullary parenchymatous-cellulosic rays.

All vascular bundles show phloem that consists of sieve tubes and companion cells and xylem formed by radiating rows of vessels, separated by cells of cellulosic parenchyma; therefore, the structure is typically primary.

On the periphery of these vascular bundles, the first additional cambium, of pericyclic origin, has already appeared, which begins to function bifacially, giving, here and there, xylem to the inside and phloem to the outside; the process of tracheogenesis is ongoing, some xylem vessels having very thin, cellulosic walls.

It should be emphasized that this first additional cambium forms a continuous, multilayered ring, which will produce vascular bundles and, in some places, medullary rays.

At the level of the middle third, in addition to the primary and secondary structures, we could also highlight a structure resulting from the activity of successive cambia, characterized by a very large amount of libriform, strongly sclerified, but little lignified, in which there are irregularly dispersed vessels of xylem; at the periphery of this thick ring of xylem, a thin ring of phloem elements is observed. We recall here the fact that some authors still use the name "tertiary structure" to denote all products of additional cambia (Hayward 1938; Şerbănescu-Jitariu and Toma 1980).

Towards the lower third, the structure remains the same, except that the central cylinder is much thicker, comprising a large amount of libriform, relatively few xylem vessels, dispersed irregularly or forming discontinuous radiating rows, and a few very small islands of libriform tissue.

Other halophyte species of the family *Chenopodiaceae* that have been analyzed (Grigore and Toma 2008) also show this interesting structure. Thus, at *Petrosimonia oppositifolia*, the phenomenon is expressed in the root (Fig. 41a), at all three analyzed levels, as well as in the stem (Fig. 41b), especially in the lower and middle third of the organ. In the root, the conducting tissues form incomplete rings or arches of phloem and xylem in the lower third of the organ; in the central position, compact secondary xylem is observed, with two islands of secondary phloem on either side; around this secondary structure areas of xylem are located, with irregularly arranged vessels, separated by much libriform strongly sclerified and intensely lignified. Towards its middle third, the root is thicker, the structure remaining the same, but there are more (three to four) concentric zones of xylem and phloem; the latter is thinner, and often discontinuous, forming wide-open arcs or islands. We also note the presence of a significant amount of libriform in this species, which has been highlighted in the central cylinder of the axial organs.

Successive cambia are also present in the root of *Atriplex tatarica* (Fig. 41c), where, in its lower third, two such additional cambia function, and in the middle third, four cambia; in all cases, their activity produces more xylem, on the inside, and less phloem, on the outside. Following the same model, the cambium forms a central, compact xylem mass, entirely sclerified and lignified.

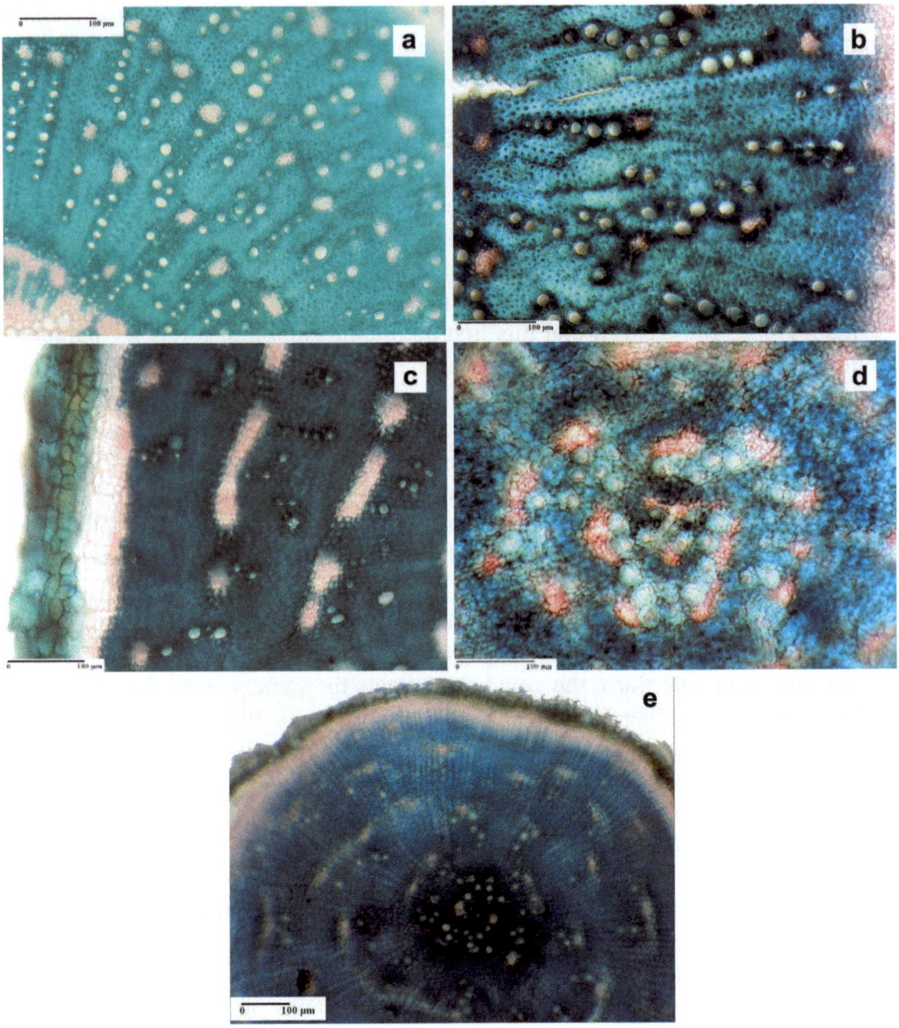

Fig. 41 Cross sections through the root (**a**) and stem (**b**) of *Petrosimonia oppositifolia*; the root of *Atriplex tatarica* (**c**); the root of *Salicornia europaea* (**d**); the root of *Suaeda maritima* (**e**) (Grigore and Toma 2010)

At *Salicornia europaea*, successive cambia affect both the root (Fig. 41d) and the unarticulated portion of the stem. At the root, successive cambia give rise to many irregularly dispersed xylem vessels and small islands of phloem. In the unarticulated stem, the structure resulting from the activity of additional cambia is represented by a very thick libriform ring, in which several small phloem islands are incorporated; on their internal face, few xylem vessels are found, the diameter of which does not differ much from that of the xylem fibers; however, they are distinguished by a slightly thinner and less lignified wall. At *Suaeda maritima*, in the root

(Fig. 41e), as a result of the activity of additional cambia, a number of five rings of conducting tissues result, each of them having the largest part occupied by libriform, with few vessels dispersed irregularly in it, and numerous islands of phloem, separated by cellulosic or weakly lignified parenchyma; these islands are not equidistant nor of the same size.

On closer analysis, several radiating rows of xylem vessels near the phloem islands can be distinguished; they are separated by a very large amount of libriform; the fibers of the latter have extremely thickened and moderately lignified walls. In the upper third of the stem, all the vascular bundles have a beginning of a secondary structure, visible at least at the level of the xylem where, among the extremely thick-walled vessels, there are also some libriform fibers. Particularly interesting is the fact that in none of the species of *Chenopodiaceae* investigated by us and other authors, have no xylem vessels with thickened walls, but weakly lignified, like those of libriform fibers.

In another halophyte species of the same family, *Camphorosma annua*, in the root structure, a number of two to three additional cambia function, resulting in a corresponding number of xylem and phloem rings, the xylem ones being much thicker and strongly lignified (Fig. 42a). The phloem forms discontinuous arches in the central area of the organ (Fig. 42b).

In the whole thickness of the root, in *Camphorosma annua*, but especially in the external xylem rings, the libriform fibers have extremely thick and intensely lignified walls. The secondary structure is recognized by the large number of vessels per unit area and the libriform elements between them, with moderately thickened and poorly lignified walls.

Regarding the structure of the stem, we highlighted an additional cambium only at its basal level. Its activity resulted in a relatively thick xylem ring and another, thinner, phloem one. In the xylem ring, the vessels are rare, and at the periphery of the phloem ring, several very thin cords of sclerenchymatous fibers are observed, with moderately thickened and poorly lignified walls (Fig. 42 c).

Interestingly, in another halophyte species we analyzed, *Bassia hirsuta*, the structure of the vegetative organs is normal, not presenting this structure of the successive cambia. The same situation, as an exception to the "rule," let us say, was also mentioned by Hérail (1885) for *Camphorosma monspeliaca*. Also as a "deviation" from the rule, we also mention the example provided by Bonnier and du Sablon (1905), which refers to a *Camphorosma* (unfortunately, the authors do not specify exactly which species), "a small shrub from the Mediterranean region," to which in the first 2 years, the cambium (we intuit that it is this one, since the authors use the expression *generative layer of phloem and xylem*) works normally and then stops its activity; in the third year, another layer (of cambium) is formed from the pericycle, which forms a new circle of secondary vascular bundles. A few years later, a third ring of secondary bundles forms, outside the second, and the process continues.

Another interesting situation is the one found in *Salicornia macrostachya* (Bonnier and du Sablon 1905), in which the normal cambium never functions; in the still young root, a cambium differentiates from the pericycle, which will give a *secondary tissue* in which the vascular bundles are embedded (Fig. 43).

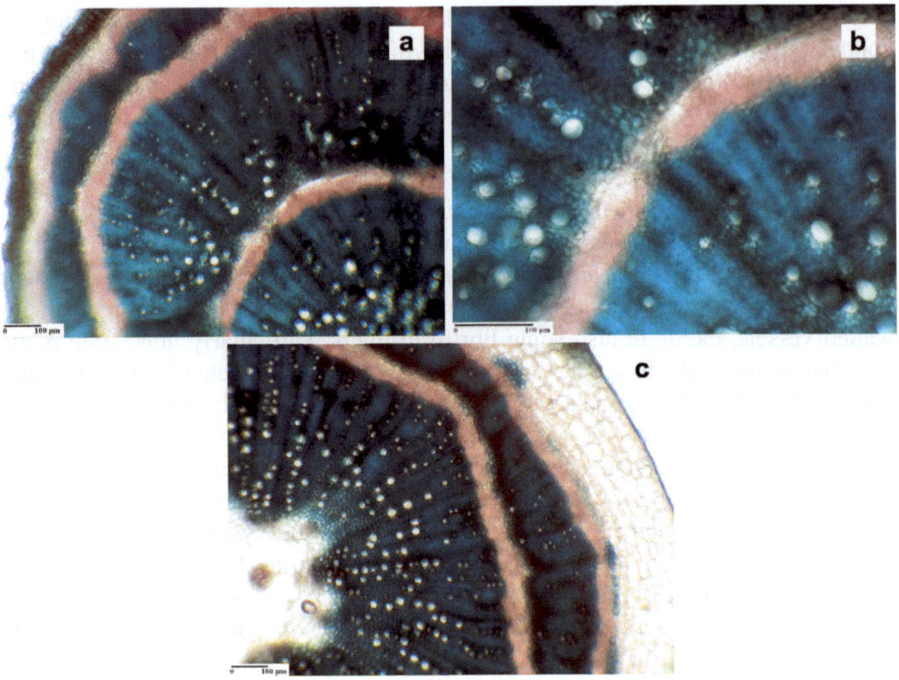

Fig. 42 Cross sections through the root (**a**, **b**) and stem (**c**) of *Camphorosma annua* (Grigore and Toma 2010)

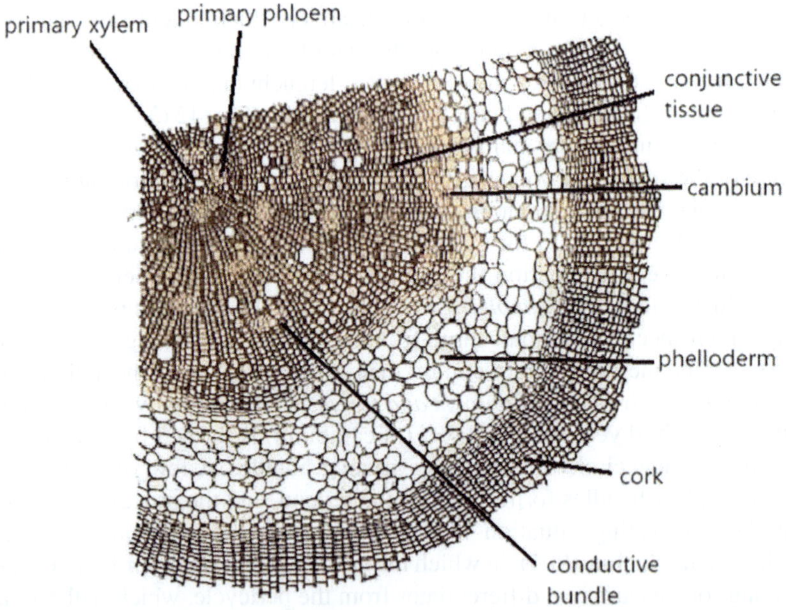

Fig. 43 Cross section through the root of *Salicornia macrostachya* (Modified from Bonnier and du Sablon 1905)

We make a parenthesis. The abovementioned authors do not use the term conjunctive tissue, but trying to explain the scheme that we also are using, they refer to a secondary parenchyma, in which the vascular bundles are embedded. We equated it with the conjunctive tissue, in the spirit of the definition and the discussions we had on the edge of this concept, which also corresponds to the tissue figured by the authors, but named differently.

In *Chenopodium murale*, according to the same authors, successive cambia form concentric circles of vascular bundles (Fig. 44).

At *Salsola kali* (Fig. 45), the additional cambia no longer have a concentric arrangement, but are arranged in a spiral, having as a starting point, towards the central part of the root, a normal cambium that was formed on the inner face of each of the two primary vascular bundles; this cambium is spiral and extends through its outer extremity, thus increasing the number of secondary bundles and implicitly contributing to the thickening of the root. The authors explain this spiral arrangement of the abnormal cambium by the fact that the tip of the radicle is compressed in the seed between the cotyledons; chenopods whose radicle is not compressed between the cotyledons present this sequence of secondary formations in a symmetrical sense.

In *Chenopodium album*, the abnormal structure is produced by a cambium that is periodically active, forming xylem in the centripetal direction and phloem in the centrifugal direction, in more limited regions (Artschwager 1926). As the phloem is formed, the cambium is gradually depleted. The continuity of the cambial ring is ensured by the formation of new portions of cambium on the outside of the phloem. The release of secondary growth zones occurs after the xylem is almost completely formed. According to the author, conjunctive tissue should not be considered as

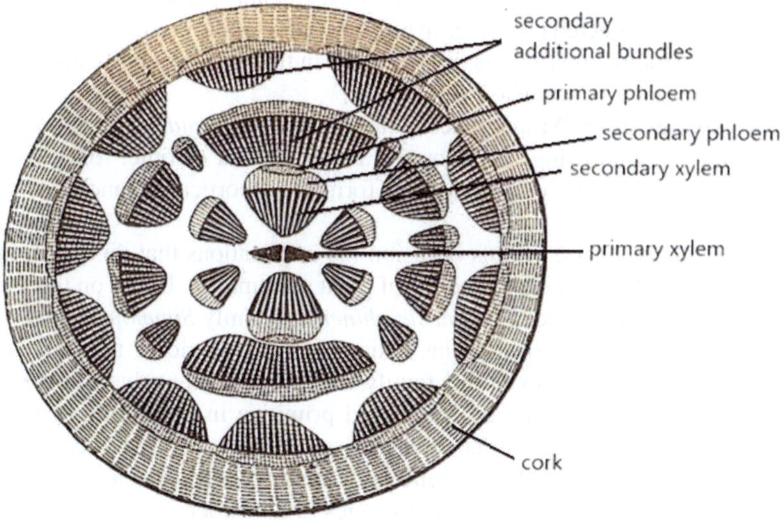

Fig. 44 Cross section through the root of *Chenopodium murale* (After Bonnier and du Sablon 1905)

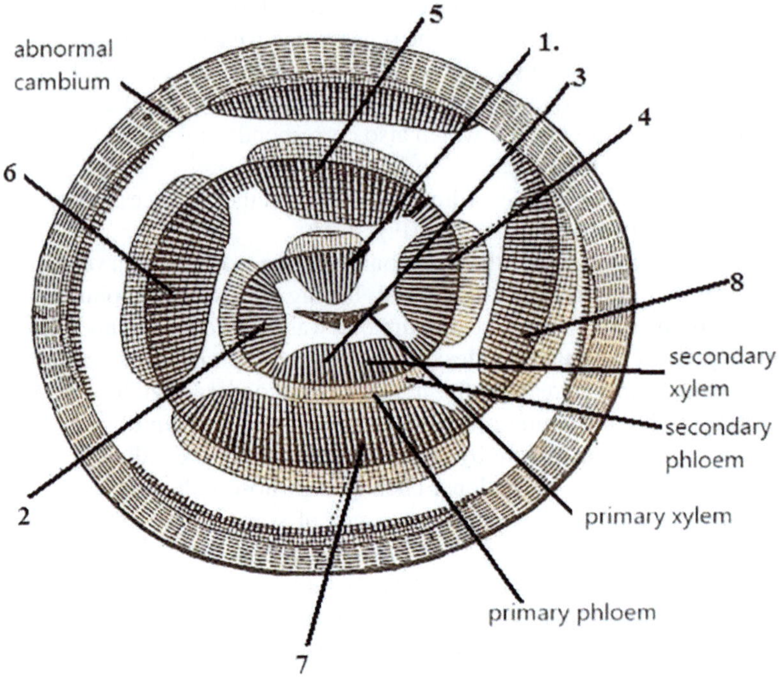

Fig. 45 Cross section through the root from *Salsola kali* (1, 2, 3, and 4, normal secondary formations, which are added, as a result of an abnormal development, to the pericyclic, "abnormal" ones, 5, 6, 7, and 8) (Modified after Bonnier and du Sablon 1905)

medullary rays in the morphological sense. Incidentally, the problem of rays in *Chenopodiaceae* (affected by successive cambia) is a controversial topic, but important, because, among other things, it intervenes in the possible clarification of some systematic problems (Carlquist 2003).

In two species of the Molluginaceae family, *Glinus lotoides* and *G. oppositifolius*, the first cambium ring functions for a short period of time, followed by the development of a second cambium ring, formed by cortical parenchyma (Rajput and Rao 2003).

It is time to emphasize on this occasion the implications that this phenomenon, of successive cambia, can have in terms of plant systematics. Based on this (and not only), an attempt to include *Simmondsia chinensis* (family *Simmondsiaceae*) in the order *Caryophyllales* has been done. This species endemic to the arid areas of California is the only species of this family. The anatomical analysis of the species revealed that it presents features considered primitive in dicotyledons: tracheids, diffuse parenchyma, and uniseriate and pluriseriate rays, but also successive cambia, which are a widespread occurrence in *Caryophyllales* species. All these are considered arguments for including the species in this order.

Synthesizing and trying to get closer to the possible implications and meanings of this phenomenon, we can state that we highlighted in the central cylinder of the axial vegetative organs of the halophyte species investigated by us a large amount of libriform, with thick, lignified walls.

Lignification—also present in the central cylinder of the roots of species affected by successive cambia—could be induced by excessive soil salinity (Neves et al. 2010). Kozlowski (1997) believed that salinity increases fiber content. We have to think that, in the context of multiple adaptations of plants to salinity, each species responds through certain metabolic changes. It would not be excluded that lignin is such a response, in the sense of increasing the resistance of the cell walls to the high osmotic pressures of the soil solution. Even so, however, the contribution played by lignin in ensuring salt resistance in halophytes must be viewed with caution, as the link between it and salinity is not always clear and relevant.

On the one hand, in some rice varieties (resistant, respectively sensitive to salinity), a relatively high salinity value (40 mM NaCl) increases the lignin content (Seung et al. 2004). In the same idea, histochemical observations in wheat varieties, tolerant and sensitive to salts, confirmed a more intense lignification in the root cells of the tolerant varieties, compared to the sensitive ones (Jbir et al. 2001).

On the other hand, high salinity can reduce the lignin content in the internodes of *Suaeda maritima* (Hagège et al. 1988). In *Atriplex prostrata*, grown in solutions with 0.5 and 1.0% NaCl, the lignified surface in the third and fourth internode decreases, compared to plants grown in 0.0% NaCl solution (Wang et al. 1997).

In addition, very interesting correlations can be made between lignin, extensin, and conditions of increased salinity. It was generally observed that in *Atriplex prostrata* (Wang et al. 1997), the extensin content increases in plants grown under salinity conditions. Thus, in the first and second internodes, plants grown in 2.0% NaCl solutions recorded the highest level of extensin in the vascular tissue, cortex, epidermis, and pith. Extensin level increased in vascular tissue as salinity increased in the third internode; it also grew in the vascular tissue and in the pith of the fourth internode. The results of the study showed that there was a greater amount of soluble extensin in the young (first and second) internodes than in the old ones (third and fourth). This is explained by the fact that extensin molecules tend to bind to each other or to other components of the cell wall. In addition, extensin is closely correlated with lignin deposition and can provide mechanical support in cells subjected to compressive or torsional stress, driven by vascular growth or vascular fluid transport (Tiré et al. 1994; Li and Showalter 1996). Therefore, both lignin and extensin serve to increase the strength of the cell wall. Lignin is generally located in the xylem, and extensin more in the phloem (Ye and Varner 1991; Showalter 1993). In young internodes, where the vascular tissue is not intensively developed, the content of soluble extensin is increased by salt stress especially in the medulla parenchyma, which has a small amount of extensin under normal conditions. In aged internodes, where the xylem and phloem are well developed, growth is not observed in the pith, but in the vascular tissue. Thus, the lignified surface (xylem) is reduced, while the phloem surface has increased under salinity conditions. This may be related to the fact that certain organic substances are synthesized to maintain the osmotic potential under saline conditions, so the phloem is more developed to translocate these organic substances. Consequently, extensin becomes abundant in the stem, but lignin content decreases. These results might suggest that extensin can substitute for lignin to enhance the strength of cell walls under salt stress conditions. Furthermore, lignin is known to be an important component in the secondary cell wall (Chapman 1987), and extensin is a common component in the primary cell wall. These observations could be related to the fact that plants subjected to salt stress may remain at juvenile stages of development, due to growth retardation, and thus the "proportion" of the primary cell wall is increased.

As a general observation, we can state, at least as a result of the analysis of the halophyte species studied by us, that lignification has a higher proportion at the root level compared to that of the stem, an aspect that was mentioned in the lines above. In our opinion, this is not an accidental fact, but can be placed in the context of the position and role played by the root in the life of the plant. Excluding perhaps the halophyte species of maritime coasts, subject to salt spray, we can say that the root is the organ most exposed to salinity. In this sense, it is logical to support the presence, at the root level, of some general mechanisms capable of controlling and managing the intrusion of salt water into the plant body and, possibly, its retention at this level. In fact, the existence of such mechanisms, such as the well-developed endodermis, capable of limiting the entering of salts into the rest of the plant organs, has already been stated in classical or more recent studies (Fahn 1963; Ginzburg 1966; Poljakoff-Mayber 1975; Schreiber et al. 1999). For example, some authors (Grigore et al. 2014; Alassimione et al. 2012; Bao et al. 2019; Wang et al. 2019) stated that certain active transport mechanisms are located either in the endodermis or in the xylem parenchyma.

All these details support the importance of the root, the role it plays in plants exposed to high salinity conditions. Our opinion is that lignin plays a major role in ensuring the strength of cell walls, in cells that must withstand quite high osmotic pressures. This reality therefore makes the root a key interface between salinity in rhizosphere and the plant as a whole. As already stated, not incidentally, the stem of many species of halophytes is not affected by successive cambia, at least not completely, so the proportion of lignin is much lower (Grigore and Toma 2006, 2007a).

However, it is not easy to find a convenient explanation regarding the ecological or ecophysiological significance of successive cambia phenomenon. Halophytes represent a heterogeneous ecological group of plants, the habitats vary widely, and correlations are therefore difficult to establish. Rather, the species taken individually can provide sources of interpretations.

Carlquist (2007) discusses the implications of the phenomenon regarding the storage and recovery of the products of photosynthesis and water. It is believed that the alternation between vascular complexes (*vascular increments*) and parenchyma, produced by successive cambia, would provide an "ideal" histological plan for water storage and recovery of the products of photosynthesis and water.

Roots from *Beta* (Artschwager 1926) and *Mirabilis* (Mikesell and Popham 1976) have thin vascular complexes separated from each other by relatively wide regions of parenchyma. Sugar is abundantly stored in the roots of *Beta* and starch in those of *Mirabilis*. In both genera, many of these recent vascular complexes are devoid of vessels; these aspects could be correlated with adaptation in the sense of recovering the products of photosynthesis, and less so with water storage.

Some genera in the *Aiozaceae*, such as *Mestoklema* and *Trichodiadema*, have thick tuberous roots but relatively thin stems. Vascular complexes ("increments") are separated in stems by wider surfaces than in roots. Unlike *Beta* and *Mirabilis*, these two genera have xylem and phloem in all these complexes. This seems primarily to be related to the storage of water in the underlying structures, although even in this case, small amounts of starch may occur in the roots.

Somewhat in the same sense, since 2006, we have issued a hypothesis regarding the ecological significance of successive cambia in halophytes (Grigore and Toma 2006). In this sense, the presence of this phenomenon in halophytes could be related to environmental modeling factors. Thus, certain regulatory mechanisms of salt content are known. One of these is salt dilution to ensure better growth (El Moukhtari et al. 2020); another is related to the retention of salts in the tissues of roots and stems (Flowers et al. 2015; Percey et al. 2016; Rahman et al. 2021; Mann et al. 2023), as well as the retransport of salts inside the roots and their elimination in the environment (Orcutt and Nilsen 2000; Rengel 2023). All this could be correlated with an increased internal surface, if we only think of the high capacity of retention, of storing water with its salts, in the root and stem. On the other hand, the corky layer on the outside of the root could delay water uptake. Therefore, salts enter the root heavily, but once there, they would be "spread" across this increased surface area. It may be that, literally, the distribution of water to the rest of the plant's organs is "delayed." Increasing the surface area would inevitably mean a space for the dispersion of salts, which are diluted at the same time, being thus, in the end, less harmful to the plant. Undoubtedly, the number and diameter of the xylem vessels can also play a role in this mechanism. Therefore, we can look at successive cambia, through its effects (many vessels, high "internal" surface), as a beneficial "compromise" for the plant between the need to limit growth (controlled by ABA) and the need to impose a salt dilution strategy in the plant, in lesser extent in the upper regions of the stem. The tip of the stem, as a growth region, would thus be protected from the harmful effects of salts, as young tissues are known to be more sensitive to salts, including flowering, as a particularly important stage in the life of the plant (Waisel 1972).

Marlothistella is another example of adaptation, through the plan of successive cambia, to the function of storing water. The stems are short and thick with numerous very short branches. The main stem is succulent. The vascular complexes are thin and scattered in the conjunctive tissue, which is formed almost completely of parenchyma having thin-walled cells. The stem of this genus represents another version of the plan of successive cambia, in which the water-storage function is amplified, and the mechanical one is reduced.

Another implication of successive cambia is that which refers to the provision of mechanical resistance and the extension of conducting capabilities. *Guapira*, *Neea*, *Pisonia*, and *Torrubia*, tropical and subtropical trees, have thin vascular complexes embedded in a fibrous tissue. This architectural configuration suggests optimal mechanical support. In addition, the rays in these genera contribute to mechanical resistance, because they have secondary cell walls, and are short in the vertical plane and narrow in the lateral plane, so that they occupy a small proportion in the stem tissues. Moreover, there are arches of parenchyma around the phloem bands in these four genera, which would suggest a slight deviation from the plane of optimal mechanical strength. But these bands would provide areas that would allow the secondary phloem volume to increase, produced outside the vascular cambium. The phloem of dicotyledonous species, with a single cambium, functions mainly for a single season; likewise, secondary xylem is functional only in the year in which it is formed. In *Guapira*, each of the numerous bands of secondary phloem, scattered in the stem (or root), is functional, a fact proven by the continuous production of secondary phloem in each of these vascular complexes; logically and correspondingly, the secondary xylem in the same complex is also functional. Consequently, a higher proportion of the stem surface becomes available for conducting, through secondary xylem and phloem, as compared to species with a single cambium.

The vascular cambium of *Charpentiera* (*Amaranthaceae*) produces not only radiating rows of vessels for a prolonged period but also associated fibers, resulting in the formation of a girdle of vessels, surrounded by fibers. The appearance of broad concentric bands of conjunctive parenchyma allows additions of secondary xylem that penetrates outwards into the conjunctive tissue. The same situation can be observed in the older parts of *Heimerliodendron*. In both genera, the addition of fibers along the xylem vessels, by the vascular cambium, ensures the increase of the mechanical resistance, as well as the prolongation of the cambial activity.

In *Aptenia*, each vascular cambium produces a lot of secondary xylem, but few fibers are produced in the last formed secondary xylem. Thus, the mechanical resistance is not increased, but the conductive xylem surface is maintained or even increased.

In any case, this phenomenon, of successive cambia, remains difficult to explain according to a single scheme; there is still quite a bit of controversy as to how these additional cambia are formed; we took over and developed the concept of master cambium. However, we remind you that most authors approached this problem from the perspective of a single type of cambium, not two types (master cambium and vascular cambium).

Much more problematic and tempting is the ecologically functional significance of this phenomenon. We specify at the end of this chapter that the perception of this phenomenon as a structural anomaly is exaggerated and somewhat unjustified. The presence of an (ancestral?) character in a heterogeneous group of plants must be seen as a particular evolutionary and adaptive response, while comparisons with the "normal" must have at most a didactic value.

Bibliography

Alassimione J., Roppolo D., Geldner N., Vermeer J. E. M., 2012 – The endodermis—development and differentiation of the plant's inner skin. *Protoplasma*, 249: 433–443

Anghel Gh., Chirilei H., 1960 – *Botanica*, Ministerul Agriculturii, Editura Agro-Silvică, București

Artschwager E., 1926 – Anatomy of the vegetative organs of the sugar beet. *J. Agric. Res.*, 33:143–176

Bao Z., Bai J., Cui H., Gong C., 2019 – A missing link in radial ion transport: ion transporters in the endodermis. *Front. Plant Sci.*, 10: 713

Barberon M., 2017 – The endodermis as a checkpoint for nutrients. *New Phytologist,* 213: 1604–1610

Belzung E., 1900 – *Anatomie et physiologie vtales*, Paris, Ancienne Librairie Germer Baillière et Cie Félix Alcan, Éditeur

Bonnier G., du Sablon L., 1905 – *Cours de Botanique. Phanérogames*, Paris, Librairie Générale de l'enseignement

Carlquist S., 1975 – Xylem anatomy of *Onagraceae*, with notes on alternative modes of photosynthate movement in dicotyledonous xylems. *Ann. Missouri Bot. Garden*, 62: 386–424

Carlquist S., 1996 –Wood. bark, and stem anatomy of Gnetales: a summary. *Int. J. Plant Sci.* 157(6 suppl.):S58-S76.

Carlquist S., 2001 – *Comparative xylem anatomy*, second ed., Berlin and Heidelberg Springer Verlag

Carlquist S. 2003 – Wood and stem anatomy of woody Amaranthaceae s. s.: ecology, systematics and the problems of defining rays in dicotyledons. *Bot. J. Linn. Soc.* 143: 1–19.

Carlquist S., 2004 – Lateral meristems, successive cambia and their products: a reinterpretation based on roots and stems of Nyctaginaceae. *Bot. J. Linn. Soc.*, 146: 129–143

Carlquist S., 2007 – Successive cambia revisited: ontogeny, histology, diversity, and functional significance. *J. Torrey Bot. Soc.*, 134(2): 301–332

Chapman A. R. O., 1987 – *Functional diversity of plants in the sea and on land*, Jones and Bartlett Publishers, Inc., Boston / Portola Valley

Cockrell R. A., 1941 – A comparative study of the xylem of several south American species of Strychnos. *Am. J. Bot.*, 28: 32–41

Crang R., Lyons-Sobaski S., Wise R., 2018 – *Plant anatomy. A concept-based approach to the structure of seed plants*, Springer Nature Switzerland AG

Doblas V. G., Geldner N., Barberon M., 2017 – The endodermis, a tightly controlled barrier for nutrients. *Curr. Opin. in Plant Biol.*, 39: 136–143

El Moukhtari A., Cabassa-Hourton C., Farissi M., Savouré A., 2020 – How does proline treatment promote salt stress tolerance during crop plant development? *Front. Plant Sci.*, 23(11):1127

Esau K., Cheadle V. I., 1969 – Secondary growth in Bougainvillea. *Ann. Bot.*, 33: 807–819

Evert R. F., Eichorn S. E., 2013 – *Raven Biology of plants*, 8th edition, W. H. Phloemman and Company Publishers, New York

Fahn A., 1963 – The fleshy cortex of articulated *Chenopodiaceae*. *J. Indian Bot. Soc.*, 42A: 39–45

Fahn A., Zimmermann M. H., 1982 – Development of the succesive cambia in *Atriplex halimus* (Chenopodiaceae). *Bot. Gaz.*, 143(3): 353–357

Flowers T. J., Munns R., Colmer T. D., 2015 – Sodium chloride toxicity and the cellular basis of salt tolerance in halophytes. *Ann. of Bot.*, 115(3): 419–431

Foster A. S., Gifford E. M., 1959 – *Comparative morphology of vascular plants*, second ed., W. H. Freeman and Company, San Francisco

Gérardin L., Guède H., 1895 – *Traité élémentaire d'histoire naturelle. Botanique. Anatomie et physiologie végétales*, Paris, Librairie J.-B. Baillière et Fils

Ginzburg C., 1966 – Xerophytic structures in the roots of desert plants. *Ann. Bot.*, 30(119): 403–418

Greguss P., 1968 – *Xylotomy of the living cycads*, Budapest Academiai Kiadó

Grigore M.-N., 2008 – *Introducere în Halofitologie. Elemente de anatomie integrativă*, Editura PIM, Iaşi

Grigore M.-N., 2012 – *Romanian salt tolerant plants. Taxonomy and ecology*, Edit. Tehnopress, Iaşi

Grigore M.-N., 2022 – *Morfologia plantelor. Aspecte ecologice*, Editura Universităţii „Ştefan cel Mare" Suceava

Grigore M.-N., Toma C., 2005 – Contributions to the knowledge of the anatomical structure of some halophytes (I). *Stud. Cerc. St., Univ. Bacău, ser. Biol.*, 10: 125–128

Grigore M.-N., C. Toma, 2006 – Evidencing the successive cambia phenomenon on some halophylous representatives among Chenopodiaceae and its possible ecological-adaptive implications. *Stud. Com. Complexul Muzeal Şt. Nat. „Ion Borcea" Bacău*, 21: 87–93

Grigore M.-N., Toma C., 2007a – Histo-anatomical strategies of *Chenopodiaceae* halophytes: adaptive, ecological and evolutionary implications. *WSEAS Transactions on Biology and Biomedicine*, 12(4): 204–218

Grigore M.-N., Toma C., 2007b – Aspects on the ecological and functional anatomy of some halophyte species from the *Chenopodiaceae* family. Proceedings of the 1st International Conference Environment – Natural Sciences- Food Industry in European Context. Baia Mare, November 16-17, 2007: 257–264

Grigore M.-N., Toma C., 2008 – Ecological anatomy of halophyte species from the *Chenopodiaceae* family. Advanced Topics on Mathematical Biology and Ecology (Proceedings of the 4th WSEAS International Conference on Mathematical Biology and Ecology – MABE '08, Acapulco, Mexico, January 25–27, 2008): 62–67

Grigore M.-N., Toma C., 2010 – *Halofitele. Aspecte de anatomie ecologică.* Editura Universităţii „Alexandru Ioan Cuza", Iaşi

Grigore M.-N., Toma C., 2017 – *Anatomical adaptations of halophytes. A review of classic literature and recent findings*, Springer International Publishing

Grigore M.-N., Toma C., 2021a – Morphological and anatomical adaptations of halophytes. A review. In: Grigore M.-N. (Ed.), *Handbook of halophytes. From molecules to ecosystems towards biosaline agriculture*, Springer, pp. 1079–1222

Grigore M.-N., Toma C., 2021b – Integrative anatomy of halophytes from Mediterranean climate. In: Grigore M.-N. (Ed.), *Handbook of halophytes. From molecules to ecosystems towards biosaline agriculture*, Springer, pp. 1293–1328

Grigore M.-N., Ana Cojocariu, 2020 – A tentative list of Romanian halophytes: taxonomy, distribution, ecology. In: Grigore M.-N. (Editor), *Handbook of Halophytes. From Molecules to Ecosystems towards Biosaline Agriculture*, Springer; https://doi.org/10.1007/978-3-030-17854-3_17-1, pp. 471–568

Grigore M.-N., Ivănescu L., Toma C., 2014 – *Halophytes. An integrative anatomical study.* Springer, Cham, Heidelberg, New York, Dordrecht, London

Grigore M.-N., Toma C., Zamfirache M.-M., Ivănescu L., Daraban I., 2013 – Anatomical and ecological observations in succulent (articulated) halophytes from *Chenopodiaceae. Lucr. Şt. (Horticultură), USAMV „Ion Ionescu de la Brad", Iaşi*, 56(2): 19–24

Hagège D., Kevers C., Boucaud J., Gaspar T., 1988 – Activités peroxydasiques, production d'éthylène, lignification et limitation de croissance chez *Suaeda maritima* cultivé en l'absence de NaCl. *Plant. Physiol. and Biochem.*, 26: 609–614

Hayward H.E., 1938 – *The structure of economic plants*, Verlag J. Cramer, New York

Hérail J., 1885 – Recherches sur l'anatomie comparée de la tige des Dicotylédones. *Ann. Sci. Nat., sér. 7, Bot.*, 12: 203–314

Jbir N., Chaibi W., Ammar S., Jemmali A., Ayadi A., 2001 – Effet du NaCl sur la croissance et la lignification des racines de deux espèces de blé différant par leur sensibilité au sel. *Comp. Rend. Acad. Sci. Paris (series III), Sci. de la Vie*, 324(9): 863–868

Kirchoff B. K., Fahn A., 1984 – Initiation and structure of the secondary vascular system in Phytolacca dioica (Phytolaccaceae). *Can. J. Bot.*, 62: 2580–2586

Kozlowski T. T., 1997 – Response of xylem plants to flooding and salinity. *The Physiology Monograph*, 1: 1–29

Li S., Showalter A. M., 1996 – Immunolocalization of extension and potato tuber lectin in carrot, tomato, and potato. *Physiol. Plant.*, 97: 708–718

Lindley J., 1861 – *The elements of botany*, London, Bradbury & Evans

Mann A., Lata C., Kumar N., Kumar A., Kumar A., Sheoran P., 2023 – Halophytes as new model plant species for salt tolerance strategies. *Front. Plant Sci.*, 14: 1137211

Metcalfe C.R., Chalk L., 1950 – *Anatomy of the Dicotyledons*, Clarendon Press, Oxford, 2

Metcalfe C.R., Chalk L., 1972 – *Anatomy of the Dicotyledons*, Clarendon Press, Oxford, 2

Mikesell J. E., Popham R. H., 1976 – Ontogeny and correlative relationship of the primary thickening meristem in four-o'clock plants (Nyctaginaceae) maintained under long and short photoperiods. *Am. J. Bot.*, 63: 427–437

Miyashima S., Nakajima K., 2011 – The root endodermis: a hub of developmental signals and nutrient flow. *Plant Signal Behav.*, 6(12):1954–1958

Moruzi C., Andrei M., 1967 – *Lucrări practice. Botanică generală*, Editura Didactică şi Pedagogică, Bucureşti

Morariu I., 1965 – *Botanică generală şi sistematică, cu noţiuni de geobotanică*, ed. a II-a, Editura Agro-Silvică, Bucureşti

Morlova I., Chirilă C., Ursu T., Baciu E., Ciocîrlan V., Cosmin S., Oprea C., Turcu Gh., Dobre F., Ungureanu L., 1966 – *Botanica. Lucrări practice*, Editura Didactică şi Pedagogică, Bucureşti

Neves G. Y. S., Marchiosi R., Ferrarese M. L. L., Siqueira-Soares R. C., Ferrarese-Filho O., 2010 – Root growth inhibition and lignification induced by salt stress in soybean. *J. Agron. & Crop Sci.*, 196: 467–473

Orcutt D. M., Nilsen E. T., 2000 – *Physiology of plants under stress. Soil and biotic factors*, John Wiley & Sons, Inc., New York, Chichester, Weinheim, Brisbane, Singapore, Toronto

Percey W. J., Shabala L., Wu Q., Su N., Breadmore M. C., Guijt R. M., Bose J., Shabala S., 2016 – Potassium retention in leaf mesophyll as an element of salinity tissue tolerance in halophytes. *Plant Physiol. Biochem.*,109: 346–354

Pfeiffer H., 1926 – Das abnorme Dickenwachstum. *Handbuch der Pflanzenanatomie* 9(2): 1–272, Berlin, FRG Gebrüder Borntraeger

Poljakoff-Mayber A., 1975 – Morphological and anatomical changes in plants as a response to salinity stress. In: *Plants in saline environments*, A. Poljakoff-Mayber, J. Gale (Eds.), pp. 97–117, Springer Verlag, Berlin, Heidelberg, New York

Rahman M. M., Mostofa M. G., Keya S. S., Siddiqui M. N., Ansary M. M. U., Das A. K., Rahman M. A., Tran L. S.-P., 2021 – Adaptive mechanisms of halophytes and their potential in improving salinity tolerance in plants. *Int. J. of Molec. Sci.,* 22(19):10733

Rajput K. S., Rao K. S, 1999 – Structural and developmental studies on cambial variant in Pupalia lappacea (Amaranthaceae). *Ann. Bot. Fennici,* 36: 137-141

Rajput K. S., Rao K. S, 2003 – Cambial variant and xylem structure in the stem of *Cocculus hirsutus* (Menispermaceae). IAWA J., 24, 4: 411–420

Răvăruţ M., Anghel Gh., Buia A., Nyarady A., Morlova I., 1967 – *Botanică*, Editura Didactică şi Pedagogică, Bucureşti

Răvăruţ M., Turenschi E., 1973 – *Botanica*, Editura Didactică şi Pedagogică, Bucureşti

Rengel Z., 2023 – *Plant responses to soil-borne ion toxicities*. In *Marschner's mineral nutrition of plants*, 4th ed, Rengel Z., Cakmak I., White P. J. (Eds.), Elsevier Academic Press, London, pp. 665–722

Sachs J., 1874 – *Lehrbuch der Botanik*, Leipzig, Verlag von Wilhelm Engelmann

Sadava D., Hillis D. M., Heller H. C., Berenbaum M. R., 2011 – *Life. The science of biology*, 9th edition, Sinauer Associates, Inc., Sunderland

Schenck H., 1893 – *Beiträge zur Biologie und Anatomie der Lianen, im Besonderen der in Brasilien einheimischen Arten*. Theil II. Beiträge zur Anatomie der Lianen, Heft V. Jena, Verlag von Gustav Fischer

Schreiber L., Hartmann K., Skrabs M., Zeier J., 1999 – Apoplastic barriers in roots: chemical composition of endodermal and hypodermal cell walls. *J. Exp. Bot.*, 50(337): 1267–1280

Seung Gon Wi, Jae Sung Kim, Jin Hong Kim, Myunghwa Baek, Daehwa Yang, Myung Chul Lee, Byung Yeoup Chung, 2004 – Effects of salinity on lignin and hydroxycinnamic acid contents in rice. *Korean J.. Crop. Sci.*, 49(5): 368–372

Showalter A. M., 1993 – Structure and function of plant cell wall proteins. *Plant Cell*, 5: 9–23

Sitte P., Ziegler H., Ehrendorfer F., Bresinsky A., 1998 – *Lehrbuch der Botanik für Hochschulen*, 34 auflage, Gustav Fischer, Jena, Lübeck, Ulm

Stevenson D. W., Popham R. A., 1973 – Ontogeny of the primary thickening meristem in seedlings of Bougainvillea spectabilis. *Am. J. Bot.*, 60: 1–9

Şerbănescu-Jitariu G., Toma C., 1980 – *Morfologia şi anatomia plantelor*, Editura Didactică şi Pedagogică, Bucureşti

Taiz L., Zeiger E., 2002 – *Plant Physiology*, third ed., Sinauer Associates

Throm G., 1996 – *Grundlagen der Botanik, 2. Überarbeitete und Erweiterte Auflage*, Nikol Verlag, Wiesbaden

Tiré C., De Rycke M., De Loose D., Inzé D., Van Montagu M., Engler G, 1994 – Extensin gene expression is induced by mechanical stimuli leading to local cell wall strenghtening in *Nicotiana plumbaginifolia. Planta*, 195: 175–181

Toma C., 1977 – *Anatomia plantelor. II. Structura organelor vegetative şi de reproducere*, Centrul de multiplicare al Universităţii „Al. I. Cuza" Iaşi

Toma C., Niţă M., Rugină R., Ivănescu L., Costică N., 2003 – *Morfologia şi anatomia plantelor. Manual de lucrări practice*, ediţia a 2-a revizuită şi îmbunătăţită, Editura Universităţii „Alexandru Ioan Cuza" din Iaşi

van Vliet G. J. C. M. 1979 – Wood anatomy of the Combretaceae. Blumea 25: 141–223

Zanoschi V., Toma C., 1985 – *Morfologia şi anatomia plantelor cultivate*, Ed. Ceres, Bucureşti

Waisel Y., 1972 – *Biology of halophytes*, Academic Press, New York, London

Wang L.-W., Showalter A. M., Ungar A., 1997 – Effect of salinity on growth, ion content, and cell wall chemistry in *Atriplex prostrata* (*Chenopodiaceae*). *Am. J. Bot.*, 84(9): 1247–1255

Wang P., Calvo-Polanco M., Reyt G., Barberon M., Champeyroux C., Santoni V., Maurel C., Franke R. B., Ljung K., Novak O., Geldner N., Boursiac Y., Salt D. E., 2019 – Surveillance of cell wall diffusion barrier integrity modulates water and solute transport in plants. *Sci. Rep.*, 9: 4227

Ye Z. H., Varner J. E., 1991 – Tissue-specific expression of cell wall protein in developing soybean tissues. *Plant Cell*, 3: 23–37

Ecological Anatomy of the Root

In a previous work (Grigore 2022), we dealt with ecology of plant organs, also including that of the root. Morphologically, it has been emphasized, since then, that the root is a key organ in the life of plants, presenting a great morphological diversity, which reflects plant ecological and functional realities.

Along with this morphological diversity, otherwise quite extensive in the world plant, the ecological anatomy of the root betrays the existence of some plant adaptations that are, of course, closely related to the environmental conditions in which they grow.

Beyond the great diversity of morphological forms of roots—and their structural peculiarities—the huge plasticity of root systems, from an ecological point of view, the general discussions regarding the root can be reduced to a few essential aspects:

1. Absorption of water with mineral salts.
2. Anchoring the plant in the soil.
3. Accumulation of nutrient reserves (substances).
4. Root growth and development in relation to ecological factor.
5. Aeration and root growth.
6. Lifetime.
7. The position of the root in the soil.
8. Aquatic roots.
9. Aerial roots (met especially in epiphytes) (Grigore 2022).

1 Ecological Anatomy of the Root of Hydrophytes (Water, Aquatic) Plants

1.1 General Characters

As we discussed in another work (Grigore 2022), the definition and classification of aquatic plants (hydrophytes, in the broad sense of the word) are very complicated and raise numerous problems of biological interpretation (see, more extensively, discussions in Grigore 2022).

M.-N. Grigore, *Plant Ecological Anatomy*,
https://doi.org/10.1007/978-3-032-01332-3_3

Morphologically speaking, as a "general" rule, the roots of some of the more specialized aquatic plants are extremely reduced or, in some cases, even completely absent, for example, in species of *Ceratophyllum*, *Aldrovanda*, *Utricularia*, *Wolffia*, *Pseudowolffia*, *Wolffiella*, and *Wolffiopsis*; the lack of roots in the last four mentioned genera represents a diagnostic character (Grigore 2022).

From an anatomical point of view, there are some general features that characterize the roots of aquatic plants (the discussion here is only about the roots of higher plants—spermatophytes; we will therefore not refer to the analogous organs found in cryptogamic species) (van Tieghem 1871; Warming 1909; Coulter et al. 1911; Barnes 1898; Schenck 1886; Snell 1907; Sauvageau 1889a, b, c; Goebel 1891–1893; Lewakoffski 1873; Batten 1918; Glück 1911; Haberlandt 1914; Schwarz 1883; Henslow 1895; Chatin 1856; Barnéoud 1848; Schimper 1903; Arber 1920; Ogden 1974; Schulze et al. 2005, 2019; Cronk and Fennessy 2001).

1. *Roots and analogous organs*. Since nutrients can be absorbed through the entire permeable surface of all submerged parts of aquatic plants, there is, in submerged plants, a reduction of those organs that normally extract mineral elements from the soil, i.e., roots (or analogous organs, in among cryptogames). As we have shown (Grigore 2022), some vascular plants, such as *Salvinia*, *Wolffia*, *Ceratophyllum*, *Utricularia vulgaris*, *Aldrovanda*, and *Genlisea*, are completely rootless; to others, such as *Azolla*, *Lemna*, *Hydrocharis*, *Pontederia*, and *Pistia*, the roots soon stop growing, do not branch, and may lose their caps. Root hairs are absent in *Lemna minor* and *L. trisulca*, *Myriophyllum* species, *Butomus umbellatus*, *Caltha palustris*, *Hippuris vulgaris*, *Nymphaea alba*, and others.

In these species, the roots and root hairs are, in many cases, only anchoring organs.

2. *The conducting elements* (which carry water with the minerals) are, for the same reasons, less required in the life of these plants; the xylem vessels and the whole xylem are consequently reduced in vascular plants. As a rule, the phloem, as a tissue specialized in conducting phloem sap, does not suffer any reduction. The conducting tissues are always grouped more towards the center of the organ, so that they constitute, as a whole, a central body.

In connection with the reduction of conducting elements in the roots of aquatic plants, van Tieghem (1871) established four types of "degenerate" roots. This French botanist, a prominent representative of the French school of botany of the late nineteenth and early twentieth centuries, interpreted the simplification of root structure in aquatic plants as a "degradation" of tissue organization within the root.

Thus, he highlighted "the increasingly simplification of structure in adventitious roots" in: *Pontederia crassipes* (Fig. 1), *Triglochin maritima* (Fig. 2), *Aponogeton distachyos* (Fig. 3), *Alisma plantago*, *Damasonium stellatum*, *Hydrocleys humboldtii* (Fig. 4), *Potamogeton lucens* (Fig. 5), and *Hydrocharis morsus-ranae* (Fig. 6). In the roots of these plants, the number of vascular bundles continues to decrease, and in each of the bundles, the vessels become less and less numerous, although they always reach normal development.

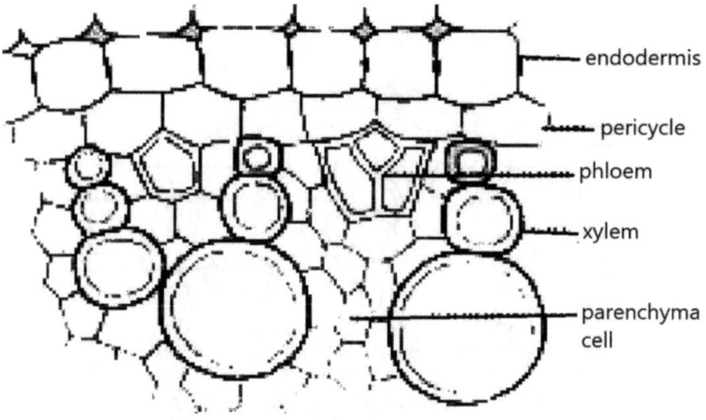

Fig. 1 Cross section through the root of *Pontederia crassipes* (van Tieghem 1871)

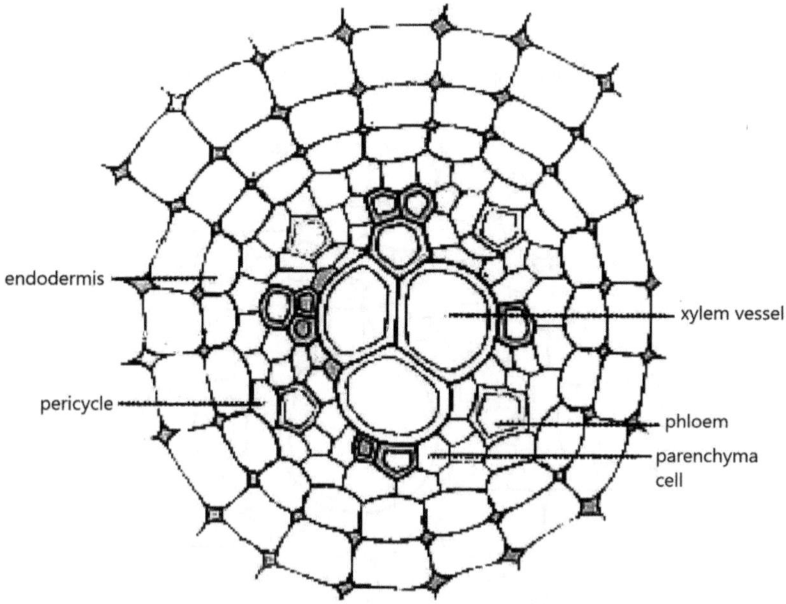

Fig. 2 Cross section through the root of *Triglochin maritima* (van Tieghem 1871)

Van Tieghem observed that, in aquatic plants, the vessels, after having—more or less—thickened their cell wall, have a marked tendency to resorb it, to become lacunae. There are, however, depending on the organ and the plant species, considerable differences regarding the developmental phase in which this resorption is expressed.

This tendency depends on the organs, because, for example, in *Aponogeton*, *Alisma*, and *Hydrocleys*, this phenomenon occurs later in the root than in the stem

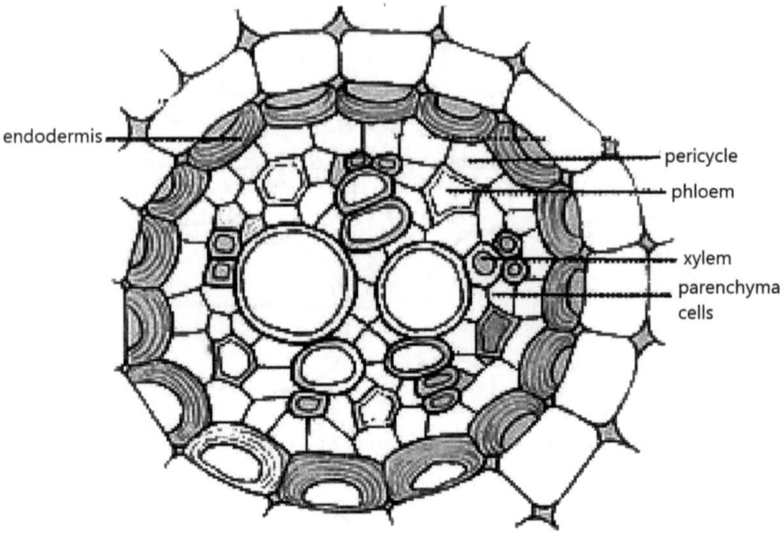

Fig. 3 Cross section through the root of *Aponogeton distachyos* (van Tieghem 1871)

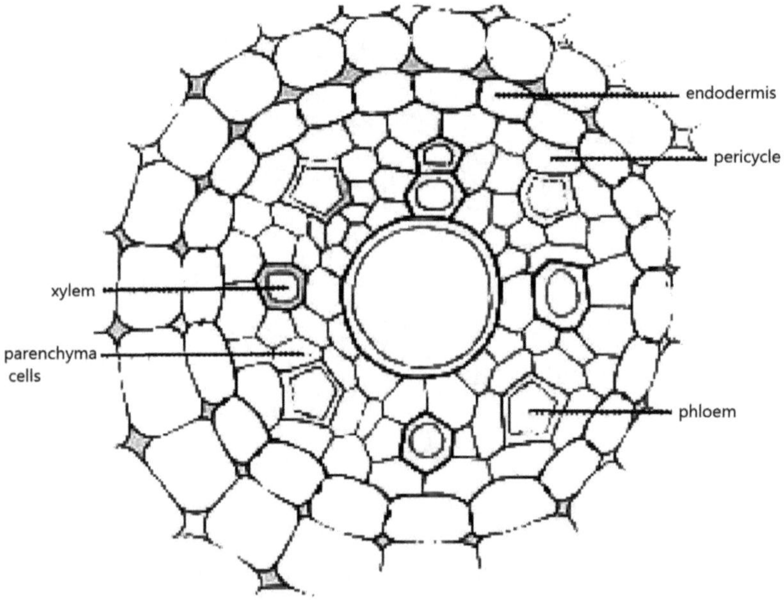

Fig. 4 Cross section through the root of *Hydrocleys humboldtii* (van Tieghem 1871)

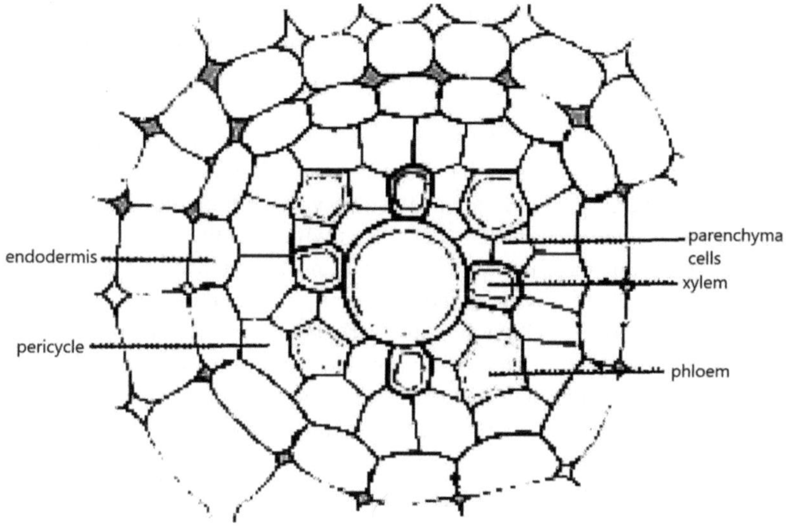

Fig. 5 Cross section through the root of *Potamogeton lucens* (van Tieghem 1871)

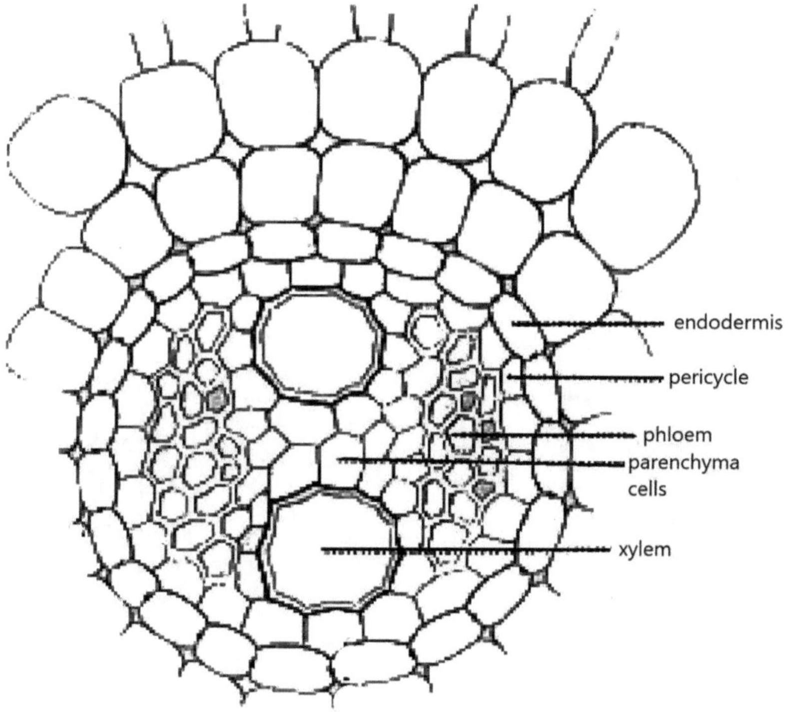

Fig. 6 Cross section through the root of *Hydrocharis morsus-ranae* (van Tieghem 1871)

and leaf and, when manifested here, first affects the large central vascular bundle; outer vessels can retain their cell walls for a very long time. Depending on the plant species, this resorption can only begin when the vessels are completely thickened, although there are other species in which resorption occurs so quickly that it is difficult to specify the moment when the vessel begins to acquire its characteristic features; it can only be supposed that the resorption reaches the wall of the vascular cell before its thickening has taken place in a very perceptible manner, as is the case, for example, in *Elodea canadensis* (Fig. 7), in which all the vessels would be very early transformed into lacunae by the resorption of their walls. A rather elaborate type of structure is thus individualized, especially if we compare it with that of the plants of the next group, in which the same early resorption is nevertheless observed, combined with a much more elementary organization of the central cylinder.

Other groups of aquatic plants in which the simplification of the root structure is observed include *Najas major* (Fig. 8), *Althenia filiformis*, and *Phucagrostis major*.

Vallisneria spiralis (Fig. 9) and *Lemna minor* (Fig. 10) give an even deeper state of "degeneration" than the preceding examples, as the central cylinder shows much earlier development.

From the study of van Tieghem (1871), we can extract some very important aspects in relation to the simplification of the anatomical structure of the root of aquatic plants and, mainly, at the level of the central cylinder.

Thus, in the roots with an increasingly simple structure, studied by van Tieghem, the cortex always has two distinct areas, limited on the outside by the rhizodermis and, on the inside, by the endodermis; the central cylinder always has a pericycle. On its internal face, the secondary changes take place (subsequent, not in the sense of passing to the secondary structure), produced by a progressive structural degradation, in which four main states can be distinguished:

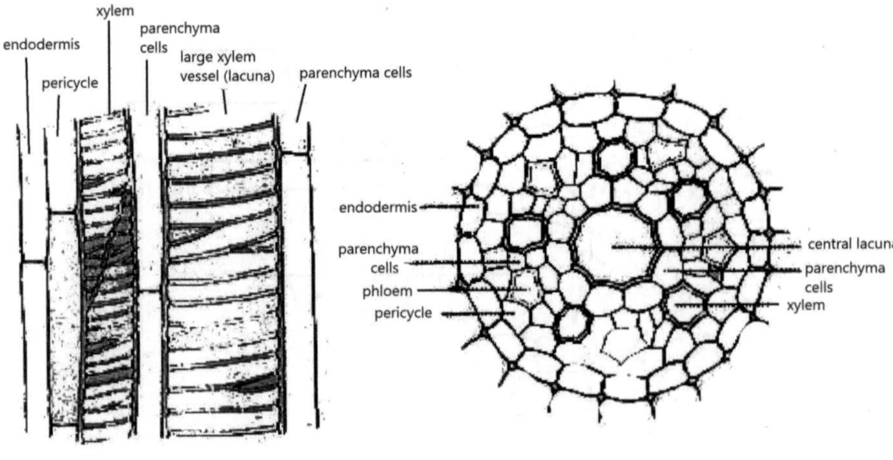

Fig. 7 Cross section through the root of *Elodea canadensis* (van Tieghem 1871)

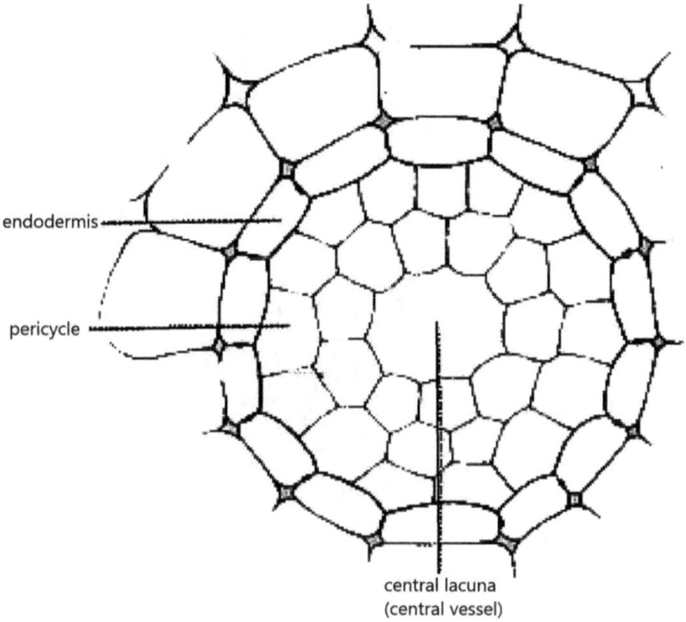

Fig. 8 Cross section through the root of *Najas major* (van Tieghem 1871)

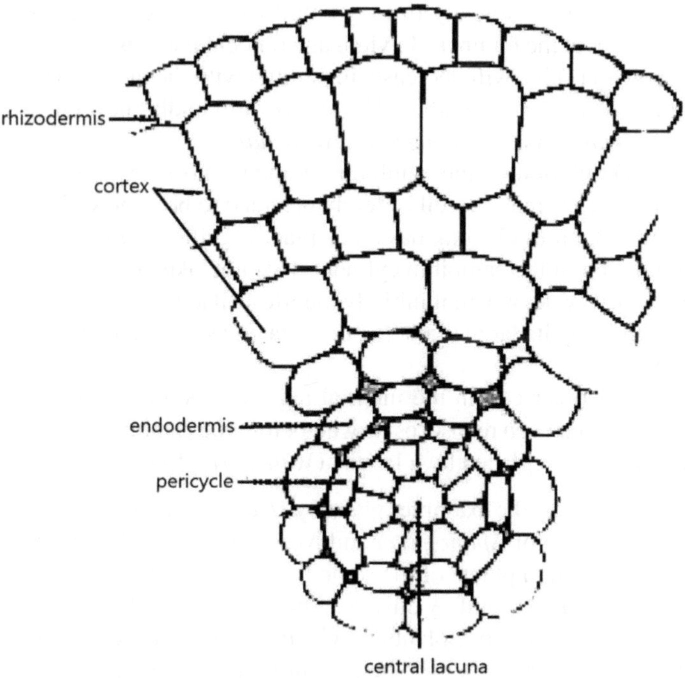

Fig. 9 Cross section through the root of *Vallisneria spiralis* (van Tieghem 1871)

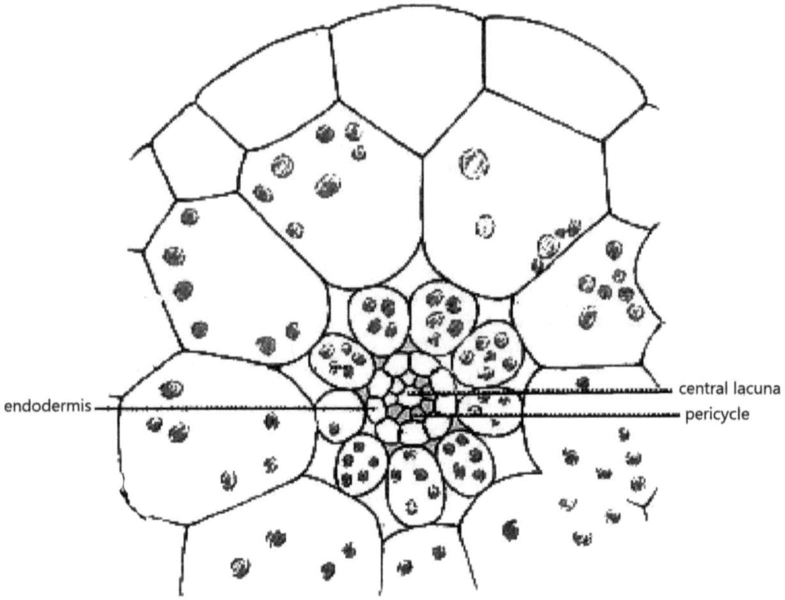

Fig. 10 Cross section through the root of *Lemna minor* (van Tieghem 1871)

(a) The mother cell of the internal region divides to give rise to a cambium cylinder, and its elements completely differentiate to form xylem, phloem, and parenchyma cells. But the number of xylem and phloem bundles, as well as the blades of parenchyma cells, will decrease, in parallel with the number of elements that constitute each of these bundles. The number of bundles never drops below two (*Alisma*, *Hydrocleys*, *Potamogeton*, *Hydrocharis*).
(b) Local differentiation of the cambium cylinder still takes place, but is incomplete, in the sense that the cells, clearly destined to become vessels, resorb their walls before their thickening has taken place significantly (*Elodea*).
(c) Differentiation of the cambium cylinder does not take place in the layer located below the pericycle, where it initially manifests itself in the usual case. Only the central cell is, by its large diameter, designated as vascular and soon becomes a canal (*Najas*).
(d) Finally, the mother cell in the internal region does not divide, but stops in its development and even reabsorbs its wall before the cambium cylinder is formed, which is entirely replaced by a lacuna (*Vallisneria*, Lemnaceae).

It can also be seen by the examples given by *Hydrocharis*, *Elodea*, and *Vallisneria*, but also through those of *Aponogeton* and *Najas*, that these different states of degradation can be found in plants of the same botanical family.

The most important secondary change, which marks the state of extreme "degradation," is manifested in those plants in which a part of the central cylinder, inside the pericycle, is resorbed before being built up and is replaced by a lacuna (*Vallisneria*, species of *Lemna*). In this case, it therefore has neither xylem vessels, nor phloem cells, nor parenchyma cells. The degradation is already less deep when the cambium cylinder is formed, but without differentiation (*Najas*).

The secondary change is still less evident when it differentiates in an incomplete manner, that is, in such a way that the cells destined to become vessels resorb their walls at the very beginning of their thickening. The root then possesses phloem bundles and parenchyma cells, but, instead of vessels, only lacunae are found there, early on (*Elodea*). The change is less pronounced in plants in which the central cylinder does not resorb the wall of its vessels until very late, after they have been completely thickened for some time, and beginning with the largest of them (*Alisma*, *Hydrocleys*). Therefore, the root has, for a certain time, vessels, phloem, and parenchyma cells, but as it ages, all its vessels can transform into lacunae. This—more or less—primitive arrest of the development of that part of the central cylinder from the inner face of the pericycle, and this more or less early resorption of its vascular elements, produces the first series of changes characteristic of plants adapted to aquatic life.

When the local differentiation of this region is complete and definitive, other sources of much more pronounced secondary variation can be identified in:

(a) The number of xylem and phloem bundles, which alternate side by side on the pericycle, a number that is never less than two, but which may exceed a hundred, which is quite fixed in the same species when small; this number becomes very variable, depending on the roots, and also along the same root as soon as it exceeds the value of five bundles.
(b) The diameter and shape, but, above all, in the number of elements that make up each xylem and phloem bundle, a number that can be reduced to one, but which can become large enough for the bundle to be dislocated, as well as in the development of more or less abundant, and in the degree of fiber formation within the parenchymatous tissue, which separates these bundles, and which seems to be abrogated in the thinnest roots.

These two sources of secondary changes are, up to a certain point, connected with each other, for both depend, to a certain extent, on the more or less considerable development in the diameter of the central cylinder.

In turn, the cortical parenchyma, by the shape and arrangement of its elements, by the absence or presence of a cork layer, laticiferous vessels or canals, mechanical elements, dissociation or resorption air lacunae, and intralacunnae hairs, is subject to several secondary variations.

Finally, one can observe how the whole series of these secondary modifications takes place, rather by comparing the main roots, as well as the adventitious roots of different species of plants of the same age, than by comparing, in the same plant, the main root and secondary roots arising from it, or of an adventitious root and the roots it produces, or, rather, by comparing different adventitious primary roots that form on the plant in different periods of vegetation.

Overall, considering the extreme "degradation" that occurs in some submerged aquatic plants, which results from premature resorption or the absence of differentiation of the inner region of the central cylinder, the same type of structure governs the root of all monocotyledonous plants. In all cases, the cortex and the central cylinder obey the same differentiation laws.

3. *Mechanical tissue* is either reduced or undeveloped because the buoyancy of water is greater than that of air. Those structural patterns, in particular, adapted to resist bending, are not developed. To resist stretching due to water movements, large, fixed aquatic plants living in turbulent waters have their mechanical tissue clustered as close to the center of the stem as possible to form a pattern adapted to withstand tensile stresses. Lignification is very little developed (in xylem vessels) or even absent. And for this reason, there are no woody plants within submerged aquatic plants. On the contrary, as we will see, lignification is intense in plants from the xerophyte group (Grigore and Toma 2021).
4. *Aerenchyma* (the totality of air spaces) is very abundant and well developed in aquatic and marsh plants (helophytes), serving partly to lower their specific gravity (as a flotation device) and partly to facilitate gas exchange and especially respiration. We will return, separately, to the ecological significance of aerenchyma in aquatic plants.
5. *Secondary growth* (in thickness, transition to secondary structure) occurs only exceptionally in the axial organs of aquatic plants. On the contrary, the submerged parts of the stems and leaves of the spermatophytes are much longer and thinner than in the ecotypes developed in contact with air, and due to the poor light, they almost acquire the appearance of some etiolated parts.

Returning to the classic study of van Tieghem (1871), Schenck (1886) also described submerged roots with also simplified structure, especially at the level of the central cylinder, in species such as *Callitriche stagnalis* (Fig. 11), *C. vernalis* (Fig. 12), *Elatine alsinastrum* (Fig. 13), *Elodea canadensis* (Fig. 14), *Lemna minor* (Fig. 15), *L. trisulca* (Fig. 16), *Myriophyllum spicatum* (Fig. 17), *M. alterniflorum* (Fig. 18), *Potamogeton crispus* (Fig. 19), *P. densus* (Fig. 20), *P. natans* (Fig. 21), *P. pectinatus* (Fig. 22), *Ranunculus aquaticus* (Fig. 23), *R. fluitans* (Fig. 24), *Vallisneria spiralis* (Fig. 25), and *Zannichellia palustris* (Fig. 26).

One of the first classic studies on the structure of vegetative organs of aquatic plants is that of Chatin (1856). Among the structural peculiarities of the root of these plants, one can note the presence of aerenchyma (air spaces), a defining anatomical feature for plants subjected to waterlogging, which creates, as we will see, conditions of hypoxia or even anoxia at the level of submerged organs. Chatin (1856) identified aerenchyma, at the root level, in the following species: *Alisma plantago* (Fig. 27), *Alisma natans* (Fig. 28), *Alisma ranunculoides* (Fig. 29), *Butomus umbellatus* (Fig. 30), *Hydrocharis morsus-ranae* (Fig. 31), *Limnocharis humboldtii* (Fig. 32), *Ottelia alismoides* (Fig. 33), *Scheuchzeria palustris* (Fig. 34), and *Stratiotes aloides* (Fig. 35).

Another classic study on the structure of vegetative organs from aquatic plants is that of Sauvageau (1889a, b, c), for whom, unlike van Tieghem's (1871) interpretations, the root structure is not so simplified ("degenerate").

From Sauvageau's observations, the central cylinder of the root of *Najas major* is composed of one to two axial vessels, representing the xylem, surrounded by a

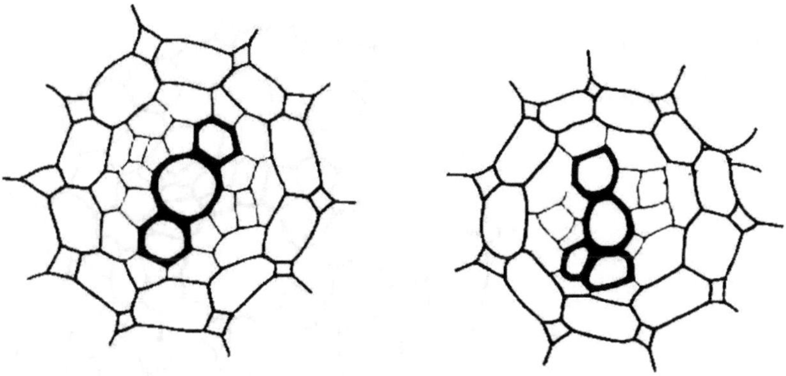

Fig. 11 Cross section through the root—central cylinder—*Callitriche stagnalis*: left, aquatic form; right, terrestrial form (Schenck 1886)

Fig. 12 Cross section through the root—central cylinder—*Callitriche vernalis* (Schenck 1886)

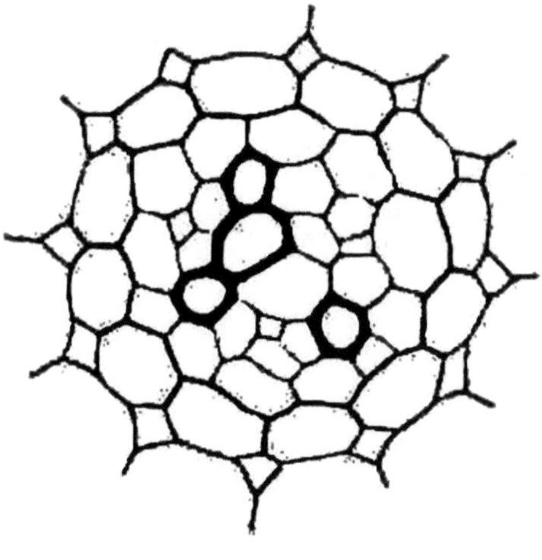

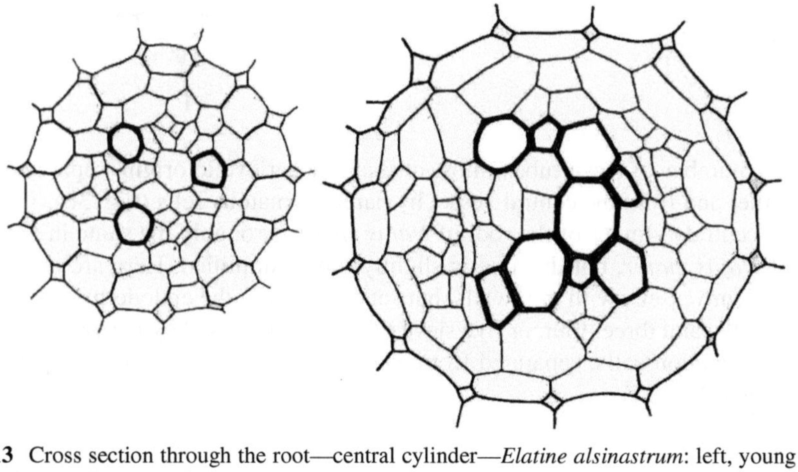

Fig. 13 Cross section through the root—central cylinder—*Elatine alsinastrum*: left, young root; right, old root (Schenck 1886)

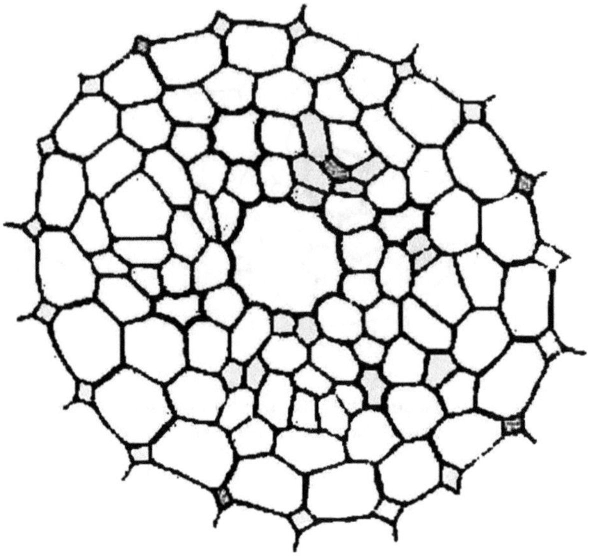

Fig. 14 Cross section through the root—central cylinder—*Elodea canadensis* (Schenck 1886)

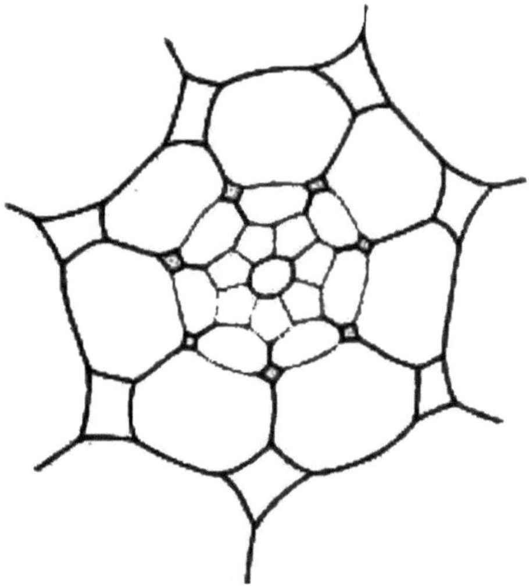

Fig. 15 Cross section through the root—the central cylinder—*Lemna minor* (Schenck 1886)

variable number of sieve tubes, most of them of pericyclic origin, separated from each other and from the central vessel by parenchymatous cells (Fig. 36).

The central cylinder of the root of *Najas minor* is roughly the same in nature as that of *Najas major*, but this one is slightly more simplified. Here are one or two axial xylem vessels, with thin walls, but separated from the endodermis by a single row of cells, and three, four, or five sieve tubes, accompanied by narrow cells, similar to companion cells, separated from each other by one, two, or three parenchymatic cells (Fig. 37).

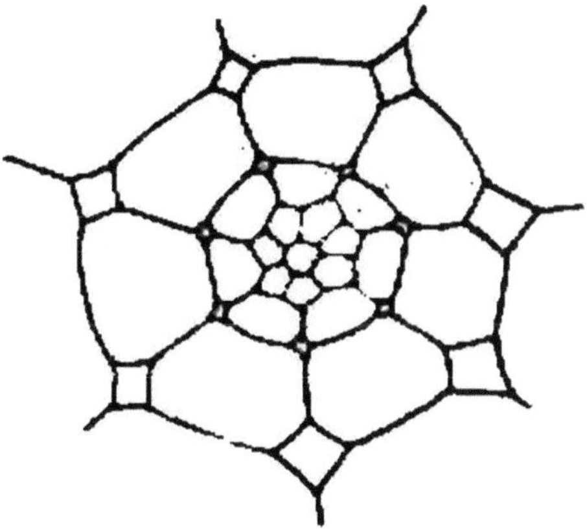

Fig. 16 Cross section through the root—central cylinder—*Lemna trisulca* (Schenck 1886)

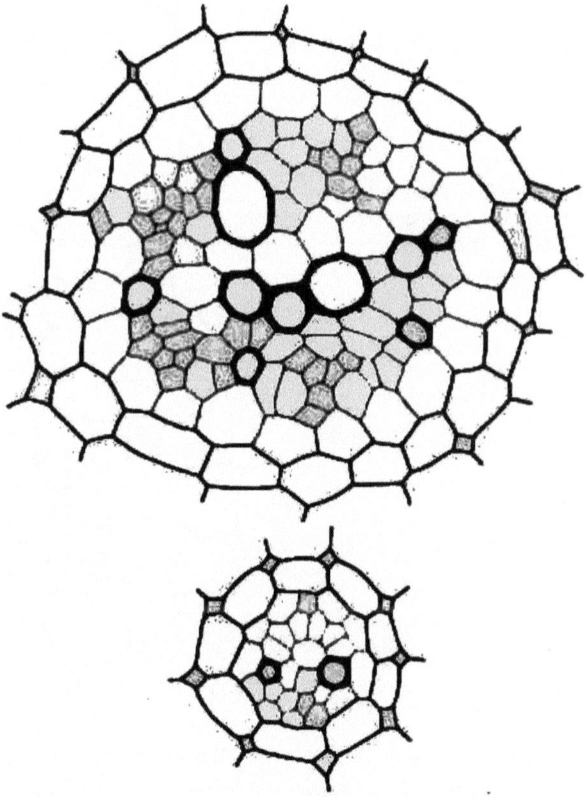

Fig. 17 Cross section through the root—central cylinder—*Myriophyllum spicatum*; top, cross section through an adventitious root; pentarchy structure; bottom, cross section through a lateral adventitious root (diarch structure) (Schenck 1886)

The central cylinder of the root of *Najas minor* and *Najas major* therefore remains entirely cellulosic and corresponds to a single bundle, with an axial vessel representing the xylem and the peripheral phloem.

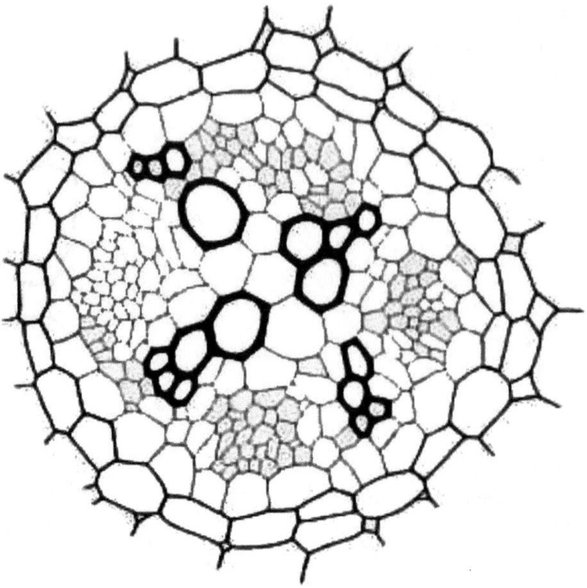
Fig. 18 Cross section through the root—central cylinder—*Myriophyllum alterniflorum* (Schenck 1886)

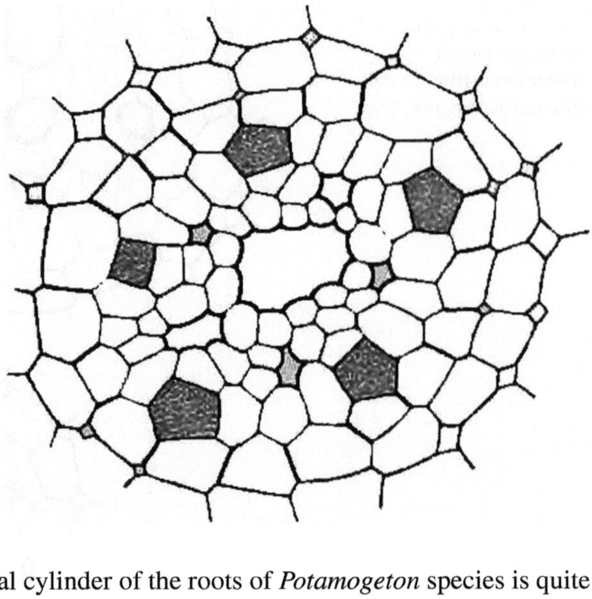
Fig. 19 Cross section through the root—central cylinder—*Potamogeton crispus* (Schenck 1886)

The structure of the central cylinder of the roots of *Potamogeton* species is quite simple. Roots of *Potamogeton plantagineus* (Figs. 38 and 39), *P. polygonifolius* (Figs. 40 and 41), *P. natans* (Fig. 42), *P. pectinatus* (Fig. 43), *P. gramineus* (Fig. 44), *P. lucens* (Fig. 45), and *P. pusillus* (Fig. 46) show many similar structural elements, especially before the appearance of changes due to sclerification.

Their central cylinder has three to seven sieve tubes, each provided with a companion cell; each xylem bundle is composed of one, two, or three vessels, of which

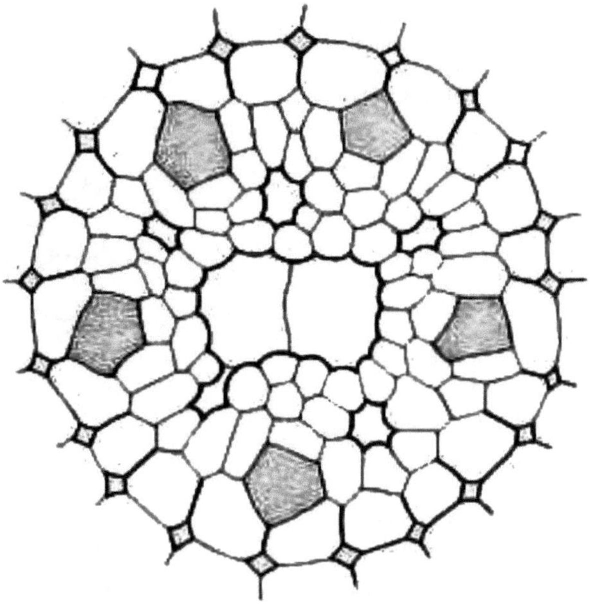

Fig. 20 Cross section through the root—central cylinder—*Potamogeton densus* (Schenck 1886)

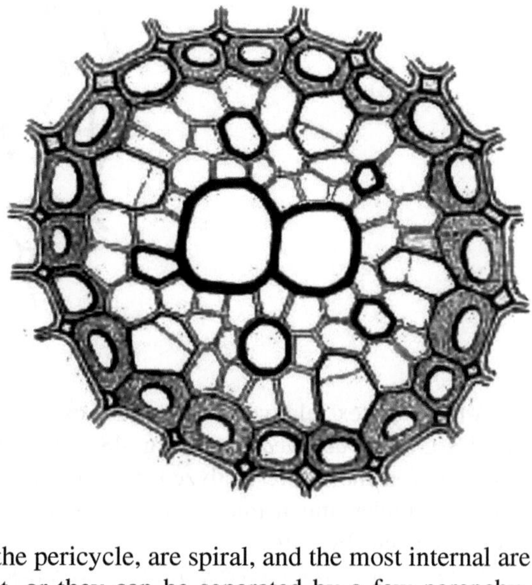

Fig. 21 Cross section through the root—central cylinder—*Potamogeton natans* (Schenck 1886)

the most external, in contact with the pericycle, are spiral, and the most internal are reticulated; they can be in contact, or they can be separated by a few parenchymal cells.

Cross sections, made in an older region, show that the endodermis, at points opposite to the sieve tubes, lignifies its cells without thickening them; the other cells of the same layer are only partially changed.

As in other aquatic species, the central root cylinder of *Zostera marina* is very narrow relative to the cortical region (Fig. 47).

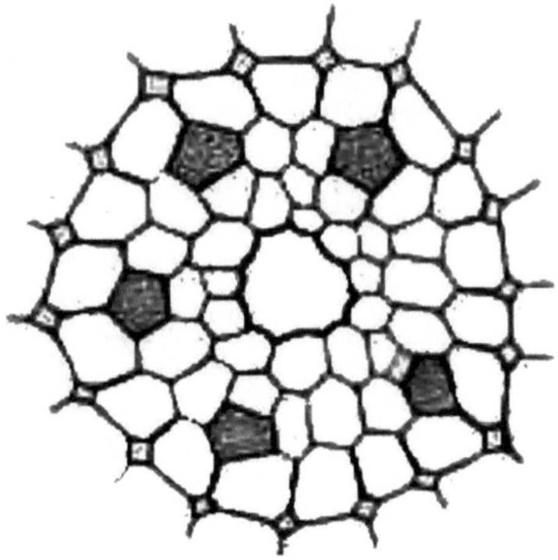

Fig. 22 Cross section through the root—central cylinder—*Potamogeton pectinatus* (Schenck 1886)

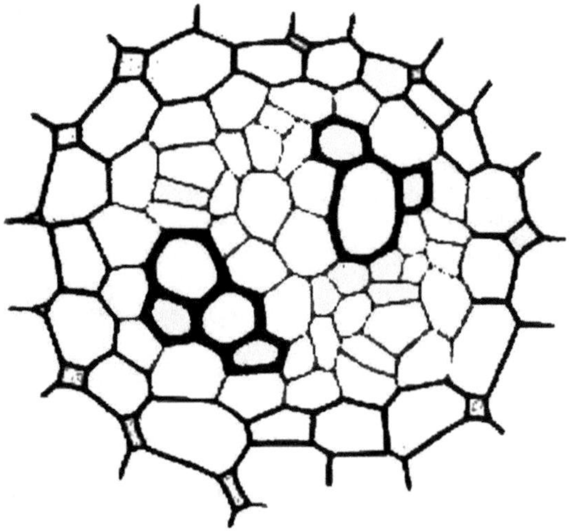

Fig. 23 Cross section through root—central cylinder—*Ranunculus aquaticus* (Schenck 1886)

A thick cortex relative to the central cylinder was also evidenced in the aquatic species *Cymodocea aequorea* (Fig. 48), in which the rhizodermis detaches easily and falls (on the figure analyzed by Sauvageau, this is not included); within the central cylinder, much thicker than that of *Zostera marina*, the pericycle is well individualized (Fig. 49).

The root of *Posidonia caulinii*, however, has a more complex structure than that of the previously presented plants. The rhizodermis, formed, in cross section, of flattened cells larger than those located below, has cells with thickened and lignified walls (Fig. 50).

In the same species, the central cylinder has an interesting structure: the cells of the endodermis have barely evident Caspary thickenings; those opposite the phloem bundles form small protective sclerenchymatic arches, sometimes reinforced by a few cells of their outer cortical layer, which are also thickened and lignified (Fig. 51).

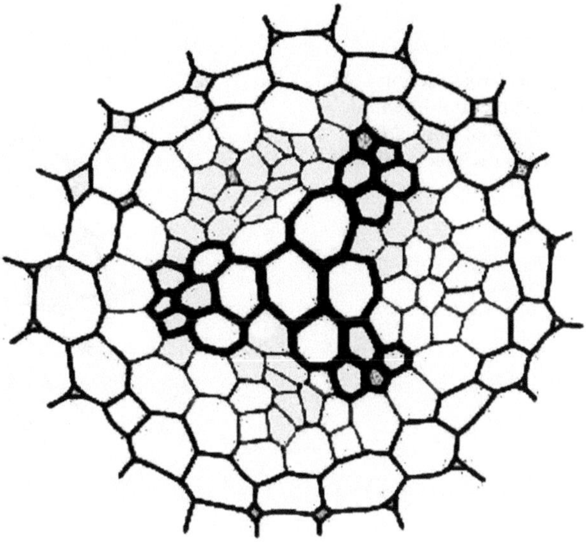

Fig. 24 Cross section through the root—central cylinder—*Ranunculus fluitans* (Schenck 1886)

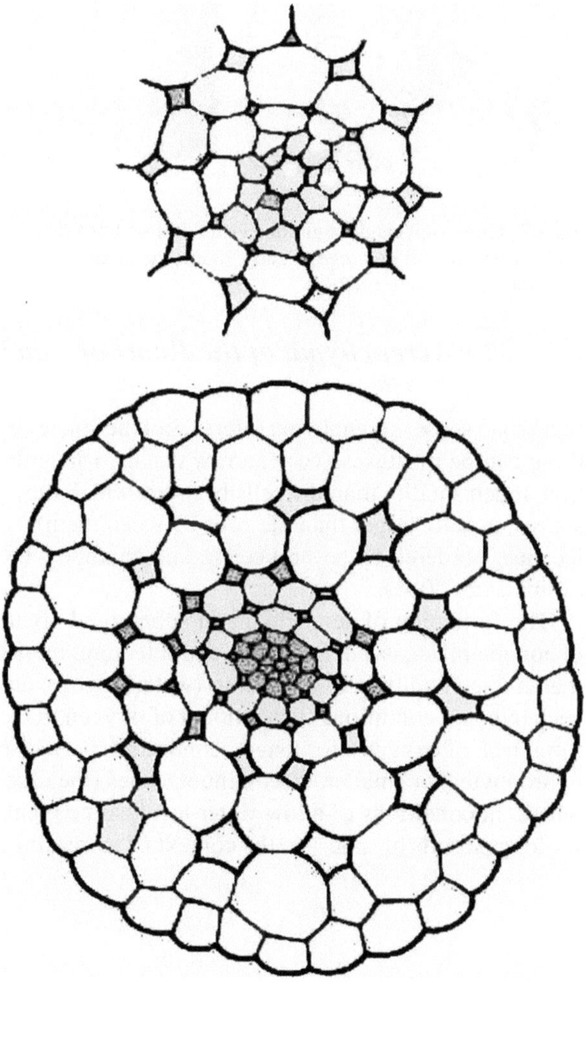

Fig. 25 Cross section through the root—central cylinder—*Vallisneria spiralis*; top, central cylinder; bottom, cross section through the adventitious root (general view) (Schenck 1886)

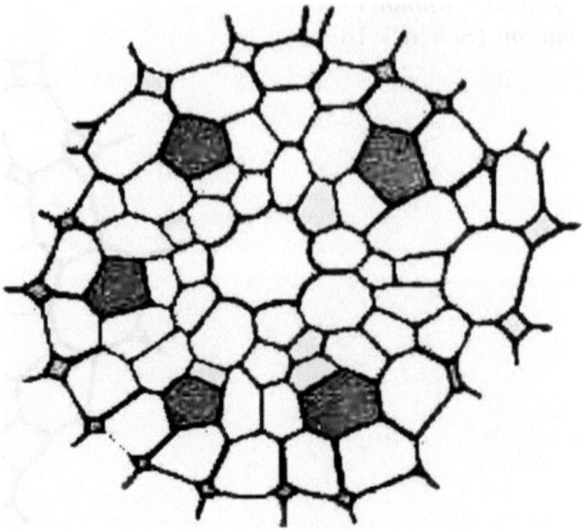

Fig. 26 Cross section through the root—central cylinder—*Zannichellia palustris* (Schenck 1886)

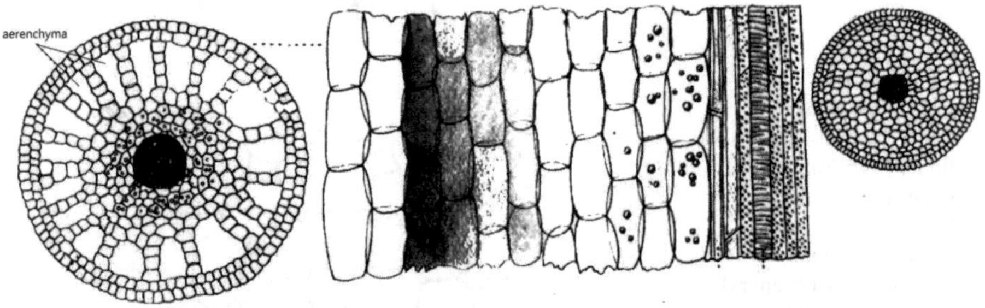

Fig. 27 Cross (left) and longitudinal (middle) sections through the main root and secondary root (right) from *Alisma plantago* (Chatin 1856); the absence of aerenchyma from the structure of the secondary root is observed

1.2 The Aerenchyma of the Roots of Aquatic Plants

In a broad sense, aerenchyma refers to all the spaces with air in the organs of a plant; these can be meatuses (very narrow canals, triangular or rhomboidal in cross section, much smaller than the cells between which they are), lacunae (isodiametric air spaces, usually larger than the cells between them), canals (long spaces, wider than lacunae, bordered by several cells), and chambers (very wide isodiametric spaces) (Toma et al. 2003).

The formation of aerenchyma in submerged organs (root, rhizome, aerial stem) of aquatic plants is a defining structural feature, related to the environmental factors that create conditions for flooding (waterlogging) and, therefore, for adaptation to such hypoxia conditions (low amount of oxygen in the environment) or even anoxia (total lack of oxygen), for longer, continuous periods of time (typical aquatic plants, hydrophytes) or smaller, discontinuous ones (the case of swamp plants, helophytes, where, in conditions of a low water level, some plant organs can remain unflooded for long periods of time—in the context of increasingly dry periods in recent years).

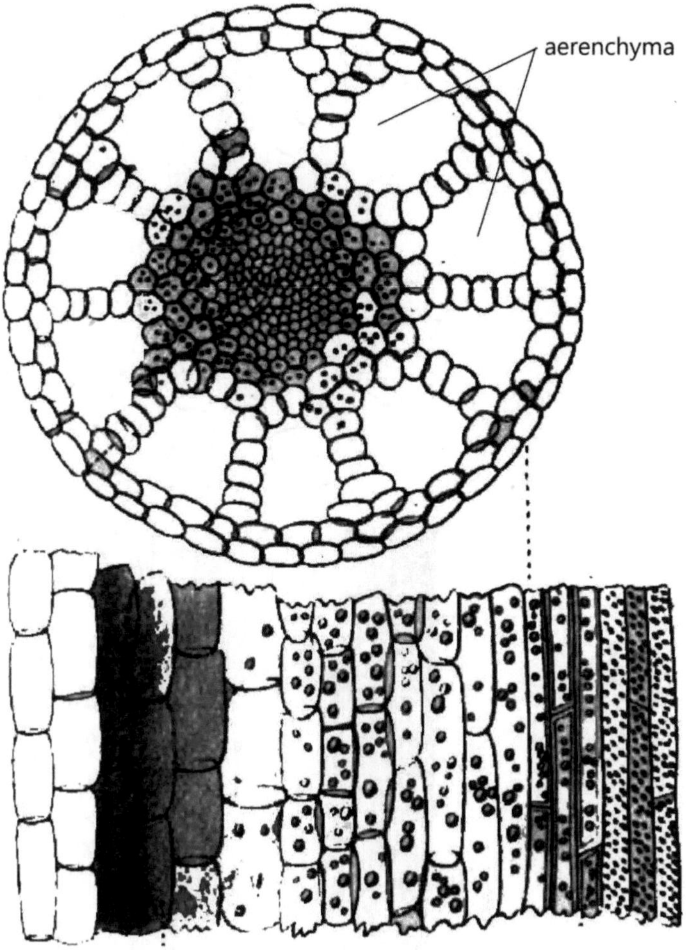

Fig. 28 Cross (top) and longitudinal (bottom) sections through the root of *Alisma natans* (Chatin 1856)

It should be noted that aerenchyma can also form in plants that are not normally aquatic, but which, exposed to waterlogging stress (flooding), induce the formation of aerenchyma in response to the action of abiotic stress (Schulze et al. 2005, 2019).

Since the formation of the aerenchyma, its functions, the chemical (hormonal) mediation of the formation of air spaces in the organs exposed to waterlogging, and the anaerobic metabolism are very complicated and intensely debated aspects, in the present work, we will limit to a few essential aspects related to the anatomy the root of some aquatic plants showing aerenchyma.

Broadly speaking, a submerged root faces two major problems: the permanence of anaerobiosis in wet, flooded soil and the presence of soluble phytotoxins. To these can be added other associated challenges, such as nutritional deficiencies

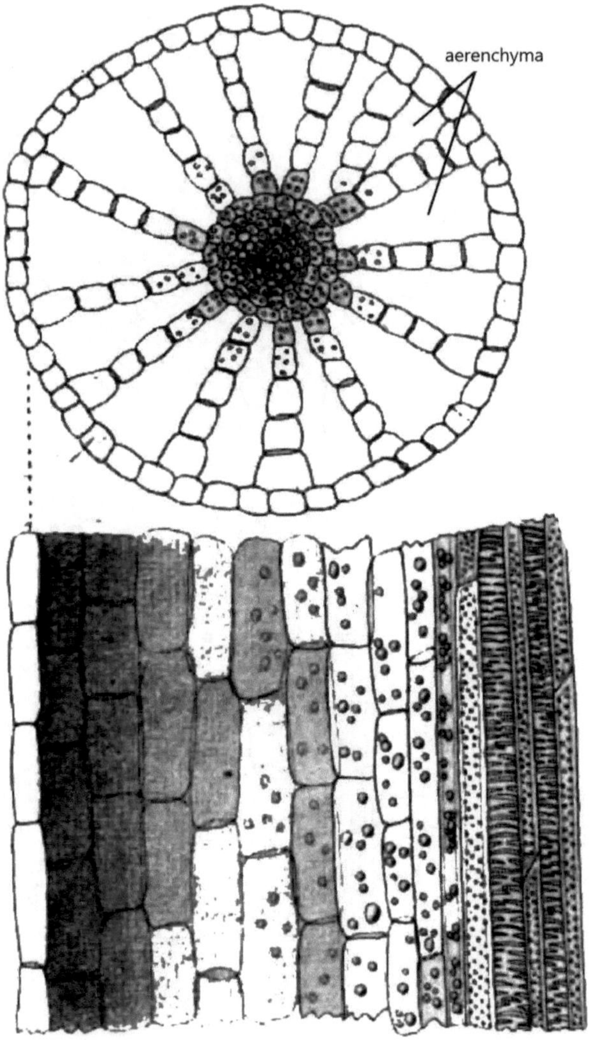

Fig. 29 Transverse (top) and longitudinal (bottom) section through the root from *Alisma ranunculoides* (Chatin 1856)

(especially in essential nutrients such as nitrates) and inactivity of aerobic bacteria and fungi. The dependence on aerobic respiration for growth and the permanence of the anoxic soil environment clearly demonstrate the need for some form of ventilation in submerged plant organs; the presence of phytotoxins, many of which can be inactivated by direct oxidation or biological destruction, also indicates the need for ventilation (Armstrong 1978; Hook and Scholtens 1978; de Wit 1978; Meet and Stolzy 1978; Cronk and Fennessy 2001; Schulze et al. 2005).

Therefore, we can consider the aerenchyma as the main structural adaptation to hypoxia or anoxia, conditions that occur, as already said, in flooded (marshy) soils.

Virtually, all rooted water plants form internal gas transport systems consisting of large gas-filled spaces called lacunae (Crawford 1993). The lacunae are held

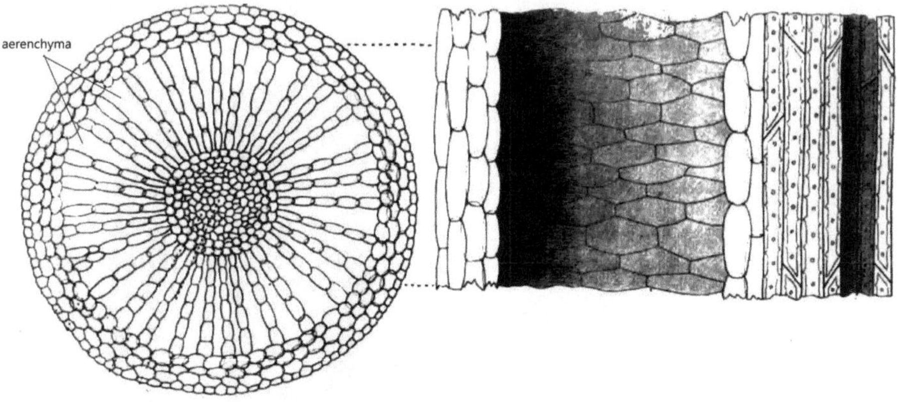

Fig. 30 Cross (left) and longitudinal (right) sections through the root from *Butomus umbellatus* (Chatin 1856)

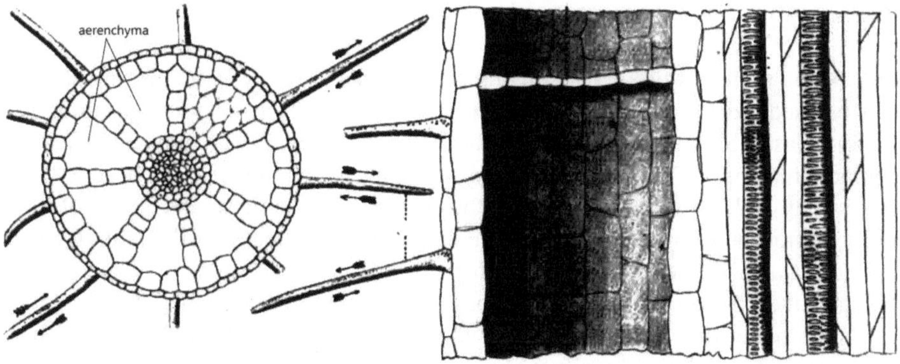

Fig. 31 Cross (left) and longitudinal (right) sections through the root of *Hydrocharis morsus-ranae* (Chatin 1856)

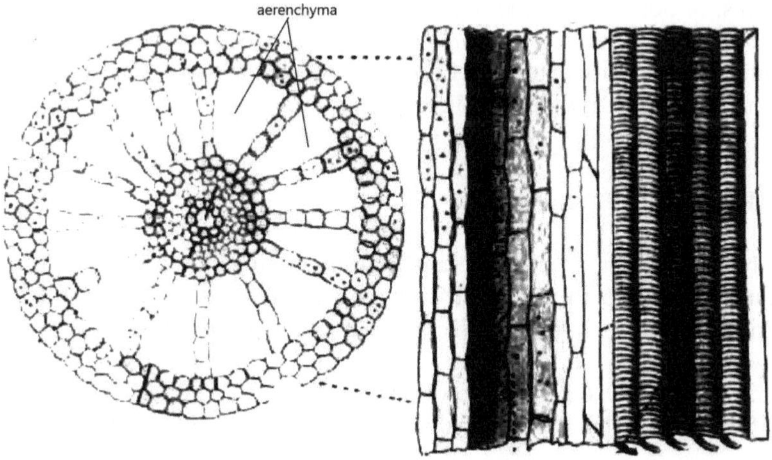

Fig. 32 Cross (left) and longitudinal (right) sections through the root of *Limnocharis humboldtii* (Chatin 1856)

Fig. 33 Cross (left) and longitudinal (right) section through the root from *Ottelia alismoides*. (Chatin 1856)

together in a porous tissue called aerenchyma (this is the generic term most commonly used worldwide, despite covering several different anatomical configurations). Gases are transported throughout the plant along channels formed by aerenchyma (Fig. 52), with little or no resistance to gas movement (Laing 1940; Armstrong 1978). In emergent plants (those that expose their organs or part of them to the atmosphere) in wetlands, oxygen enters the aerial parts of the plant through the stomata in the leaves and through the lenticels in the stem or xylem tissue. Oxygen then moves to the roots through the aerenchyma, usually by diffusion. Carbon dioxide follows the opposite route, moving upwards from the roots, where it is produced as a derived product of respiration, through the aerial part of the plant, where it is released into the atmosphere through the stomata (Armstrong 1978; Topa and McLeod 1986).

Aerenchyma forms in both new and old tissues, roots, rhizomes, stems, petioles, and leaves, in both woody and herbaceous wetland plants (Jackson 1989; Arteca 1997). In some species, such as *Cladium mariscus* and *Spartina alterniflora*, there is a continuous air space extending from the leaves to the roots (Teal and Kanwisher 1966; Smirnoff and Crawford 1983).

1.2.1 Formation of Aerenchyma

Aerenchyma forms in species tolerant to waterlogging (flooding), but to a lesser extent in many intolerant species (Armstrong 1978, 1979; Crawford 1982; Justin and Armstrong 1987). In flood-intolerant to flooding plants, air spaces may occupy 10–12% of the total root cross section, but, in flood-tolerant plants, the total air space may be over 50–60% of the root surface (Smirnoff and Crawford 1983; Smits et al. 1990). The volume of aerenchyma varies considerably among species, but

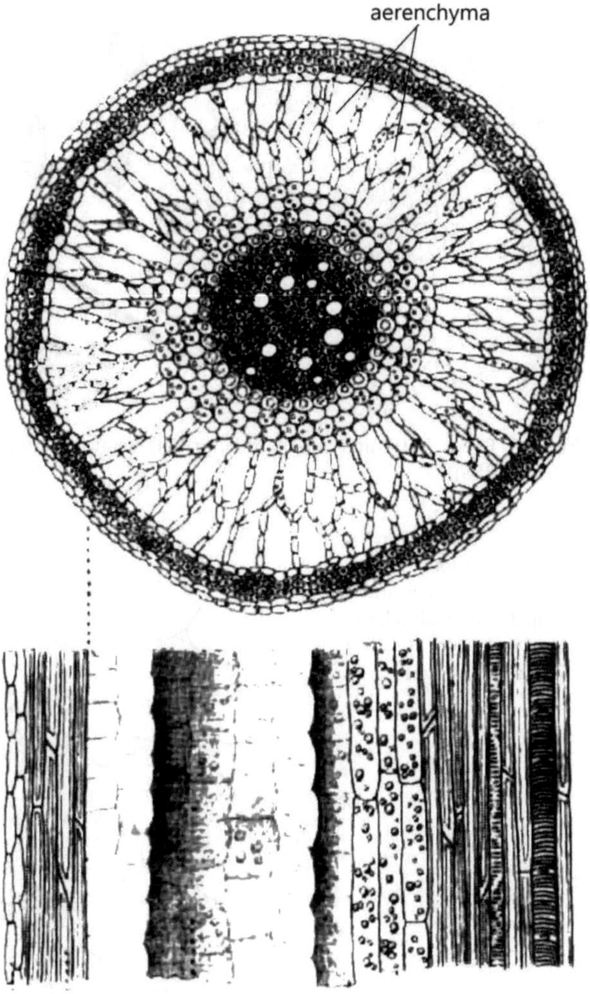

Fig. 34 Cross (top) and longitudinal (bottom) sections through the root of *Scheuchzeria palustris* (Chatin 1856)

porosity is generally greater in emergent plants than in submerged plants (Sculthorpe 1967).

Aerenchyma is formed in two ways:

1. By cell wall separation and cell collapse, known as the lysigeny pathway.
2. By cell enlargement and separation (without collapse), a process known as schizogeny.

In lysigeny, the cells disintegrate and the total number of cells in the air spaces is reduced. Lysigeny is more common than schizogeny (Arteca 1997). Smirnoff and Crawford (1983) observed lysigeny in *Mentha aquatics*, *Ranunculus flammula*, *Potentilla palustris*, *Juncus effusus*, *Narthecium ossifragum*, *Glyceria maxima*, and *G. stricta* and in some members of the family Cyperaceae, including *Eriophorum*.

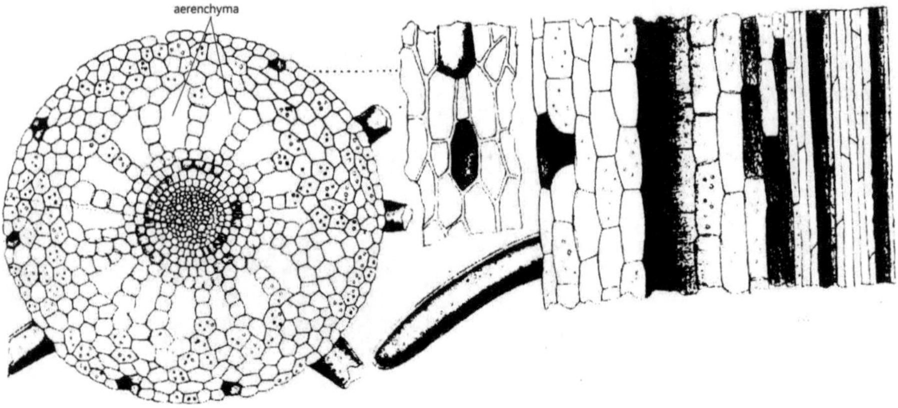

Fig. 35 Cross (left) and longitudinal (right) sections through the root of *Stratiotes aloides* (Chatin 1856)

Fig. 36 Cross section through the root of *Najas major*; sieved tubes are hatched; the axial xylem vessel is double (Sauvageau 1889a)

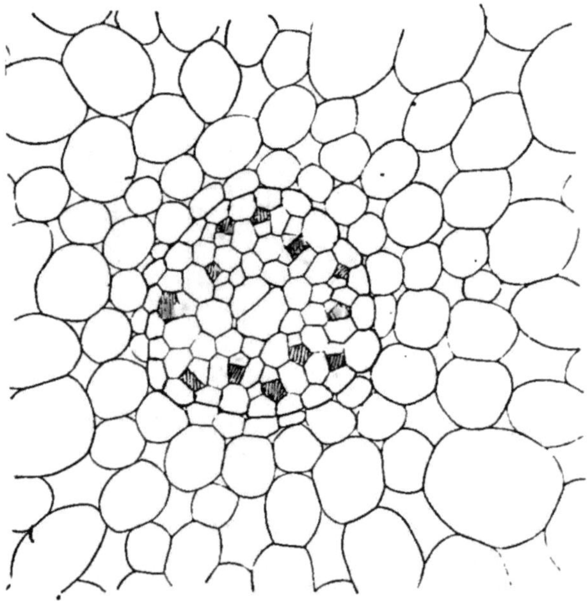

In plants in which the aerenchyma is formed by schizogeny, the number of cells is not reduced, but a honeycomb-shaped structure is produced, by enlarging the intercellular spaces. The cells move further from each other, thus creating space, but they do not disintegrate (Arteca 1997). Schizogeny was observed in *Caltha palustris*, *Filipendula ulmaria* (Armstrong 1978; Smirnoff and Crawford 1983), and *Rumex maritimus* (Laan et al. 1989).

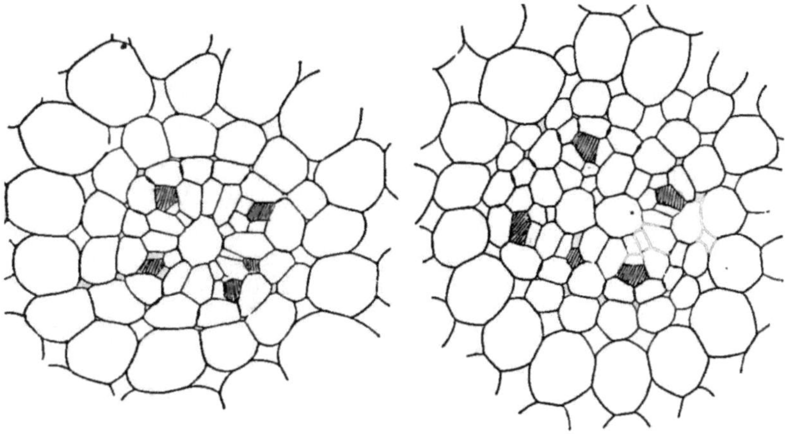

Fig. 37 Cross section through the root of *Najas minor*; right, the axial xylem vessel is double (Sauvageau 1889a)

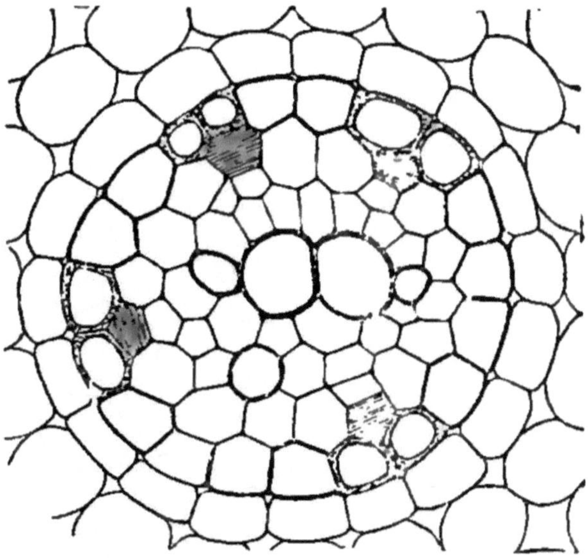

Fig. 38 Cross section through the young root—central cylinder—*Potamogeton plantagineus* (Sauvageau 1889b)

The precise mechanism of aerenchyma formation is not fully understood, but it is known for certain that there is a gaseous plant hormone, ethylene, which is involved in this process. When a chemical inhibitor is used to stop ethylene production, aerenchyma formation is also stopped (Arteca 1997). Low oxygen levels stimulate the production of the enzyme 1-aminocyclopropane-1-carboxylate (ACC) synthetase, which in turn leads to increased levels of another enzyme, ACC oxidase. ACC oxidase is directly responsible for the production of ethylene, which requires oxygen. ACC oxidase diffuses throughout the plant, and ethylene is produced in the aerated parts of the plant (Jackson 1994; Arteca 1997). Ethylene normally diffuses

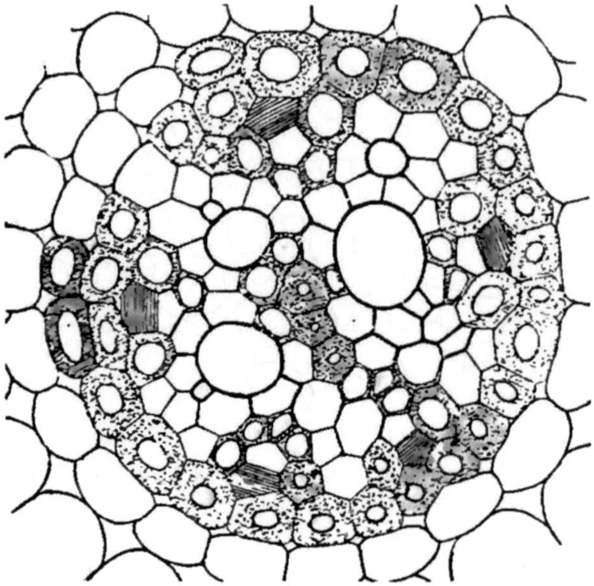

Fig. 39 Cross section through the old root—central cylinder—*Potamogeton plantagineus* (thickened cells of the endodermis are observed in "O" shape) (Sauvageau 1889b)

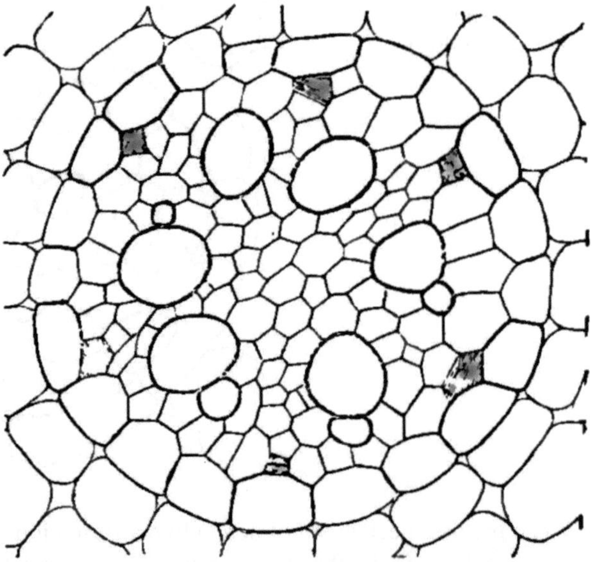

Fig. 40 Cross section through the young root—central cylinder—*Potamogeton polygonifolius* (Sauvageau 1889b)

away from plants, but diffusion is inhibited when the plant is surrounded by water. As ethylene accumulates, it induces, through a very complicated mechanism, the programmed cell death (apoptosis) (Fig. 53) (Drew et al. 2000; Schulze et al. 2005), cell wall degeneration, and an increase in the activity of cell wall-degrading compounds (Vartapetian and Jackson 1997).

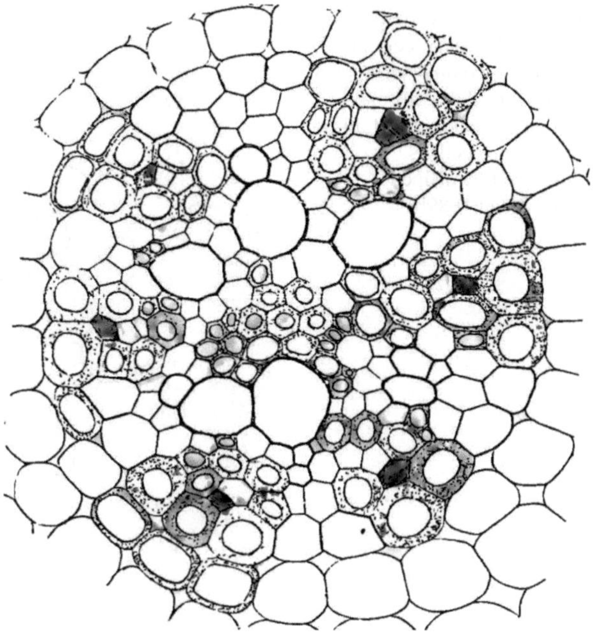

Fig. 41 Cross section through the old root—central cylinder—*Potamogeton polygonifolius*; the advanced degree of sclerification (thickened cells) is observed (Sauvageau 1889b)

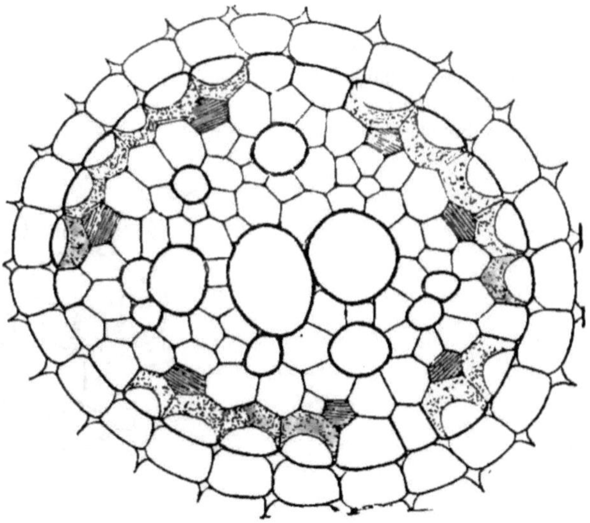

Fig. 42 Cross section through the root—central cylinder—*Potamogeton natans*; the uneven degree of thickening of the endodermis cells is observed ("U-shaped") (Sauvageau 1889b)

Smirnoff and Crawford (1983) observed that several flood-tolerant species formed aerenchyma at the onset of flooding and that porosity increased as soil redox potential decreased. Plants collected from different habitats in a given area had porosity from 1.2 to 33.6% after 11 weeks of waterlogging, and as the stress exposure time increased to 32 weeks, the percentage of porosity increased with up to 50% in some species (*Eriophorum vaginatum* and *E. angustifolium*).

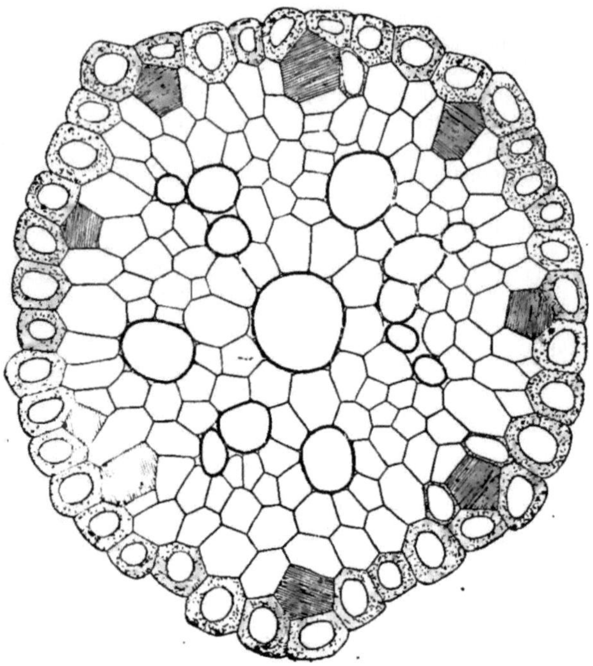

Fig. 43 Cross section through the root—central cylinder—*Potamogeton pectinatus*; the uniform degree of thickening of all endodermis cells is observed ("U-shaped") (Sauvageau 1889b)

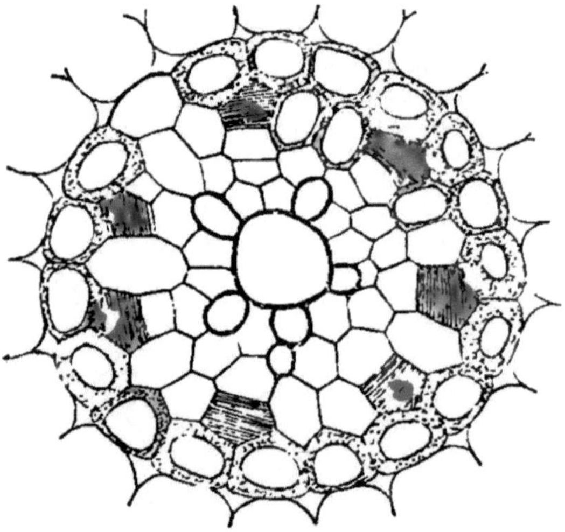

Fig. 44 Cross section through the root—central cylinder—*Potamogeton gramineus* (Sauvageau 1889b)

As soil water content increased from 70 to 90%, *Senecio aquaticus* root porosity increased from 10 to 35%. Air space increases with increasing sediment anaerobiosis in other herbaceous plants, notably *Zostera marina* (Penhale and Wetzel 1983), *Salicornia virginica* (Seliskar 1987), *Oryza sativa* (Kludze et al. 1993), *Spartina patens* (Burdick and Mendelssohn 1990; Kludze and De Laune 1994), *Cladium jamaicense*, and *Typha domingensis* (Kludze and De Laune 1996); *Taxodium*

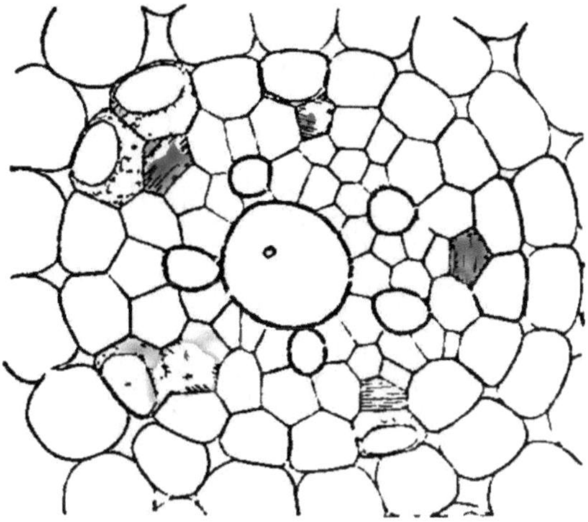

Fig. 45 Cross section through the root—central cylinder—*Potamogeton lucens* (Sauvageau 1889b)

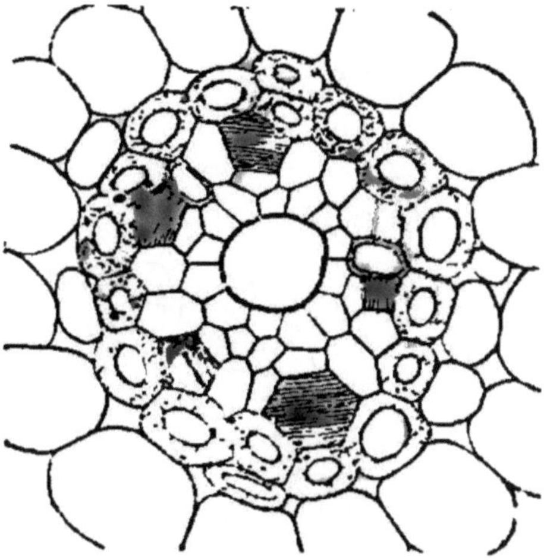

Fig. 46 Cross section through the root—central cylinder—*Potamogeton pusillus* (Sauvageau 1889b)

distichum (bog cypress) also forms more aerenchyma under increasingly reduced oxygen conditions (Kludze et al. 1994).

Some plants, such as *Oryza sativa*, *Schoenus nigricans*, and some species of *Juncus*, form aerenchyma even in well-aerated soils as part of normal root development. This suggests that aerenchyma formation in these plants has a genetic component and does not require ethylene accumulation (John et al. 1974; Jackson et al. 1985; Justin and Armstrong 1987; Jackson 1990).

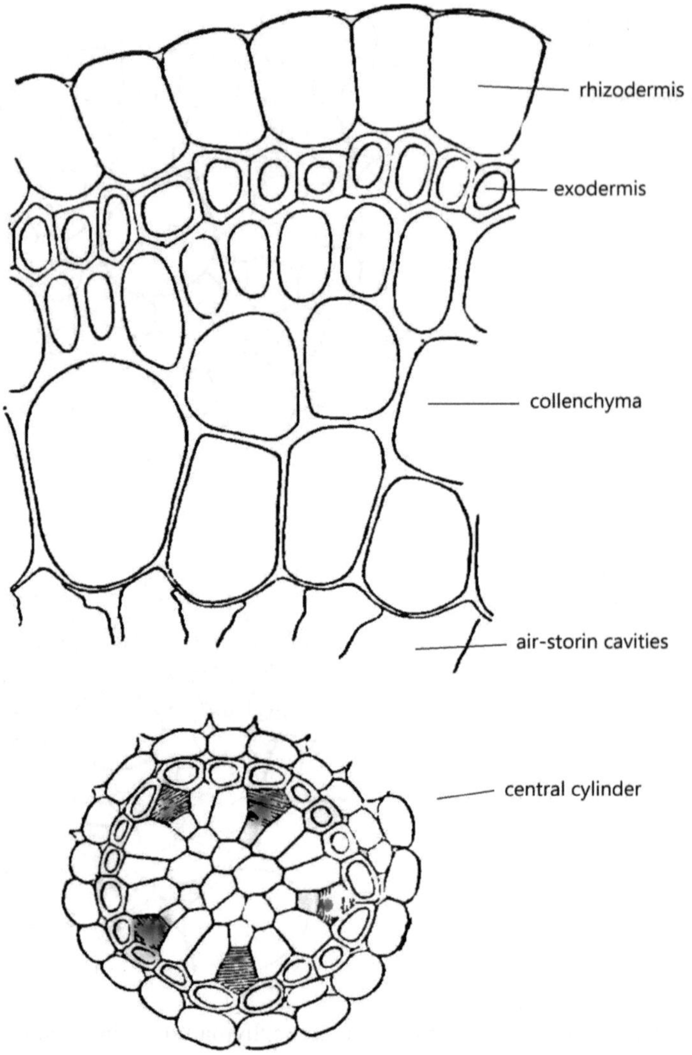

Fig. 47 Cross section through the root of *Zostera marina* (Sauvageau 1889c)

1.2.2 Functions of Aerenchyma

Aerenchyma lowers the resistance to flow encountered by oxygen and other gases in the plant tissue, thereby allowing oxygen to reach the submerged portions of the plant relatively easily (Vartapetian and Jackson 1997). At the same time, the aerenchyma allows gases produced by plants, such as carbon dioxide and ethylene, to be released into the atmosphere (Visser et al. 1997). Naturally, aerenchyma is effective in aerating the roots and rhizomes of aquatic plants; however, aeration is often incomplete. Even when there is a developed aerenchyma, roots can suffer some

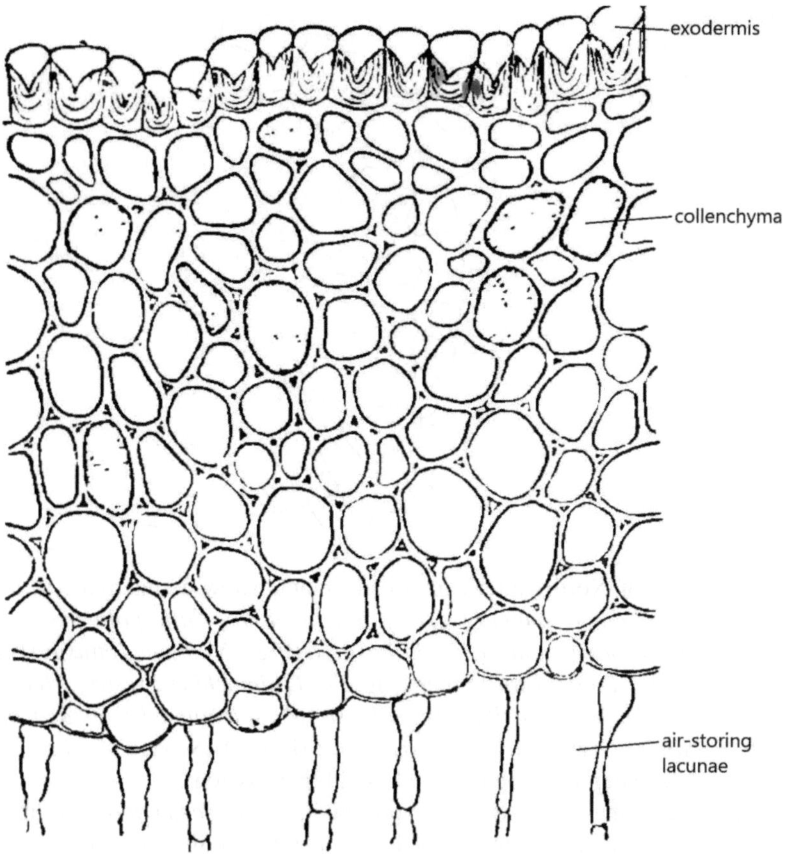

Fig. 48 Cross section through the root of *Cymodocea aequorea* (Sauvageau 1889c)

degree of anoxic stress and switch to anaerobic metabolism (Saglio et al. 1983). At the beginning of the growing season, the roots and rhizomes of some emergent species suffer from oxygen deficiency until their new shoots emerge and connect them to the atmosphere (Burdick and Mendelssohn 1990; Koncalova 1990; Naidoo et al. 1992; Weber and Brandle 1996).

Another important function of aerenchyma is gas storage. However, the gas storage capacity of herbaceous plants is limited and can be depleted within minutes to hours. New supplies from the atmosphere are needed to support the plant's oxygen needs. In general, the more air space there is within the plant, the greater its storage capacity, and monocots tend to have greater porosity and storage capacity than dicots (Crawford 1993). At *Typha latifolia*, approximately half of the total leaf volume is occupied by air spaces, and the internal concentration of CO_2 in the leaves is up to 18 times higher than atmospheric levels (Constable et al. 1992). Internal CO_2 is assimilated by leaves and provides the plant with a significant additional carbon pool (Constable and Longstreth 1994).

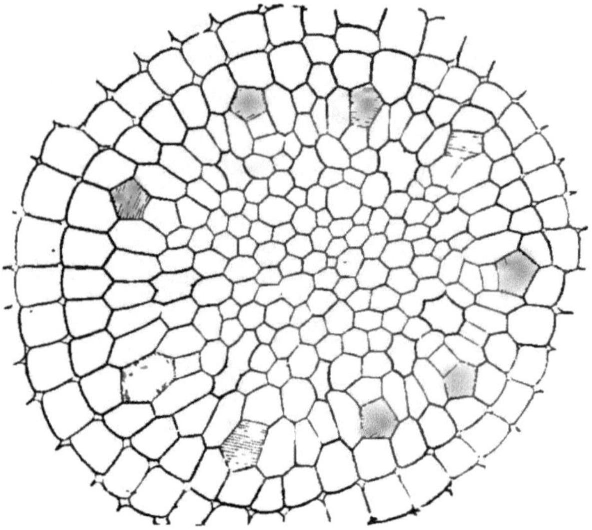

Fig. 49 Cross section through the central root cylinder of *Cymodocea aequorea* (Sauvageau 1889c)

Many species of North American aquatic plants show aerenchyma at the root level: *Sparganium americanum* (Fig. 53), *Bidens beckii*, *Calla palustris* (Fig. 54), *Caltha palustris*, *Elodea nuttallii*, *Gratiola aurea* (Fig. 55), *Juncus effusus* (Fig. 56), *Justicia americana*, *Lobelia dortmanna* (Fig. 57), *Lysimachia terrestris*, *Menyanthes trifoliata* (Fig. 58), *Myriophyllum brasiliense*, *M. spicatum*, *Nymphaea odorata* (Fig. 59), *Peltandra virginica*, *Sagittaria latifolia* (Fig. 60), *Scirpus validus* (Fig. 61), and *Trapa natans* (Fig. 62) (Ogden 1974—with permission).

Aerenchyma, at the root level, is also present in *Cymodocea nodosa*, a submerged aquatic species, which grows in the Mediterranean Sea or in some areas adjacent to the Atlantic Ocean (Grigore et al. 2014). In this species, the air channels are found in the middle zone of the root cortex, being separated by uniseriate blades formed by elongated cells (Figs. 63 and 64).

2 Ecological Anatomy of the Roots of Marsh Species (Helophytes)

Aquatic plants include all plants whose photosynthetic organs are submerged or, at most, on the surface of the water.

As discussed previously (Grigore 2022), the definition and classification of aquatic plants, as well as the types of aquatic plants, are very difficult topics from a biological point of view.

The boundary between typical aquatic plants and, for example, marsh plants (helophytes) on the one hand, and that between helophytes and terrestrial plants, on the other hand, is very fragile; genuine aquatic plants may have organs that are not

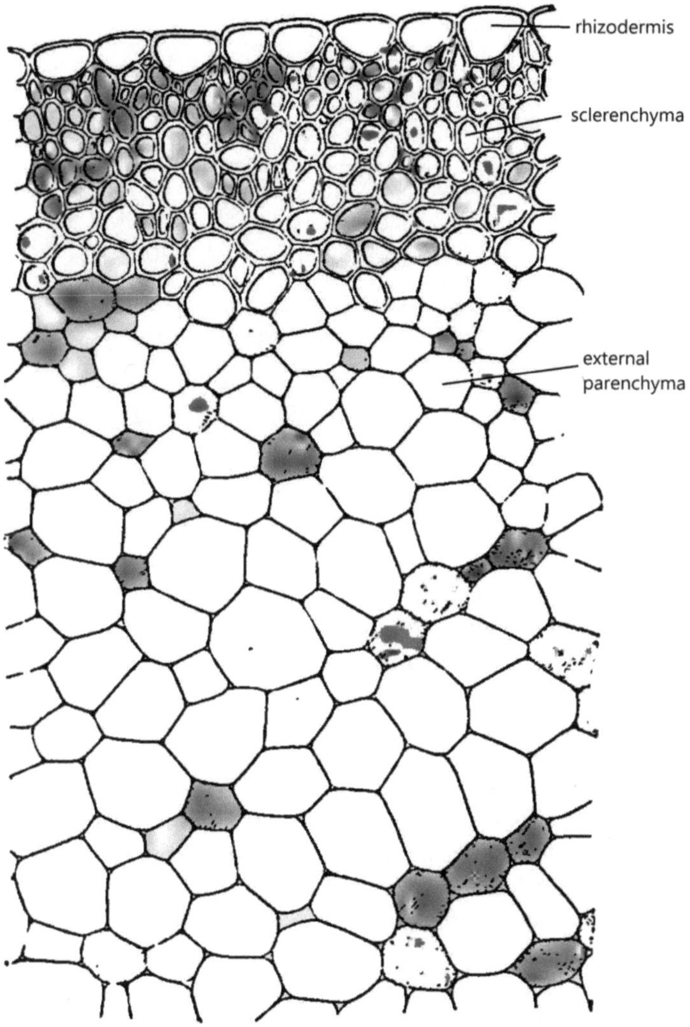

Fig. 50 Cross section—aged root of *Posidonia caulinii* (Sauvageau 1889c)

in contact with water, while the organs of some marsh plants may be, for greater or lesser periods of time, continuously or not, immersed in water.

The presence of water in a habitat depends on many factors and, for some aquatic habitats, is not a permanent factor.

Marsh plants include all those species that normally have their roots under water or in marshy soil, but, like land plants, raise their aerial parts above the surface of the water.

Moreover, many marsh plants are more or less amphibious (they have organs that are located in two life environments—aquatic and aerial) and plastic, so they can change their structure depending on the level of immersion in water.

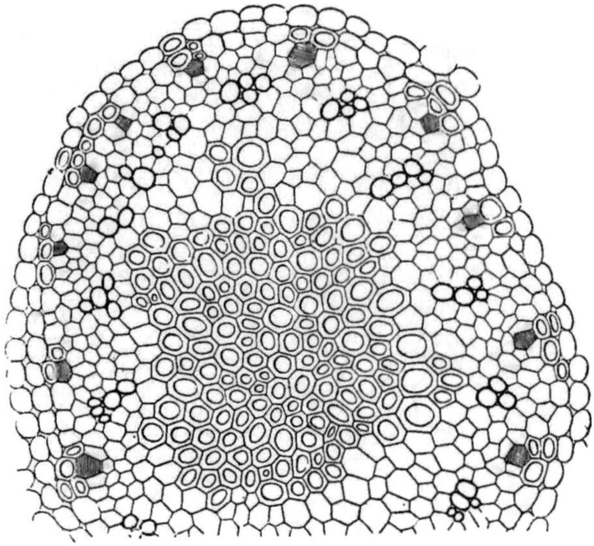

Fig. 51 Cross section—central cylinder—root of *Posidonia* caulinii (Sauvageau 1889c)

Marsh plants are generally restricted to shallow, stagnant, or gently flowing water, fixed in a soil that contains a large amount of water, at least for an extended period. The soil is loose, often very loose and soft, and is also rich in humus in the form of peat or mud.

For these reasons, there are morphological and anatomical adaptations common to aquatic plants (hydrophytes) and marsh plants (helophytes).

Of these, as we have seen for aquatic plants, the aerenchyma, present at the level of the underground, submerged organs, the root, and the rhizome, represents an anatomical feature specific to many species of marshes (and, in our example, salty): *Glaux maritima* (Figs. 65 and 66), *Triglochin maritima* (Figs. 67 and 68), *Plantago crassifolia* (Fig. 69), and *Bassia sedoides* (Fig. 70), to list here only some species of salt marsh halophytes.

3 Ecological Root Anatomy of Mangrove Species

The term "mangrove" is very versatile, being able to have multiple definitions (Grigore 2008; Ivănescu et al. 2020; Grigore and Toma 2021); in this subsection, we will briefly refer to other anatomical adaptations of mangroves, in addition to discussing the anatomy of aerial respiratory roots (pneumatophores) (Børgesen and Paulsen 1898, 1900; Børgesen 1909), in order to have a more comprehensive picture of this fascinating ecological group of plants (see also Grigore 2022).

Davis (1940) refers to mangroves as a general term applied to plants living in marshy, loose, moist soils of tropical tidal waters (Fig. 71). According to Macnae (1968), mangroves are trees or shrubs growing between the high water mark of

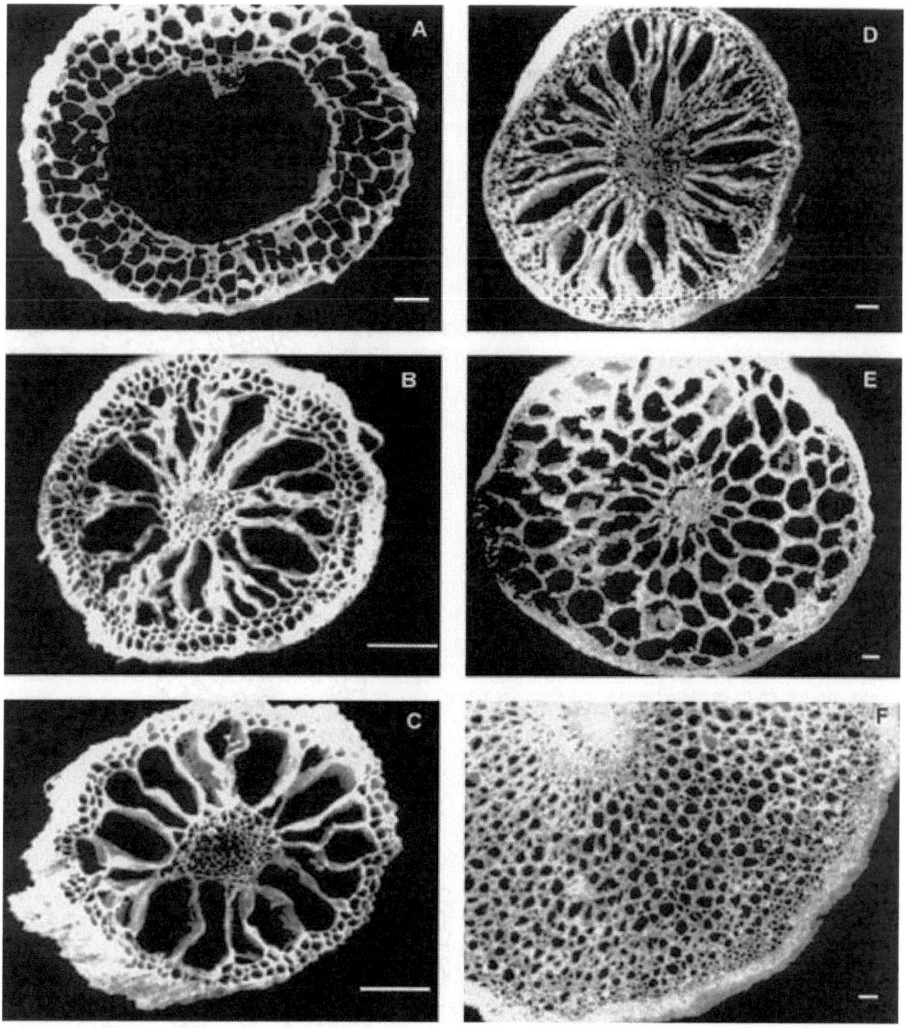

Fig. 52 Cross sections through the root of some species of aquatic plants: large air spaces—aerenchyma: (**a**) *Isoetes lacustris*; (**b**) *Littorella uniflora*; (**c**) *Luronium natans*; (**d**) *Nymphoides peltata*; (**e**) *Nymphaea alba*; (**f**) *Nuphar lutea* (Smits et al. 1990)

spring tide and a level close to but above mean sea level. These species are circumtropical, on sheltered shores, and often grow along river banks as far inland as the sea can penetrate (Walsh 1974).

Børgesen (1909) studied the mangroves of the Danish West Indies as a formation of evergreen, treelike plants growing on dry shores, partly in shallow, salt, or brackish water, partly on low-lying soils, which are relatively rare, and sometimes never covered by salt or brackish water.

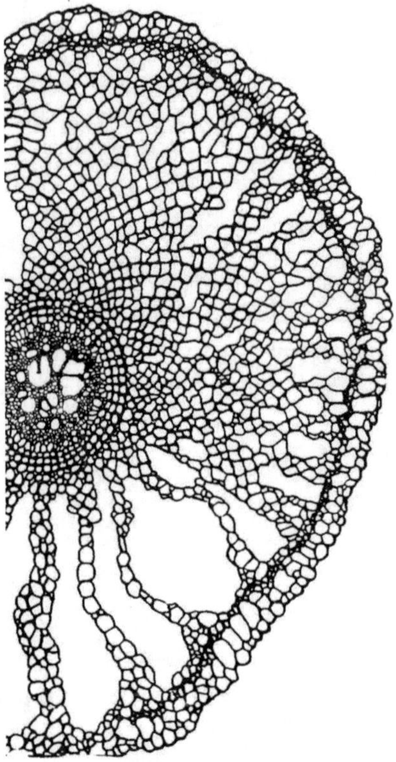

Fig. 53 Cross section through the root of *Sparganium americanum*—aerenchyma (Ogden 1974, with permission)

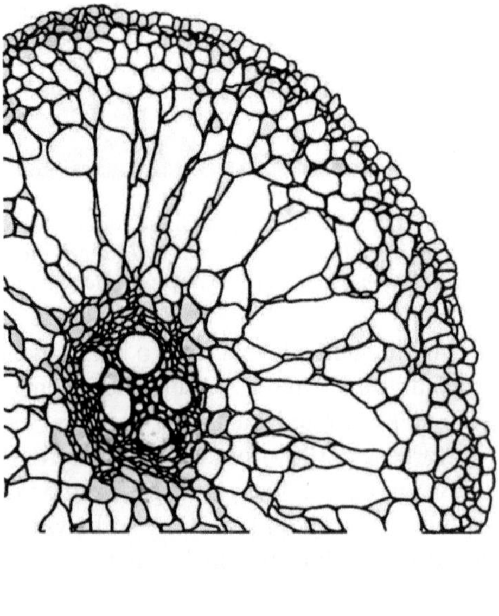

Fig. 54 Cross section through the root of *Calla palustris*—aerenchyma (Ogden 1974, with permission)

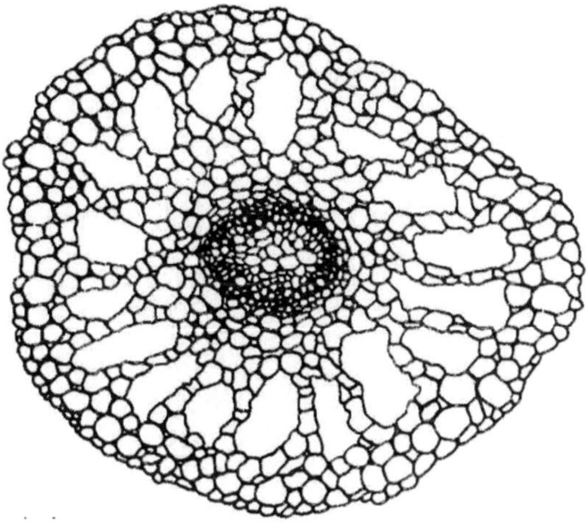

Fig. 55 Cross section through the root of *Gratiola aurea*—aerenchyma (Ogden 1974, with permission)

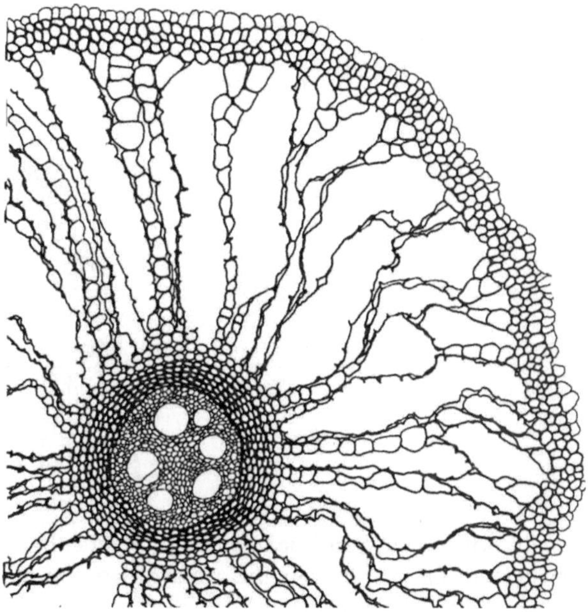

Fig. 56 Cross section through the root *Juncus effusus*—aerenchyma (Ogden 1974, with permission)

In very general terms, there appear to be five basic requirements for extensive mangrove forest development: tropical temperatures, fine-grained alluvium, shorelines free of strong waves and tidal action, salt water, and a large tidal range (Walsh 1974).

Warming (1877, 1883, 1909) described very well the environmental conditions under which mangrove species grow. They are found in all tropical seas, especially on flat, marshy shores where the water is relatively calm, such as in lagoons, small islands, and estuaries, but not where rocky soil predominates. The substrate is also

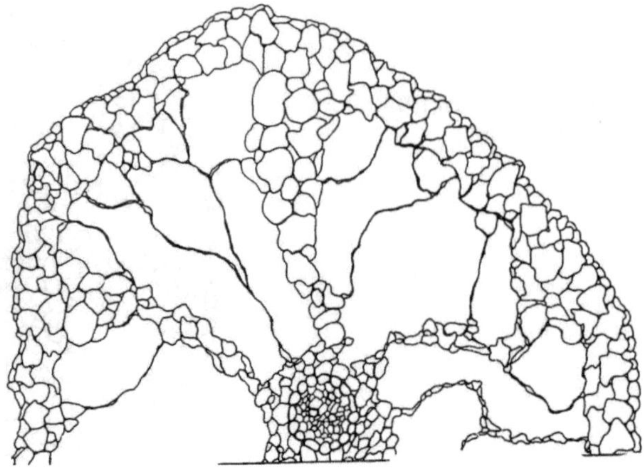

Fig. 57 Cross section through the root of *Lobelia dortmanna*—aerenchyma (Ogden 1974, with permission)

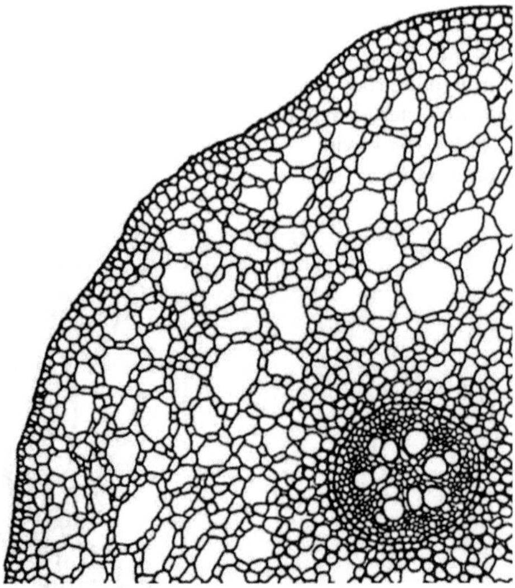

Fig. 58 Cross section through the root of *Menyanthes trifoliata*—aerenchyma (Ogden 1974, with permission)

flooded with water, permanently, or at high tide. In many places, mangrove vegetation extends far inland along the banks of large rivers. The water is usually more or less brackish as the tide extends.

The soil, which in places is not covered with water at ebb, is soft, deep, muddy black, and full of decaying organic bodies in which bacteria abound. The water between the trees can be covered with a dirty film, and gas bubbles rise from below to the surface.

The morphological and anatomical adaptations of mangrove species are well documented in classic studies (Warming 1877, 1909; Liebau 1914; Bowman 1917; Schimper 1891, 1884, 1903; Børgesen and Paulsen 1898, 1900; Børgesen 1909; Walsh 1974; Grigore 2008; Ivănescu et al. 2020; Grigore and Toma 2021).

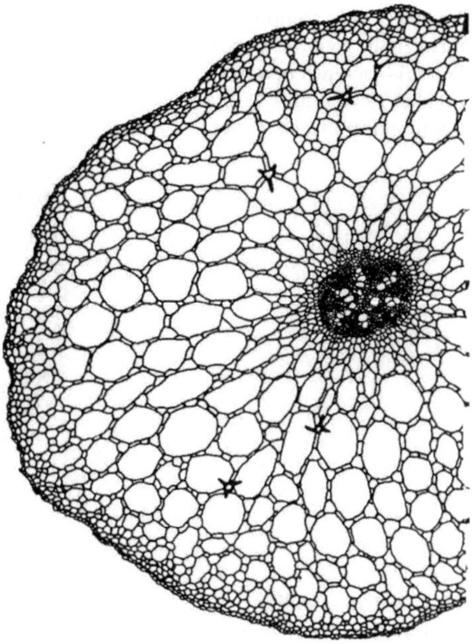

Fig. 59 Cross section through the root of *Nymphaea odorata*—aerenchyma (Ogden 1974, with permission)

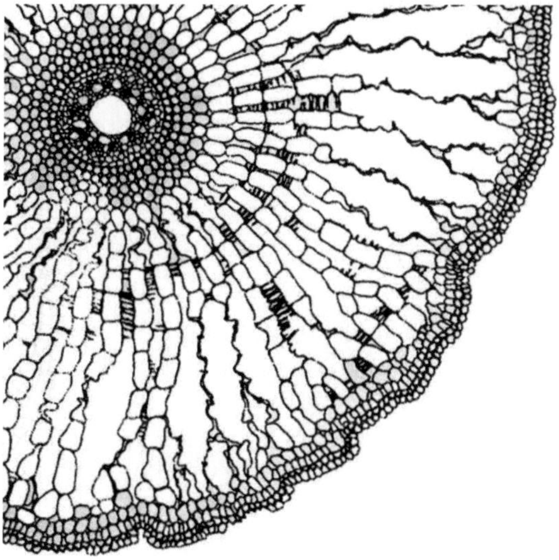

Fig. 60 Cross section through the root of *Sagittaria latifolia*—aerenchyma (Ogden 1974, with permission)

3.1 Fixation to the Substrate with the Help of Prop Aerial Roots

The morphology of metamorphosed roots from mangrove species was discussed in a previous paper (Grigore 2022). In the continuation of the discussion, it must be

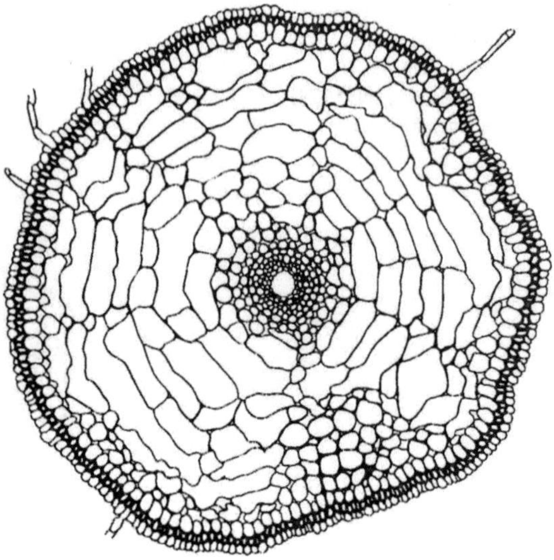

Fig. 61 Cross section through the root of *Scirpus validus*—aerenchyma (Ogden 1974, with permission)

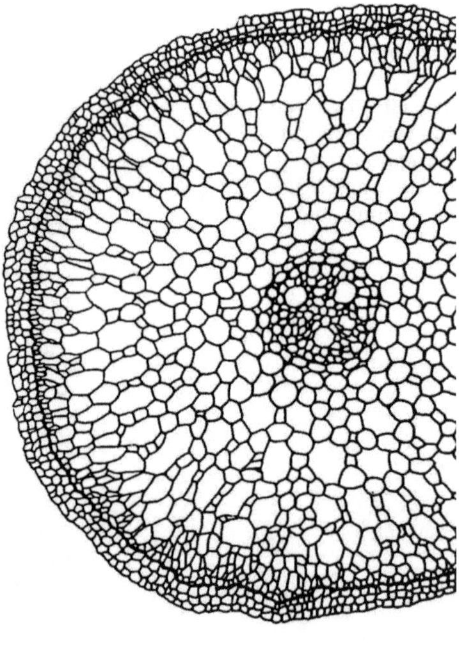

Fig. 62 Cross section through the root of *Trapa natans*—aerenchyma (Ogden 1974, with permission)

said that the degree of soil compaction and the disposition of water in depth induce certain differences and determine that a mangrove species assumes a zonal distribution within plant associations.

On the landward side, mangrove plants disappear and are replaced by other shrubs and trees. The species nearest to the sea are those that are most able to fix themselves in deep water, namely, the species of *Rhizophora*; within them, in

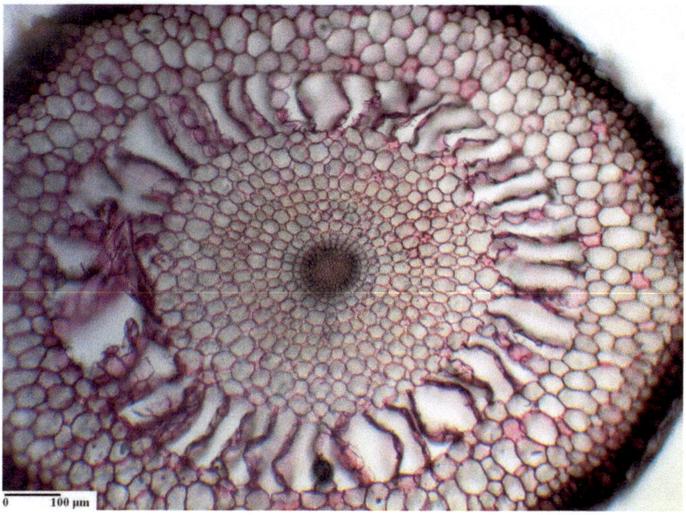

Fig. 63 *Cymodocea nodosa*—aerenchyma—cross section through the root (Grigore et al. 2014)

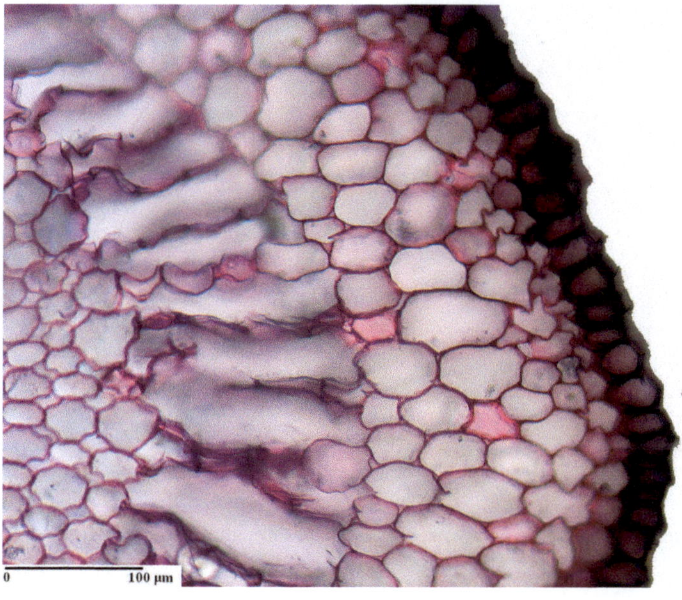

Fig. 64 *Cymodocea nodosa*—aerenchyma—cross section through the root (Grigore et al. 2014)

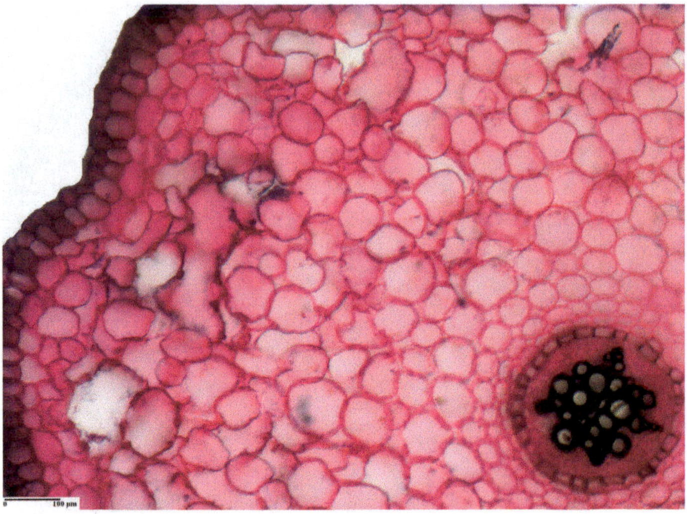

Fig. 65 *Glaux maritima*—aerenchyma; cross section through the root (Grigore et al. 2014)

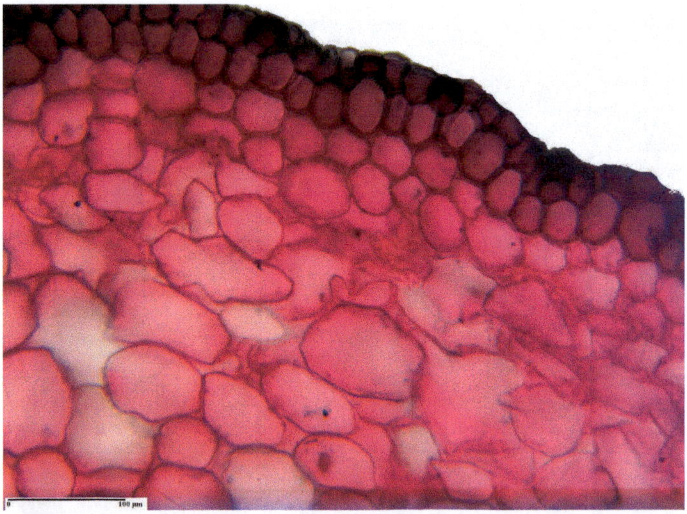

Fig. 66 *Glaux maritima*—aerenchyma; cross section through the root (Grigore et al. 2014)

shallower waters or on drier land, follow those species that have a more limited capacity in this sense, such as the species of *Avicennia*, *Bruguiera*, *Aegiceras*, and *Carapa*.

Rhizophora species are fixed by prop roots (Grigore 2022), which are aerial roots that are formed from the main trunk and form a kind of arched "pillars" that penetrate the soil. Sometimes aerial roots emerge accidentally from the lower branches

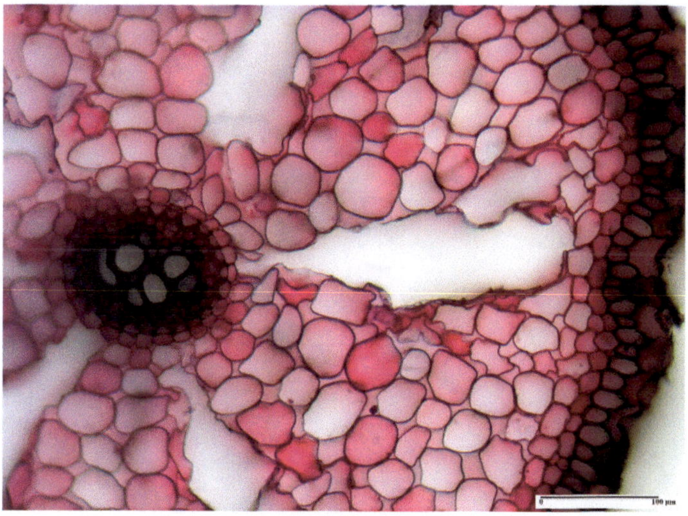

Fig. 67 *Triglochin maritima*—aerenchyma; cross section through the root (Grigore et al. 2014)

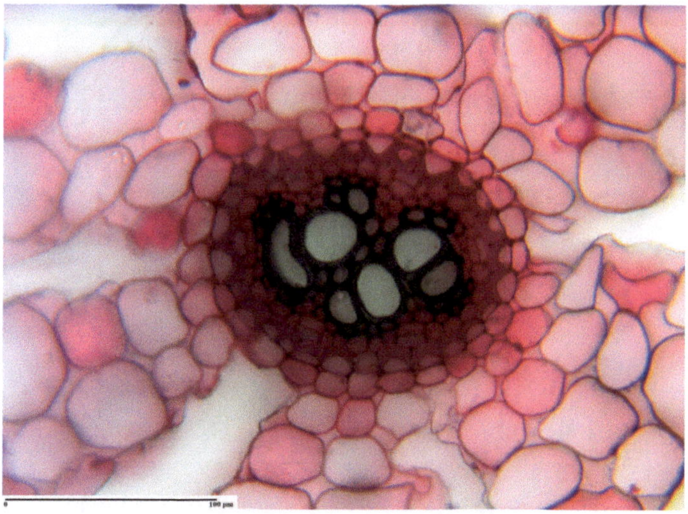

Fig. 68 *Triglochin maritima*—aerenchyma; cross section through the root (Grigore et al. 2014)

of the tree and then curve down to the ground in an arching fashion. These arc-shaped roots, on which the tree comes to depend, are very numerous; in this way, the foundation of the tree is firmer, and its power of resisting bending, which may be caused by the movements of wind or waves, is greater than if the stem were the sole support (Figs. 72 and 73).

The anatomical construction of these roots is consistent with the unusual nature of the stresses to which they are exposed, as supporting structures, and differs from

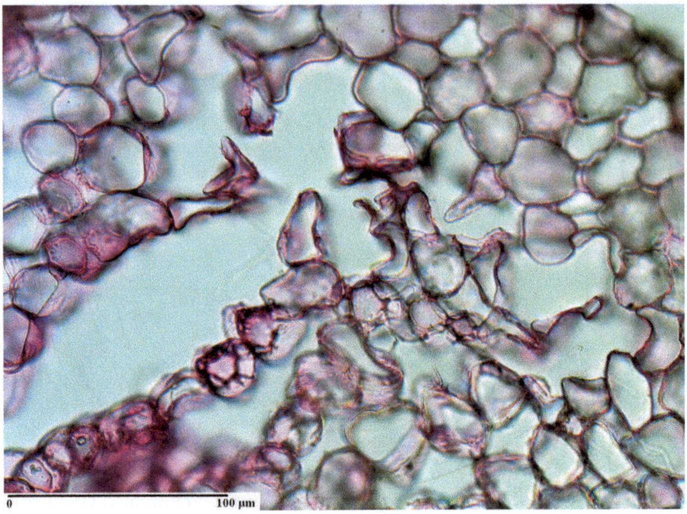

Fig. 69 *Plantago crassifolia*—aerenchyma; cross section through the root (Grigore et al. 2014)

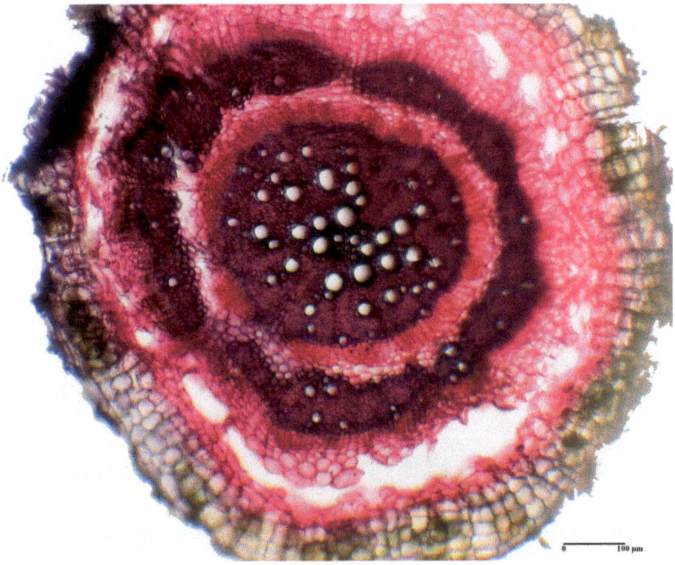

Fig. 70 *Bassia sedoides*—aerenchyma; cross section through the root (Grigore et al. 2014)

that of most other roots in that the mechanical tissue is made to provide a tubular arrangement around the medulla of size large (Fig. 74). Roots with a similar mechanical support function are found in *Ceriops* and *Acanthus ilicifolius*. The lower parts of these propitiatory roots are often covered with algae.

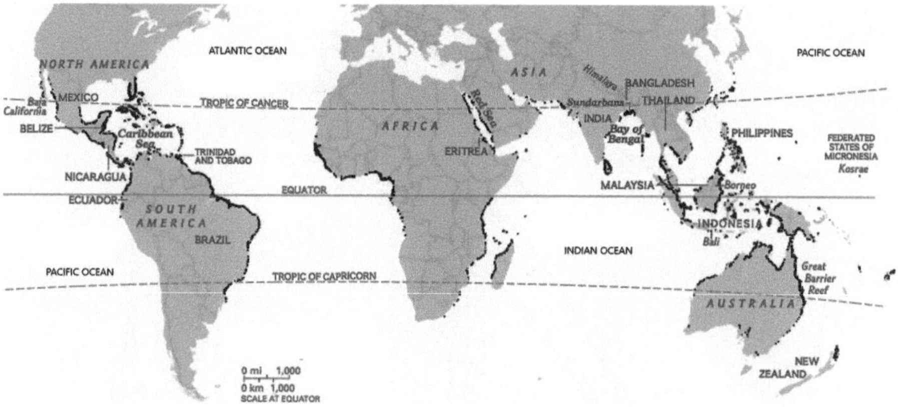

Fig. 71 Distribution of mangroves (marked with the black hashes) on the globe (adapted and modified after UNEP WORLD CONSERVATION MONITORING CENTRE AND INTERNATIONAL SOCIETY FOR MANGROVE ECOSYSTEMS NGM MAPS)

Fig. 72 General aspect of the prop aerial roots of *Rhizophora mangle* (Warming, taken from Børgesen and Paulsen 1900)

3.2 Respiratory Roots (Pneumatophores)

As we have seen in the lines above, respiration is a very difficult process in flooded (swampy) soil rich in organic matter and poor in oxygen. For this reason, all mangrove plants have a strongly developed aerenchyma system; the parts immersed in water are very spongy and soft in structure; the unusually large stomata and lenticels on the organs that develop above water provide communication of the atmosphere with the intercellular spaces.

Rhizophora prop roots serve as respiratory roots. Other species possess distinct respiratory roots (pneumatophores). *Avicennia* has such erect, unbranched roots, 30–40 cm high; these are arranged in very long rows, which “radiate” around the

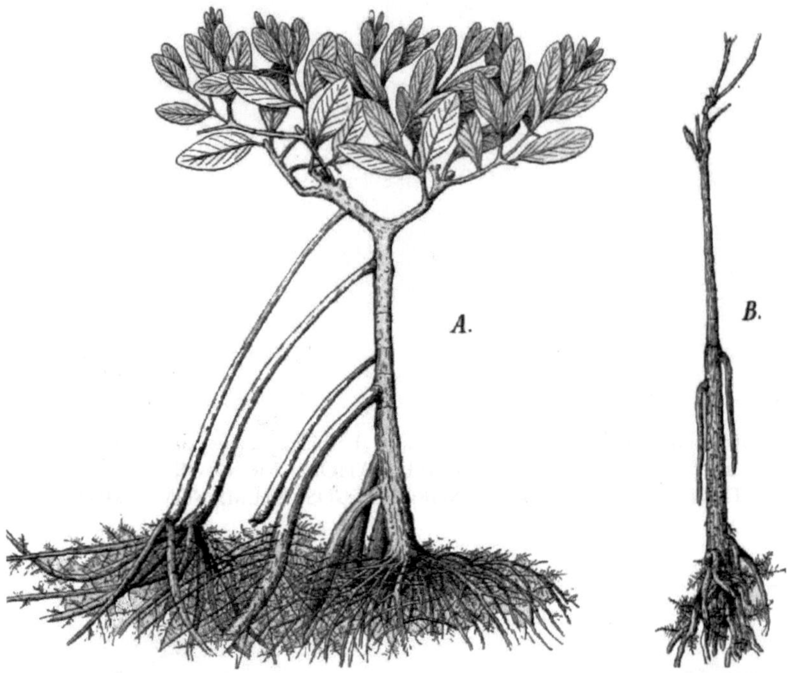

Fig. 73 Arch-shaped aerial roots (**a**) and almost vertical (**b**) roots in *Rhizophora mangle* (Børgesen and Paulsen 1900)

tree and indicate the position of the horizontal roots from which they emerge (Figs. 75, 76, and 77). Their respiratory role is revealed by the anatomical structure (Fig. 78).

Similar respiratory roots are found in *Sonneratia acida*, *Laguncularia racemosa* (Figs. 79 and 80), *Bruguiera caryophyllaeoides* (Fig. 81a), *B. gymnorrhiza* (Figs. 81b and 84), and *Carapa obovata* (Fig. 81c); roots in the form of bent knees, with the knee projecting above the water, occur at *Bruguiera* and, to a lesser extent, at *Lumnitzera*, while comb-like extensions, connected to the root, occur at *Carapa*. The anatomical configuration supports this function.

At *Laguncularia racemosa*, the primary cortex initially develops as air storage tissue, and the inner part of the cortex is bounded by a distinct endodermis; in its growth, the primary cortex shrinks, and the large surface of the phloem acquires an air storage function (Fig. 82). Idioblasts (trichoblasts), like those shown in *Rhizophora*, are also found in *Sonneratia* (Fig. 83) and *Bruguiera gymnorrhiza* (Fig. 84).

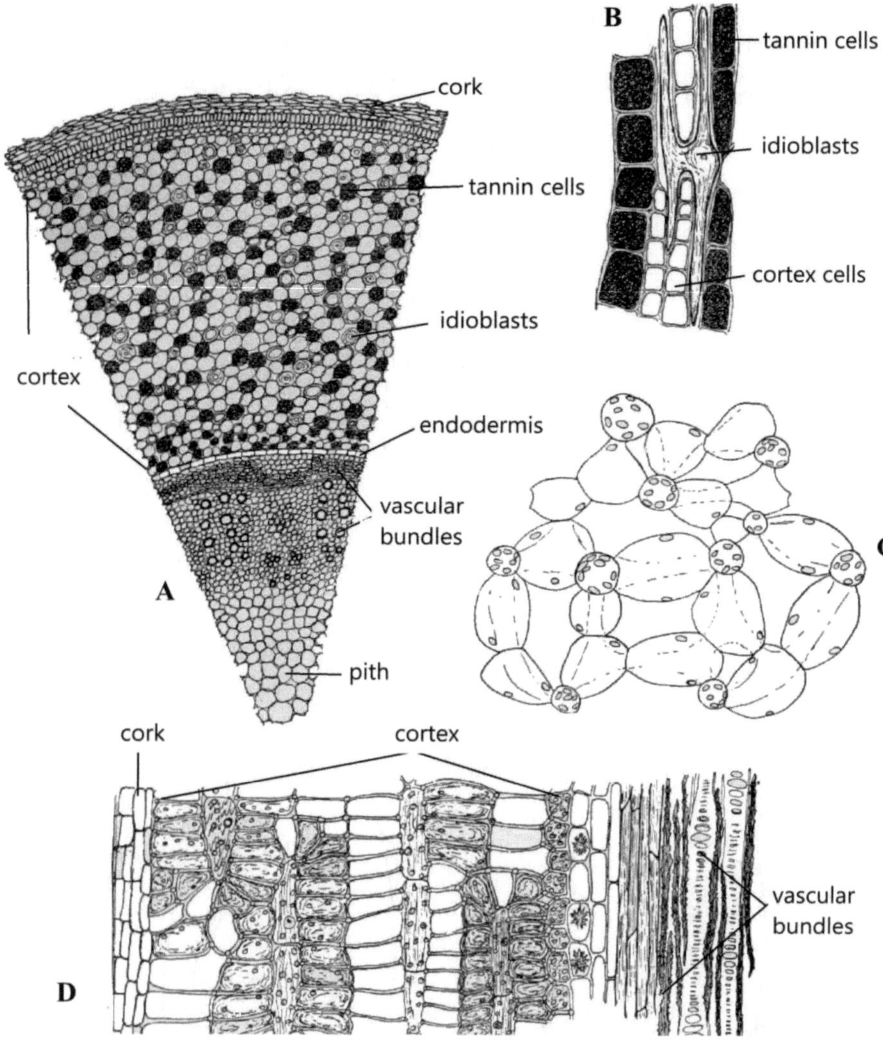

Fig. 74 *Rhizophora mangle*: (**a**), cross section through the prop aerial root; (**b**), longitudinal section through the prop aerial root; (**c**), cross section through the cortex of the absorbing root; (**d**), longitudinal section through the absorbing root (Bowman 1917)

4 Ecological Anatomy of Roots of Epiphytic Species

Epiphytes are plants that germinate and grow on other plants, but without obtaining food from them (from the support plants); in this way, they differ from parasitic plants, with which they are often confused. The definition of epiphytes, but especially their ecological classification, is not always easy; there are several classification systems for epiphytes. The biological approach to epiphytes is complicated (Benzing 1990; Grigore 2022).

Fig. 75 *Avicennia nitida*—isolated plant—with pneumatophores (Børgesen and Paulsen 1900)

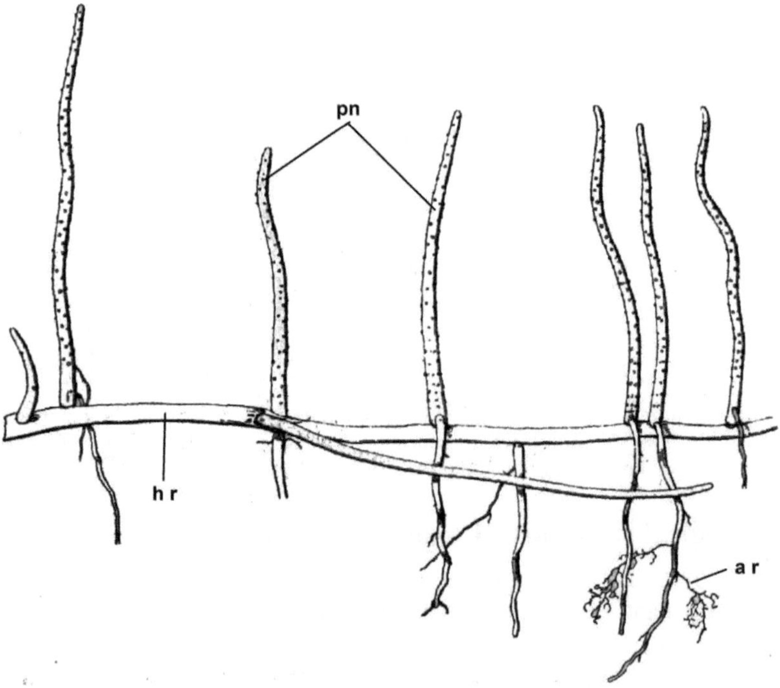

Fig. 76 *Avicennia nitida*—horizontal root (creeping) (h r), with emergent pneumatophores (pn) and absorbing roots (a r) (Børgesen and Paulsen 1900)

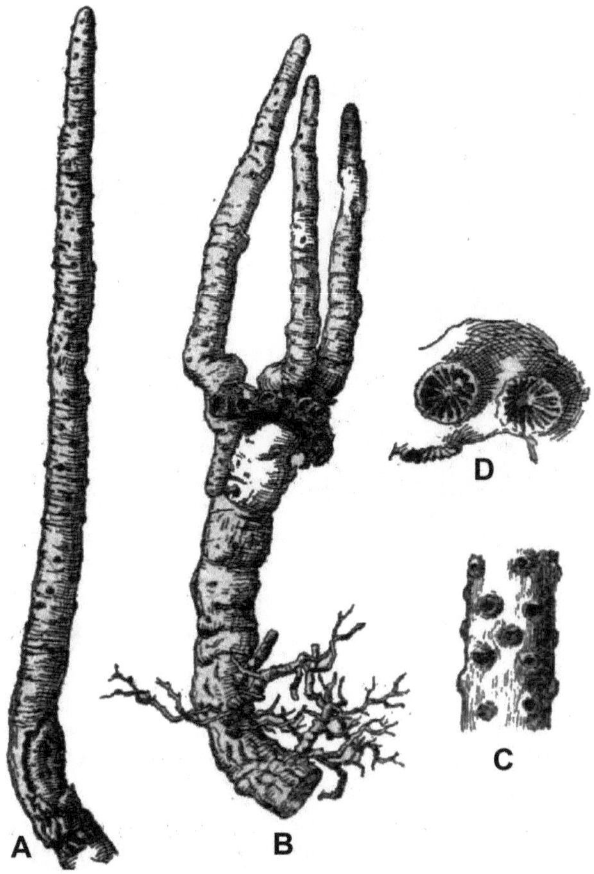

Fig. 77 *Avicennia nitida*—pneumatophores: (**a**), unbranched root; (**b**), branched root; (**c**, **d**), enlarged details of pneumatophores—pneumatodes (lenticles) visible (Børgesen and Paulsen 1900)

In another work, we dealt with the morphology of the roots found in epiphytes, especially tropical ones (Grigore 2022); as seen, a special characteristic of epiphytes is aerial roots, adventitious in origin, which play a role in water absorption (most often in the form of vapors) and sometimes in photosynthetic assimilation.

Epiphytes native to all forests, perhaps less so those of wetter or deeply humus-anchored forests, must either create considerable water reserves within their own bodies or meet their water needs during occasional contact with fluids at canopy level (see the morphological adaptations of epiphytes—Grigore 2022). Some epiphytes possess characteristic absorption devices. Although different organs and tissues are involved in this absorption mechanism, there is basically only one type of structural architecture for water uptake, which is based on a biphasic system: a superficial, nonliving, absorbent tissue that sits on top of a deeper absorbent tissue, containing transfer cells. The two best-known aerial organs of absorption are the aerial roots with velamen (found in many orchids) and the leaf scales (trichomes) of bromeliad epiphytes (which we will deal with in the chapter dedicated to the leaf).

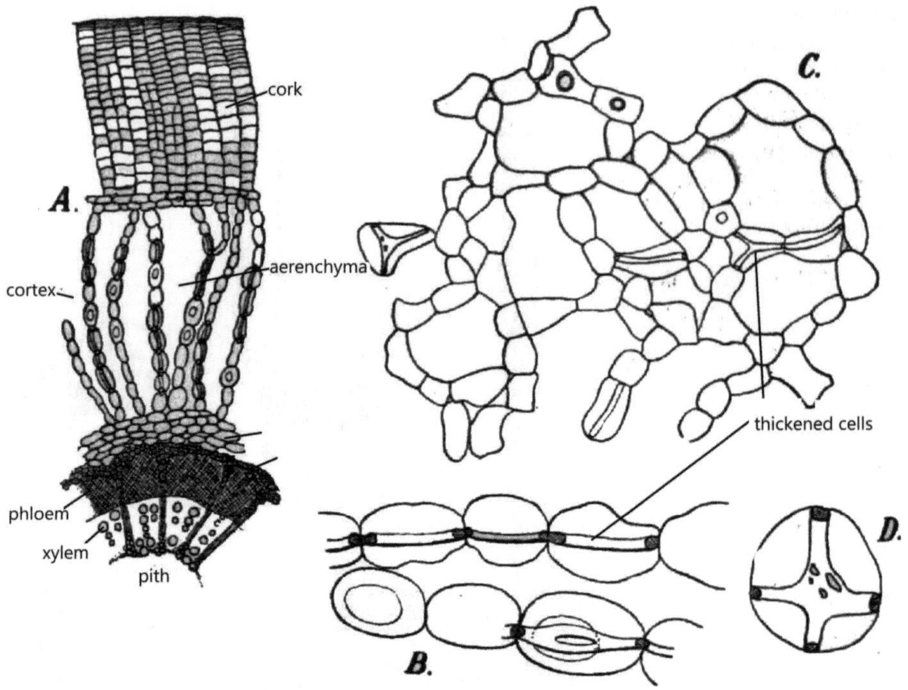

Fig. 78 *Avicennia nitida*—cross section through a pneumatophore; (**a**), general aspect; (**b**), thickened cells within the aerenchyma; (**c**), cross section at the aerenchyma level; (**d**), thickened cell within the aerenchyma (Børgesen and Paulsen 1900)

The aerial roots of many orchids and species of the Araceae family, tropical epiphytes, are characterized by the presence of a special covering (layer), which has long been known as velamen (*velamen radicum*; *velamen*—Latin language—"sheath or covering, used to cover the body or other objects") (Figs. 85, 86, 87, 88, 89, 90, 91, 92, 93, 94, 95, and 96); the anatomy and development of this tissue were studied most closely and thoroughly in classical studies (Leitgeb 1865).

It should be noted that the interpretation of the name of some tissues that make up the structure of the aerial roots of epiphytes can raise some aspects of differentiation and naming of these tissues, from author to author; for example, the velamen, as a tissue located on the outside of the aerial root, is interpreted by some authors as a multilayered epidermis (Coulter et al. 1911) or as a silvery parchment-like sheath of variable thickness. Other times, the layer on the outside of the velamen—when present—is called the epivelamen (Benzing 1990).

Studies of the development of the velamen show that it is a product of the protodermis (see Grigore 2022), which, near the tip of the root, becomes stratified by tangential division. In some exceptional cases (*Vanilla planifolia*, *V. aphylla*, *Dendrocolla teres*, etc.), the protoderm divides only by anticlinal walls, and consequently the velamen remains unilayered.

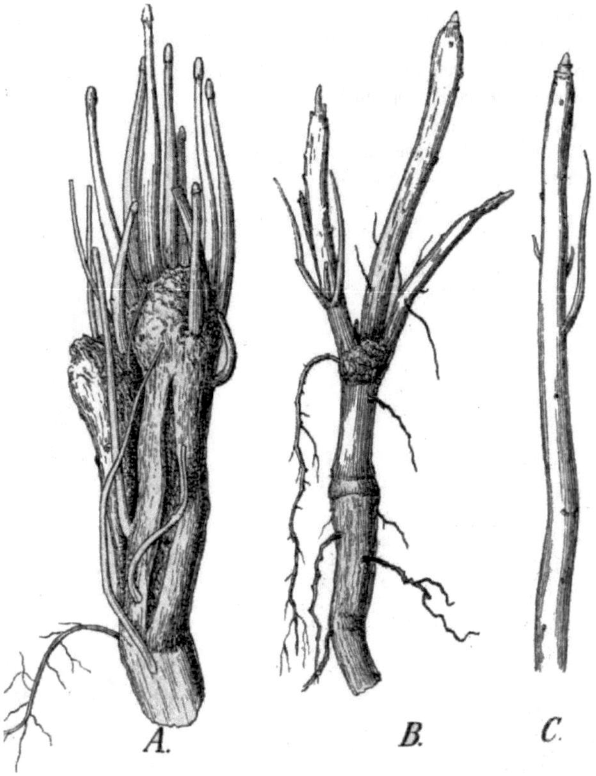

Fig. 79 *Laguncularia racemosa*—pneumatophores (**a–c**, varying degrees of branching) (Børgesen and Paulsen 1900)

The number of layers contained in the velamen varies (Figs. 85, 86, 87, 88, and 89) from 1 to 18 (*Cyrtopodium* spp.), or 24, but generally remains at a fairly constant value in each species (Haberlandt 1914). The cells of the velamen are in continuous contact with each other, and their form varies widely. Sometimes, the cells are roughly isodiametric, but somewhat elongated radially, as seen in cross section; in other cases, their long axes are parallel to the axis of the root itself.

The walls of the cells of the velamen are thickened in various ways, but most often by means of spiral mechanical fibers (idioblasts) (Figs. 85, 86, and 88); these fibers may be parallel to each other, or may form meshes of variable width, or finally may be grouped together to form wide ribbons or bands.

Reticulate thickenings are less common than the spiral type, but occur, for example, in *Dendrocolla teres* and *Vanda furva*. Rarely, the walls are uniformly thickened and provided with more or less numerous depressions (*Angraecum subulatum*); in a few cases, finally, the cell wall remains uniformly thin (*Trichotosia ferox*). All these various forms of thickening sometimes occur side by side within the same velamen, while the types of thickening may differ even on different parts of one and the same cell (*Renanthera matutina*). Between the thickening fibers, the cell walls are often perforated by holes; these perforations are not limited to the septums between adjacent cells, but also occur in the outer walls of the superficial layer.

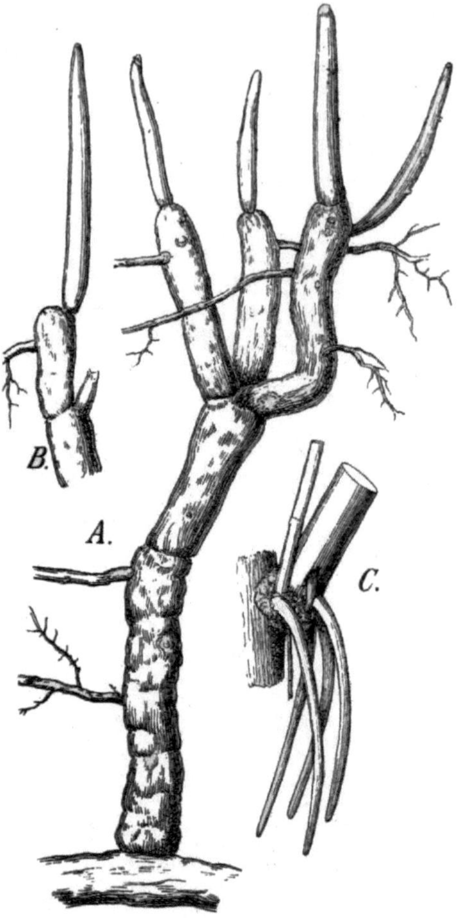

Fig. 80 *Laguncularia racemosa*—pneumatophores; (**a–b**), with negative geotropism; (**c**), with positive geotropism (Børgesen and Paulsen 1900)

The walls of the cells that make up the velamen are ornamented and show small depressions, sometimes being differentiated into several anatomically distinct areas. The membranes near the depressions break throughout the system, allowing massive infiltration with microorganisms.

In their adult state, the cells of the velamen almost invariably contain nothing but air, which gives, macroscopically, the silvery appearance to such aerial roots.

On contact with fluids, the velamen is instantly filled by the capillary flow of water. Only the pneumatodes remain covered with air, repelling moisture through a mechanism still incompletely elucidated. Fluids from a saturated velamen are located horizontally at the level of the outermost layer of the cortex—a unilayered, suberized exodermis—impermeable to water, except for its transfer cells (passage cells). After moisture passes through the cortical parenchyma, it encounters another tissue—a robust endodermis that also contains passage cells. From there, the distance to the conducting tissue is quite short—a few layers of parenchyma cells.

Fig. 81 Pneumatophores—*Bruguiera caryophyllaeoides* (**a**); *B. gymnorrhiza* (**b**); *Carapa obovata* (**c**); *Avicennia officinalis* (**d**) (Schimper 1903)

An embolized velamen (hollow inside) provides protection against water loss due to hydrophilic substrates.

Therefore, the cortical parenchyma is separated from the velamen by a very characteristic intermediate layer, the exodermis, which, in many ways, resembles a typical endodermis. As we have seen, the exodermis comprises two different types of cells; some are more or less elongated, and their external walls are often thickened, but show no perforations. Among these long cells, there are shorter ones, rounded, and almost invariably thin-walled, with abundant protoplasmic contents.

The long cells are the exodermic elements in the strict sense; they have suberized walls that are therefore relatively impermeable to water. These cells consequently serve to prevent drying of the aerial roots during periods of prolonged drought, the velamen itself being of course unable to assume the protective functions usually performed by a dermal tissue. Small thin-walled cells (passage cells, transfer cells),

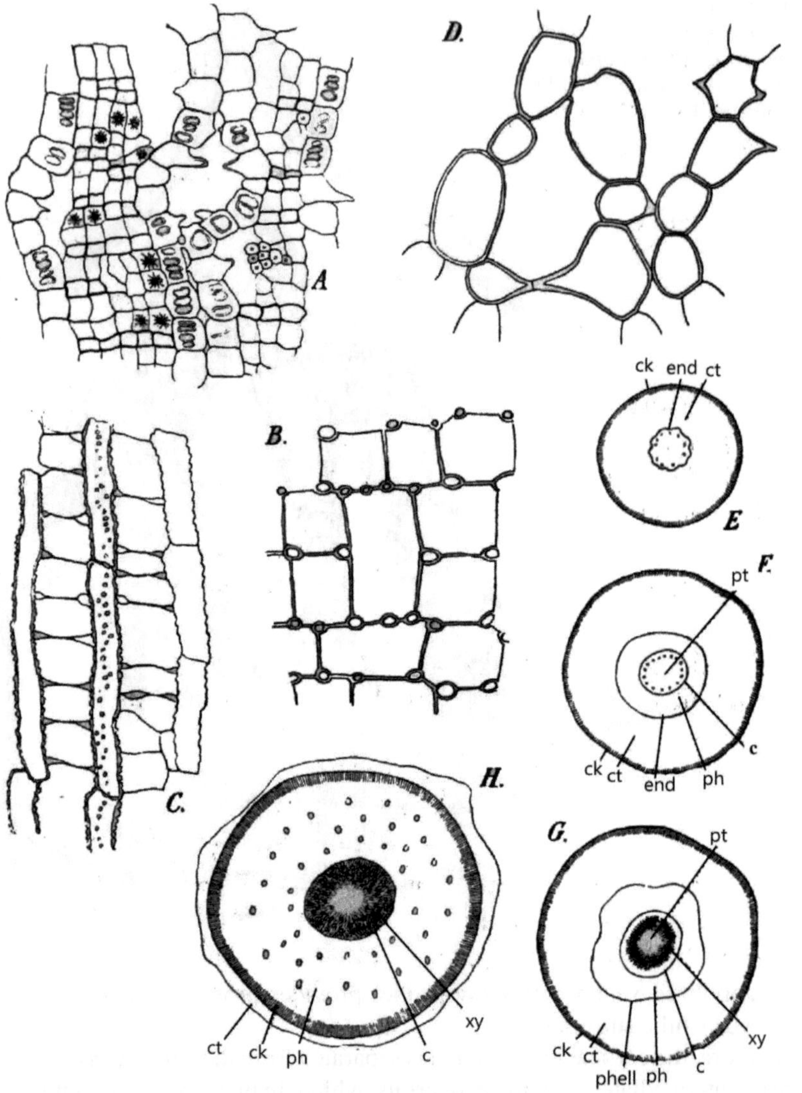

Fig. 82 *Laguncularia racemosa*—(**a**), cross section through an aerial respiratory root (pneumatophore), at the phloem level; (**b**), longitudinal and radial section through the phloem and through the primary cortex (**c**); (**d**), cross section through the primary cortex; (**e–h**), various developmental stages of a young aerial root, showing phloem formation and removal of primary cortex; *c* cambium, *ct* cortex, *xy* xylem, *ph* phloem, *ck* cork, *phell* phellogen, *pt* pith, *end* endodermis (Børgesen and Paulsen 1900)

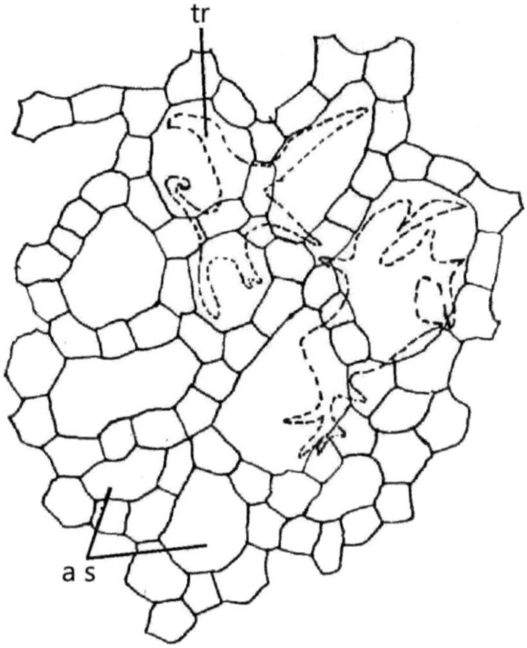

Fig. 83 *Sonneratia acida*, cross section through aerial respiratory root (tr, trichoblasts (idioblasts); a s, air-containing spaces) (Liebau 1914)

on the other hand, act as channels for the inward passage of water collected by the velamen.

In the aerial roots of certain orchids, the elements of the velamen covering the passage cells are distinguished by the presence of very characteristic disk-shaped or spherical local thickenings on their inner walls, thickenings called by some older authors "fibrous bodies" (Figs. 86, 88, 97a). The function of these fibrous formations is not fully understood; they appear to be involved in absorbing and retaining the water that reaches them and subsequently transferring the liquid gradually to the underlying passage cell (hence a sort of "condensing organ").

5 Ecological Anatomy of Roots in Xerophytes

The morphological adaptations of xerophytes, i.e., plants that grow in dry (xeric) environments, were extensively discussed in a previous work (Grigore 2022), on which occasion it was established that xerophytes represent such a heterogeneous ecological group and that the term "xerophyte" came to have more of a didactic, orientated value than a purely scientific one.

We have repeatedly emphasized that there is no precise boundary between the various ecological groups ("classes") of plants; for example, as we discussed in the previous lines, there is a group of plants, *marsh plants* (*helophytes*), which, like water plants (aquatic, hydrophytes), develop their lower parts (roots, rhizomes, and,

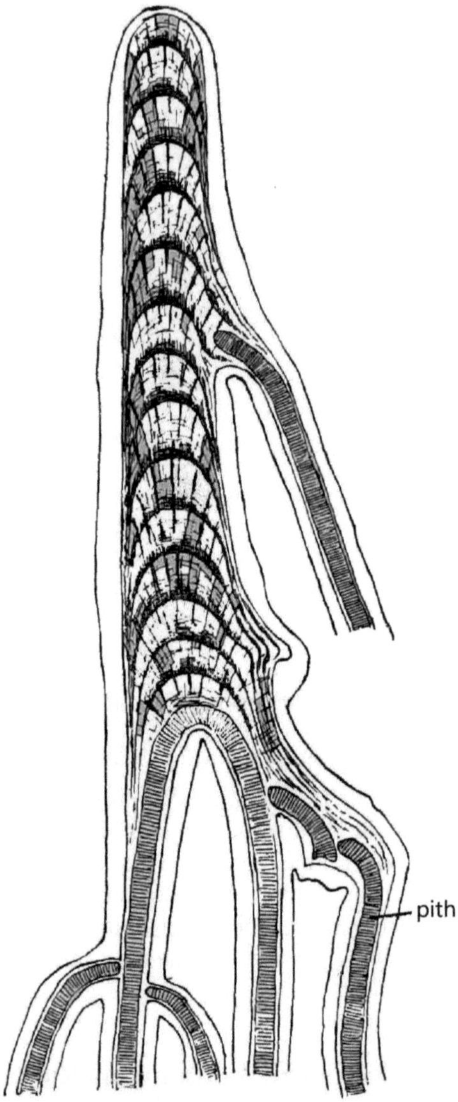

Fig. 84 *Bruguiera gymnorrhiza*, longitudinal section through aerial respiratory root (Liebau 1914)

to some extent, the leaves) in water, or at least in moist soil, but they have similar organs adapted especially for the aerial environment (life in the atmospheric environment), as is the case with terrestrial plants, to which they are closely related.

Helophytes give rise to special forms of communities, and terrestrial plants show many degrees of adaptation to their way of life, in contact with the air.

Returning to xerophytes (plants of dry, xeric places, deserts), these plants encounter the greatest difficulties in terms of providing water supply (although in this case too, there are many nuances and exceptions—see Grigore 2022—rather than an absolute statement).

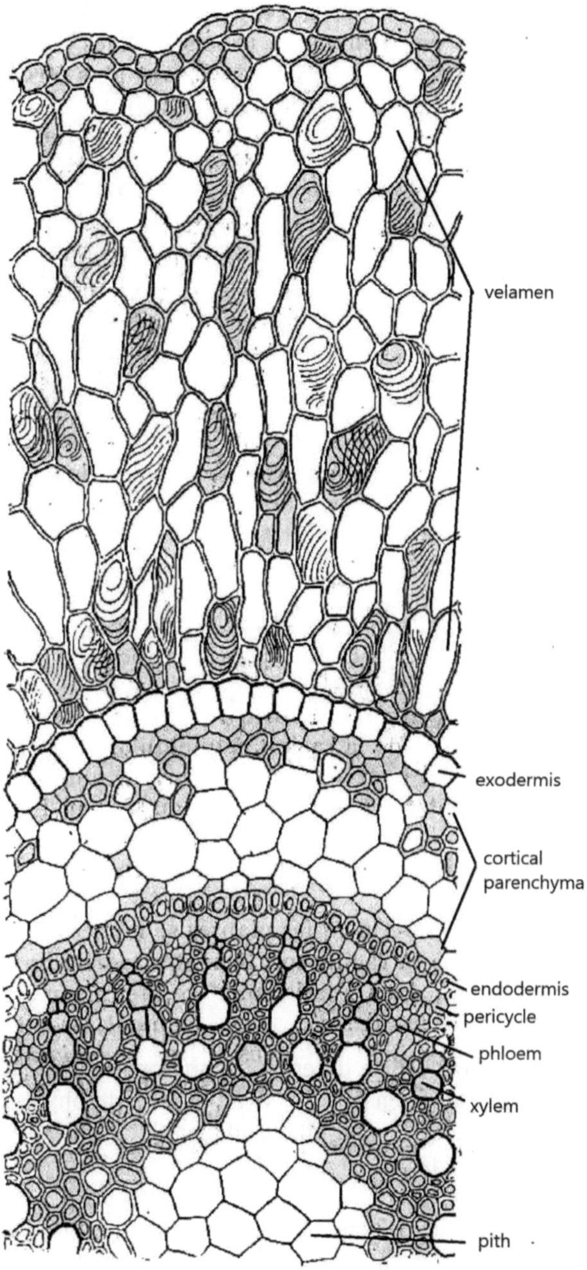

Fig. 85 Cross section through the aerial root of *Oncidium* sp (Gérardin and Guède 1895)

Fig. 86 Cross section through the aerial root of *Stanhopea oculata* (Haberlandt 1914)

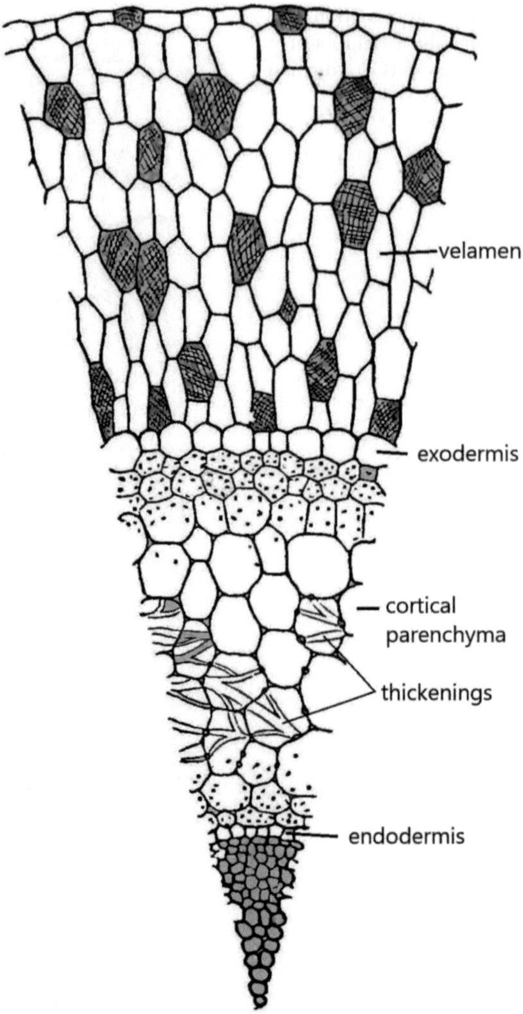

Other species are described as mesophytes because in some respects, they are "halfway" between the two "extremes," hydrophytes and xerophytes.

The differentiation of the land plant in one direction or another is decided by the ecological, edaphic, and climatic factors, which prevail in the station or habitat of the plant.

But edaphic and climatic factors cannot be viewed separately: the plant community is always the product of both. The nature of a soil is also influenced by climate, and it is undeniable that climate (e.g., rainfall) causes great differences between, for example, a desert and a rainforest.

Since xerophytes represent, probably, the most heterogeneous ecological group of plants, being very difficult to approach even from a didactic point of view, some

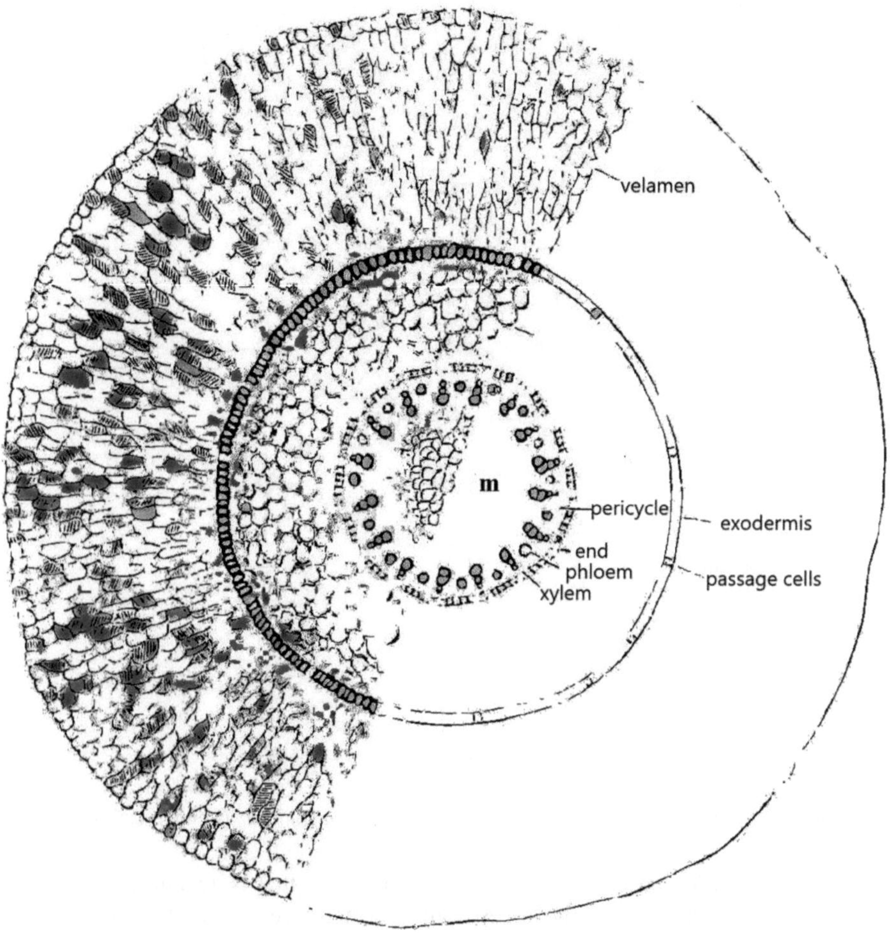

Fig. 87 *Dendrobium nobile*—cross-section through aerial root (*end*, endodermis; *m*, medulla) (Schimper 1903)

notions regarding the classification of plants into ecological classes (groups) of plants are required here (Warming 1909; Schimper 1903). Thus, plants can be classified according to certain factors, as follows (Warming 1909). When:

A. *The soil (in the broadest sense) is very wet*, and abundant water is available to the plant, the formations are therefore more or less *hydrophilous*:

    Class 1. *Hydrophytes* (aquatic, water formations).
    Class 2. *Helophytes* (marsh formations).

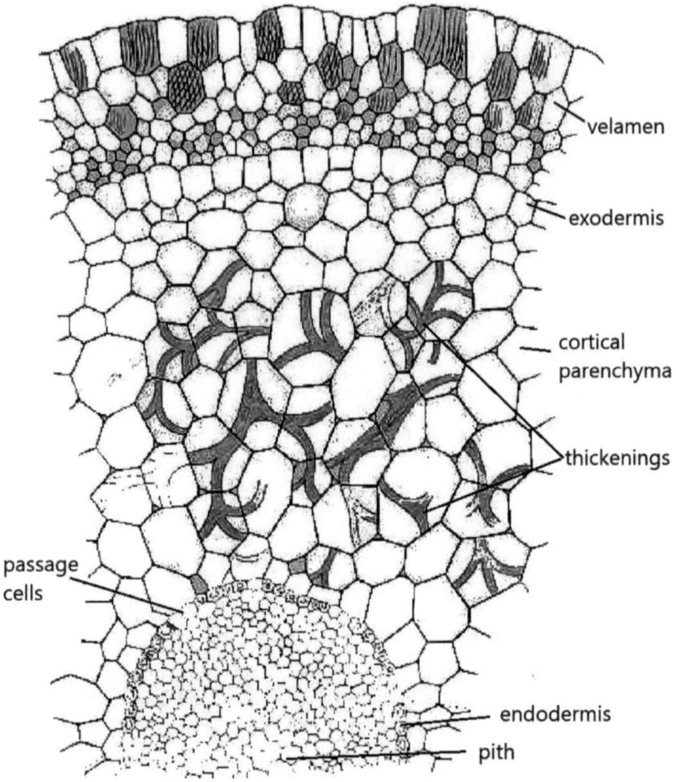

Fig. 88 Cross section through the aerial root of *Oncidium sanguineum* (Leitgeb 1865)

B. *The soil is physiologically dry*—it contains water that is available to the plant only to a small extent; the formations are therefore essentially composed of *xerophilous* species:

Class 3. *Oxylophytes* (formations on acid soils).
Class 4. *Psychrophytes* (of cold soil formations).
Class 5. *Halophytes* (formations on saline soils).

C. *The soil is physically dry,* and its poor water-holding capacity determines the vegetation, the climate being of secondary importance; the formations are therefore *xerophilous*:

Class 6. *Lithophytes* (rock formations).
Class 7. *Psammophytes* (formations on sand and gravel).
Class 8. *Chersophytes* (of waste land formations).

D. *The climate is very dry and decides the character of the vegetation*; soil properties are dominated by climate; the formations are *xerophilous*:

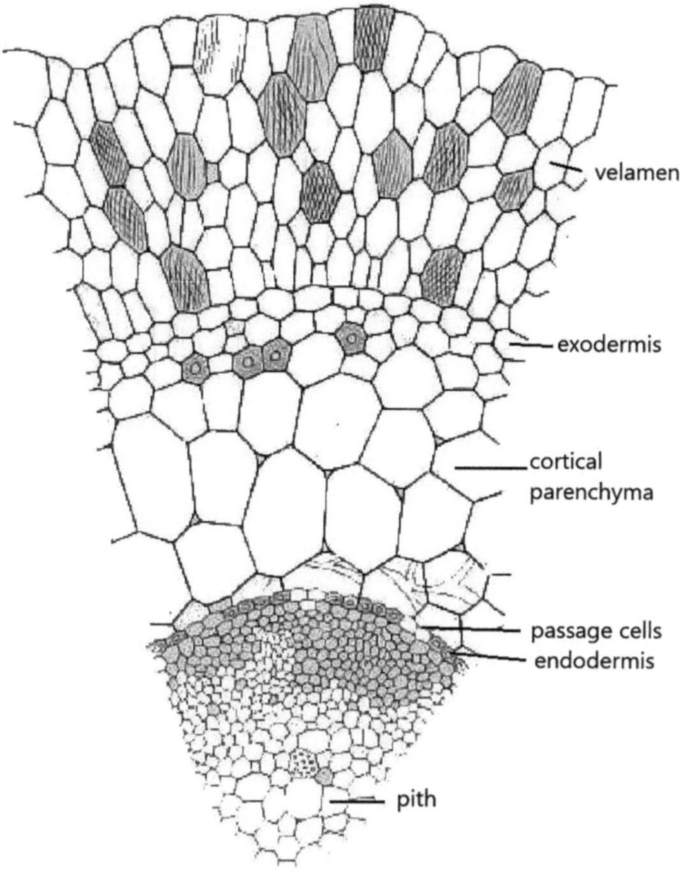

Fig. 89 Cross section through the aerial root of *Oncidium flexuosum* (Leitgeb 1865)

Class 9. *Eremophytes* (desert and steppe formations).
Class 10. *Psilophytes* (savanna formations).
Class 11. *Sclerophyllous* formations (bush and forest).

E. *The soil is physically or physiologically dry*:

Class 12. *Coniferous formations* (coniferous forests).

F. *The soil and climate favor the development of mesophilic formations*:

Class 13. *Mesophytes* (Warming 1909).

Xerophytes (*xeros*, dry; *phyton*, plant) can be divided into different groups, depending on the nature of the soil or climate.

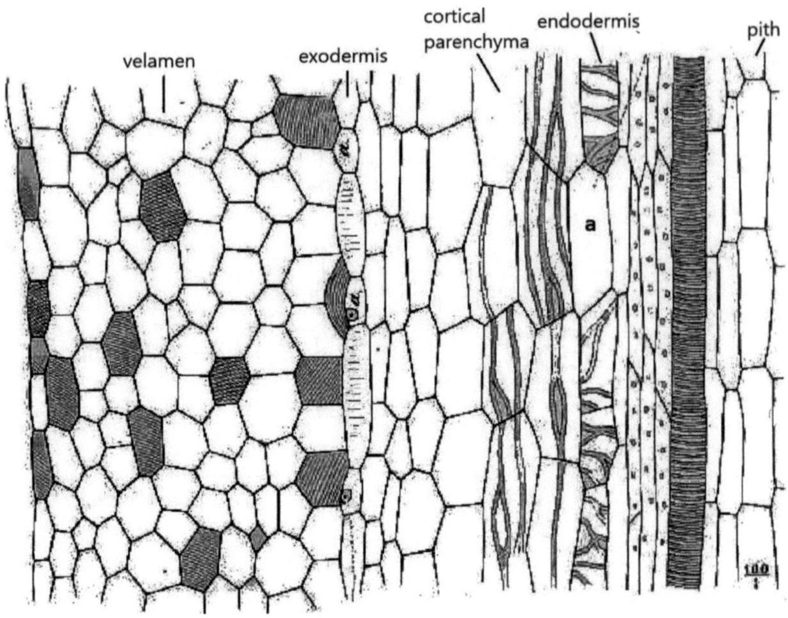

Fig. 90 Tangential section through the aerial root of *Gongora jaenischii*; a, passage cells (Leitgeb 1865)

Fig. 91 Cross section through the aerial root of *Sarcanthus rostratus* (Leitgeb 1865)

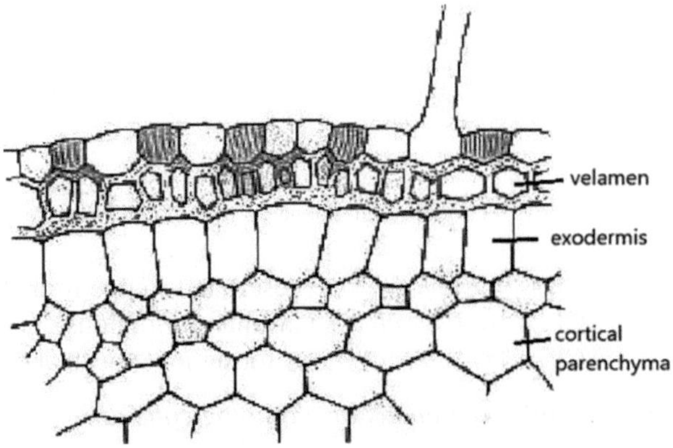

Soil can be physically or physiologically dry (affected by physical or physiological drought).

The soil is physically dry when it contains very little free water; this situation occurs on the following:

1. The surface of rocks or stones—occupied by plants that make up the *lithophilous* formations (lithophytes).
2. Sandy soil that is so high above the permanent ground water that it does not influence it and that is very dry in dry seasons due to rapid drainage and drought;

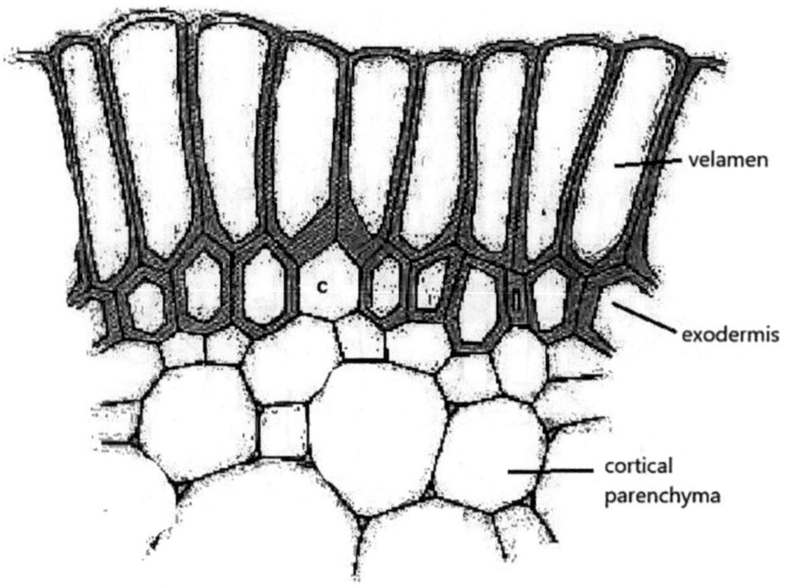

Fig. 92 Cross section through the aerial root of *Sarcopodium lobii*; c, passage cells (Leitgeb 1865)

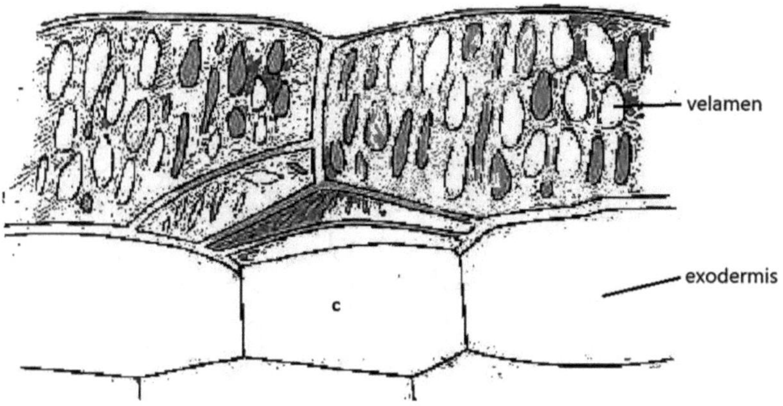

Fig. 93 Cross section through the aerial root of *Dendrocolla teres;* c, passage cells (Leitgeb 1865)

there are *psammophilous* formations on the sands (psammophytes, arenicola plants), which are related to the plants on the gravels.

Epiphytes must also be included here, which almost all have precise adaptations for preserving water (Grigore 2022).

The soil is affected by physiological drought (Schimper 1903; Warming 1909; Grigore and Toma 2011; Grigore 2021) when it contains a considerable amount of water, but which is available to the plant only *to a small extent or can be absorbed only with difficulty*, either because the soil firmly holds a large amount of water or

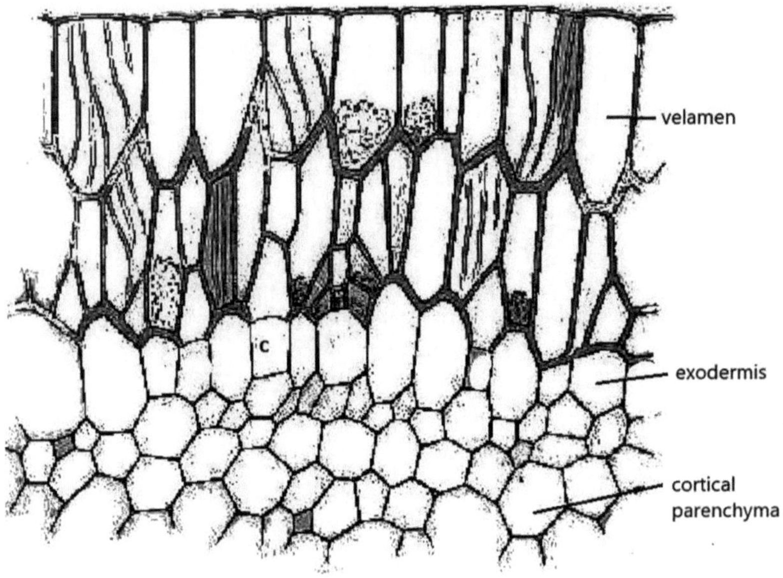

Fig. 94 Cross section through the aerial root of *Renanthera matutina*; c, passage cells (Leitgeb 1865)

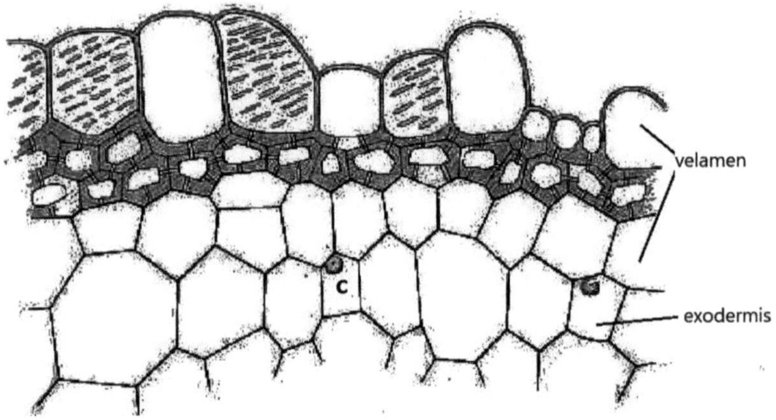

Fig. 95 Cross section through aerial root from *Angraecum subulatum*; c, passage cells (Leitgeb 1865)

because the root osmotic pressure is not high enough to overcome that of the concentrated salt solution in the soil.

This theory of the physiological drought of the soil explains very well why groups of plants—apparently—so diverse from the point of view of the habitats in which they grow, show common morpho-anatomical characters (convergences) of xeromorphosis; e.g., desert plants (typical xerophytes, where there is soil and

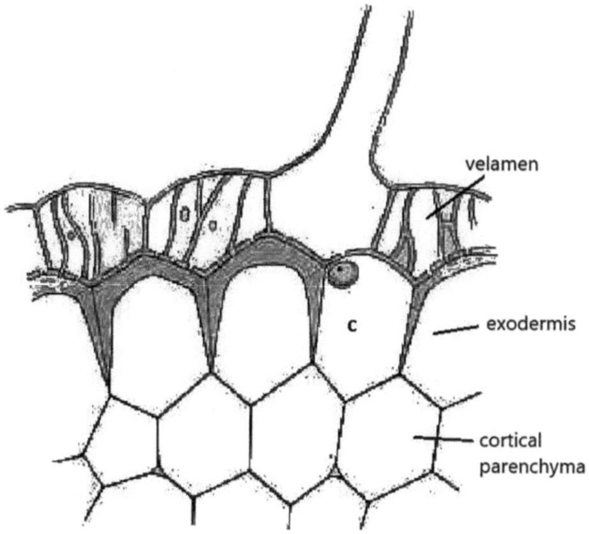

Fig. 96 Cross section through the aerial root of *Vanilla aphylla*; c, passage cells (Leitgeb 1865)

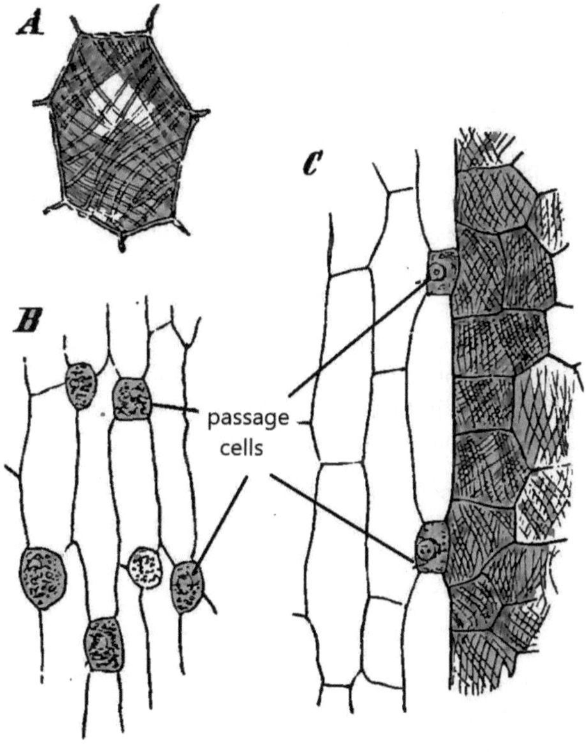

Fig. 97 Cells in the velamen of the aerial root of *Stanhopea oculata*: A, isolated cell—the thickening of the cell wall can be observed; B, exodermis sector (front view); C, aerial root sector—longitudinal section (Haberlandt 1914)

atmospheric dryness), bog plants (peats, where there is sufficient soil moisture but the environment is acidic and rich in humic acids), salt marsh plants (halophytes, where there is water at ground level, but it is difficult to absorb due to the high concentration of the soil solution, due to soluble salts dissolved in the water), and arctic or tundra plants (where there is water, theoretically, at ground level, but it is frozen or the soil temperature is very low, which makes its absorption by plants very difficult, or even impossible).

Summing up, physiological soil drought occurs when:

1. The soil is rich in free humic acids, or in chemical compounds that, through their particular action on the plant, induce xerophilia; those plant formations that grow on acid soil (*oxylophytes*) result.
2. Soil rich in soluble salts, which induces a form of xerophilia found in halophilic formations (halophytes). A halophyte is actually a special form of xerophyte (Grigore 2019; Grigore 2021; Grigore and Toma 2011).

Therefore, the habitats affected by physical and physiological dryness are very diverse; to form an image, here are the most important natural regions affected by this phenomenon (Schimper 1903):

(a) Deserts, steppes, characterized by a dry substrate, dry air, as well as very high temperatures, sporadic or permanent, so with intense lighting.
(b) The rhytidome of trees, rocks, where rapid drying occurs on the surface, because there is no deep substratum.
(c) Sandy soils, gravel, which dry quickly on the surface, due to high permeability.
(d) Seashores, volcanic surfaces that contain a high amount of soluble salts.
(e) Peatlands, due to the humic acids in the soil.
(f) Polar areas, or near glaciers in high mountains, where the soil temperature is very low.

In xerophytes, physiological drought is caused by external factors that either reduce water absorption or promote transpiration, or, most commonly, there is a combination of these influences.

Factors that reduce water absorption:

1. Low amount of free-soil water.
2. The abundance of soluble salts in the soil.
3. Abundance of humic acids in soil.
4. Low soil temperature.

Factors that promote transpiration:

1. Dry atmosphere.
2. High air temperature.
3. Air rarefaction.
4. Light.

Returning to the root anatomy of xerophytes (therefore viewed *lato* sensu), being such a diverse group, from the point of view of the habitats in which these plants grow, but also of the histological characters of the botanical family, it is practically impossible to find such "clear" adaptations, as is the case to other ecological groups of plants (we previously discussed the case of aquatic plants).

Lignification is perhaps one of the most obvious anatomical features common to all xerophytes; by contrast, it is greatly reduced in aquatic plants.

However, the ecological value of lignification is not very clear; it seems that this adaptation gives the plant mechanical resistance.

It is worth noting that lignification is directly related to the environment, as it becomes more pronounced the drier the habitat (except for succulents). Botanical families such as Apiaceae, Caryophyllaceae, Linaceae, Labiatae, Rubiaceae, and Dipsacaceae, as well as genera that in temperate countries are rich in herbaceous species, become much richer in woody plants in tropical, warm temperate, or even Mediterranean countries. Lignification is particularly developed in xerophytes that contain only little xylem sap. In these, the wood is dense and hard, but at the same time often fragile; it resembles the so-called autumn wood because the diameters of the vessels and cells are narrow, and the reason for the similarity is probably that some conditions of development are identical.

In this sense, a relevant example of massive lignification, at the root level, is that of halophytes, whose organs are affected by the phenomenon of successive cambia (see the previous discussion and Figs. 30, 31, 32, 33, 34, 35, 36, 37, 38, 39, 40, 41, and 42 in chapter "Root"); we recall, for complete clarity, that halophytes represent a special case of xerophytes (Grigore 2008; Grigore and Toma 2010; Grigore 2021; Grigore and Toma 2021).

When the roots develop, the secondary structure, the primary cortex, which dries out, usually exfoliates and disappears. In desert and steppe plants, however, the cortex often persists. The peripheral layers can become rough and thus protect the plants against the pressure of hard soils and drying (as in *Haloxylon articulatum*) (Killian 1941). Instead, the inner cortex can become spongy (by puncturing the disintegrating cell walls), forming dead protective coverings. This is the case with many species of South African grasses (*Gerbera*, *Commelina*) studied by Henrici (1929), who describes the absorption and storage function of water, an observation confirmed by the experiments of Killian (1942) and Lemée (1954). Killian argued that in the case of *Bromus rubens*, such roots immediately absorbed on contact with liquid water (12% of their weight by capillary suction); Lemée showed that dead roots are strongly hygroscopic, absorbing 100% of their initial weight in moist air. Small mosses conduct liquid water easily, both peripherally and in the central conducting bundle.

The roots of plants from sandy habitats in dry regions often possess very dense and long absorptive hairs, which agglutinate the sand by more or less mucilaginous secretions of pectic substances.

Killian (1941) considers that such permanent root hairs, which appear near the tip of the roots in *Haloxylon articulatum*, are functional even in very dry soil after prolonged droughts.

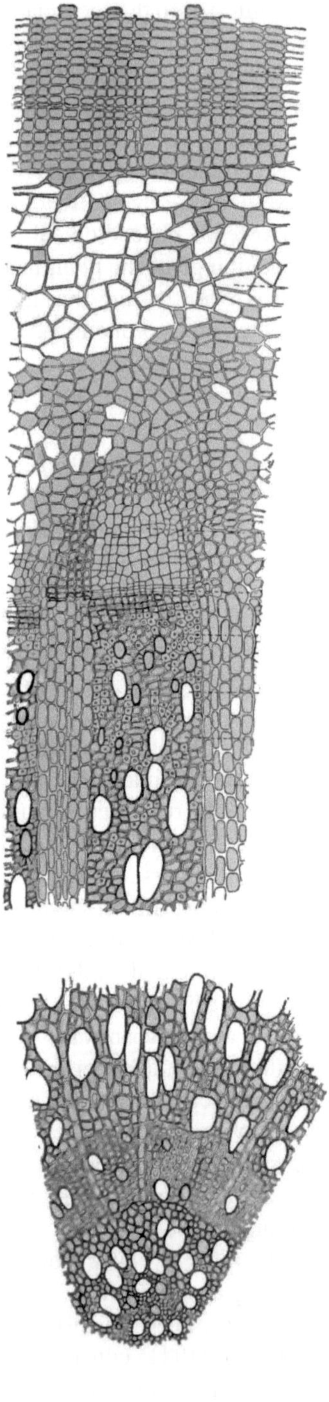

Fig. 98 Cross section through the root of *Potentilla arguta* (Hayden 1919)

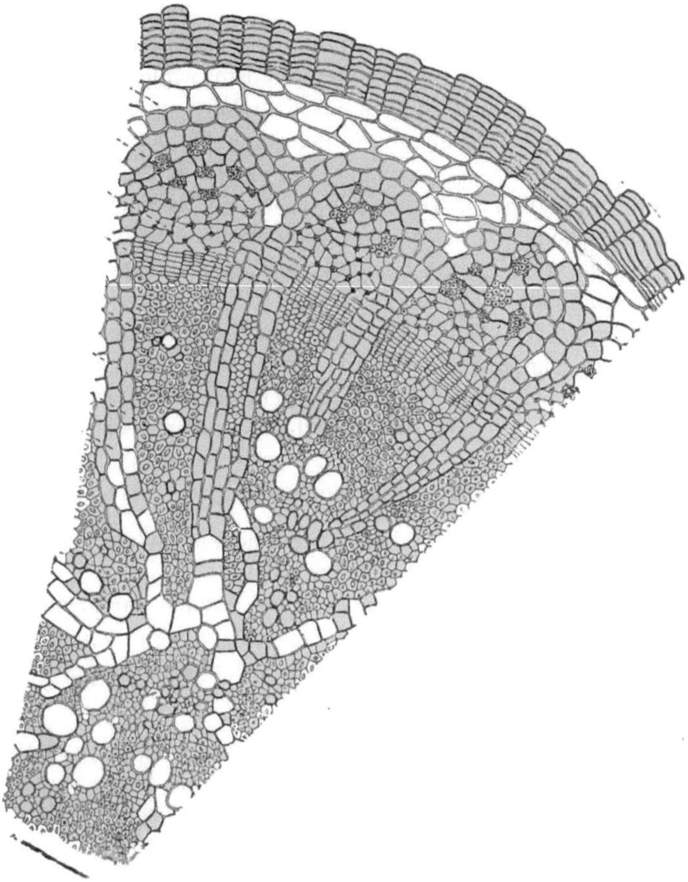

Fig. 99 Cross section through the root of *Petalostemum candidum* (Hayden 1919)

6 Ecological Anatomy of Roots in Eremophytes (Here, Steppe—Prairie Species)

Prairie species represent a subgroup of xerophytes, and sometimes, broadly ecologically speaking, and for the sake of simplifying, the American prairie could be interchanged with Euro-Asiatic steppe (see larger discussion on page 443). There are some studies that confirm, from an anatomical point of view, the xeromorphic nature of prairie plants (Shimek 1915; Hayden 1919; Urcelay 1923); it must be said that within all organs, the roots were studied lesser, and the leaf being the most intensely analyzed organ, in relation to water conservation (see the section dedicated to ecological anatomy of the leaf).

Hayden (1919) conducted an intensive study on the ecological subterranean anatomy of plants vegetating in a prairie province in Central Iowa. She found out that every species has established, in different degrees, its strategy in economy of water relations; as anatomical indicators of this, she considered the parenchyma (storage and aerenchyma), mechanical tissue, and conductive tissue. The prominence of the latter two tissues suggests xerophytism, while that of the first tissue indicates the opposite condition. In her study, the relative proportion of these tissues in each stem or root has been utilized to determine the degree of xerophytism in each species.

Within the seven studied root structures, six were of upland plants, and one of an alluvial basin plant; some of them recorded a prominent development of mechanical tissue: *Potentilla arguta* (Fig. 98), *Desmodium illinoense*, *Petalostemum candidum* (Fig. 99), and *Solidago canadensis.*

In *Potentilla arguta*, Hayden (1919) found that the protective cork is well developed; the tracheae are large and numerous; and the mechanical tissue in the form of wood fibers is prominent (Fig. 98); *Desmodium illinoense* shows a prominent cork, few parenchyma, and mechanical tissue in the form of phloem and xylem fibers, forming a tough, compact root.

In summary, this study reveals some conclusions:

1. There is a tendency to the production of prominent mechanical tissue in plants of dry habitats and reduction of parenchyma tissue.
2. In moist habitats, the proportion of parenchyma tissue is prominent. Aerenchyma is abundant in swamp plants.
3. The vascular tissue is variable in quantity, seemingly more or less subject to systematic variation.

The subterranean stem is predominant as an equivalent of the primary root, especially in moist lowland regions. It is more efficient than the root in propagation. Primary roots that show secondary thickening resemble stems in their concentric manner of expansion.

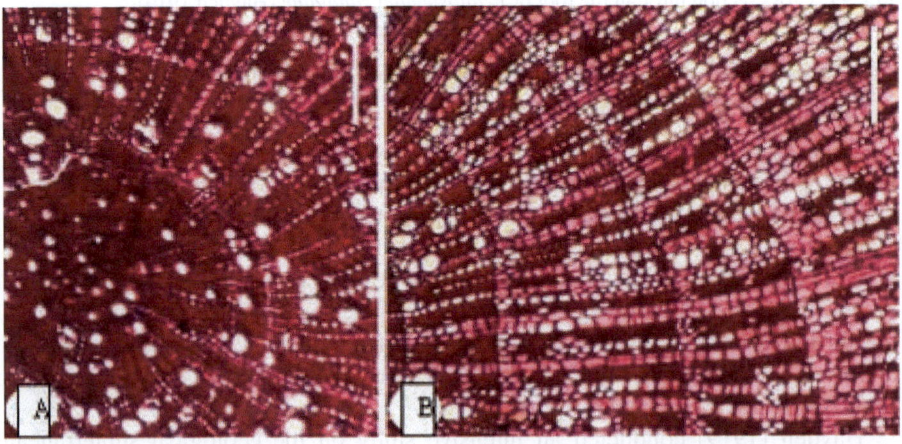

Fig. 100 Cross section through the xylopodium of *Jacaranda ulei*: (**a**), diffuse porous vessels; (**b**), confluent paratracheal axial parenchyma (Miranda et al. 2018, with permission)

7 Ecological Anatomy of Xylopodia ("Lignotubers")

As previously discussed, some shrubs or small trees are included in the group of "pyrophytes," i.e., fire-resistant (fire, pyrophytic) plants, which show different strategies for resistance (survival) to fires; among these are xylopodia (lignotubers, woody lignified underground roots or stems) (Grigore 2022). In this ecological group of plants, the formation of lignified underground organs (xylopodia, lignotubers) is a typical phenomenon of arid areas, especially of Brazilian savannas of the "Cerrado" type (Grigore 2022).

The definition of xylopodium first appeared in 1906 (Lindman and Löfgren 1906, cited by Miranda et al. 2018) and refers to a rigid, sclerified region in the transition zone between the stem and the taproot. Contrary to the general perception that the bud-producing organs are soft, the xylopodium shows a high capacity to produce shoots and is more commonly found in shrubs and sub-shrubs (Rizzini and Heringer 1962). Some species show xylopodia and can generate trees, such as *Brosimum gaudichaudii* (Palhares et al. 2006).

For example, the underground system of *Jacaranda ulei*, a species of "Cerrado," shows an axial taproot and a xylopodium, which can be recognized as a rigid thickened fusiform portion between the root and the stem. Xylopodium has a great capacity to produce buds and aerial shoots, despite being very tough. Unlike many other "Cerrado" (Brazilian savannah) species, the aerial stem is not resistant to fire, so the survival and persistence of individuals depend on the ability of the xylopodium to produce shoots (Miranda et al. 2018).

The anatomy of a xylopodium (Fig. 100) betrays, in the spirit of the previous discussion, a massive lignification, in connection with the adaptations of xeromorphosis.

The anatomy of the xylopodia was also studied in two other species that grow in the open areas of the Cerrado range, *Aldama tenuifolia* and *A. kunthiana* (Asteraceae) (Da Silva et al. 2014), found especially in "campos sujos," "campos limpos," and "campos rupestres," which are characterized by intense solar irradiation, frequent fires, and water deficit during autumn and winter (Ratter et al. 1997); the specific names assigned to these types of ecosystems are very useful to remember in the native form, because perfectly valid equivalences in other languages are extremely problematic, if not impossible to do. In the mentioned species, the xylopodia are located in the upper layers of the soil and are responsible for the production of tuberous and non-tuberous adventitious roots.

Due to its high budding potential, the xylopodium generates (aerial) shoots during favorable periods of development. This process increases the anatomical complexity of the organ, since "self-grafting" occurs at the base of the branches in the proximal region.

The covering tissue of the xylopodium is similar to that of tuberous roots; it resembles the stratified cork, originating in the subepidermal layers. As the organ thickens, the cortical parenchyma cells show periclinal divisions, followed by suberization of the inner cell layers. In the later stages of development, the outer layers of the cortex are removed, as the secondary phloem is located immediately below the covering tissue. The secondary xylem is well developed, contributing to the lignification of the organ. The primary xylem of endarch type occurs at the periphery of the pith, which is undeveloped and may sclerify.

In the cortex (if persistent), secondary phloem, and medullary parenchyma, there are numerous secretory structures of lipophilic substances.

The underground system from another Brazilian xerophytic species, *Wunderlichia azulensis*, presents morpho-anatomical features that have been interpreted as forming a xylopodium (Trindade and Feitoza 2022).

Beyond these very interesting discussions about the structure and functions of xylopodia, we retain the idea that they are considered as an effective adaptation to xeromorphic and disturbed environments, including fire-prone ecosystems (Pausas et al. 2018); the aforementioned genus, *Wunderlichia,* is endemic to Brazil, and the geographic distribution of its species is limited to xeromorphic environments.

Bibliography

Arber A., 1920 – *Water plants. A study of aquatic angiosperms*, Cambridge at the University Press

Armstrong, W. 1978 – Root aeration in the wetland condition. In *Plant life in anaerobic environments*, D. D. Hook, R. M. M. Crawford (Eds.), Ann Arbor, MI, Ann Arbor Science Publishers, pp. 269–297

Armstrong, W. 1979 – Aeration in higher plants. In *Advances in Botanical Research*, H. W. Woolhouse (Ed.), London Academic Press, pp. 225–332

Arteca, R.N., 1997 – Flooding. In *Plant Ecophysiology*, M. N. V. Prasad (Ed.), pp. 151–171, New York. John Wiley & Sons

Barnes C. R., 1898 – *Plant life considered with special reference to form and function*, New York, Henry Colt & Company

Barnéoud F. M., 1848 – Mémoire sur l'anatomie et l'organogénie du *Trapa natans* (Linn.). *Ann. Sci. Nat., 3 ème sér., Bot.*, 9: 222–244

Batten L., 1918 – Observations on the ecology of *Epilobium hirsutum*. *J. of Ecol.*, 6: 162–177

Benzing D. H., 1990 – *Vascular epiphytes. General biology and related biota*, Cambridge University Press, Cambridge, New York, Port Chester, Melbourne, Sydney

Børgesen F., 1909 – Notes on the shore vegetation of the Danish West Indian islands. *Botanisk Tidsskrift*, 29: 201–259

Børgesen F., Paulsen O., 1898 – *Om vegetationen paa de Dansk-Vestindiske øer*, Kjøbenhavn/Bogforlaget: Det Nordiske Forlag/Ernst Bojesen

Børgesen F., Paulsen O., 1900 – La végétation des Antilles danoises. *Revue gén. de Bot.*, 12(135): 99–107; 12(136): 138–153; 12(138): 224–245; 12(139): 289–297; 12(141): 344–354; 12(143): 434–446; 12(144): 480–510

Bowman H. H. M., 1917 – Ecology and physiology of the Red Mangrove. *Proc. of the Am. Phil. Soc. Philadelphia*, 56: 589–672

Burdick D. M., Mendelssohn I. A., 1990 – Relationship between anatomical and metabolic responses to waterlogging in the coastal grass *Spartina patens*. *J. of Exp. Bot.*, 41: 223–228

Chatin G.-A., 1856 – *Anatomie comparée des végétaux. Plantes aquatiques*. Paris, J.-B. Baillière, Londres, New-York

Constable J. V. H., Grace, J. B., Longstreth D. J., 1992 – High carbon dioxide concentrations in aerenchyma of *Typha latifolia*. *Am. J. of Bot.*, 79: 415–418

Constable J. V. H., Longstreth D. J., 1994 – Aerenchyma carbon dioxide can be assimilated in *Typha latifolia* L. leaves. *Plant Physiol.*, 106: 1065–1072
Coulter J. M., Barnes C. R., Cowles H. C., 1911 – *Textbook of botany for colleges and universities, vol. II. Ecology*, New York, Cincinnati, Chicago, American Book Company
Crawford, R. M. M., 1982 – Physiological responses to flooding. In *Physiological Plant Ecology, II.* O. L. Lange, P. S. Nobel, C. B. Osmond, H. Ziegler (Eds.), Berlin. Springer-Verlag, pp. 453–477
Crawford, R. M. M., 1993 – Root survival in flooded soils. In *Mires: Swamp, Bog, Fen and Moor. Ecosystems of the World*, Vol. 4A, A. Gore (Ed.), Amsterdam. Elsevier Science, pp. 257–283
Cronk J. K., Fennessy M.S., 2001 – *Wetland plants. Biology and ecology*, Lewis Publishers, Boca Raton, London, New York, Washington, D.C.
Da Silva E. M. S., Hayashi A. H., Appezzato-da-Glória B., 2014 – Anatomy of vegetative organs in *Aldama tenuifolia* and *A. kunthiana* (*Asteraceae*: *Heliantheae*). *Braz. J. Bot.*, 37(4): 505–517
Davis J. H., 1940 – The ecology and geologic role of mangroves in Florida. *Pap. from the Tortugas Lab.*, Vol. 32, Carnegie Institution of Washington Pub., 517, 303–412
de Wit M. C. J., 1978 – Morphology and function of roots and shoot growth of crop plants under oxygen defficiency. In *Plant Life in Anaerobic Environments.* D. D. Hook, R. M. M. Crawford (Eds.), pp. 333–350, Ann Arbor, MI. Ann Arbor Science Publishers
Drew M. C., He Ch.-J., Morgan P. W., 2000 – Programmed cell death and aerenchyma formation in roots. *Trends Plant Sci.*, 5: 123–127
Gérardin L., Guède H., 1895 – *Traité élémentaire d'histoire naturelle. Botanique. Anatomie et physiologie végétales*, Paris, Librairie J.-B. Baillière et Fils
Glück H., 1911 – *Biologische und morphologische Untersuchungen über Wasser- und Sumpfgewächse*, Verlag von Gustav Fischer in Jena
Goebel K., 1891–1893 – *Pflanzenbiologische Schilderungen*, zweiter teil, erste lieferung, zweiter teil, zweite lieferung, Marburg N. G. Elwert'sche Verlagsbuchhandlung
Grigore M.-N., 2008 – *Introducere în Halofitologie. Elemente de anatomie integrativă*, Editura PIM, Iași
Grigore M.-N., 2019 – Defining halophytes – A conceptual and historical approach in an ecological frame, In *Halophytes and climate change: Adaptive mechanisms and potential uses*, Hasanuzzaman M., Sergey Shabala and Masayuki Fujita (Eds.), CABI, pp. 3–18
Grigore M.-N., 2021 – Definition and classification of halophytes as an ecological group of plants. In: Grigore M.-N. (Ed.), *Handbook of halophytes. From molecules to ecosystems towards biosaline agriculture*, Springer, pp. 3–50
Grigore M.-N., 2022 – *Morfologia plantelor. Aspecte ecologice*, Editura Universității „Ștefan cel Mare" Suceava
Grigore M.-N., Toma C., 2010 – *Halofitele. Aspecte de anatomie ecologică.* Editura Universității „Alexandru Ioan Cuza", Iași
Grigore M.-N., Toma C., 2011 – Halofitele, o categorie ecologică polimorfă. Între seceta fiziologică a solului și stresul salin. *Revista Botanică (Chișinău)*, 2(3): 38–46
Grigore M.-N., Toma C., 2021 – Morphological and anatomical adaptations of halophytes. A review. In: Grigore M.-N. (Ed.), *Handbook of halophytes. From molecules to ecosystems towards biosaline agriculture*, Springer, pp. 1079–1222
Grigore M.-N., Ivănescu L., Toma C., 2014 – *Halophytes. An integrative anatomical study.* Springer, Cham, Heidelberg, New York, Dordrecht, London
Haberlandt G., 1914 – *Physiological plant anatomy*, Macmillan and Co, London
Hayden A., 1919 – The ecologic subterranean anatomy of some plants of a prairie province in central Iowa. *Am. J. Bot.*, 6: 87–105
Henrici M., 1929 – Structure of the cortex of grass roots in the more arid regions of South Africa. *Pretoria (Union of Souty Africa, Department pf Agriculture, scientific bulletin no 85)*
Henslow G., 1895 – *The origin of plant-structures by self-adaptation to the environment*, London, Kegan Paul, Trench, Trübner, & Co, Ltd, Paternoster House, Charing Cross Road

Hook D. D., Scholtens J. R., 1978 – Adaptations and flood tolerance of tree species. In *Plant Life in Anaerobic Environments*, D.D. Hook, R.M.M. Crawford (Eds.), pp. 299–332, Ann Arbor, MI. Ann Arbor Science Publishers

Ivănescu L., Grigore M.-N., Toma C., 2020 – Ce sunt mangrovele? *Columna*, 9: 61–72

Jackson M. B., 1989 – Regulation of aerenchyma formation in roots and shoots by oxygen and ethylene. In *Cell Separation in Plants: Physiology, Biochemistry and Molecular Biology*, D. J. Osborne, M.B. Jackson (Eds.), pp. 263–274, Berlin. Springer-Verlag

Jackson M. B., 1990 – Hormones and developmental change in plants subjected to submergence or soil waterlogging. *Aquat. Bot.*, 38: 49–72

Jackson M. B. 1994 – Hormone action and plant adaptations to poor aeration. *Proc. of the Royal Soc. of Edinburgh*, 102B: 391–405

Jackson M. B., Fenning T. M., Jenkins W., 1985 – Aerenchyma (gas-space) formation in adventitious roots of rice (*Oryza sativa* L.) is not controlled by ethylene or small partial pressures of oxygen. *J. of Exp. Bot.*, 36: 1566–1572

John C. D., Limpinuntana V., Greenway H., 1974 – Adaptation of rice to anaerobiois. *Austral. J. of Plant Physiol.*, 1: 513–520

Justin S., Armstrong W., 1987 – The anatomical characteristics of roots and plant response to soil flooding. *New Phytol.*, 106: 465–495

Killian C., 1941 – Sols et plantes indicatrices dans les parties non irriguées des oasis de Figuig et de Beni Ounif'. *Bull. Soc. Hist. nat. Afr. N.*, 32: 301–314

Killian C., 1942 – *Bromus rubens* L. Contribution à l'étude des plantes annuelles xérophvtiques du désert. *Bull. Soc. bot. Genève*, 52: 215–238

Kludze H. K., De Laune R.D., Patrick W.H., Jr., 1993 – Aerenchyma formation and methane and oxygen exchange in rice. *Soil Sci. Soc. of Am. J.*, 57: 386–391

Kludze H. K., De Laune R. D., 1994 – Methane emissons and growth of *Spartina patens* in response to soil redox intensity. *Soil Sc. Soc. of Am. J.*, 58: 1838–1845

Kludze H. K., De Laune R. D., 1996 – Soil redox intensity effects on oxygen exchange and growth of cattail and sawgrass. *Soil Sci. Soc. of Am. J.*, 60: 616–621

Kludze H. K., Pezeshki S. R., De Laune R. D., 1994 – Evaluation of root oxygenation and growth in baldcypress in response to short-term soil hypoxia. *Can. J. of Forest Res.*, 24: 804–809

Koncalova H., 1990 – Anatomical adaptations to waterlogging in roots of wetland graminoids: limitations and drawbacks. *Aquatic Bot.*, 38: 127–134

Laan P., Berrevoets M. J., Lythe S., Armstrong W., Blom C. W. P. M., 1989 – Root morphology and aerenchyma formation as indicators of the flood-tolerance of *Rumex* species. *J. of Ecol.*, 77: 693–703

Laing H. E., 1940 – The composition of the internal atmosphere of *Nuphar advenum* and other water plants. *Am. J. of Bot.*, 27: 861–868

Leitgeb H. H, 1865 – Die luftwurzeln der Orchideen. *Denkschrift. der Kaiser. Akad. Der Wissenschaft. Mathematisch-Naturwissenschaftliche*, 24: 179–221

Lemée G., 1954 – L'économie de l'eau chez quelques Graminées vivaces du Sahara septentrional. *Végétation*, 5–6: 534–541

Lewakoffski N., 1873 – Ueber den Einfluss des Wassers auf das Wachsthum der Stengel und Wurzeln einiger Pflanzen (Gelehrte Schriften der k. Universitat in Kasan), Abstract in *Just's Bot. Jahresbericht, Jahrg.*, p. 594

Liebau O., 1914 – Beiträge zur Anatomie und Morphologie der Mangrove-Pflanzen, insbesondere ihres Wurzelsystems. *Beiträge zur Biologie der Pflanzen*, 12, 181–213

Lindman C. A., Löfgren A., 1906 – *A vegetação no Rio Grande do Sul*, Porto Alegre: Livraria Universal de Echenique

Macnae W., 1968 – A general account of the fauna and flora of mangrove swamps and forests in the Indo-West-Pacific region. *Adv. in Marine Biol.*, 6: 73–270

Meet B. D., Stolzy L. H., 1978 – Short-term flooding. In *Plant life in anaerobic environments*, D. D. Hook, R. M. M. Crawford (Eds.), pp. 351–374, Ann Arbor, MI. Ann Arbor Science Publishers

Miranda T. D., Palhares D., Cury N. F., Pereira L. A. R., Dos Santos Silveira C. E., 2018 – Comparative xylem and bark anatomy of stem, root and xylopodium of *Jacaranda ulei* (*Bignoniaceae*). *Balduinia*, 64: 1–18

Naidoo G., McKee, K. L., Mendelssohn I. A., 1992 – Anatomical and metabolic responses to waterlogging and salinity in *Spartina alterniflora* and *S. patens* (Poaceae). *Am. J. of Bot.*, 79: 765–770

Ogden E. C., 1974 – *Anatomical patterns of some aquatic vascular plants of New York*, BULLETIN 424, New York State Museum and Science Service

Palhares D., Paula J. E., Silveira C. E. S., 2006 – Morphology of stem and subterranean system of *Brosimum gaudichaudii* Trécul (Moraceae). *Acta Bot. Hung.*, 48(1–2): 89–101

Pausas J. G., Lamont B. B., Paula S., Appezzato-Daglória B., Fidelis A., 2018 – Unearthing below-ground bud banks in fire-prone ecosystems. *New Phytologist*, 217: 1435–1448

Penhale P.A., Wetzel R. G., 1983 – Structural and functional adaptations of eelgrass (*Zostera marina* L.) to the anaerobic sediment environment. *Can. J. of Bot.*, 61: 1421–1428

Ratter J. A., Ribeiro J. F., Bridgewater S., 1997 – The Brazilian cerrado vegetation and threats to its biodiversity. *Ann. Bot.*, 80: 223–230

Rizzini C.T.; Heringer E.P., 1962 – Studies on the underground organs of trees and shrubs from some southern Brazilian savannas. *Anais da Academia Brasileira de Ciências*, 34(2): 235–247

Saglio P. H., Raymond P., Pradet A., 1983 – Oxygen transport and root respiration of maize seedlings. *Plant Physiology*, 72: 1035–1039

Sauvageau C., 1889a – Sur la racine du *Najas*. *J. de Bot.*, 3: 3–11

Sauvageau C., 1889b – Contribution a l'étude du système mécanique dans la racine des plantes aquatiques. Les*Potamogeton*. *J. de Bot.*, 3: 61–72

Sauvageau C., 1889c – Contribution a l'étude du système mécanique dans la racine des plantes aquatiques. Les *Zostera*, *Cymodocea* et *Posidonia*. *J. de Bot.*, 3: 169–181

Schenck H., 1886 – Vergleichende Anatomie der submersen Gewächse. *Bibliotheca Bot.*, 1: 1–67

Schimper A. F. W., 1884 – Ueber bau und Lebensweise der Epiphyten Westindiens. *Botanisches Zentralblatt*, 17: 192–195; 223–227; 253–258; 284–294; 319–326; 350–359; 381–389

Schimper A. F. W., 1891 – Die indo-malaysche Strandflora. *Bot. Mit. Trop.*, 3: 1–204

Schimper A. F. W., 1903 – *Plant geography upon a physiological basis*, Oxford: Clarendon Press

Schulze E.-D., Beck E., Müller-Hohenstein K., 2005 – *Plant Ecology*, Springer Verlag, Berlin, Heidelberg

Schulze E.-D., Beck E., Buchmann N., Clemens S., Müller-Hohenstein K., Scherer-Lorenzen M., 2019 – *Plant ecology*, 2nd ed., Berlin: Springer Nature

Sculthorpe, C. D., 1967 – *The biology of aquatic vascular plants*, London, Edward Arnold Publishers

Schwarz F., 1883 – *Die Wurzelhaare der Pflanzen Ein Beitrag zur Biologie und Physiologie dieser Organe*, Breslau, Druck der Breslauer Genossenechafts-Buchdruckerei

Seliskar D. M., 1987 – The effect of soil moisture on structural biomass characteristics of four salt marsh plants. *J. of Exp. Bot.*, 38: 1193–1202

Shimek E., 1915 – The ecological histology of prairie plants. *Proc. of the Iowa Acad. of Sci.*, 22(1), 121–126

Smirnoff N., Crawford R. M. M., 1983 – Variation in the structure and response to flooding of root aerenchyma in some wetland plants. *Annals of Botany*, 51: 237–249

Smits A. J. M. Laan P., Thier R. H., van der Velde G., 1990 – Root aerenchyma, oxygen leakage patterns and alcoholic fermentation ability of the roots of some nymphaeid and isoetid macrophytes in relation to the sediment type of their habitat. *Aquatic Bot.*, 38: 3–17

Snell K., 1907 – Untersuchungen über die Nahrungsaufnahme der Wasserpflanzen. *Flora oder Allgem. Bot. Zeit.*, 98(2): 213–249

Teal J. M., Kanwisher J. W., 1966 – Gas transport in the marsh grass, *Spartina alterniflora*. *J. of Exp. Bot.*, 17: 355–361

van Tieghem Ph., 1871 – Recherches sur la symétrie de structure des plantes vasculaires. *Ann. Sci. Nat. Bot.*, 13: 6–314

Toma C., Niţă M., Rugină R., Ivănescu L., Costică N., 2003 – *Morfologia şi anatomia plantelor. Manual de lucrări practice*, ediţia a 2-a revizuită şi îmbunătăţită, Editura Universităţii „Alexandru Ioan Cuza" din Iaşi

Topa, M. A., McLeod K.W. 1986 – Aerenchyma and lenticel formation in pine seedlings: a possible avoidance mechanism to anaerobic growth conditions. *Physiol. Plant.*, 68: 540–550

Trindade F. G., Feitoza R. B. B., 2022 – Morphoanatomy of the vegetative organs of *Wunderlichia azulensis* Maguire & G. M. Barroso (*Asteraceae*) from a rocky outcrop in northern Rio de Janeiro, Brazil. *Biotemas*, 35(4): 1–14

Urcelay J. C., 1923 – Contribución al estudio de la adaptación de las plantas para diminuir la transpiración. Estudio anatómico de algunas especies de las estepas españolas. *Trab. Museo Nacional de Ciencias Nat. de Madrid, Ser. Bot.*, 18: 5–75

Vartapetian B. B., Jackson M. B., 1997 – Plant adaptations to anaerobic stress. *Ann. of Bot.*, 79: 3–20

Visser E. J. W., Nabben R. H. M., Blom C. W. P. M., Voesenek L. A. C. J., 1997 – Elongation by primary lateral roots and adventitious roots during conditions of hypoxia and high ethylene concentrations. *Plant, Cell and Environ.*, 20: 647–65

Walsh G. E., 1974 – Mangroves: A review. In R. J. Reimold, W. H. Queen (Eds.), *Ecology of halophytes*, pp. 51–174, New York/London: Academic

Warming E., 1877 – Om *Rhizophora mangle* L. *Botaniska Notiser*, 13–21

Warming E., 1883 – Tropische fragmente. II. *Rhizophora mangle* L. *Bot. Jahrb. fur Syst., Pflanzengesch. und Pflanzengeographie*, 4: 519–548

Warming E., 1909 – *Ecology of plants. An introduction to the study of plant communities* (English edition prepared by P. Groom, I. B. Balfour), Oxford: Clarendon Press

Weber M., Brandle R., 1996 – Some aspects of the extreme anoxia tolerance of the sweet flag, *Acorus calamus* L. *Folia Geobotanica Phytotaxonomica*, 31: 37–46

Stem

Stem

1 Primary Structure (Growth) of the Stem

This structure characterizes the stem of all cormophytes, i.e., plants whose body is organized in a cormus, thus presenting well-differentiated organs, with complex organization, in tissues.

It should be reminded that stems are usually aboveground organs that grow towards the light (positive phototropism) and away from the soil (negative gravitropism). They provide support for the plant's aerial organs and perform multiple other functions in the plant's life. Stems play an important role in the transport—often over long distances—of sap and hormones through the vascular tissues (xylem and phloem). The evolution of the vascular system represents, along with other morpho-anatomical acquisitions, one of the major advances necessary for plants to colonize (dry) land.

The epidermis arises from the protoderm, the cortex and pith from the ground meristem, and the vascular bundles from the procambium (Fig. 1), all three meristems originating from the tunica and corpus, being visible at a certain distance from the shoot apex; the procambium cords differentiate later (first giving phloem outwards and then xylem inwards), after the epidermis, cortex, and pith have resulted from the other two meristems (Toma et al. 2003; Crang et al. 2018). It should be noted that, in reality, the structure of the shoot apex is more complex, and its functioning is much more complicated.

The procambium gives rise, first, to a primary phloem to the outside and then to a primary xylem to the inside, thus resulting in conductive (vascular) vascular bundles, which can be of several types (Figs. 2 and 3):

M.-N. Grigore, *Plant Ecological Anatomy*,
https://doi.org/10.1007/978-3-032-01332-3_4

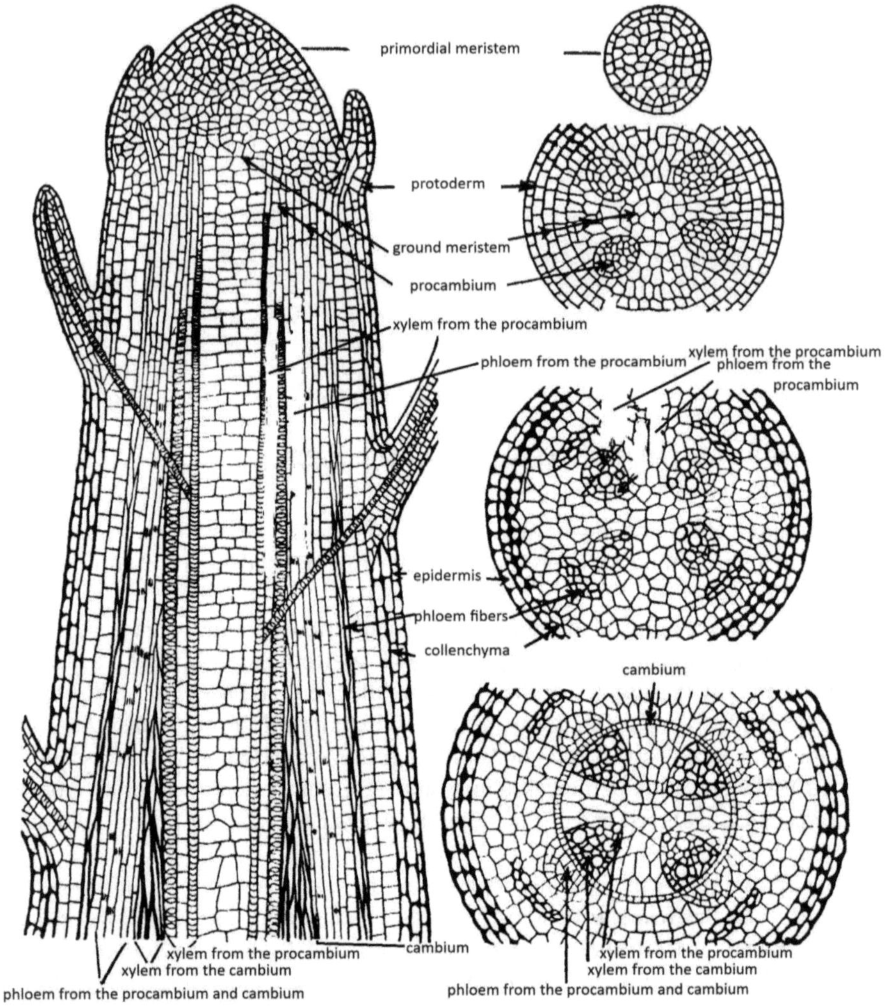

Fig. 1 Diagram showing the evolution of tissues from the primordial meristem down to the beginning of cambial activity (Stevens 1924)

(a) Collateral type:

- Closed, in which all procambial cells differentiate into conductive elements, as in monocotyledons (see also Fig. 1 in chapter "Primary Structure (Growth) *Versus* Secondary Structure (Growth): General Considerations"—corn).
- Open, in which a blanket of procambial cells is maintained between the phloem and the primary xylem, as in dicotyledons and gymnosperms.
- Bicollateral open, in which the xylem is framed by the phloem externally and internally, a type found in plants from the families Cucurbitaceae,

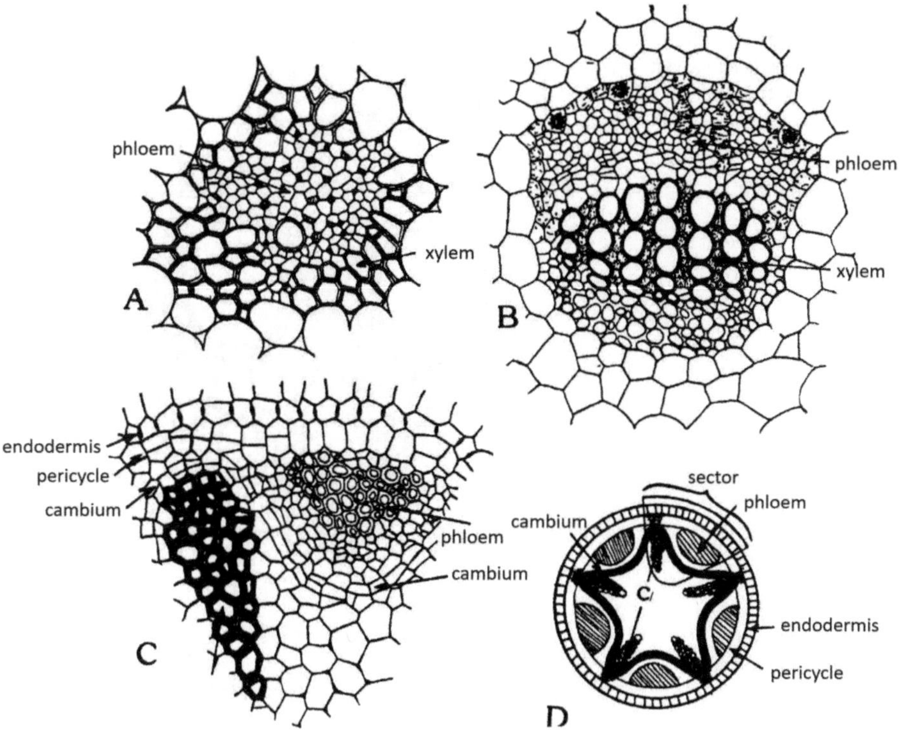

Fig. 2 Types of vascular bundles: A, concentric leptocentric (*Polypodium* sp.); B, open collateral with details in C and complete structure in D (Stevens 1924)

Solanaceae, etc.; thus, one area of the phloem is located towards the cortex and the other, towards the pith.

(b) Concentric type:

- Leptocentric (amphivasal), in which the phloem is in the center, surrounded by the xylem, a type found in the rhizomes of many monocots.
- Hadrocentric (amphicribral), in which the xylem is in the center, surrounded by the phloem, a type of bundle found in the rhizomes of many pteridophytes (Toma 1977; Crang et al. 2018).

It should be noted that the stems with primary structure can have vascular bundles of closed or open collateral type.

1.1 The Anatomical Areas of the Stem with Primary Structure

In dicotyledons (Figs. 4, 5, 6, 7, 8, 9, 10), the following anatomical areas can be described:

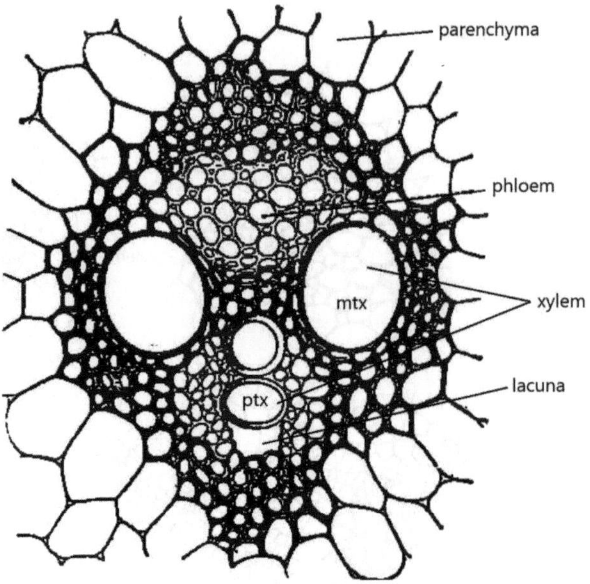

Fig. 3 Types of vascular bundles: closed vascular bundle in corn (*Zea mays*); mtx, metaxylem; ptx, protoxylem (Barnes 1900)

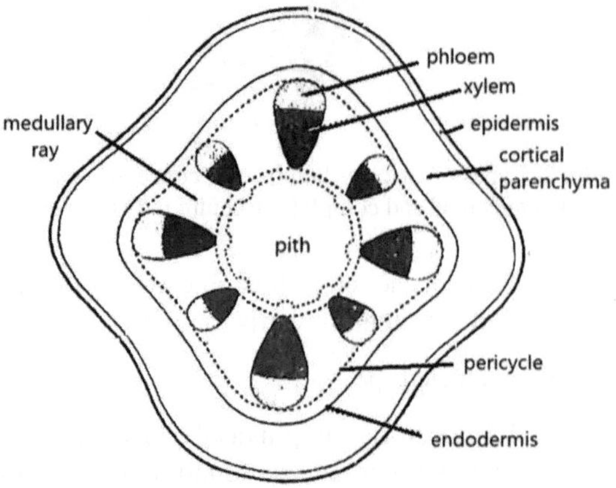

Fig. 4 The primary structure of the stem, in *Mercurialis annua* (dicotyledonous), cross section, diagram—general view (Bonnier and du Sablon 1905)

1. The epidermis is single-layered and often has stomata, and trichomes can form at its level (protecting, secretory-glandular). The epidermal cells have more convex outer wall and thicker than the others, being cutinized; sometimes it can be covered with wax and is rarely mineralized and lignified. In cross section, the epidermal cells have a square or rectangular outline, the cells being tangentially elongated.
2. The cortex is always multilayered; thicker in ferns, gymnosperms, and dicotyledons than in monocotyledons; and thicker in underground and aquatic stems than in aerial ones, but usually thinner than the central cylinder.

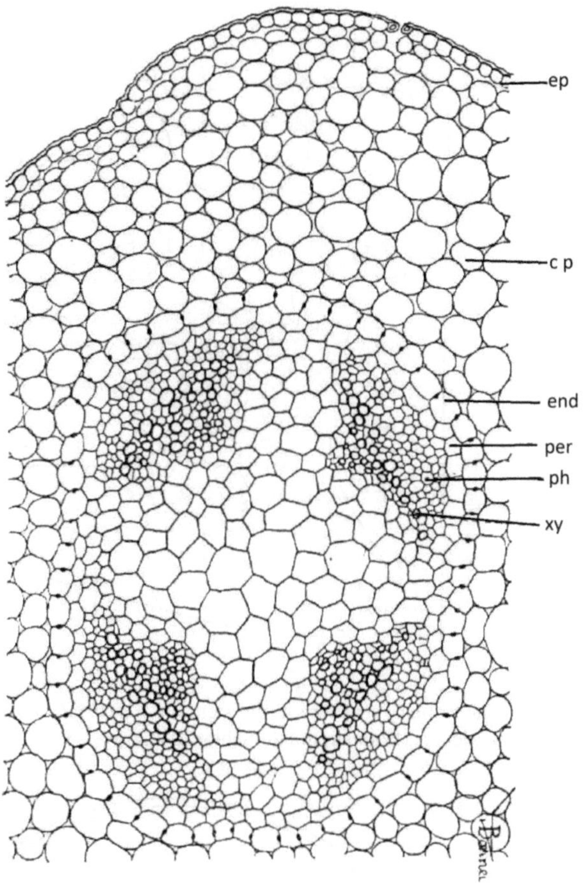

Fig. 5 The primary structure of the stem, in *Lupinus* sp. (dicotyledonous), cross section, diagram—general view (Gérardin and Guède 1895); ep, epidermis; c p, cortical parenchyma; end, endodermis; per, pericycle; ph, phloem; xy, xylem

Cortex cells are often parenchymatous, with thin, cellulosic walls, leaving meatus, rarely air gaps between them. The outer layers of the cortex are smooth, therefore presenting chlorophyll cells (which, in some species with stems without leaves or with reduced leaves, can even be of palisade shape—see, in this case, the part of the ecological anatomy of the stem), and the deep ones can accumulate reserve substances (especially starch) (Toma 1977; Toma et al. 2003).

In several situations, many particular elements can be differentiated in the cortex, some having ecological significance or being characters of the botanical family:

- Supporting (mechanical) elements of collenchyma or sclerenchyma (fibers, sclereids).
- Cells with calcium oxalate (sand, crystals).
- Secretory elements (cells with oils), secretory channels, secretory cavities, cells and channels with mucilage, cells with tannin, laticiferous vessels, etc.
- Conductive elements that form foliar and branch bundles (traces), which connect the conductive tissues of the stem to those of the leaves and branches, respectively.

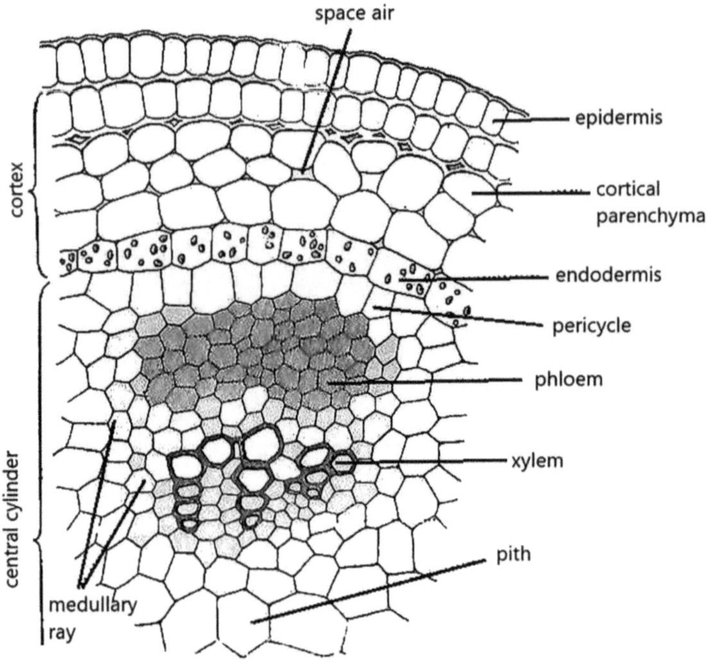

Fig. 6 The primary structure of the stem, in *Mercurialis annua* (dicotyledonous), cross section, diagram—details (Bonnier and du Sablon 1905)

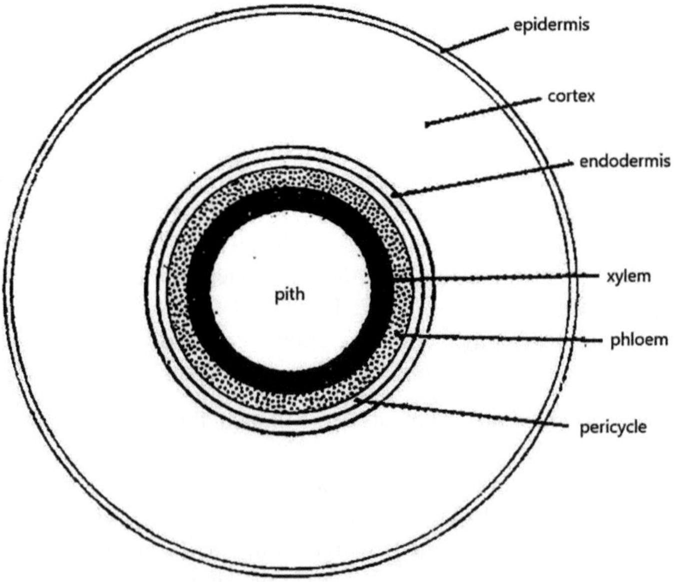

Fig. 7 The primary structure of the stem, in *Veronica beccabunga* (dicotyledonous), cross section, diagram—general view (Bonnier and du Sablon 1905)

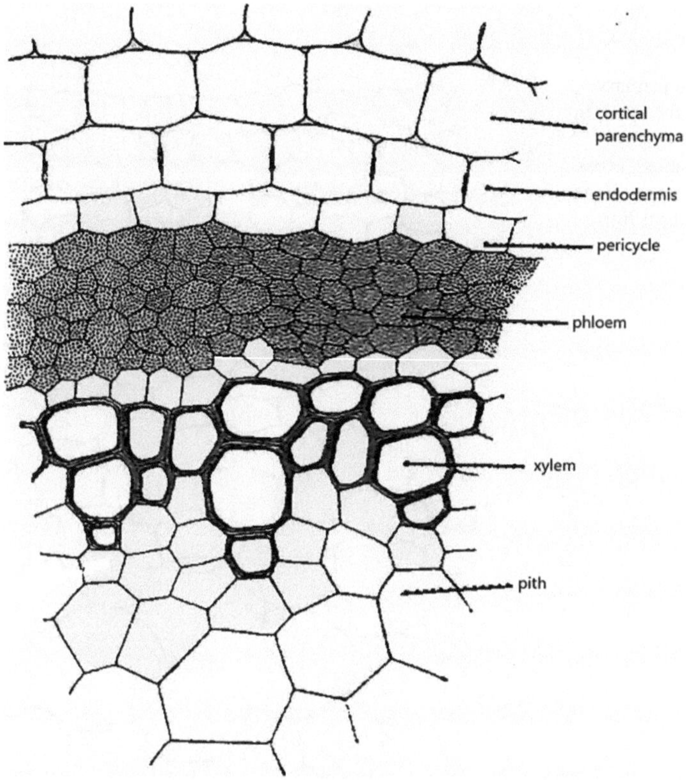

Fig. 8 The primary structure of the stem, in *Veronica beccabunga* (dicotyledonous), cross section, diagram—details (Bonnier and du Sablon 1905)

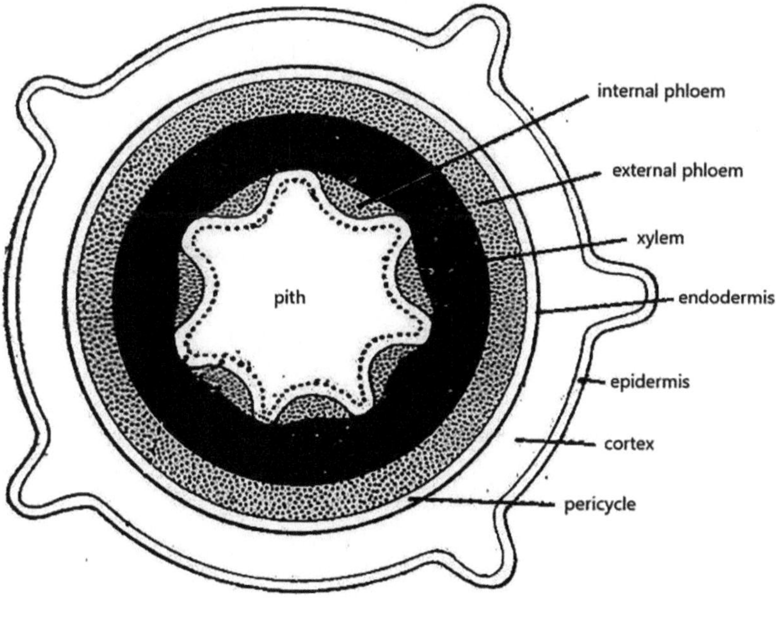

Fig. 9 The primary structure of the stem, in *Convolvulus arvensis* (dicotyledonous), cross section, diagram—general view (Bonnier and du Sablon 1905)

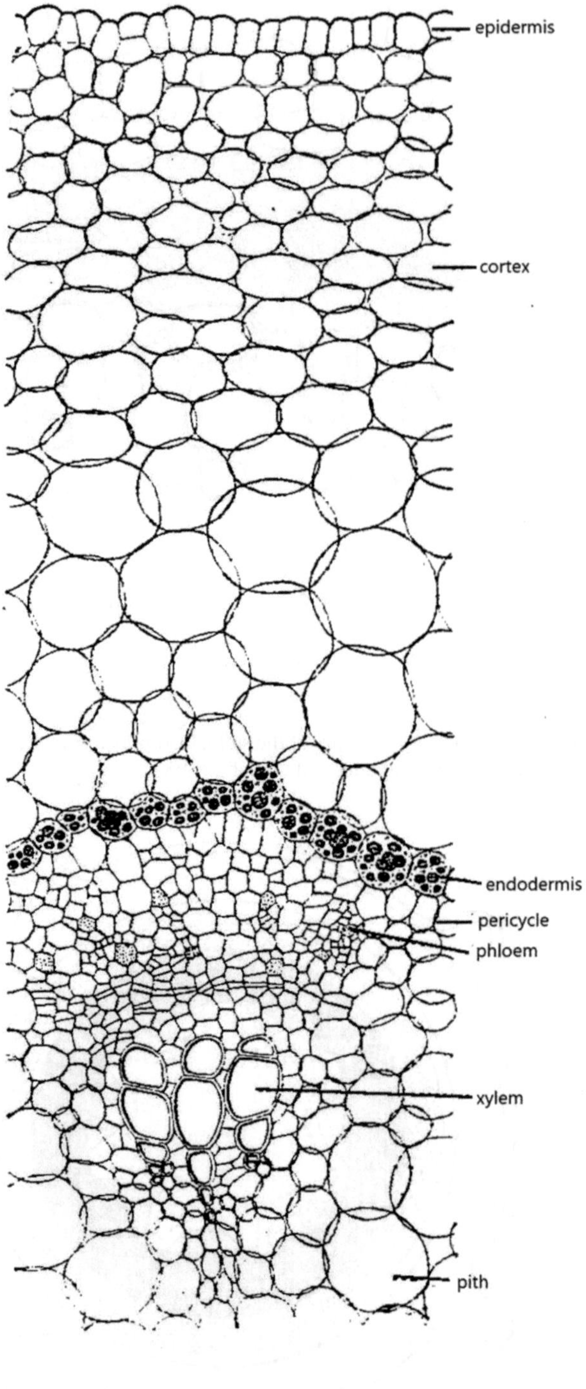

Fig. 10 The primary structure of the stem, in *Ricinus* sp. (dicotyledonous), cross section, diagram—general view (Belzung 1900)

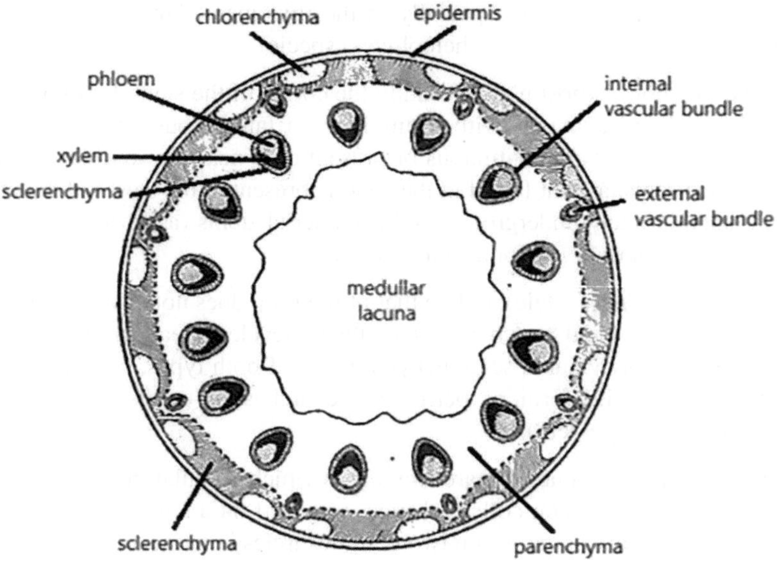

Fig. 11 The primary structure of the stem, in *Hordeum murinum* (monocot), cross section, diagram—general view (Bonnier and du Sablon 1905)

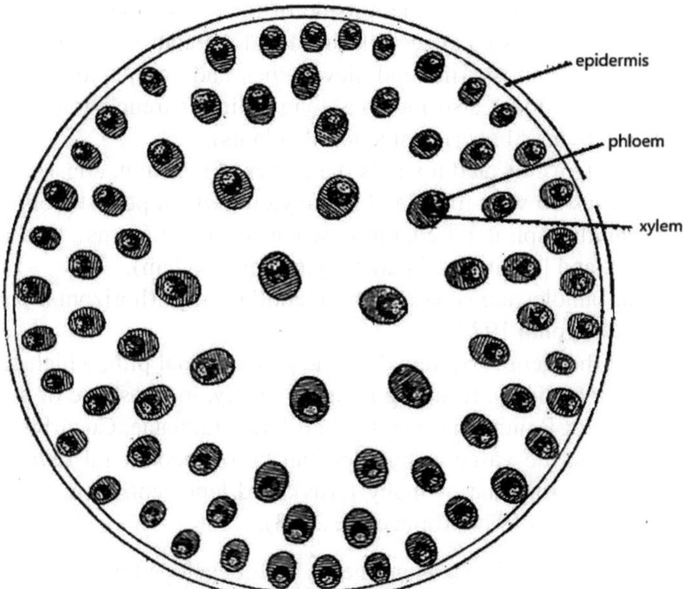

Fig. 12 The primary structure of the stem, in *Zea mays* (monocot), cross section, diagram—general view (Bonnier and du Sablon 1905)

– Aerenchyma (air spaces), especially in the rhizomes of many aquatic (hydrophytes) and marshes (marsh—helophytes) species.

Unlike the root, in most plants, usually the cortex of the stem does not end with a typical endodermis, but only with an amyliferous, tanniferous, or crystalline endodermoid or starch sheath, continuous or present only near the vascular bundles. A typical endodermis, as that found in the root, is present in the stems of aquatic and marsh plants, in many underground and even aerial stems (as in Scrophulariaceae and Labiatae) (Toma 1977; Toma et al. 2003).

3. The central cylinder (stele), unlike that of the root, does not always begin with a pericycle (Figs. 4 and 8); when present, the pericycle appears unilayered or multilayered, parenchymatic, sclerenchymatic, or of both types, continuous or discontinuous (in the form of periphloemic arches or cords—on the outside/periphery of the phloem).

It consists of a fundamental parenchyma, in which vascular bundles are embedded, especially of the open collateral type, arranged on a single circle in gymnosperms and monocotyledons, or on two to three circles in monocotyledons (eustelic structure); sometimes, they are scattered (apparently) irregularly in the fundamental parenchyma, as in corn, sugarcane, etc. (atactostelic structure). The xylem has centrifugal development, while the phloem has centripetal development (Toma et al. 2003).

The phloem, located towards the outside of the stem, consists of protophloem (which will flatten and disorganize with time) and metaphloem, consisting of sieve cells in ferns and gymnosperms and sieve tubes and companion cells in angiosperms; rarely, the phloem also includes longitudinal parenchyma cells (phloem) and even sclerenchyma fibers (in some textile plants).

The xylem (xylem), located towards the center of the stem, consists of protoxylem (which will resorb with time) and metaxylem (which persists longer), formed by tracheids (annular, spiral, pitted, rarely scalariform, as in ferns), tracheae (reticulated and pitted), and longitudinal parenchyma cells (xylem).

The vascular bundles are separated by medullary rays (horizontal parenchyma) of different width (Toma 1977).

The center of the stem is occupied by the parenchymal pith, which can disorganize and resorb sometimes, resulting in an air cavity, in the shape of a channel (in species belonging of Ranunculaceae, Umbelliferae, Labiatae, Cucurbitaceae, etc.).

In some species, the vascular bundles can be of bicollateral (*Cucurbitaceae*), hadrocentric (in the rhizome of many ferns), and leptocentric (in the rhizome of some monocots) (Toma 1977; Toma et al. 2003).

4. Pith is always present in the stem, being well developed and parenchymatic. It consists of large, non-chlorophyllous cells, rich in starch. With age, the cell walls can lignify, from the periphery to the center of the pith. The pith is very thick and amyliferous in the potato tubers (*Solanum tuberosum*) and also well developed in the elder stem (*Sambucus nigra*). Sometimes, it includes oxaliferous cells, secretory channels, etc.

In fistulous stems, the pith of the internodes is resorbed, resulting in an air-storing medullary cavity (Poaceae, Apiaceae, Compositae, Lamiaceae, Ranunculaceae, Cucurbitaceae).

The medullary rays occupy the spaces between the pith, the pericycle, and the vascular bundles, being generally of parenchymatic type (Toma 1977; Toma et al. 2003).

In reality, the pith and the cortex present itself as fundamentally continuous—and almost homogeneous—parenchymatic tissues in which the vascular bundles are located.

Distinction between the cortex and pith is somewhat relative, more in terms of position in relation to the bundles, than anatomical (Morariu 1965; Toma et al. 2003).

In monocots (Figs. 11 and 12), the following anatomical areas of the stem with primary structure can be highlighted:

1. The epidermis is cutinized and has stomata, with or without trichomes, having cells with an external wall that shows some secondary chemical changes: it can be covered with wax, mineralized or lignified, which therefore makes the wall thicker.
2. The cortex is very thin in normal, aerial stems and difficult to separate from the central cylinder because the endodermis and pericycle are missing. It is the reason why the entire fundamental tissue, regardless of its location in the stem, is also called conjunctive tissue, due to the lack of a clear demarcation between the cortex and pith (as it is, on the contrary, in the stem of dicotyledons) (Crang et al. 2018).

In corn, for example, the cortex consists of one to three layers of cells with strongly lignified walls forming an interrupted ring (hypodermis) next to the stomata, where there is a parenchymatic tissue. In grasses, in which the stem is a culm (straw), sclerenchyma strands alternate with collenchyma bands in the cortex (see also the ecological implications of this type of histological arrangement below).

As already stated, in rhizomes (metamorphosed underground stems), the endodermis is well differentiated, thus clearly marking the boundary between the cortex and the central cylinder (in grasses, *Agropyron* species, and water lily, *Convallaria majalis*).

3. The central cylinder consists of a fundamental parenchyma, made up of cells with cellulosic walls, in which the vascular bundles are arranged, most often of the closed collateral type. At the periphery of the stem, the vascular bundles are small and numerous, and towards the center, they are larger and rarer (the case of wheat). In general, the bundles can be arranged on a single circle (rarely), and on two or more concentric circles (more often, forming an eustelic structure), or dispersed (apparently) disorderly throughout the mass of the central cylinder (atactostelic structure, as in corn, *Zea mays*; sugar cane, *Saccharum officinarum*; and palm trees, species of *Palmae*).

In the straw of grasses (see the morphology of aerial stems—Grigore 2022), where the vascular bundles are distributed on two circles at the periphery of the stem, in the center of the stem the parenchyma is resorbed, resulting in a wide medullary (air-storing) lacuna (Fig. 3).

Although the straw is practically hollow inside, the peripherally located sclerenchyma allows its erect port (Fig. 11); sclerenchyma is usually located in a deeper hypodermic position, in the form of cords or a continuous ring.

As in dicotyledons, in each vascular bundle, the xylem and phloem are superimposed, proto and metaxylem having centrifugal differentiation, and proto and metaphloem having centripetal differentiation; the xylem is arranged in the form of the arms of the letter “V” (Fig. 3) (Toma 1977).

1.2 The Origin and Evolution of the Stele (Central Cylinder)

Plants colonized the Earth about 480 million years ago. The first land plants were small and leafless and lacked a vascular system like their aquatic ancestors. Life on land involved the development of a number of unique structures, mainly lignin, to provide mechanical support (mechanical strength); in addition, a cuticle was needed to limit water loss, stomata to allow gas exchange, leaves to increase the photosynthetic surface, and, last but not the least, a vascular system to move and transport water, minerals, and products of photosynthesis throughout the plant.

Stems evolved first, and the first stem vascular systems, like the first land plants, were simple in organization. As plants grew in size and complexity, so did the vascular system in their stems. The later evolution of the leaf, in particular, had a great impact on the vasculature of the stem, because the vasculature of the leaf must be directly connected to the vasculature of the stem through a cord of xylem and phloem called the leaf trace.

The stele (central cylinder) comprises the vascular tissues in the center of a root or stem, and there is a wide diversity in the arrangement of xylem and phloem in tracheophytes. Several classification systems exist to define and classify the various arrangements of conducting tissues within the central cylinder.

The proposal presented below (Crang et al. 2018) describes two basic (main) types of stele: the protosteles (without a central pith) and the siphonosteles (with a central pith), each with several variants (Table 1).

2 Secondary Structure (Increase in Thickness) of the Stem

As can also be seen from Fig. 16, the secondary structure (growth) complements the primary one, contributes to the thickening (increase in thickness and diameter) of the stem, and is due to the two secondary (lateral) meristems: cambium and phellogen. It therefore characterizes all woody plants (gymnosperms and

Table 1 Organization of steles types in plants (Crang et al. 2018)

Protosteles (no central pith)	*Haplostele* (Fig. 13)	The central cylindrical body of xylem is surrounded by phloem	
	Actinostel	The central cylindrical xylem body is lobed and surrounded by phloem	
	Plectostele	Flattened (interconnected) regions of xylem surrounded by and embedded in phloem	
Siphonosteles (with central pith)	*Solenostele* (Fig. 14)	*Ectophloic*	The phloem is on the outside of the xylem
		Amphiphloic	Phloem is located on both sides of the xylem
	Dictyostele (Fig. 15)	The leaf gaps create a net-like arrangement of the bundles	
	Eustele	Conductive bundles are discretely arranged in one area of the pith; eudicots have a eustele with a circle of bundles	
		The atactostele is a subtype of the eustele in which the conducting bundles are apparently scattered in the pith; found mainly in monocots	

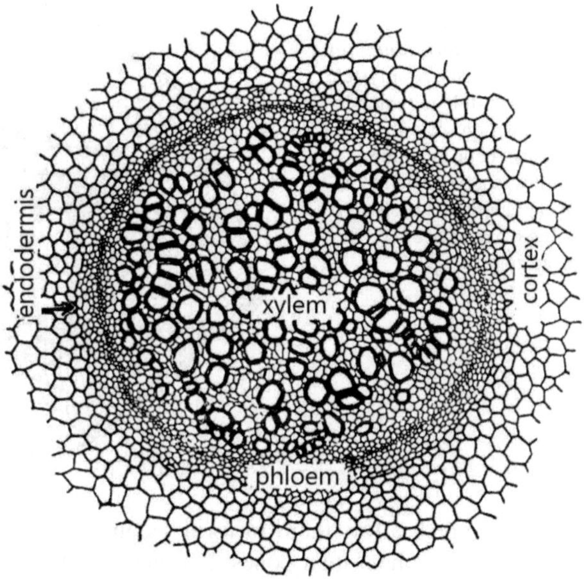

Fig. 13 Protostele—cross section through the rhizome of *Gleichenia pubescens* (Stevens 1924)

dicotyledonous angiosperms), some perennial herbaceous plants, and even some annual plants (*Helianthus annua*—sunflower).

The stem of monocots and some annual dicots, which have in their anatomical structure closed vascular bundles, remains for the rest of its life only with the primary structure. The other dicotyledons and gymnosperms have open vascular

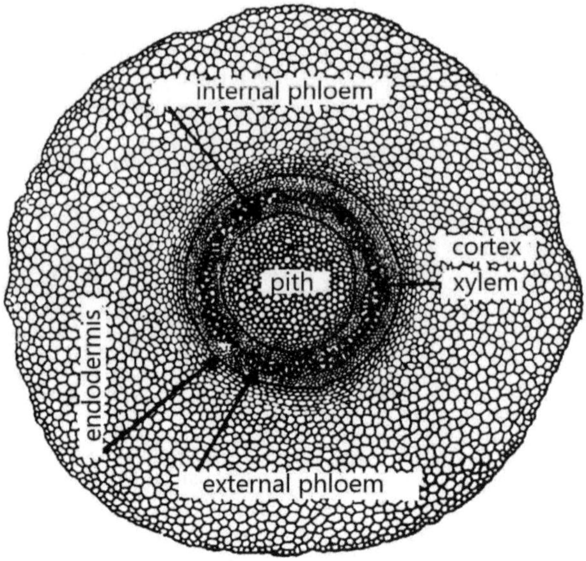

Fig. 14 Amphiphloic siphonostele—cross section through the rhizome of *Adiantum pedatum* (Stevens 1924)

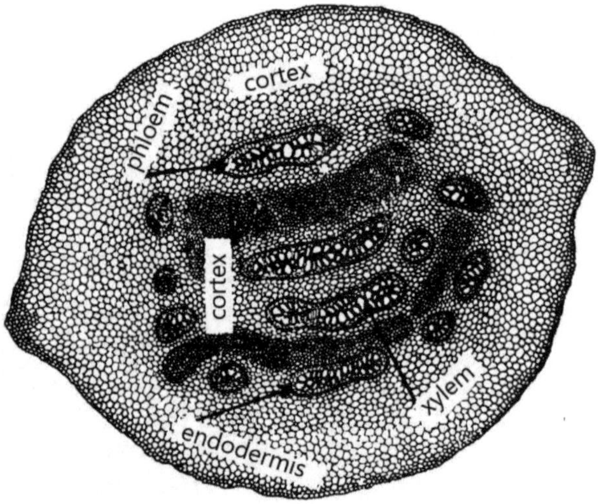

Fig. 15 Polystele—cross section through the rhizome of *Pteris aquilinia* (Stevens 1924). A polystele is a dictyostele in which in cross section several xylems occur as single strands, each surrounded by a phloem and each unit being a haplostele type of protostele

bundles, which allow them to form secondary tissues, which are added or interspersed among the primary ones. This new reality complicates the anatomical structure quite a bit. Usually, stems with the primary structure remain simple or very little branched, even if they live for many years. The secondary structure appeared and evolved as an adaptation to the biological need to support a plant body usually branched and with continuously growing biomass. This adaptation is, of course, more pronounced in woody plants, where new branches (with foliage) are added to the stem in an increased amount every year, where, during maturity, the fruits are

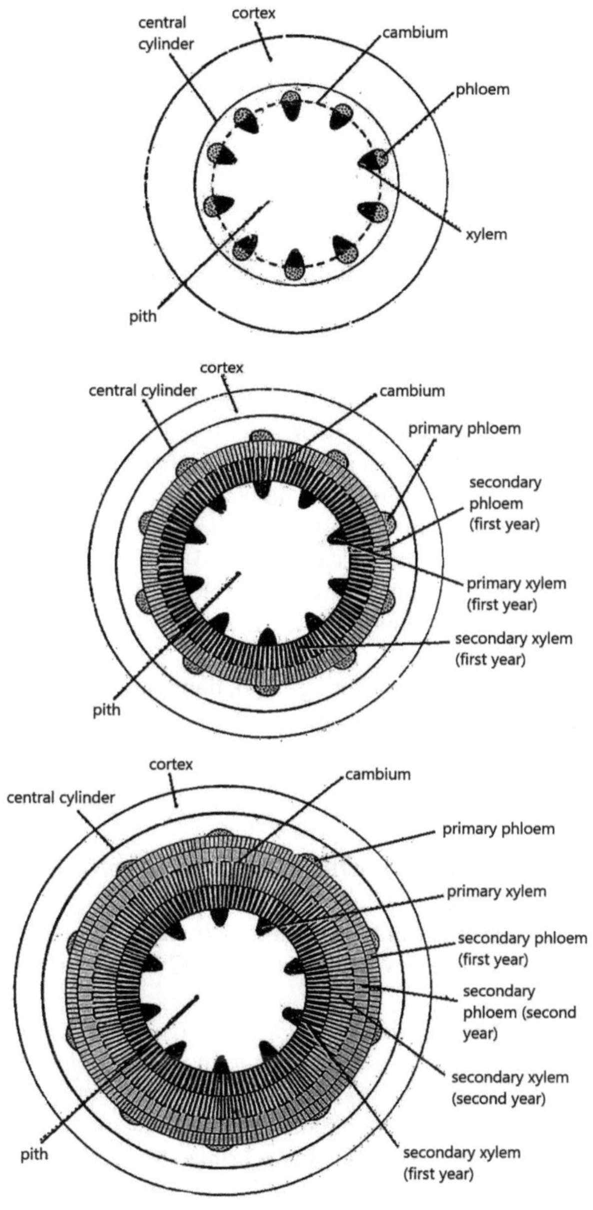

Fig. 16 Secondary stem structure (growth): (**a**), start of cambium activity; (**b**), xylem and phloem production at the end of the first year; (**c**), end of the second year (Bonnier and du Sablon 1905)

also formed. For these reasons, the progressive increase in diameter and the mechanical strengthening of the stem appeared as a natural necessity (Morariu 1965).

In addition to the mechanical function (support) that a stem with secondary growth (structure) plays, there is another one, very important in the life of a plant in continuous growth and development, the transport function. On the one hand, the expanded foliage demands increased the amounts of water, and on the other hand, the substances synthesized as a result of photosynthesis must be transported to other regions of the plant, often far away from the site of synthesis. There is therefore a need for a more efficient circulation of fluids, through more numerous conducting systems.

Therefore, with the growing age of the plant, they also grow its functional (nutritional) needs, and, to satisfy them, it is necessary to produce new secondary tissues in the stem, which cause thickening and its annual elongation. Moreover, it appears that herbaceous angiosperms arose, phylogenetically, from woody forms.

As stated, the two secondary meristems that generate the secondary structure of the stem are the cambium and phellogen.

The cambium appears first and originates from meristematic tissues (procambium), or from the procambium and from definitive tissues (through dedifferentiation), having the most important role in increasing the stem diameter (Toma 1977). The phellogen appears later, always by dedifferentiation (often due to the outer cortical layers).

Along with secondary growth, flattening, destruction, and exfoliation of external tissues of primary origin (epidermis, cortex, primary phloem) occur; they are replaced by tissues that play the same role, but of secondary origin (Toma 1977).

2.1 *Formation and Activity of Cambium*

2.1.1 Transition from Primary to Secondary Structure: Types (Patterns) (at the Level of the Central Cylinder).

1. The *Aristolochia* type (Figs. 17, 18, 19, and 20) is found in herbaceous plants and liana species belonging to the *Aristolochia*, *Vitis*, and *Clematis* genera. The cambium begins to function as intrafascicular arches (resulting from the fascicular procambium). Later, the interfascicular cambium is also formed, through dedifferentiation, due to the primary medullary rays. Later, the two types of cambial strings connect, resulting in the cambial ring, which produces secondary phloem outwards and secondary xylem inwards, next to the primary xylem phloem bundles; in the interfascicular sectors, the cambial ring produces secondary, parenchymatous medullary rays, which intercalate in the primary medullary rays, causing their elongation (Toma 1977; Crang et al. 2018).

Overall, the secondary structure remains fascicular, like the primary one. This type is found in many lianas, whose stem must retain some degree of flexibility (for

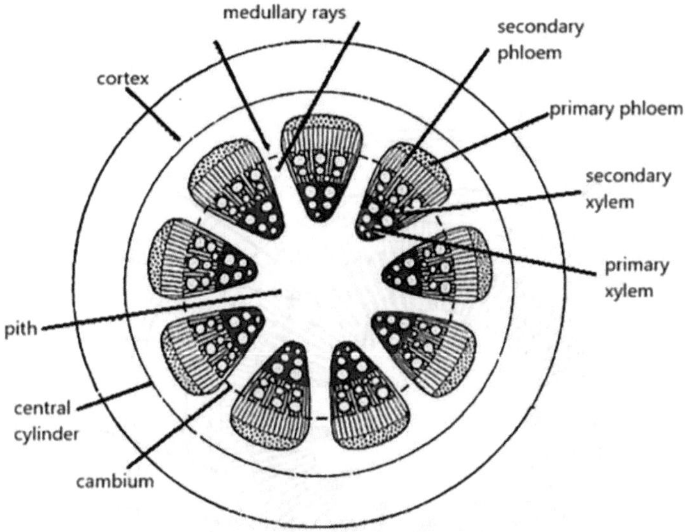

Fig. 17 Types of transition from the primary to the secondary structure of the stem: cross section through the stem of *Aristolochia* sp. (Bonnier and du Sablon 1905)

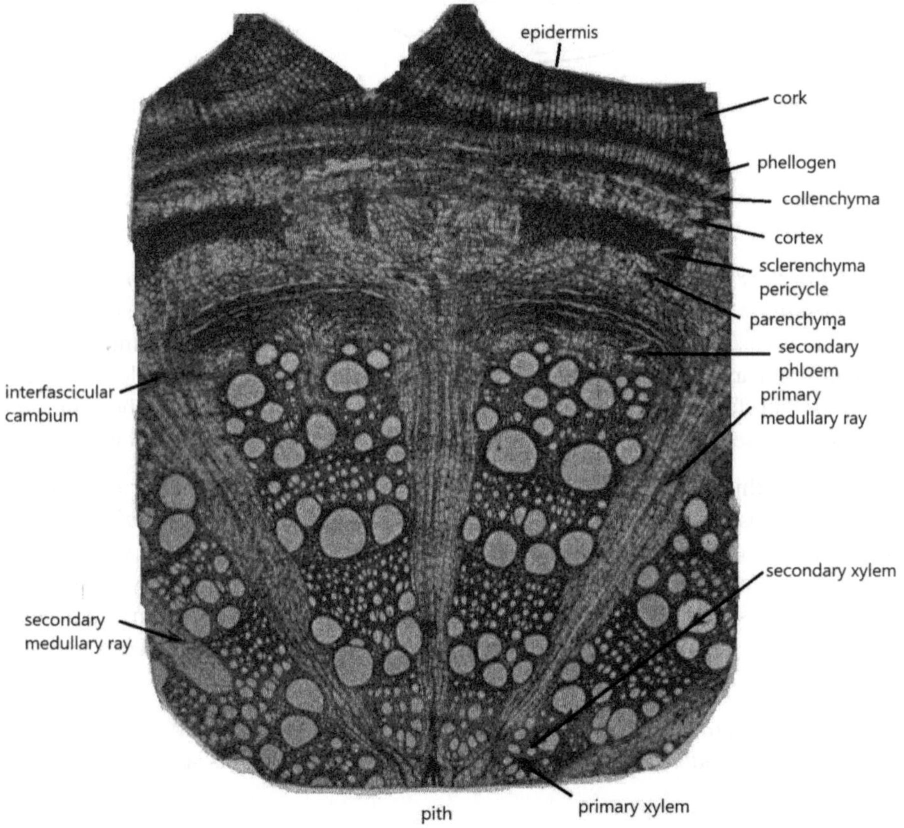

Fig. 18 Portion of cross section of 4-year-old stem of *Aristolochia sipho*, as shown by the rings of growth in the secondary wood (Stevens 1924)

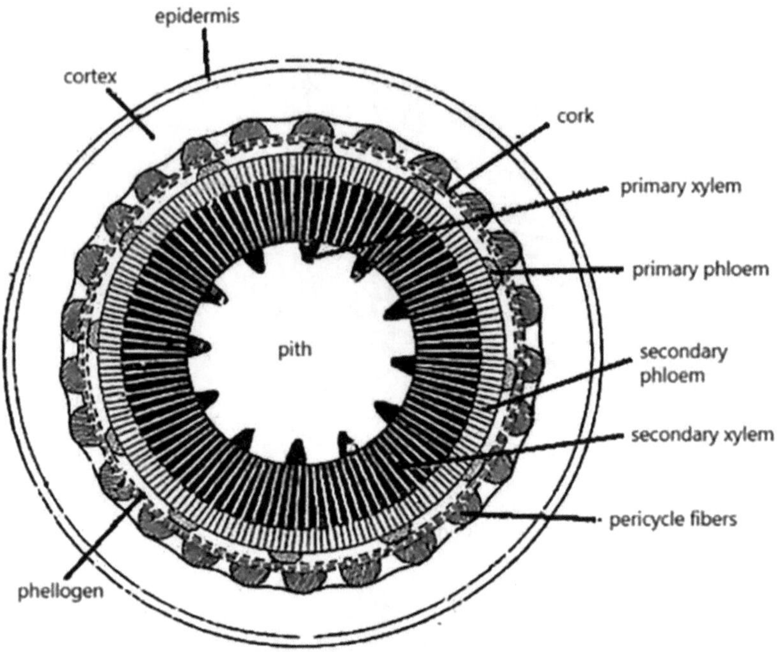

Fig. 19 The secondary structure of the 1-year stem of *Vitis* sp.—general drawing (Bonnier and du Sablon 1905)

the discussion of lianas, from a morphological and ecological point of view, see Grigore 2022).

2. *Helianthus* type (Fig. 21), found in some herbaceous dicots (*Helianthus*, *Vicia*, *Phaseolus*). In the primary medullary rays, due to the meristem resulting here, smaller vascular bundles arise, each of them with intrafascicular cambium. Ring cambium arises from the sum of all intrafascicular cambia and, to a lesser extent, from interfascicular cambia (resulting from dedifferentiation). The cambium produces secondary phloem and xylem as well as secondary medullary rays.

Broadly, this type resembles the *Aristolochia* type, the secondary structure being also of the fascicular type, like the primary one, but, unlike the *Aristolochia* type, there are two categories of vascular bundles: large, initial ones, which have in their composition primary conductive tissues (primary phloem and primary xylem) and secondary vascular tissues (secondary phloem and secondary xylem), and small ones, entirely secondary, having in their composition secondary phloem and secondary xylem only (Toma 1977).

3. Type *Coniferae* (Fig. 22) is found in conifers (e.g., *Pinus* sp.), some woody dicots (*Acer, Salix, Prunus, Sambucus*) and even herbaceous species.

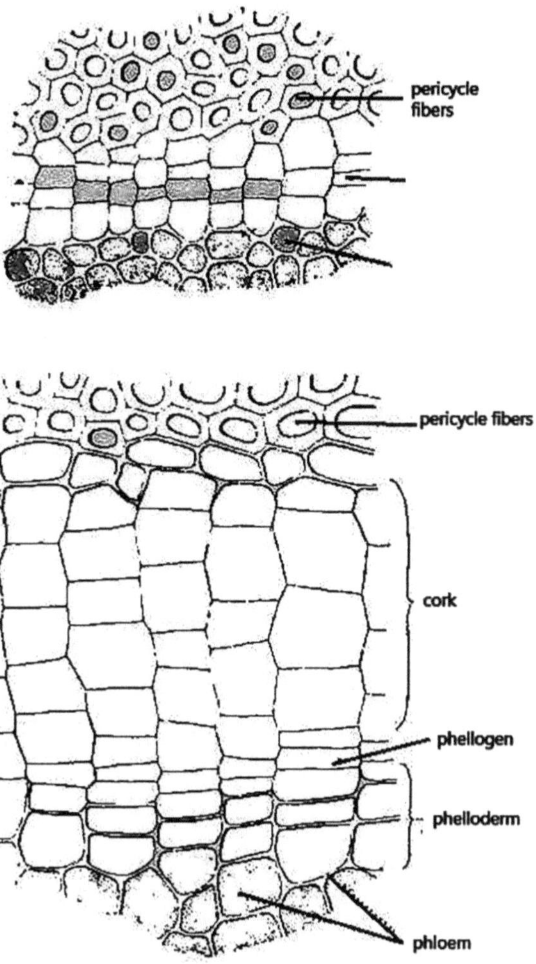

Fig. 20 The secondary structure of the 1-year stem of *Vitis* sp.—details on formation and functioning of phellogen (Bonnier and du Sablon 1905)

Procambium and the primary conducting tissues constitute cords (bundles) separated by very narrow medullary rays; the ring cambium mostly originates from the sum of the procambia of all the bundles of the open collateral type.

Secondary tissues each form a continuous ring (cylinder): secondary xylem on the inside and secondary phloem on the outside, resulting from the bifacial activity of the interfascicular cambium.

In this way, it goes from a primary structure of fascicular type to a secondary structure of ring type; annual rings of secondary xylem are formed, while the secondary phloem ring thickens from year to year. The age of the trees can be estimated by the number of annual rings, formed due to the seasonal activity of the cambium (Toma 1977; Toma et al. 2003).

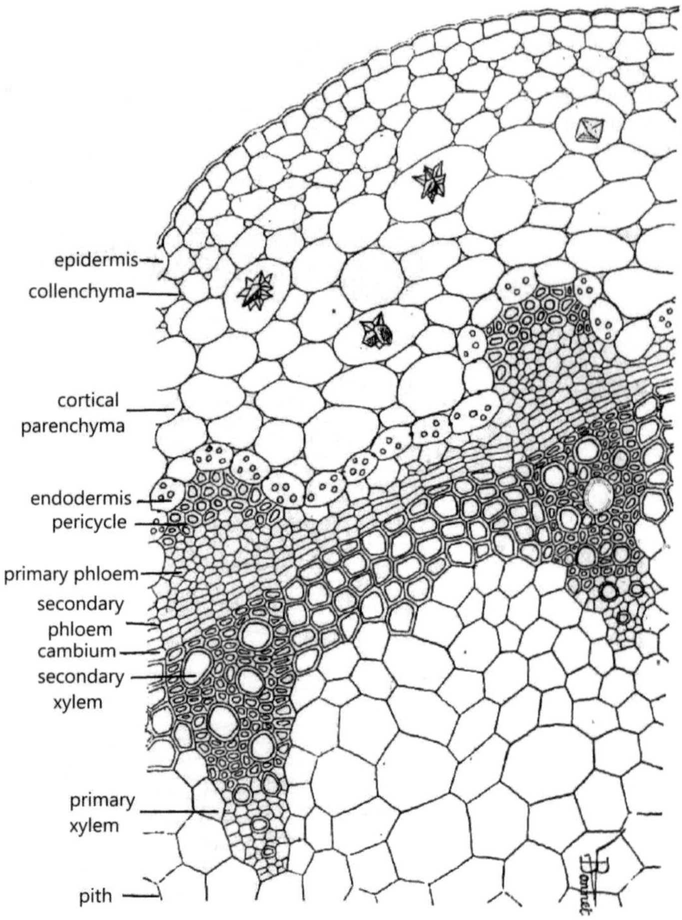

Fig. 21 Cross section through the stem of *Begonia* sp.—start of the secondary thickening (Gérardin and Guède 1895)

4. *Tilia* type (Fig. 23), found in woody (*Tilia*, *Syringa*) and herbaceous plant species (*Nicotiana*, *Veronica beccabunga*). In these plants, the procambium is from the beginning annular (cylindrical) and will therefore produce even from the primary structure rings of primary phloem and primary xylem, crossed only by narrow medullary rays.

Then, the procambium will become a cambium ring, which produces rings of phloem and secondary xylem, with narrow rays (at least at the level of the secondary xylem) and somewhat wider at the level of the secondary phloem, where it forms a veritable expansion parenchyma, of the form of triangles with the apex on the cambium in cross section. The limit between primary and secondary conducting tissues is difficult to establish. In this way, the secondary phloem ring is fragmented into trapezoidal formations in cross section (Toma 1977; Toma et al. 2003) (Fig. 8 in chapter "Ecological Anatomy of the Stem").

Fig. 22 The secondary structure of the 2-year stem of *Pinus* sp. (Gérardin and Guède 1895)

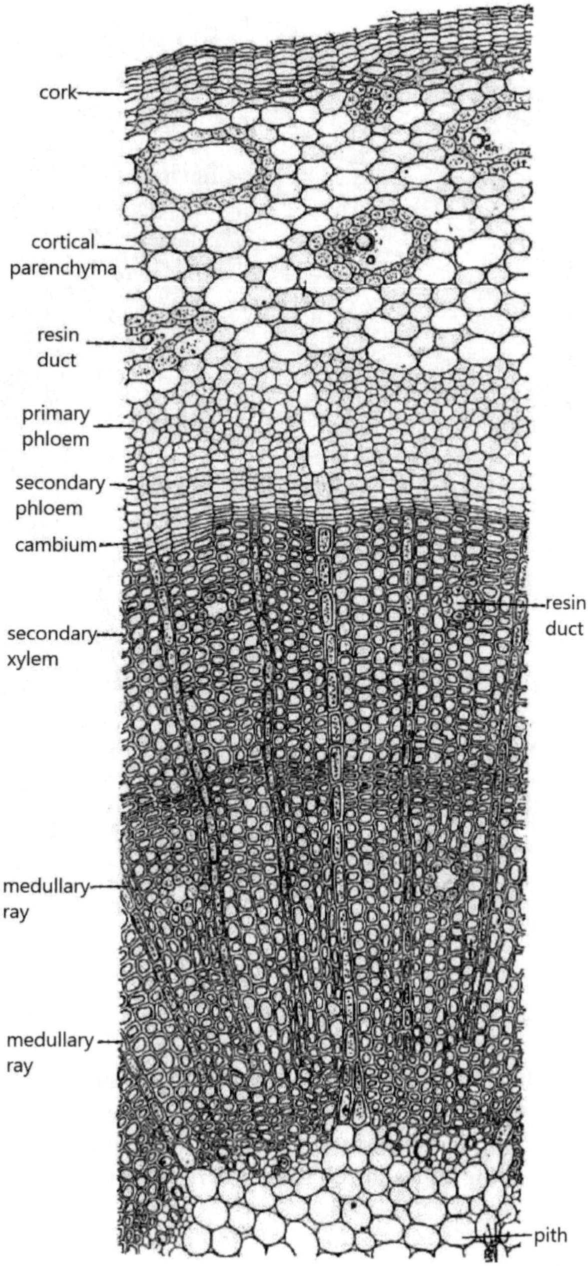

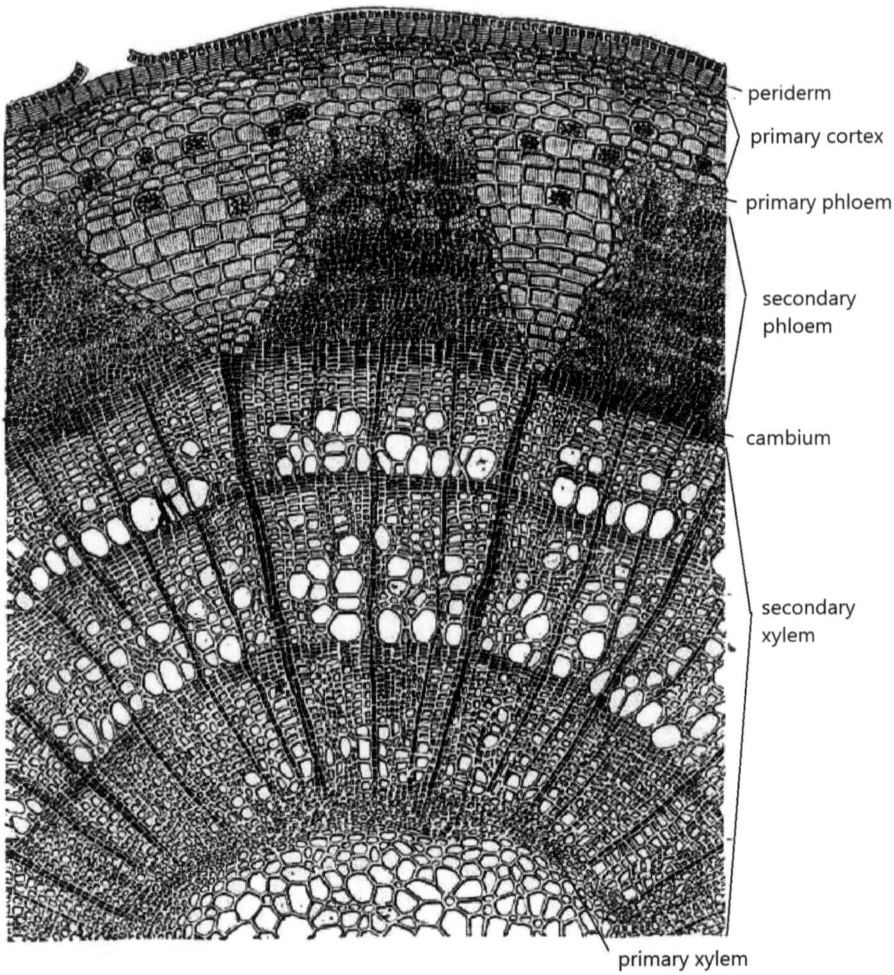

Fig. 23 Secondary structure of the 3-year stem of *Tilia* sp (Gérardin and Guède 1895)

Thus, it goes from a ring-type primary structure to a ring-type secondary structure. Over time, the ring structure of the phloem (primary and secondary) changes, appearing in the form of characteristic traps, while the secondary xylem will still have a ring appearance (Toma 1977; Toma et al. 2003).

2.2 *Formation and Activity of Phellogen*

The phellogen appears later than the cambium and is always formed following a process of dedifferentiation, at various levels: the inner layer of the epidermis, or deeper, up to the endodermis and even the pericycle (this is considered, by some authors, in general, a meristematic tissue—Crang et al. 2018).

From the bifacial activity of the phellogen, cork outwards and phelloderm inwards result, all these three secondary tissues forming the periderm or secondary cortex. In some plants, especially woody ones, a new periderm is formed every year, due to the areas of phellogen formed further inside the organ. The complex of successive periderms is called the rhytidome (Grigore 2022).

The newly formed cork pushes the older cortical layers out, and they eventually die, dry, and come off in the form of scales or plates. Sometimes, the rhytidome persists and cracks, as in the poplar (*Populus*); other times, it detaches periodically (*Platanus*, plane tree; *Eucalyptus*, eucalyptus) or progressively (*Betula*, birch).

As is known, the cork plays a protective role, and the lenticels formed at the level of the periderm ensure gas exchanges with the external environment. The phelloderm plays the same role as the primary cortical parenchyma.

Further details on the structure (composition) of secondary tissues will not be provided; it should be noted that secondary xylem includes spring (early) xylem, consisting mainly of wide vessels (to ensure an increased transport of sap, so necessary in spring) and summer-autumn (late) xylem, with smaller diameter elements (the plant prepares for the rest period before winter) (Grigore 2022). For this reason, in many plants from the temperate zone (therefore also from Romania), especially woody ones, annual (growth) rings can be distinguished in the thickness of the secondary xylem, due to the interruption of cambium activity at the end of summer; in each ring, on the inner part, there are spring vessels, and on the external part, there are summer-autumn vessels. The boundary between last year's late xylem and next year's early xylem marks the boundary between two annual rings. The contrast between the xylem formed at the end of the growing season and that formed the following spring makes it possible to measure for each year the increase in thickness of the stems and to determine the age of the trees by counting these rings (the science that has this object of study is called dendrochronology). This counting is usually done at the level of the trunk, where the annual rings are well distinguished with the naked eye.

Regarding the secondary phloem, we only say that in its thickness, no annual rings can be distinguished, although in some species (*Tilia*), hard phloem and soft phloem zones can be distinguished (Morariu 1965; Toma 1977).

2.3 Heartwood and Sapwood

In some woody species, with the aging of the wood, a differentiation occurs, distinguishing two areas: a central one of a darker color, called *heartwood* (hard wood, the "heart" of the xylem), and another lighter in color, peripheral, called *sapwood* (Fig. 24). The formation of the heartwood consists of a process of impregnating the cell walls with organic substances (gums, resins, tannins, dyes), and sometimes also mineral salts, accompanied by the formation of tyloses (a kind of "hernias"—invaginations of parenchymal cells, passing through punctures in the lumen of the vessels) in the trachea. All of heartwood elements, including parenchyma and medullary

Fig. 24 Cross section through the trunk of beech tree (*Fagus* sp.) (original)

rays, are dead, but the substances they contain give them, for example, resistance to rotting or other alterations. The heartwood therefore no longer has a conducting role in the plant; instead, it has a high economic and technological value (Morariu 1965).

The age at which heartwood formation begins varies with the species: in *Robinia pseudoacacia* at 5 years, in oak at 20, and in ash at 40, but in poplars and willows, it is not produced at all.

Naturally, species without heartwood have soft wood and short life span, because the trunk is attacked by fungi in the central part and grinds, becoming fragile. Only in the peripheral areas the young annual rings last for a while, which they maintain—contrary to expectations—the tree is alive (scurvy willows).

The sapwood (soft wood) consists of the youngest annual rings, and it keeps its conducting function (conducts water and stores reserve substances in the living elements of the parenchyma and medullary rays) (Morariu 1965; Mason et al. 2011; Evert and Eichorn 2013).

Depending on the predominance of heartwood or sapwood within the secondary wood, there are trees with "white wood" (*Populus*, poplar; *Betula*, birch; *Acer*, maple; *Tilia*, linden trees), which do not form heartwood; these plants are also called "soft essences," with few xylem fibers and many vessels. "Strong essences" (*Quercus*, oak; *Ulmus*, elm; *Robinia*, acacia; *Juglans regia*, walnut) have a developed heartwood (dark wood), the vessels are rare, and the xylem fibers are numerous.

A mention should be made in respect with the language designating some tissues (or complexes of tissues) within the trunk of a tree (therefore, it is a stem with secondary, woody structure); this language is somewhat problematic and often even inconsistent from author to author. For example, the *outer bark* refers to the periderm (hence, phellogen assemblage, phelloderm, cork), plus possibly dead secondary phloem; the *inner bark* usually includes the living secondary phloem (sometimes the phellogen is also included here). Generically, the bark consists of the "internal bark" and the "external bark."

3 Abnormal Structures of the Stem

Compared to usual situations (at least, as they are treated didactically), the anatomical structure of the stem may suffer some "deviations" from the "normal" structure, as we have seen can also be the case with the root. Structural anomalies of the stem can mainly refer to the formation of the perimedullary (internal) phloem (from species of Cucurbitaceae) and, above all, to structural anomalies resulting from the abnormal functioning of the cambium.

Among the anomalies resulting from the abnormal functioning of the cambium, two situations can be mentioned:

(a) In certain cases, the cambium, located normally, can form in some places a larger quantity of phloem than of xylem. For example, a cross section through the stem of *Bignonia* shows phloem cones that appear to be embedded in the layers of secondary xylem (except the central region); this is due to the fact that, after functioning normally at the beginning, it then, at certain points, stops producing secondary xylem and generates, on the outside, a more abundant phloem; the cambium is then no longer circular, but follows the contour of the xylem and becomes sinuous. For the same reasons, winged stems and flattened stems can result (Toma 1977; Crang et al. 2018).

(b) Another type of anomaly regarding the functioning of the cambium determines the formation of islands of intraxylary phloem (which should not be confused with the intraxylem phloem that is formed in the cellulosic secondary xylem parenchyma of some roots). In order to produce these islands, small segments of cambium give birth to the phloem on both its faces and then resume their normal activity, giving inside xylem that includes from all sides the phloem island thus formed (*Combretum*, *Entada*). At *Strychnos*, the mechanism is slightly different: in some places, the cambium stops dividing towards the inside, giving rise only to the phloem; then new cambium segments are formed above these phloem islands and, functioning normally, connect to the edges of the general cambium segments that continued their activity in the meantime, thus encompassing the phloem islands (Toma 1977; Crang et al. 2018).

Bibliography

Barnes C. R., 1900 – *Outlines of plant life with special reference to form and function*, Henry Colt and Company, New York

Belzung E., 1900 – *Anatomie et physiologie vgtales*, Paris, Ancienne Librairie Germer Baillière et Cie Félix Alcan, Éditeur

Bonnier G., du Sablon L., 1905 – *Cours de Botanique. Phanérogames*, Paris, Librairie Générale de l'enseignement

Crang R., Lyons-Sobaski S., Wise R., 2018 – *Plant anatomy. A concept-based approach to the structure of seed plants*, Springer Nature Switzerland AG

Evert R. F., Eichorn S. E., 2013 – *Raven Biology of plants*, 8th edition, W. H. Phloemman and Company Publishers, New York

Gérardin L., Guède H., 1895 – *Traité élémentaire d'histoire naturelle. Botanique. Anatomie et physiologie végétales*, Paris, Librairie J.-B. Baillière et Fils

Grigore M.-N., 2022 – *Morfologia plantelor. Aspecte ecologice*, Editura Universității „Ștefan cel Mare" Suceava

Mason K. A., Losos J. B., Singer S. R., 2011 – *Biology*, 9th edition, McGraw-Hill, New York

Morariu I., 1965 – *Botanică generală și sistematică, cu noțiuni de geobotanică*, ed. a II-a, Editura Agro-Silvică, București

Stevens W. C., 1924 – *Plant anatomy from the standpoint of the development and functions of the tissues and handbook of micro-technic*, London J. & A. Churchill

Toma C., 1977 – *Anatomia plantelor. II. Structura organelor vegetative și de reproducere*, Centrul de multiplicare al Universității „Al. I. Cuza" Iași

Toma C., Niță M., Rugină R., Ivănescu L., Costică N., 2003 – *Morfologia și anatomia plantelor. Manual de lucrări practice*, ediția a 2-a revizuită și îmbunătățită, Editura Universității „Alexandru Ioan Cuza" din Iași

Ecological Anatomy of the Stem

We reviewed in a previous paper (Grigore 2022) some ecological aspects related to stem morphology. The anatomical structure of the stem also suggests the close connection of the plant with environmental factors, aspects that are visible, as we will see, especially in some ecological groups of plants (e.g., aquatic plants, hydrophytes, and marsh plants, helophytes).

1 Aspects of Ecological Anatomy of the Stem in Aquatic Plants (Hydrophytes)

The aquatic environment imposes on the plants some morphological characters (see also the morphological adaptations of the stem and the root—Grigore 2022), but also very particular anatomical ones (see, above, also the anatomical adaptations of the roots of aquatic plants).

We discussed the ability of aquatic plants' underground organs to form aerenchyma, in particular, but also of other plants, in general, when they are exposed to hypoxic (or even anoxic) stress, in waterlogged (flooding) conditions. In addition to the ability of aquatic plant stems to form aerenchyma, they also show other adaptations to avoid lack of oxygen. Total immersion (immersion) in water stimulates the stems of some aquatic plants to grow rapidly towards the surface of the water to reach atmospheric oxygen (Barnes 1898; Schimper 1903; Warming 1909; Coulter et al. 1911; Arber 1920; Cronk and Fennessy 2001; Schulze et al. 2005, 2019). Stems of herbaceous and woody plants sometimes swell at the base due to increased porosity (hypertrophy). The aerenchyma in the stems of many submerged and floating-leaved plants allows them to float near or on the surface of the water, where they have access to oxygen, light, and carbon dioxide.

Next, we present some general biological features of aquatic plant stems.

M.-N. Grigore, *Plant Ecological Anatomy*,
https://doi.org/10.1007/978-3-032-01332-3_5

1.1 Rapid Elongation of Submerged Stems

Rapid extension (elongation) of submerged stems has long been observed in many aquatic plants (Barnes 1898; Warming 1909; Coulter et al. 1911), such as in *Sagittaria pygmaea*, *S. latifolia*, *Nymphaea alba*, *Nymphoides peltata*, *Oryza sativa*, *Potamogeton distinctus*, *Victoria amazonica*, and species of *Ranunculus* and *Rumex* (Ridge 1987; Arteca 1997; Vartapetian and Jackson 1997). Elongation of stems (shoots) brings the plant close to or above the surface of the water, where it has access to more light, oxygen, and carbon dioxide (Jackson 1990; Laurentius et al. 1996). Rapid shoot elongation is more pronounced when species that are normally emergent—or have floating leaves—are submerged (Ridge 1987).

Jackson (1982) showed that shoots of *Callitriche platycarpa* grew an average of 7.3 cm in 4 days when submerged (as opposed to 2.5 cm for floating shoots). It must be said that, in such conditions, the leaf petioles also grow (elongate), not just the submerged stems; for example, petioles of *Regnellidium diphyllum* grew in length by an average of 4.5 cm per day when submerged, and only 0.6 cm per day when emergent (in the atmosphere) (Ridge 1987).

Stem elongation usually begins within about 30 min of flooding. It can be stimulated by an accumulation of ethylene that causes stem cells to elongate (Jackson 1990, 1994; Vartapetian and Jackson 1997; Schulze et al. 2005, 2019).

Rapid elongation of stems originating from hibernating roots, rhizomes, and tubers was observed early in the growing season in *Scirpus lacustris*, *Scirpus maritimus*, *Typha latifolia*, *Acorus calamus*, and *Potamogeton pectinatus*. Because stem growth is rapid, the plant comes into contact with oxygen and carbon dioxide before the plant's winter reserves are depleted (Vartapetian and Jackson 1997; Summers et al. 2000). The mechanism for the rapid growth of shoots from the underground parts of plants can be explained by the increased rates of glycolysis during anaerobiosis (known as the "Pasteur effect") (Summers et al. 2000).

In reality, the phenomenon of the elongation of submerged stems (and petioles) is much more complex and delicate; in fact, at least four hormones participate in the elongation of the two organs, a process triggered by hypoxia: ethylene, abscisic acid, auxin, and gibberellic acid (Schulze et al. 2005); we recall the involvement of ethylene also in the formation of the aerenchyma within the submerged organs (above, we discussed the case of the roots).

1.2 Hypertrophy

Hypertrophy is the swelling of the stem base that occurs in response to flooding in both herbaceous and woody plants. It has been observed in a number of flood-intolerant crop plants as well as in many flood-tolerant plants. Swelling is due to accelerated cell expansion caused by cell separation and rupture that occurs as aerenchyma forms and is stimulated by ethylene (Jackson 1990). Hypertrophy increases the porosity of the stem base and improves aeration. In some species, hypertrophy allows the development of adventitious roots (Arteca 1997).

Wetland trees often have swellings at the base of the trunk, called buttresses, such as, for example, in the swamp cypress (*Taxodium distichum*); this "buttress" strengthens the trunk and increases the stability of the plant in water by widening its base (Cronk and Fennessy 2001).

1.3 Stem Buoyancy

As we have shown, the aerenchyma of submerged plants not only serves as a channel for the diffusion of gases but also provides buoyancy, so that the stems remain upright in general, but in the optimal position to take up oxygen and dioxide from carbon on the water surface (Kemp et al. 1986). Viewed in this way, the aerenchyma forms the aeration system of the root and stem. We will return to aerenchyma in a separate chapter (below).

Submerged plants such as *Myriophyllum spicatum*, *Hydrilla verticillata*, and *Lagarosiphon major* form a floating network of leaf-bearing stems, also concentrating much of the biomass at the water surface (Grace and Wetzel 1998; Yeo et al. 1984; Howard-Williams 1993).

Plants with floating (natant) leaves have elongated stems that are also floating (Figs. 16 and 18) and help keep the leaves afloat. The stems are supported by the surrounding water and thus relieved of the task of keeping the plant's leaves straight. The long stems allow the leaves to spread over the surface of the water. The floating leaves can be interpreted as an adaptation to the low levels of oxygen, light, and carbon dioxide in the water column (Sculthorpe 1967).

At *Neptunia oleracea* (*Desmanthus natans*) (Fig. 1), this spongy mass that ensures the buoyancy of the stem (at the level of mature internodes) is nothing more than an aerenchyma developed from phellogen. This floating portion of the stem helps support the bipinnate leaves and flowers at the surface of the water.

Although the young internodes do not macroscopically show these floating formations, in cross section, they still have aerenchyma (Fig. 2), thus also contributing to the stem's buoyancy.

An example of stem buoyancy is also found in *Nesaea verticillata* (Fig. 3), which has a very interesting biology, to which the formation of this floating tissue is also related. Thus, in July and August, the plant bends under its own weight until the tip of the stem touches the water and then curves upwards again.

In the region of contact between the stem and the water, a swelling is formed, and from this region, roots also form, anchoring the floating part of the stem to the soil (Fig. 3). The epidermis of the swollen region becomes fissured, revealing a whitish, soft, elastic, spongy tissue that arises from a pericyclic phellogen. The contraction of the roots draws the swollen part into the water, and the spongy layer gradually expands over the submerged regions. In the fall, the long, thin stems die, except for those portions that have produced floating tissue around them and rooted in the

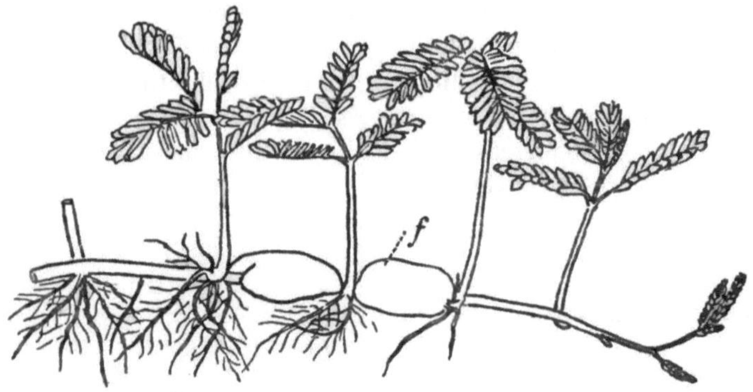

Fig. 1 *Neptunia oleracea* (*Desmanthus natans*), floating internodes (f—"internodes lined with spongy tissue") (Rosanoff 1871); on the left (old internodes) and on the right (young internodes)—these floating formations are not here observed

Fig. 2 Cross section through a young internode from *Neptunia oleracea* (*Desmanthus natans*)—the presence of aerenchyma is observed (Rosanoff 1871)

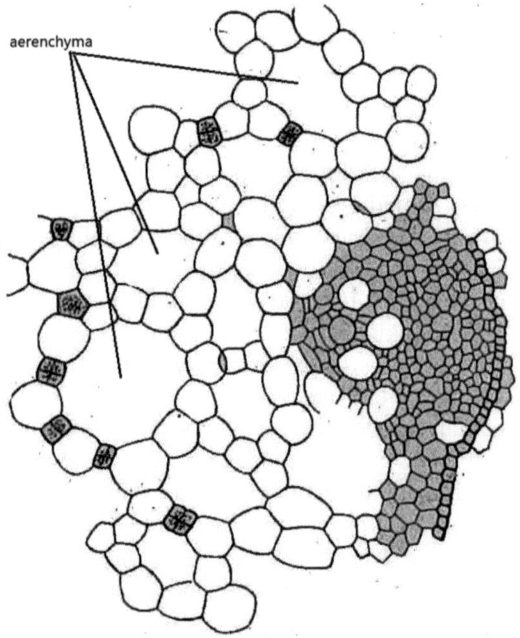

mud. In this way, a new plant develops vegetatively, sometimes at a considerable distance from the mother plant.

In the species *Jussiaea peruviana* (Fig. 4), the submerged parts of the stems are provided with aerenchyma, which also develops in normal roots that are planted in mud, and in certain erect roots that seem to serve entirely for aeration. Special respiratory roots also occur in the *Jussiaea repens*. They present, in cross section, in turn, a very small central cylinder, surrounded by a voluminous aerenchyma. And in this case, we are dealing with a modification of these roots, directly related to the aquatic

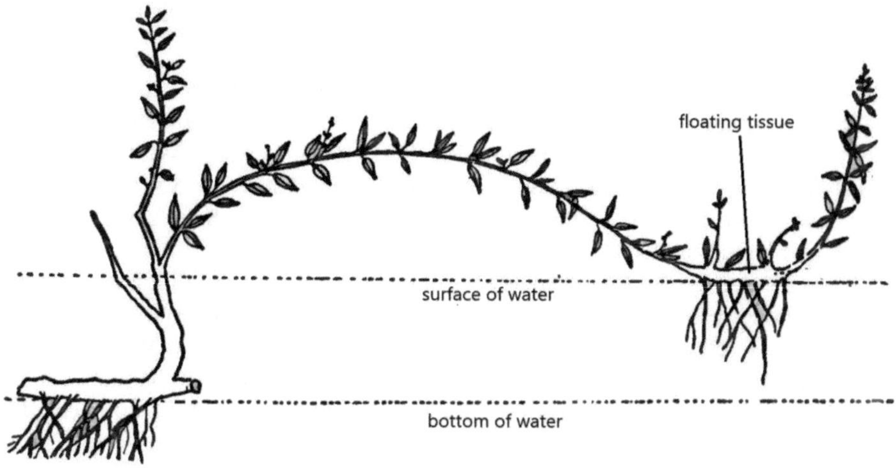

Fig. 3 Formation of floating (floating) tissue in *Nesaea verticillata* (Schrenk 1889)

Fig. 4 *Jussiaea peruviana.* (**a**) General appearance of the plant; (**b**) cross section through the submerged part of the stem, showing the development of the aerenchyma (Schenck 1889)

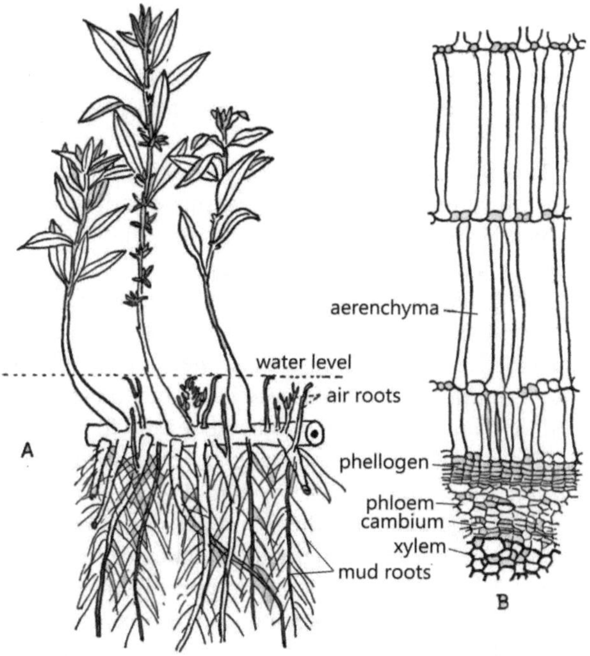

environment. In this sense, it has been demonstrated since the nineteenth century that *the Jussiaea grandiflora*, when cultivated for several years in the botanical garden at Marburg, as a terrestrial plant, produced only normal adventitious roots, but when transferred to the aquatic environment, developed roots with aerenchyma (Goebel 1891–1983).

1.4 Anatomy of Stems of Aquatic Plants: General Features

The anatomy of the vegetative organs of aquatic plants has attracted the attention of many researchers (van Tieghem 1871; Warming 1909; Coulter et al. 1911; Barnes 1898; Schenck 1886; Snell 1907; Goebel 1891–1983; Lewakoffski 1873; Batten 1918; Glück 1911; Haberlandt 1914; Schwarz 1883; Henslow 1895; Chatin 1856; Barnéoud 1848; Schimper 1903; Arber 1920; Ogden 1974; Schulze et al. 2005; Cronk and Fennessy 2001).

The vascular system of submerged stems shows certain modifications—compared to the terrestrial type, the clearest difference being that the xylem tends to be quantitatively reduced, while the lignification is often very weak. We recall that, although it is not a general, absolute rule, plants from xeric (dry) habitats tend to present organs (especially the root) with an increased level of lignification (see the case of successive cambia), while those from humid (aquatic) habitats tend to show very weak lignification (Fig. 5—general view).

The spiral or annular vessels, when present in the vicinity of the apex, may in some cases be completely destroyed by the elongation of the internodes and may survive only at the nodes, e.g., in *Potamogeton lucens*, *Zannichellia palustris*, and *Althenia filiformis*, while in *Elodea canadensis*, the tracheal thickenings do not even persist in the nodal tissues (Arber 1920). *Ceratophyllum* (Fig. 6) affords an example of a further degree of reduction, as here lignification is entirely absent, even in the apical region.

This reduction in the degree of lignification has sometimes been regarded as a generalization of the idea that transpiration flow does not exist in submerged plants (Arber 1920). In reality, it appears that the functional importance of xylem is not directly proportional to its degree of lignification.

Elongated, submerged stems, unless growing in completely stagnant water, must be subjected to some tension from water currents. It is likely—more than mere coincidence—that the vascular system of aquatic plants is so densely condensed into a central cord, reminiscent of the central cylinder of roots and climbing stems, both of which are organs subject to tensile forces.

The structure of the central axis, even if it is extremely simple, as in species of *Callitriche*, *Hydrillus*, *Aldrovandia* (Fig. 7), *Najas*, and *Hippuris*, does not represent a single conducting bundle, but represents an entire vascular system, in which the bundles are not individually differentiated. According to Arber (1920), the reduction of xylem does not represent, in itself, the effect of uniting the isolated bundles in an axial cord. This statement can also be deduced from the comparison with the stems of saprophytic or colorless parasitic plants; in these, there is little transpiration flow and no photosynthetic assimilation, and the xylem is proportionately reduced. But the simplified bundles retain their ancestral position and do not merge into an axial cord (Arber 1920).

In some aquatic species of *Ranunculus* (Figs. 8 and 9), the bundles remain perfectly separated, but in most species, a certain degree of their condensation can be observed.

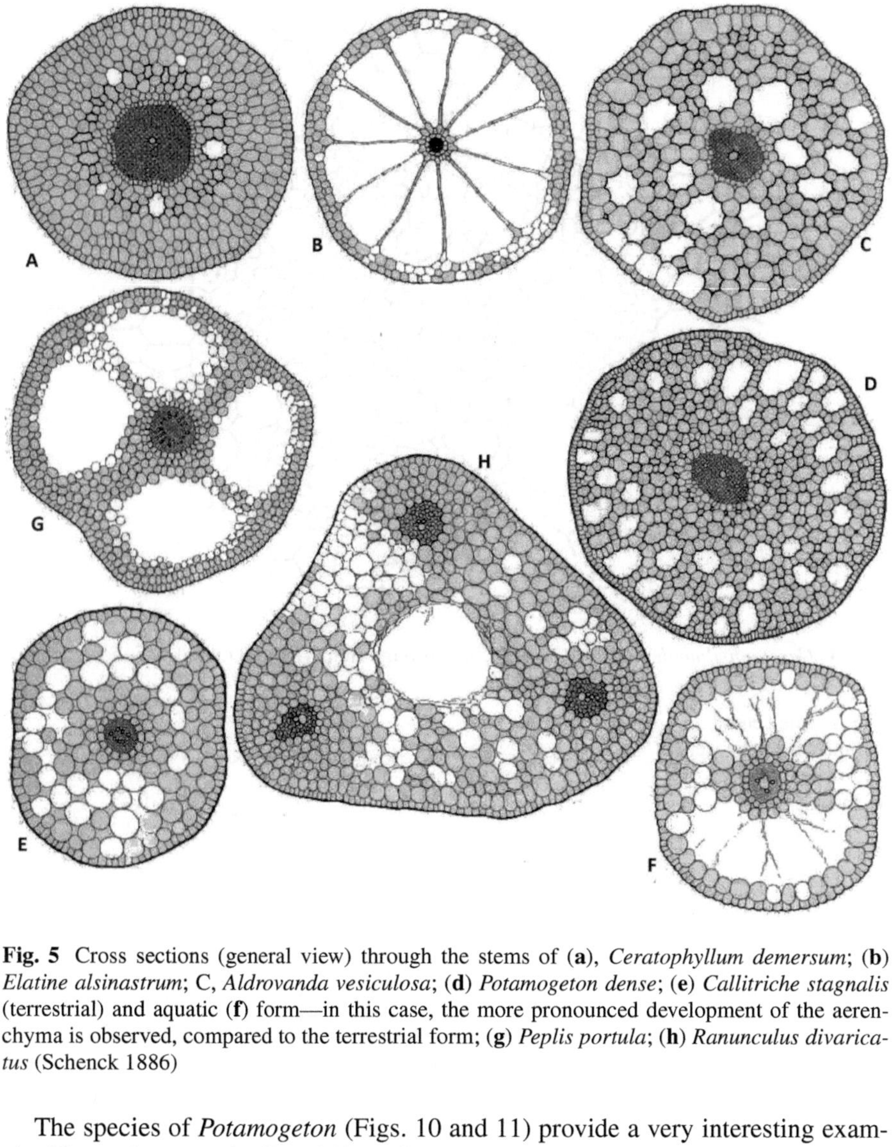

Fig. 5 Cross sections (general view) through the stems of (**a**), *Ceratophyllum demersum*; (**b**) *Elatine alsinastrum*; C, *Aldrovanda vesiculosa*; (**d**) *Potamogeton dense*; (**e**) *Callitriche stagnalis* (terrestrial) and aquatic (**f**) form—in this case, the more pronounced development of the aerenchyma is observed, compared to the terrestrial form; (**g**) *Peplis portula*; (**h**) *Ranunculus divaricatus* (Schenck 1886)

The species of *Potamogeton* (Figs. 10 and 11) provide a very interesting example, illustrating, practically within a single genus, the stages of condensation of the vascular cylinder. But there are other typical examples, which show different degrees in the reduction of the vascular system.

At *Peplis portula*, there is a well-differentiated axial cord, in which the individual bundles can no longer be distinguished (Fig. 12). In cross section, an outer ring of disconnected phloem groups can be seen surrounding a ring of xylem, consisting of short radial rows of vessels separated by rows of parenchyma. The internal phloem characteristic of species of the Lythraceae family is developed within the xylem, also forming a pith. A cambial layer appears, but with little activity.

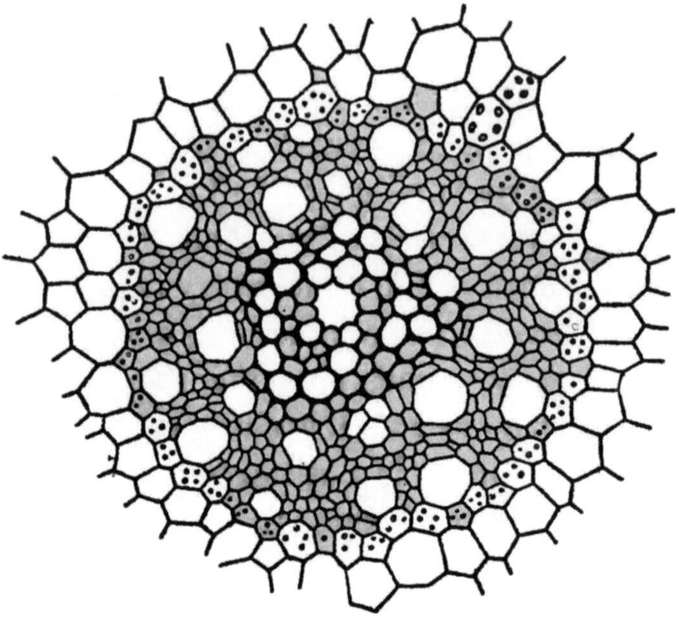

Fig. 6 *Ceratophyllum demersum*. Cross section through stem—central cylinder (Schenck 1886)

Fig. 7 The central cylinder from the stem of *Aldrovanda vesiculosa* (Schenck 1886)

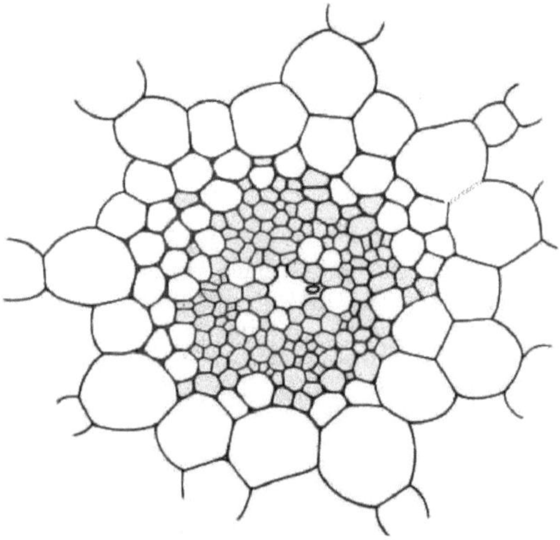

The next stage of reduction may be illustrated by the stem of *Callitriche* (Fig. 13), showing in cross section a small ring of xylem surrounded by phloem; there is no cambium. In the aquatic forms (Fig. 13a, b), the pith is resorbed at an early stage and is represented by a medullary lacuna.

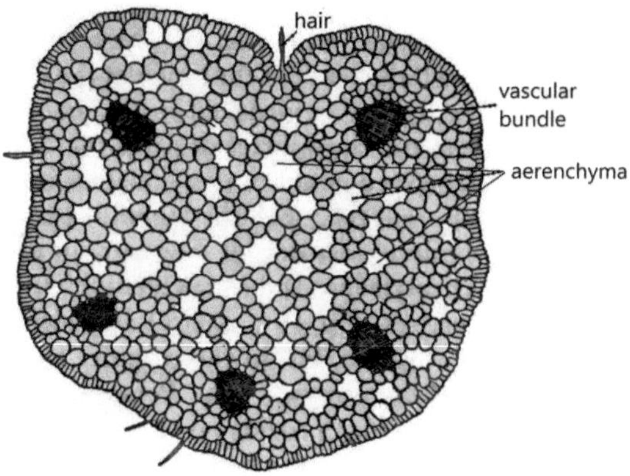

Fig. 8 Cross section through the stem of *Ranunculus trichophyllus* (Arber 1920)

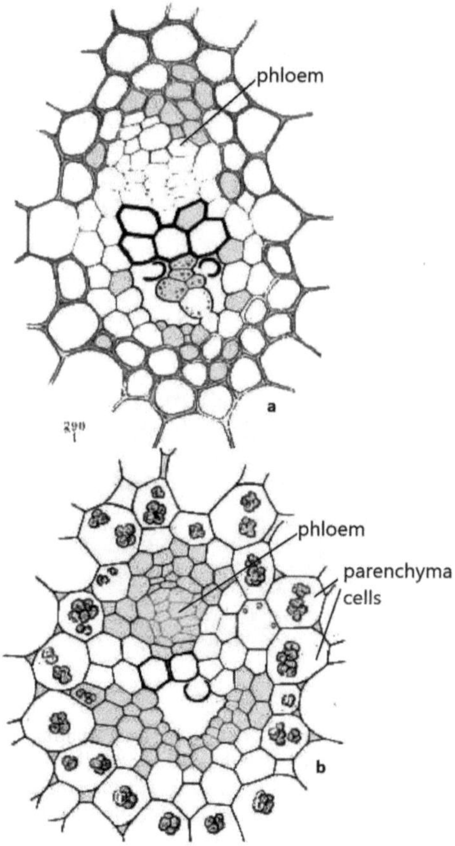

Fig. 9 The central cylinder from the stem of *Ranunculus aquatilis* (**a**) and *Ranunculus divaricatus* (**b**) (Schenck 1886)

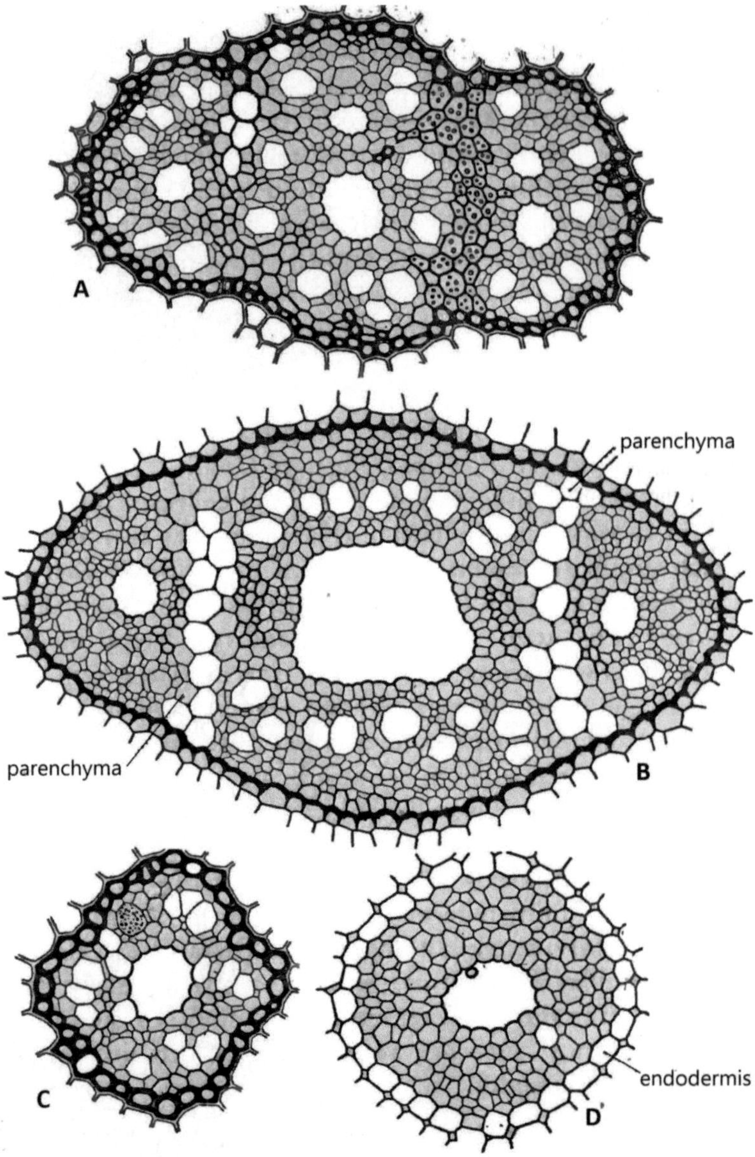

Fig. 10 Series of the reduction of the central cylinder in the stem of *Potamogeton*: (**a**) *P. crispus*; (**b**) *P. lucens*, in which the fusion has gone further, so that each of the three groups of bundles has only one xylem; (**c**) *P. pusillus*, in which the xylems of all individual bundles form a single central lacuna; (**d**) *P. pectinatus*, completely concentric structure in which any trace of the component bundles is lost (Schenck 1886)

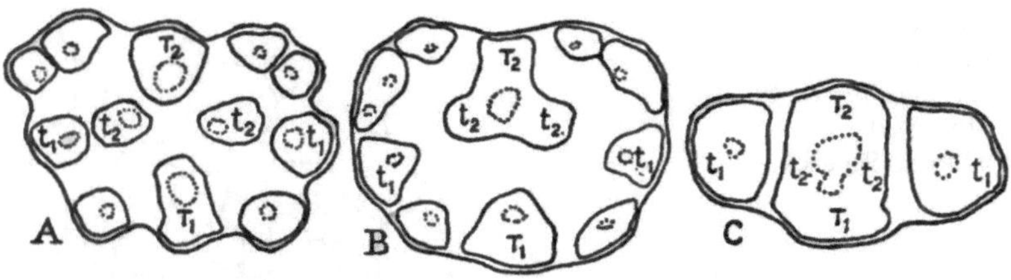

Fig. 11 Diagram (cross section)—the central cylinder of the stem in three species of *Potamogeton*, to show the reduction and fusion of groups of vascular bundles; t1, T1, t1, leaf traces of the following (upper) leaves; t2, T2, t2, traces of the second upper leaf (**a**, *Potamogeton pulcher*; **b**, *P. natans*; **c**, *P. crispus*) (Chrysler 1907)

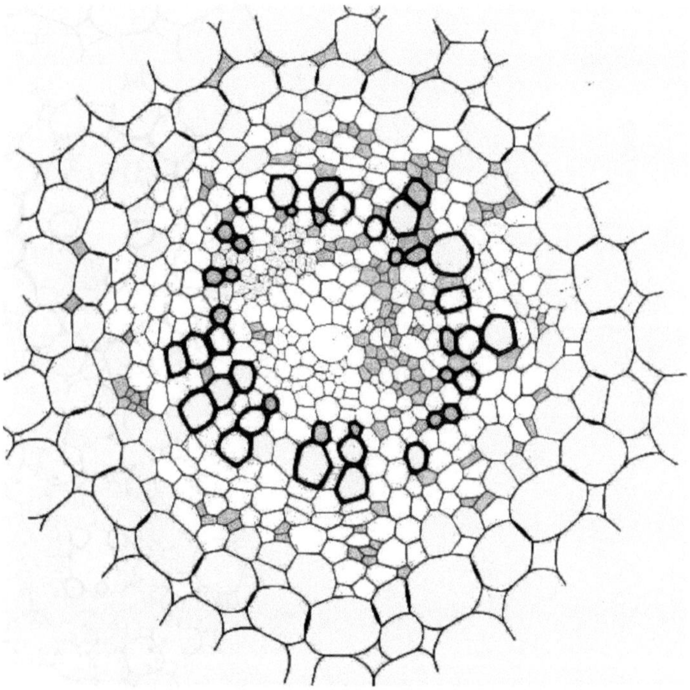

Fig. 12 The central cylinder from the stem of *Peplis portula* (Schenck 1886)

The genus *Hippuris* registers an even more advanced stage in the direction of this specialization. The vascular tissue is concentrated in a well-defined cylinder with external phloem and internal xylem, enclosing what appears at first sight to be a pith.

Myriophyllum (Fig. 14) closely resembles *Hippuris* in vascular anatomy and has the same tracheal elements in the pith, but the xylem is more reduced.

The genus *Ceratophyllum* (Fig. 15) may be regarded as representing the most advanced stage in the characteristic simplification of the stem anatomy of dicot water plants. There is a central canal, surrounded by elements whose walls are

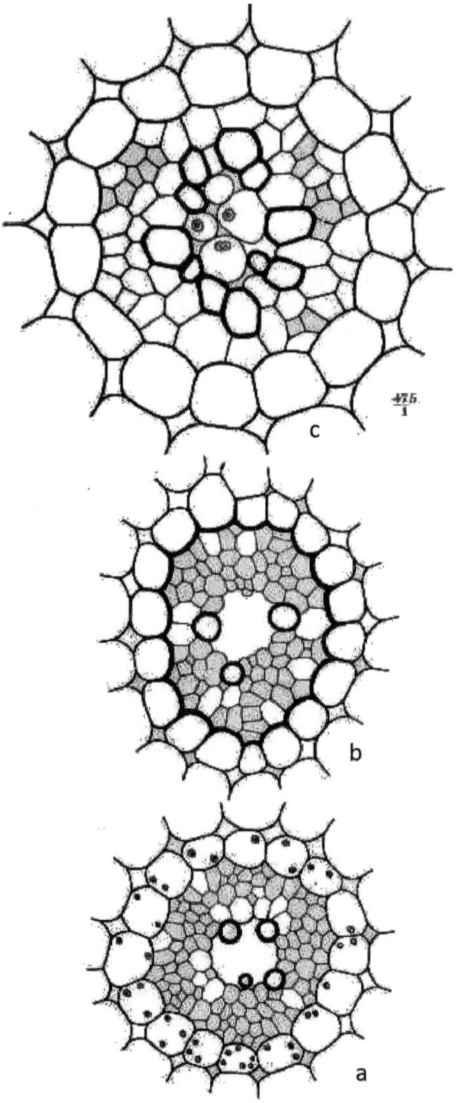

Fig. 13 The central cylinder from the stem of *Callitriche stagnalis*. (**a**, **b**) Aquatic form; (**c**) terrestrial form (Schenck 1886)

somewhat thickened, but are only cellulosic. These thick-walled cells are again surrounded by a wide area of phloem.

Therefore, one can speak of the strong tendency shown by aquatic plants to condense the vascular system to a single central cylinder (single central cylinder), devoid of secondary thickening, and in which the individual bundles cannot be distinguished (Fig. 15).

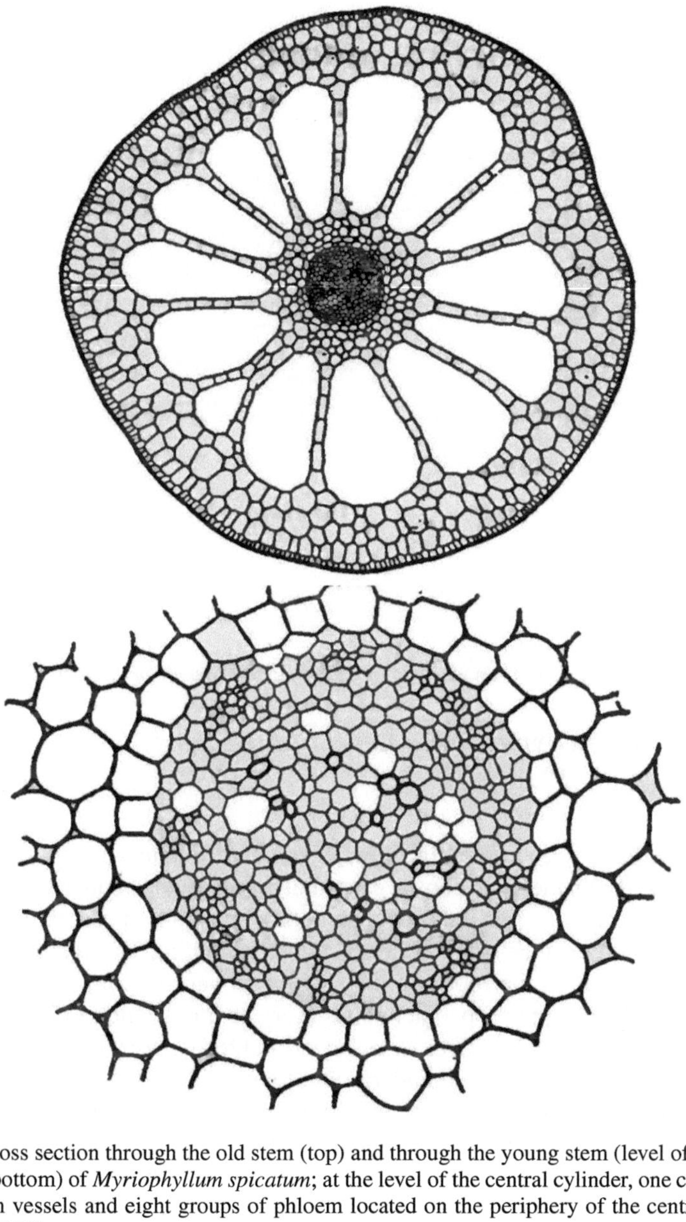

Fig. 14 Cross section through the old stem (top) and through the young stem (level of the central cylinder) (bottom) of *Myriophyllum spicatum*; at the level of the central cylinder, one can see scattered xylem vessels and eight groups of phloem located on the periphery of the central cylinder (Vöchting 1872)

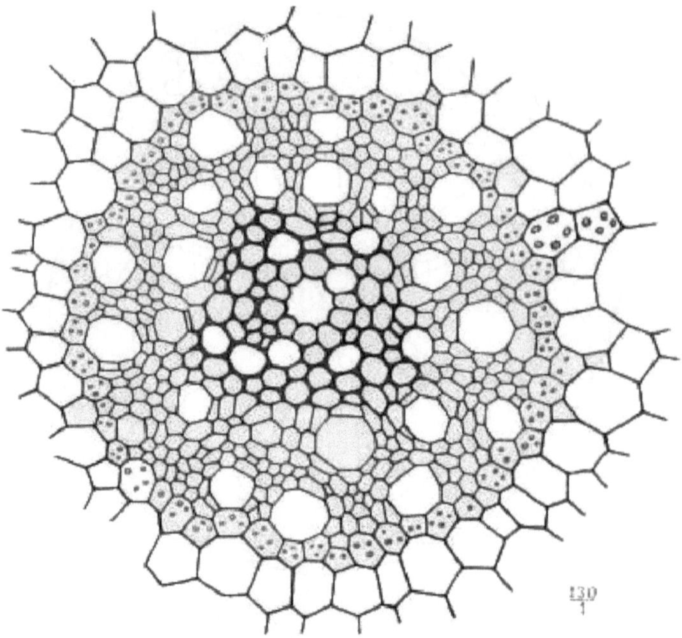

Fig. 15 The central cylinder from the stem of *Ceratophyllum demersum* (Schenck 1886)

1.5 The Aerenchyma at the Level of the Stems of Aquatic Plants

As we have already seen, the aerenchyma of submerged plants not only serves as an internal network for the diffusion of gases but also provides buoyancy, so that the stems remain upright, in general, but in an optimal position to take oxygen and carbon dioxide to the surface of the water (Kemp et al. 1986). The aerenchyma forms the aeration system of the root and the stem (it is very important in this case to consider both the normal proper stem and—very characteristic of plants from wet, aquatic environments, thus proper water plants and for swampy environments— the rhizomes, those metamorphosed underground stems).

The existence of a highly developed system of intercellular spaces therefore represents one of the most striking anatomical characters of aquatic plants (Arber 1920; Cronk and Fennessy 2001). This system of lacunae, which serves to store oxygen and transport it to the parts of the plant's body that have a special need for it, is found mainly in roots and rhizomes that are submerged in mud that provides hypoxia/anoxia conditions. Also, the mesophyll of the leaf, the basal tissue of the petiole, and the cortex of the stem and root are the regions where aerenchyma reaches its greatest development.

In the stem, cortex, which is generally large in relation to the central cylinder, is pierced by lacunae, which may be so numerous as to render the whole organ extremely fragile in texture (Arber 1920).

There are, however, two features in the arrangement of the cells of the cortex, which contribute to confer a certain degree of strength.

In the first place, the air spaces are mainly confined to the middle cortex, while the outer one consists, in many cases, of elements that are closer together and thus form a firmer (harder) peripheral layer; the septa, again, are arranged radially and are thus able to withstand pressures acting at right angles to the axis, which might otherwise break the stem.

Secondly, mechanical support is also obtained through the so-called diaphragms (Fig. 16), which appear mainly at the nodes, which divide the air spaces into sections (Fig. 17); these diaphragms are not air-tight, but are more or less water-tight, so as to form a protection against the flooding of the whole ventilation system in case of accidental injury.

As in the case of the root, the formation of the aerenchyma can be schizogenous (through the separation of cells) or lysigenic (destruction, death of the cells). In the case of schizogeny, the air spaces can be arranged in a single ring (layer) (as in *Myriophyllum*—see Fig. 14) or several rings, giving the cross section a lacy appearance (as in *Hippuris vulgaris*—Fig. 18).

At *Stratiotes aloides*, the result of the process of formation of air gaps is that the middle cortex, as it appears in cross section, consists of radial plates of cells, like the

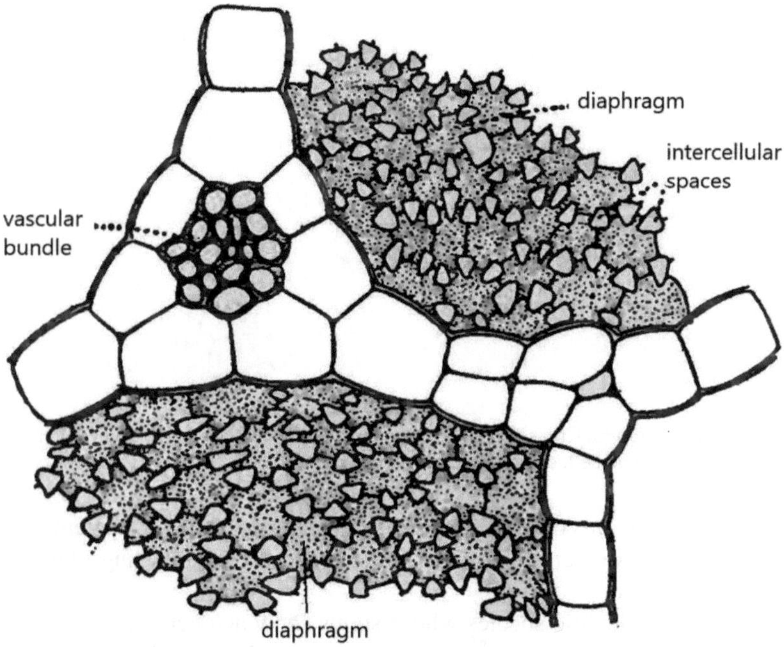

Fig. 16 Cross section—portion of stem of *Potamogeton natans* (Blanc 1912)

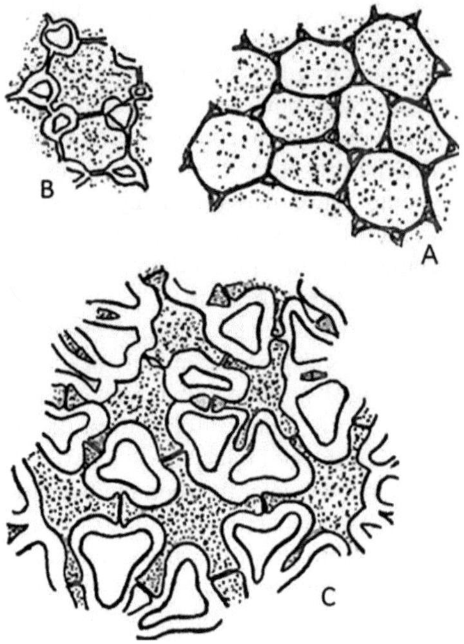

Fig. 17 *Hippuris vulgaris*. Three stages in the development of stem nodal diaphragms, seen in cross section. (**a**) Young stem, small intercellular spaces, and barely thickened walls; (**b**) the same strain as in **c**, but from a region 35 mm in diameter, indicating intermediate traits; (**c**) old stem (7 mm in diameter); the intercellular spaces are so widened that the cells are stellate and the walls greatly thickened (Arber 1920)

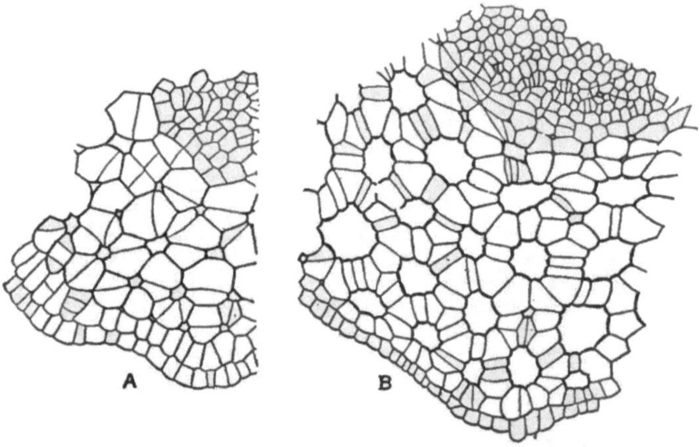

Fig. 18 *Hippuris vulgaris*. Cross sections through a younger stem (**a**) and an older stem (**b**) showing the origin of air lacunae located in the cortex (Barratt 1916)

spokes of a wheel, in contact or separated by gaps, while in tangential section (Fig. 19), the plates meet at certain intervals with the neighboring plates, so that they form a network.

The number of aquatic species whose stems show aerenchyma is very high; an extensive study of the anatomy of aquatic plants in New York State revealed the presence of aerenchyma in the following species: *Acorus calamus* (aerial stem and

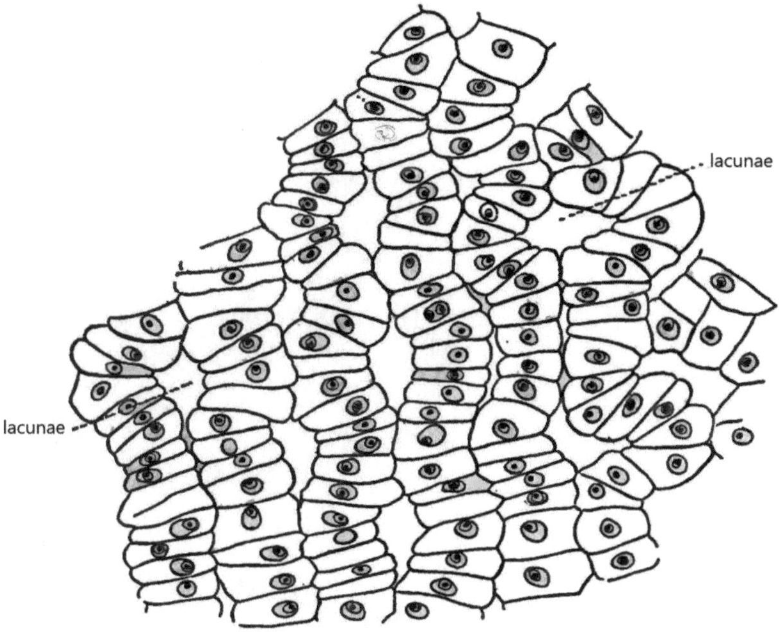

Fig. 19 Cross section through the aerial stem of *Stratiotes aloides* (Arber 1920)

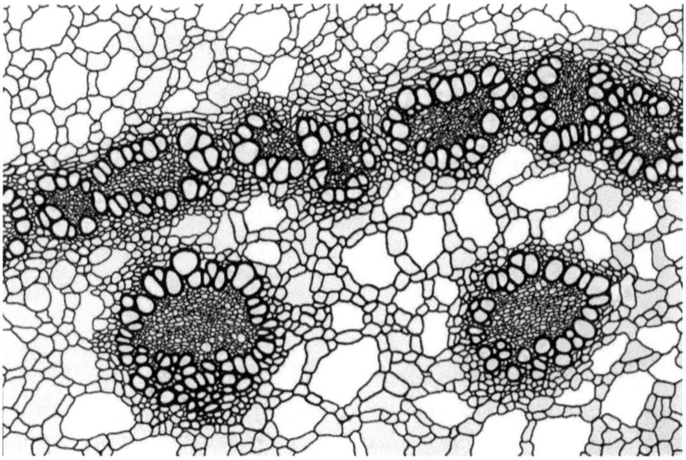

Fig. 20 Cross section through the rhizome of *Acorus calamus* (Ogden 1974)

rhizome—Fig. 20), *Armoracia aquatica*, *Bidens beckii* (Figs. 21 and 22), *Brasenia schreberi* (Fig. 23), *Cabomba caroliniana* (Fig. 24), *Calla palustris* (Fig. 25), *Caltha palustris*, *Carex comosa*, *Ceratophyllum demersum*, *Cladium mariscoides*, *Dulichium arundinaceum*, *Eleocharis obtusa* (Fig. 26), *E. palustris*, *Equisetum hyemale*, *E. variegatum*, *E. scirpoides*, *Eriocaulon septangulare* (Fig. 27),

Fig. 21 Cross section through the stem (above water) of *Bidens beckii* (Ogden 1974)

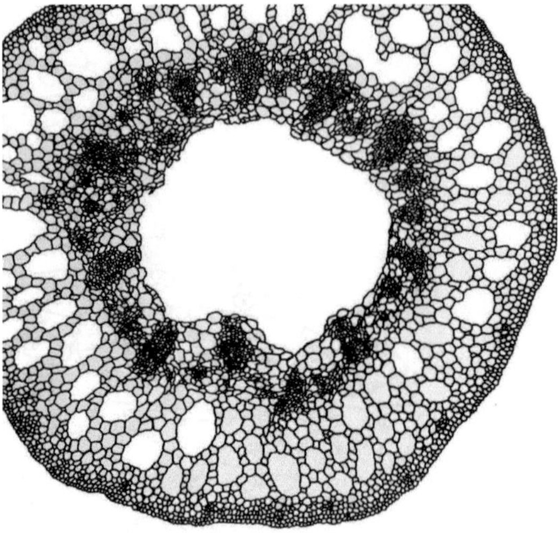

Fig. 22 Cross section through the submerged stem of *Bidens beckii* (Ogden 1974); in this case, the more important development of the aerenchyma is observed

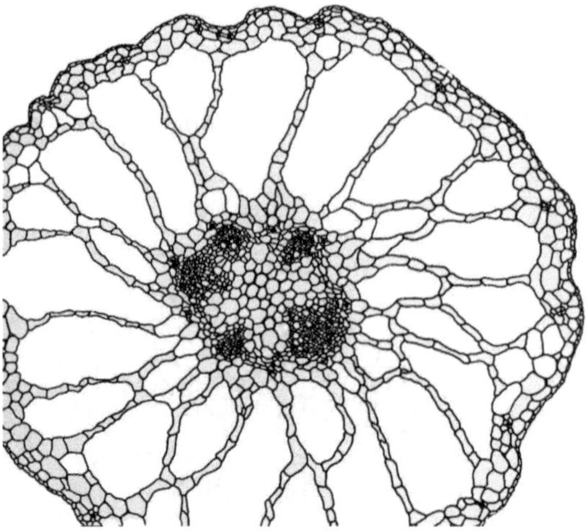

Eriophorum vaginatum, *Fuirena squarrosa*, *Gratiola aurea* (Fig. 28), *Heteranthera dubia*, *Hippuris vulgaris*, *Hydrocotyle umbellata* (Fig. 29), *Hypericum ellipticum*, *Juncus articulatus*, *Juncus effusus*, *Justicia americana*, *Lindernia dubia* (Fig. 30), *Ludwigia palustris*, *Lysimachia terrestris* (aerial stem and rhizome—Fig. 31), *Marsilea quadrifolia* (Fig. 32), *Menyanthes trifoliata*, *Myriophyllum brasiliense* (Fig. 33), *M. spicatum*, *Najas flexilis* (Fig. 34), *Nasturtium officinale*, *Nymphoides cordatum* (Fig. 35), *N. peltatum* (Fig. 36), *Pontederia cordata*, *Potamogeton crispus*, *P. foliosus*, *P. gramineus*, *P. natans*, *P. pectinatus*, *P. perfoliatus*, *P. pulcher*,

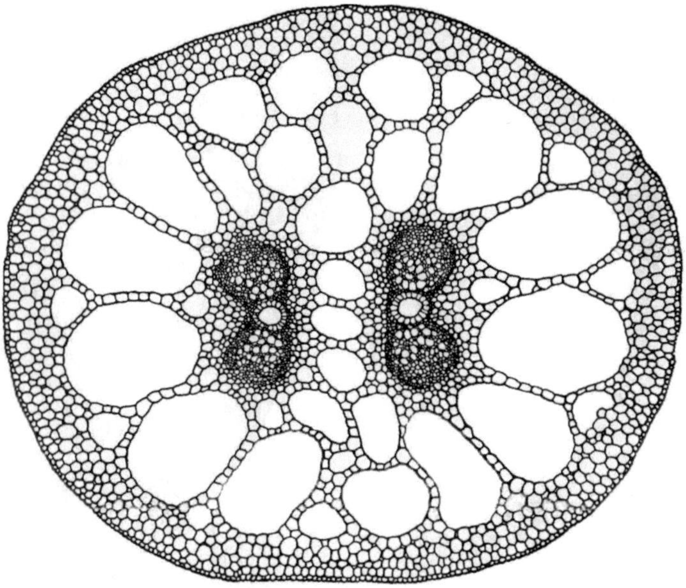

Fig. 23 Cross section through the stem of *Brasenia schreberi* (Ogden 1974)

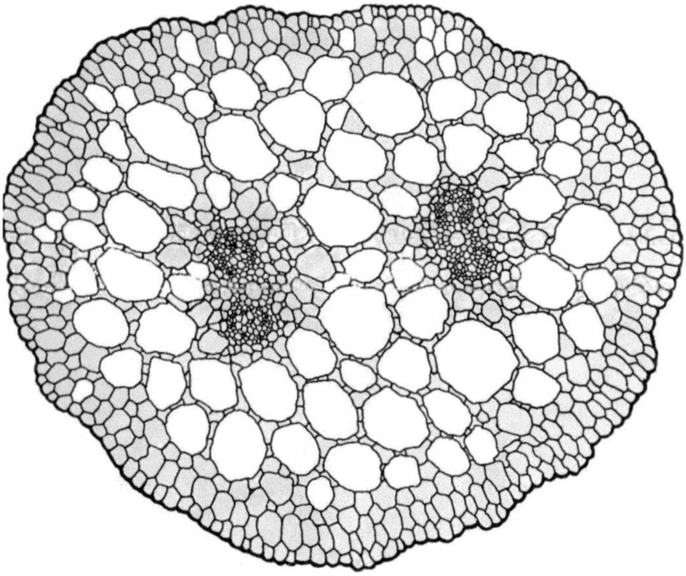

Fig. 24 Cross section through the stem from *Cabomba caroliniana* (Ogden 1974)

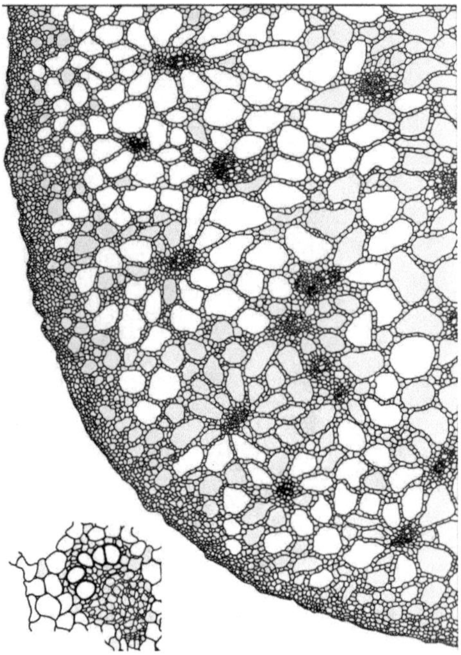

Fig. 25 Cross section through the stem of *Calla palustris*; left, bottom—detail of vascular bundle (Ogden 1974)

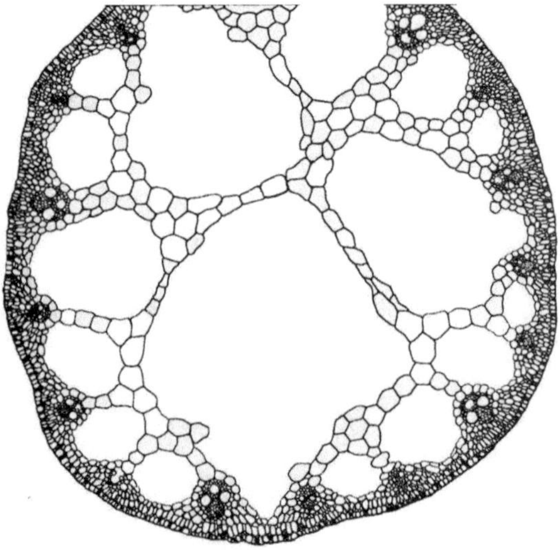

Fig. 26 Cross section through the stem of *Eleocharis obtusa* (Ogden 1974)

P. zosteriformis, *Proserpinaca palustris*, *Rhynchospora alba*, *Ruppia maritima*, *Sagittaria latifolia* (rhizome—Fig. 37), *Scheuchzeria palustris*, *Scirpus americanus* (aerial stem and rhizome), *Sparganium americanum*, *S. angustifolium*, *Trapa natans* (Fig. 38), *Triadenum fraseri*, *Utricularia vulgaris*, *Vallisneria americana*, and *Zannichellia palustris* (Ogden 1974) (Fig. 39).

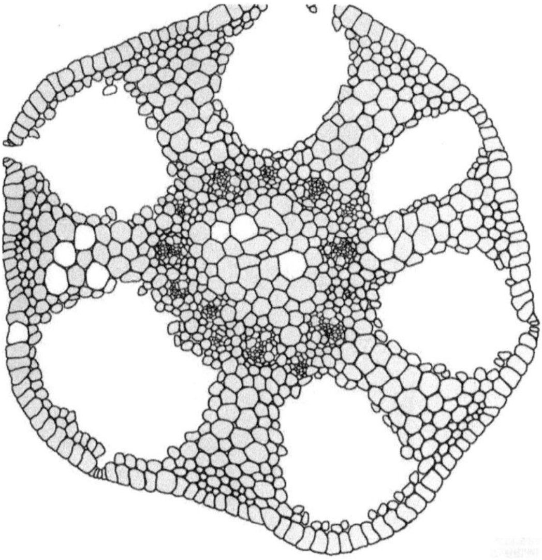

Fig. 27 Cross section through the stem of *Eriocaulon septangulare* (Ogden 1974)

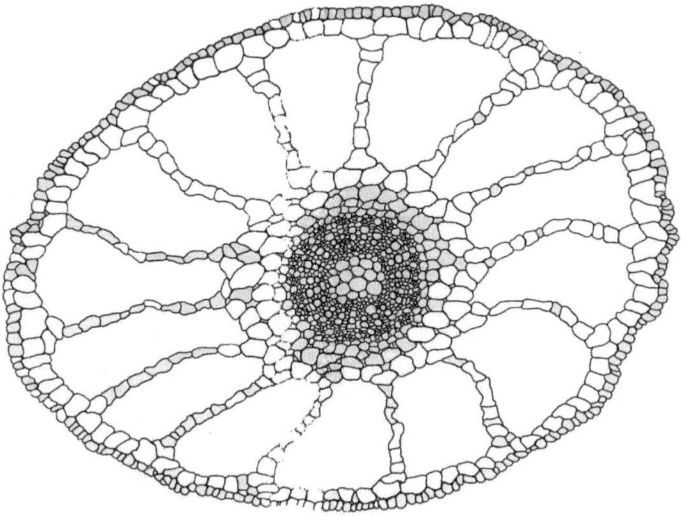

Fig. 28 Cross section through the stem of *Gratiola aurea* (Ogden 1974)

The presence of aerenchyma at the stem level has also been confirmed in some salt marsh species such as *Crithmum maritimum* (Fig. 40), *Glaux maritima* (Fig. 41), *Aster tripolium* (rhizome, Figs. 42 and 43, and aerial stem, Fig. 44), *Triglochin maritima* (Fig. 45), and *Juncus gerardii* (Fig. 46), but also in *Cymodocea nodosa* (Fig. 47), a proper aquatic plant (Grigore et al. 2014).

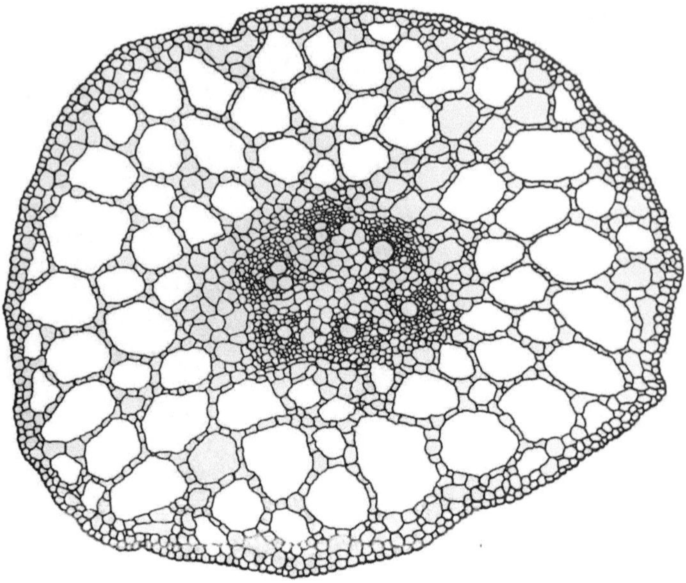

Fig. 29 Cross section through the stem of *Hydrocotyle umbellata* (Ogden 1974)

Fig. 30 Cross section through the stem of *Lindernia dubia* (Ogden 1974)

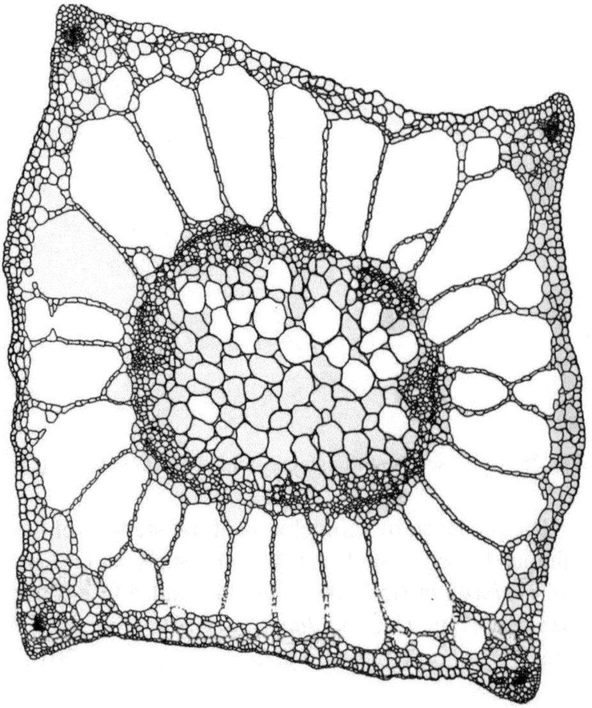

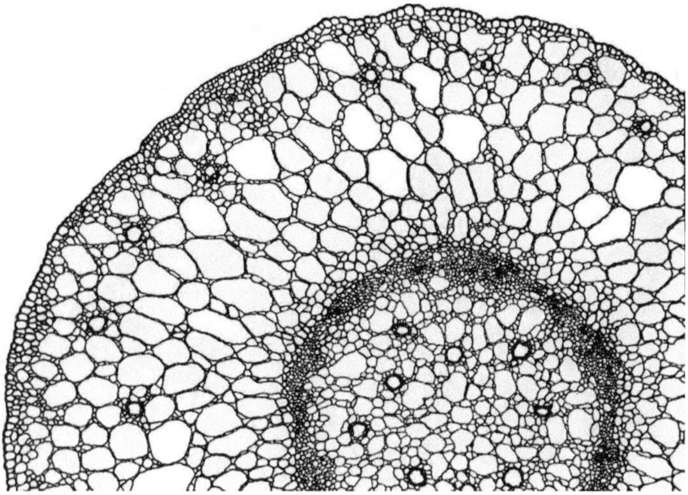

Fig. 31 Cross section through the rhizome of *Lysimachia terrestris* (Ogden 1974)

Fig. 32 Cross section through the rhizome of *Marsilea quadrifolia* (Ogden 1974)

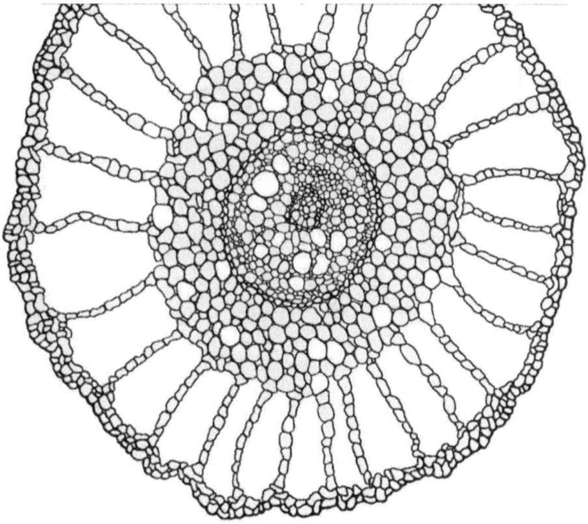

2 Ecological Anatomy of the Stem in Lianas Species

Lianas' stem morphology—along with lianas biology in general—was discussed in a previous paper (Grigore 2022). As then stated, the term "liana" is very versatile; the term liana has been used and still used here in the broadest sense, referring to voluble plants as well as various other types of climbing plants. Lianas owe their origin to the grouping of plants in communities in the form of forests and bushes; the shade due to dense vegetation has caused them in the past to elongate, to

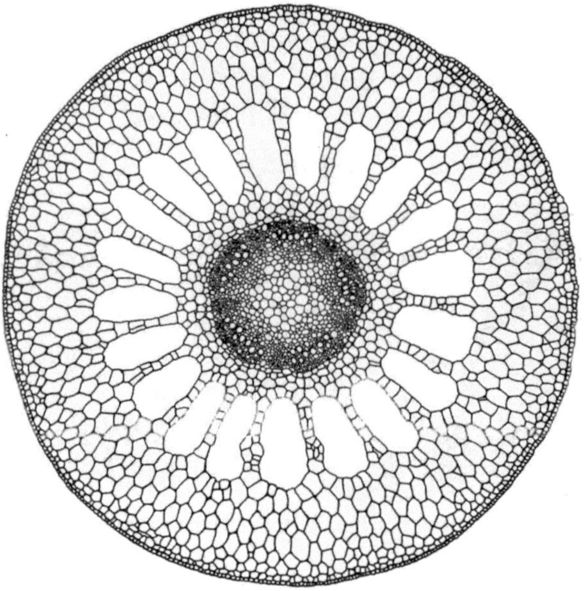

Fig. 33 Cross section through the stem of *Myriophyllum brasiliense* (Ogden 1974)

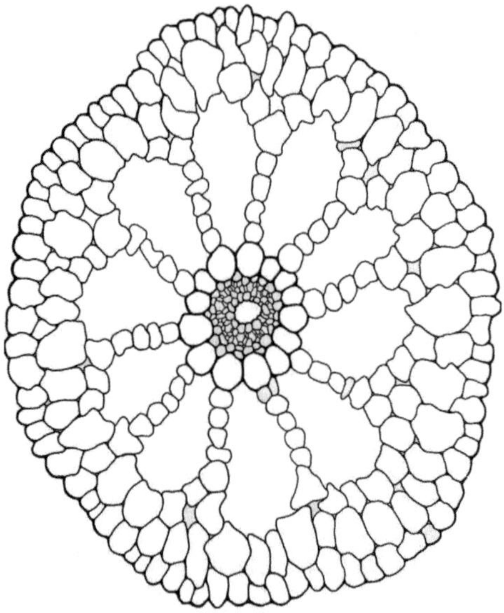

Fig. 34 Cross section through the stem of *Najas flexilis* (Ogden 1974)

produce long internodes, and in the course of time to adapt in various ways not only to fix themselves firmly on their supports but also—by an adequate internal structure—to resolve the problem of conducting sap over long distances, as well as to answer other problems due to the length and thinness of their stems (Grigore 2022).

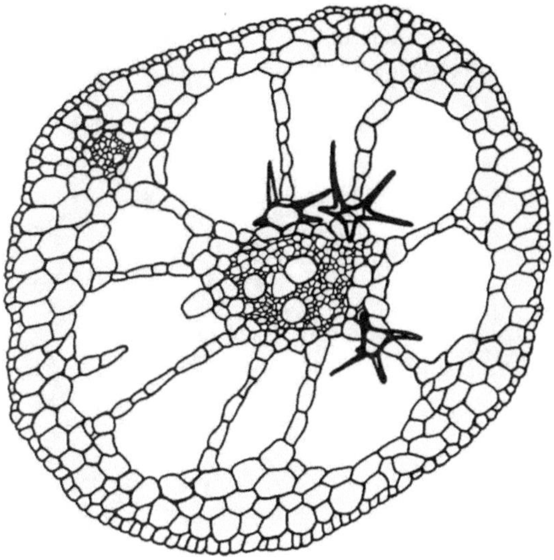

Fig. 35 Cross section through the stem of *Nymphoides cordatum* (Ogden 1974)

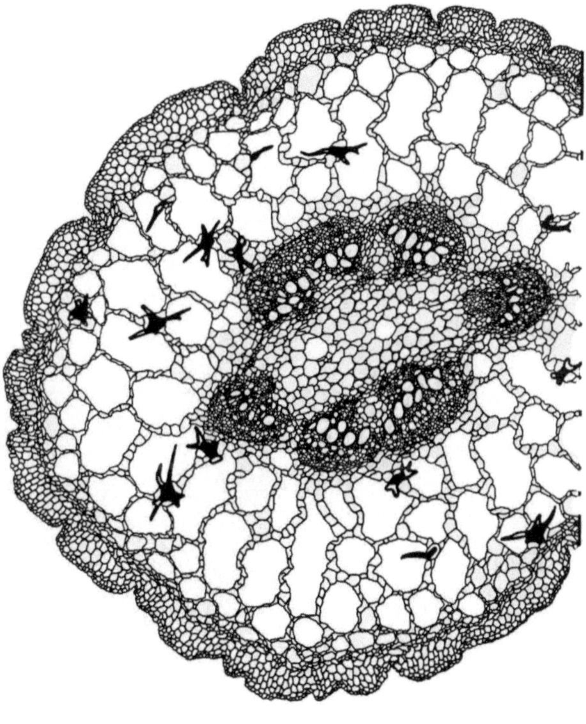

Fig. 36 Cross section through the stem of *Nymphoides peltatum* (Ogden 1974)

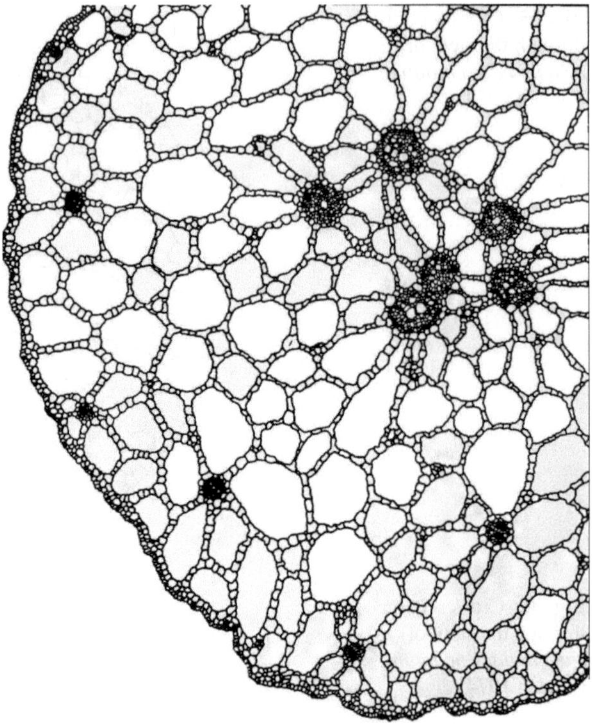

Fig. 37 Cross section through the rhizome of *Sagittaria latifolia* (Ogden 1974)

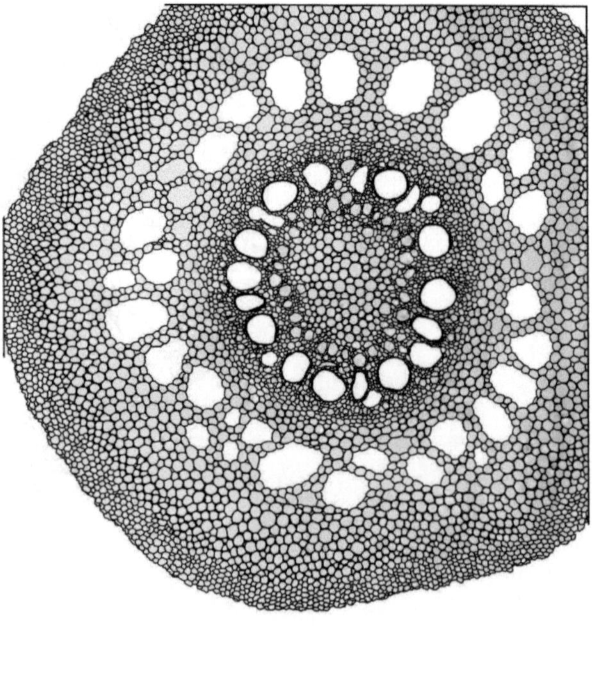

Fig. 38 Cross section through the stem of *Trapa natans* (Ogden 1974)

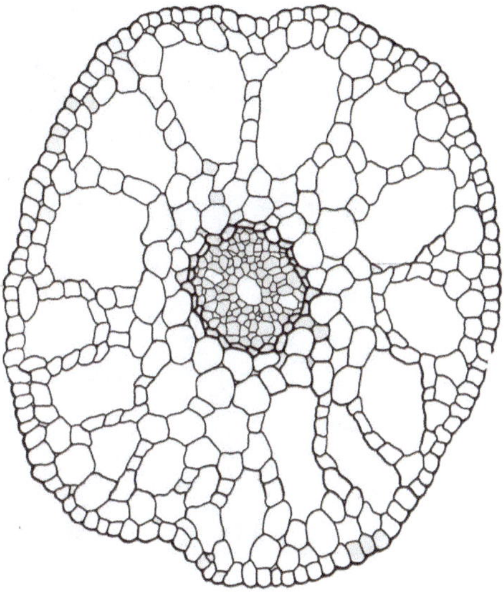

Fig. 39 Cross section through the stem of *Zannichellia palustris* (Ogden 1974)

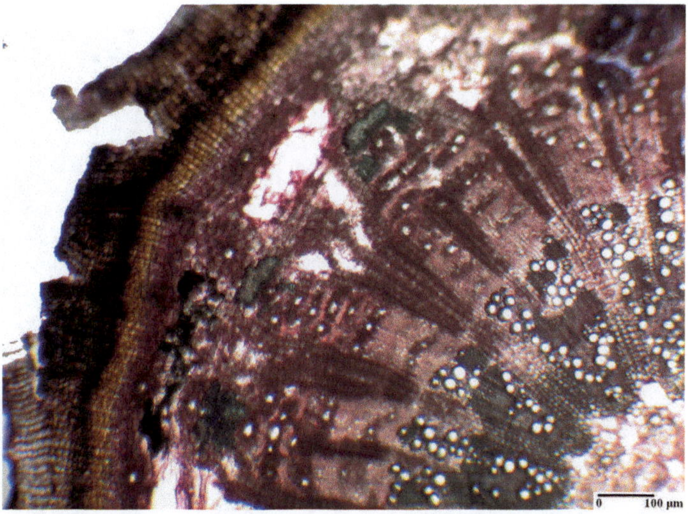

Fig. 40 Cross section through the rhizome of *Crithmum maritimum* (Grigore et al. 2014)

In a narrower but operational sense, lianas comprise, in general terms, perennial plants with extremely long stems that normally need mechanical support on other species (Ellenberg et al. 1991).

From an ecological point of view, broadly speaking, the stem of a liana is still built on the same principle: its xylem is not compact and uninterrupted as in the stem of a tree, but rather cracked in various ways or even subdivided into isolated cords (sectors).

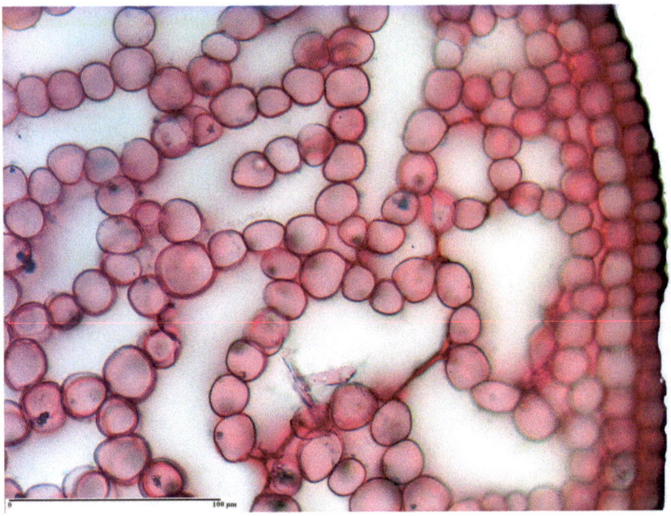

Fig. 41 Cross section through the stem of *Glaux maritima* (Grigore et al. 2014)

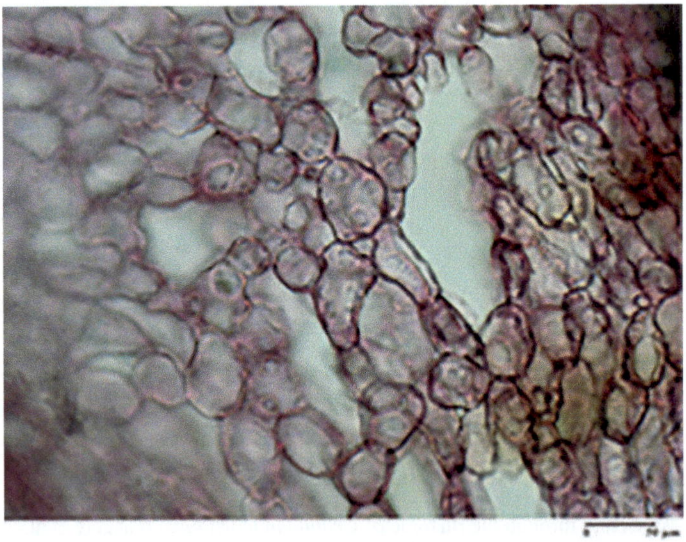

Fig. 42 Cross section through the rhizome of *Aster tripolium* (Grigore et al. 2014)

Therefore, as discussed above, the anatomical structure of the stem of lianas undergoes various "anomalies."

Lianas are characterized, in addition, by the great length and width of the conducting elements (xylem vessels and sieve tubes), by means of which the transport of the sap along the stem, often extremely long (Fig. 48), is facilitated (van Tieghem 1871; Schenck 1893; von Marilaun 1902; Solereder 1908; Haberlandt 1914;

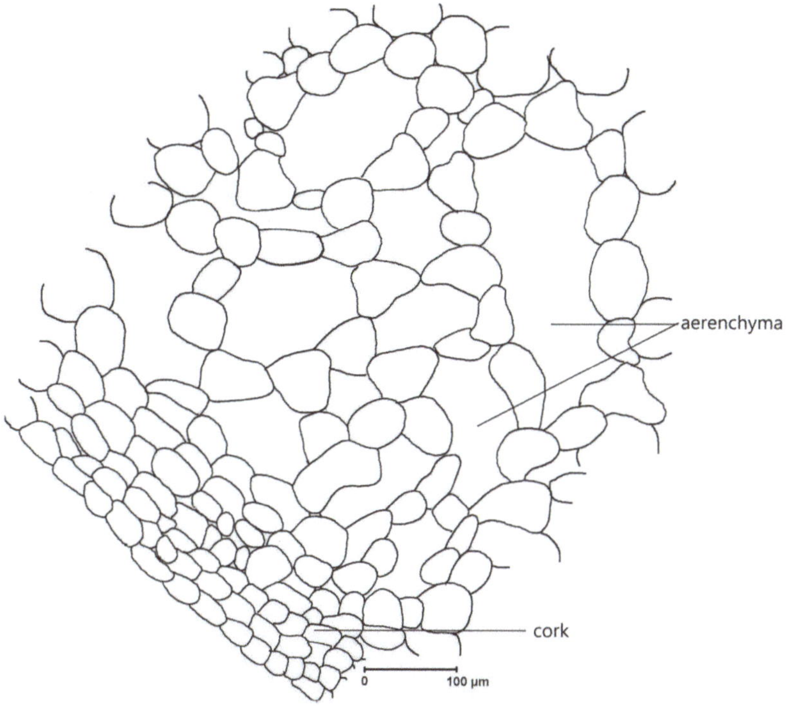

Fig. 43 Cross section (general view) through the rhizome of *Aster tripolium* (Grigore et al. 2014)

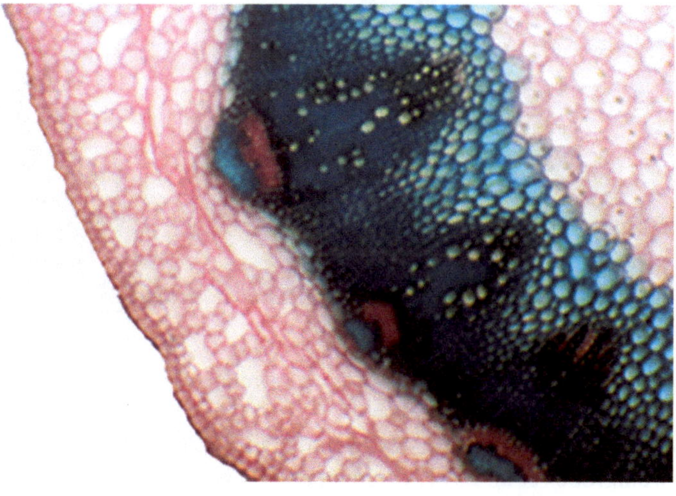

Fig. 44 Cross section through the aerial stem (basal part) of *Aster tripolium* (Grigore et al. 2014)

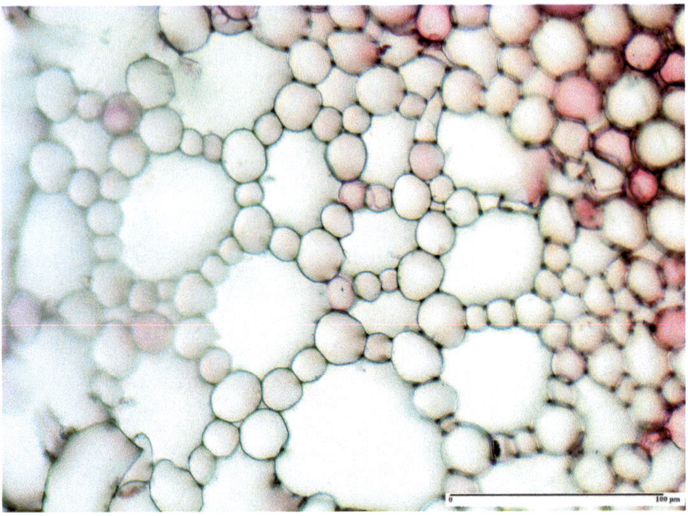

Fig. 45 Cross section through the stem of *Triglochin maritima* (Grigore et al. 2014)

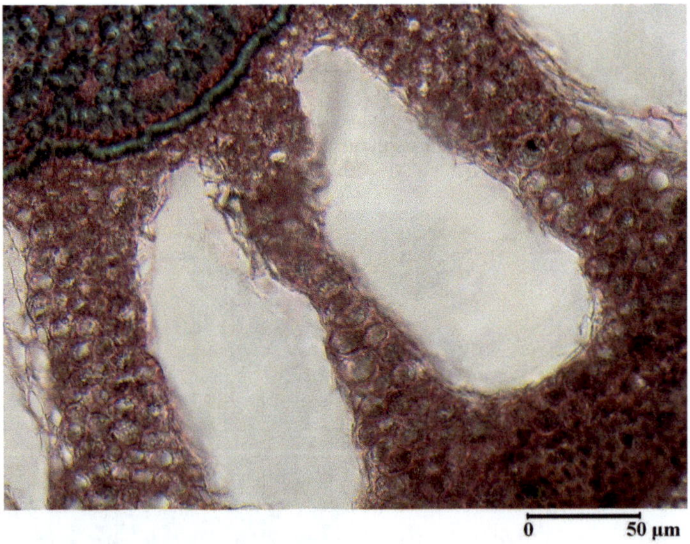

Fig. 46 Cross section through the rhizome of *Juncus gerardii* (Grigore et al. 2014)

Strasburger et al. 1894; Walter 1974; Schulze et al. 2019; Holzman 2008; Lüttge 2008; Schweingruber et al. 2011; Crang et al. 2018).

The mechanical requirements of stems of lianas are quite different from those of the usual trunks and their branches, which are, from an anatomical point of view, always built up in the direction of inflexibility (rigidity). On the one hand, a climbing stem requires, above all, to be inextensible (see Grigore 2022, for the diversity of climbing stems); when hang loose, it must support its own weight; then when it

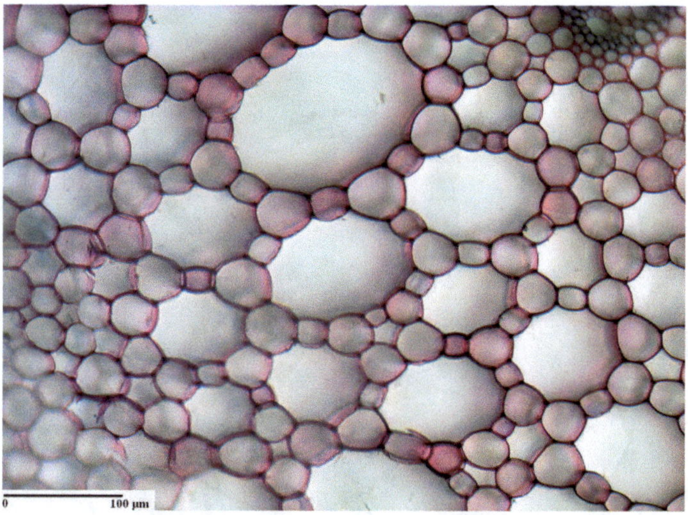

Fig. 47 Cross section through the rhizome of *Cymodocea nodosa* (Grigore et al. 2014)

is furl around another living stem (or then when it is attached to it through tendrils), it is exposed to traction and shear, as a result of the expansion of the support through increase or agitation produced by the wind. On the other hand, the lianoid long stems must be flexible to avoid the risk of breaking, when they themselves or their supports are shaken by the violent winds. If branches of the supporting tree are broken or if the entire tree falls, the flexibility of liana stems is put to an even more intense mechanical stress.

In addition, each climbing stem amends constantly the relationship with its mechanical support; as the older supporting organs become useless, stem glide slowly in the down and may be finally arrived to lie down in some measure on the ground; as a result, it can often become folded or twisted into loops. Intertwined stems are, moreover, often exposed to radial compression, because of growth in thickness of the supporting body. These very special mechanical conditions are responsible for some of the most remarkable types of abnormal secondary growth (thickening) (as already seen, in part, in the subchapter dedicated to structural anomalies of the stem).

In other words, the anatomical structure of the lianas stem brings together, as we will see, a multitude of structural "abnormalities," typical for this ecological group of plants.

Synthesizing all the abnormalities of stem structure in relation to lianas, discussed at length above—some of them—or not mentioned at all, they can be grouped as follows (Crang et al. 2018):

1. Normal location of cambium, but its abnormal activity. The cambium is in normal position, but its activity is abnormal in that it is inactive in certain areas, produces only parenchyma, or is unifacial—producing only xylem or only phloem. Here we can include several situations:

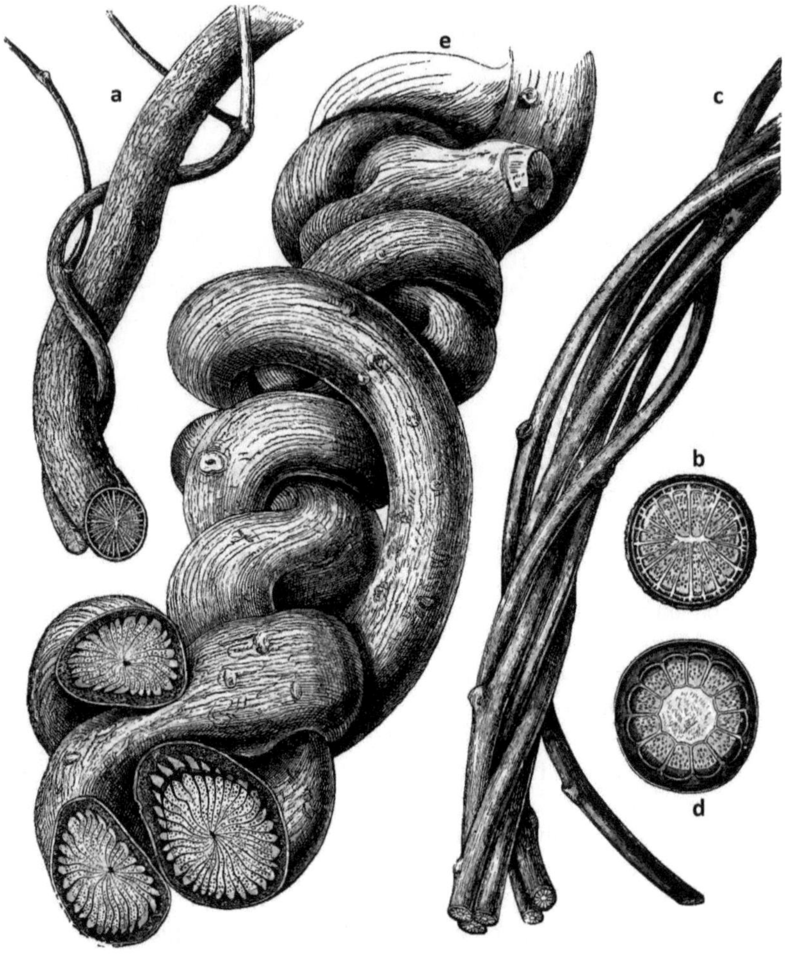

Fig. 48 (**a**) Portion of the stem of a tropical species of *Aristolochia*; (**b**) cross section through the stem of an *Aristolochia* species; (**c**) stem of *Menispermum carolinianum*; (**d**) cross section through the stem of *Menispermum* ssp.; (**e**) portion of a liana (unspecified species—probably a tropical species of *Asclepias*) (von Marilaun 1902)

(a) Interfascicular cambium—generates parenchyma, resulting in medullary rays, such as in the stem of *Aristolochia* (about which we spoke before when we discussed the types of transition from the primary structure of the stem to the secondary one—the "*Aristolochia*" type— Here Fig. 17 and 18 refer to the Fig. 17 and 18 in the Stem chapter under Formation and Activity of Cambium). In this case, bifacial fascicular cambium produces xylem and phloem tissues normally. In addition, an interfascicular cambium is formed, as described above; however, instead of producing inner and outer xylem and phloem tissues, respectively, the bifacial interfascicular cambium only generates parenchyma cells in both directions. The result is a discontinuous cambium with large rays separating sectors (like woody wedges) of conducting tissues.

(b) Formation of xylem/medullary lobes due to differentiated activity, as in *Bauhinia* stems. In this case, certain areas of the cambium become inactive, while others remain active or even accelerate their mitotic activity. The result is a cross-shaped or lobed xylem with interspersed areas of phloem. In some cases, the stem acquires, in outline, a lobed or star shape. These arrangements of xylem appear to create a structurally flexible stem that can bend and wrap around a substrate as it grows (Crang et al. 2018).

2. Abnormal location of the cambium, but its normal activity. The cambium is in an abnormal position or is not ring-shaped, but still has normal, bifacial activity. Here we can include several situations:

(a) Successive cambium within a vascular bundle, as in *Menispermum*. In this situation, successive cambia are produced in a single vascular bundle and in a single growing season. The initial vascular cambium develops normally and generates xylem on the inside and phloem on the outside. This cambium ceases its activity, and a new (successive) bifacial cambium is formed from the parenchyma cells on the inner face of the first phloem. This cambium forms xylem (inside) and phloem (outside) for a short time and then goes dormant. The process is repeated several times in a single growing season. The same development of medullary rays as in *Aristolochia* is present in these species.
(b) Double cambium, as in *Campsis*. These species have a normal vascular cambium that generates phloem on the outside and xylem on the inside. However, they also have a second abnormal vascular cambium on the inside, which is reversed; i.e., it produces external xylem and internal phloem, resulting in the reverse arrangement of vascular tissues.
(c) Included phloem, as in the *Bougainvillea* stem. In this situation, the bifacial vascular cambium initially develops in a normal way, producing secondary xylem on the inside and secondary phloem on the outside. However, this cambium eventually ceases to function, and a separate (abnormal) secondary cambium is formed in the phloem tissue of each vascular bundle. Like the first cambium, it also generates xylem and phloem, but the internally generated xylem grows around—and completely surrounds—the previous phloem; this phloem, thus surrounded, is also called included phloem.
(d) Successive "winged" cambia in certain areas along the stem, as in *Dillenia*. We have already noted that many lianas contain successive cambia (additional, supernumerary—see also the discussion about them at the root level, including their possible adaptive-ecological explanation—Grigore and Toma 2010) that work for a limited time and are then replaced by a new cambium. When the cambium ceases to function, the parenchyma cells in the phloem differentiate into a new cambium and generate xylem on the inside and phloem on the outside. Several stages in the successive development of the cambium, directed to each side of the stem, result in the "winged" growth pattern seen in stems of *Dillenia* (Crang et al. 2018).
(e) Multiple vascular cylinders (polystelic), as in *Serjania* (Fig. 52). In this case, in some genera, such as *Paullina* and *Serjania*, several vascular cambia develop in

one stem. Primary growth gives rise to a normal stem of dicots, but then individual islands of parenchyma cells in the pith differentiate into cambial cells and form their own circular cambia that develop xylem, phloem, and pith. The result is a stem with multiple, separate vascular cylinders, all surrounded by a single periderm. In *Serjania polyphylla*, for instance, five vascular bundles, star-shaped, separated, can be found. This structural arrangement is also called by some authors "polystelic" (i.e., having several eusteles—central cylinders).

In reality, compared to the examples discussed so far in a synthetic manner, the diversity and complexity of building such structures are much more complicated. In the lines below, some particularizing details for certain situations, genres, or species will be also included.

The classic study by Schenck (1893) on the anatomy of the stem of some liana species is worth mentioning, because it includes a very large number of investigated species, belonging to most of the botanical families that include liana (Figs. 49, 50, 51, 52, 53, 54, 55, 56, 57, 58, 59, and 60), but it also highlights many delicate histological aspects involved in defining these structures.

All the long, inextensible stems that characterize lianas are constructed following the pattern of a twisted rope or cable. This structure, reminding of a cable, which combines, however, flexibility with a high degree of tensile strength, cannot theoretically work if the central cylinder (xylem) would be developed in the form of a solid mass. On the contrary, the wood must to be made up of separate cords, which can "slip" one beside the other one; this essential condition cannot be fulfilled if the xylem body is cleaved in more or less separated cords, by interpolation of soft tissues.

Actually, this xylem splitting is a characteristic anatomical feature common for all lianas stems, but it can be expressed in many different modes (see the classic study of Schenck 1893).

In the little bit specialized lianoid forms of stems, the cambium is normal from the beginning, but soon produces xylem and phloem in different proportions, at various points of its circumference. This condition of simplicity prevails, for example, among climbing plants in the family Bignoniaceae, as well as in different species of Malpighiaceae (*Tetrapterys*, *Banisteria*, *Stigmaphyllon*) and *Apocynaceae* (*Phytocrene* genus).

From bignoniaceous species, production of secondary xylem is relatively small, from one in four points, which are arranged in the form of a cross, in time what the secondary phloem production in these points increases accordingly. As a result, the xylem body anatomically appears as being provided with four longitudinal grooves.

These grooves are constantly deeper, as the growth in thickness progresses; but, because these grooves are filled with secondary phloem, the outline of the overall central cylinder, in the cross section, is not affected appreciably of this unequal development of secondary xylem. The cambium itself becomes broken, as a result of his particular behavior; in a sequence of strips (bands), those four (those widest) are being located in front of the four prominent edges of xylem protrusions, while the narrow portions are located at the base of grooves.

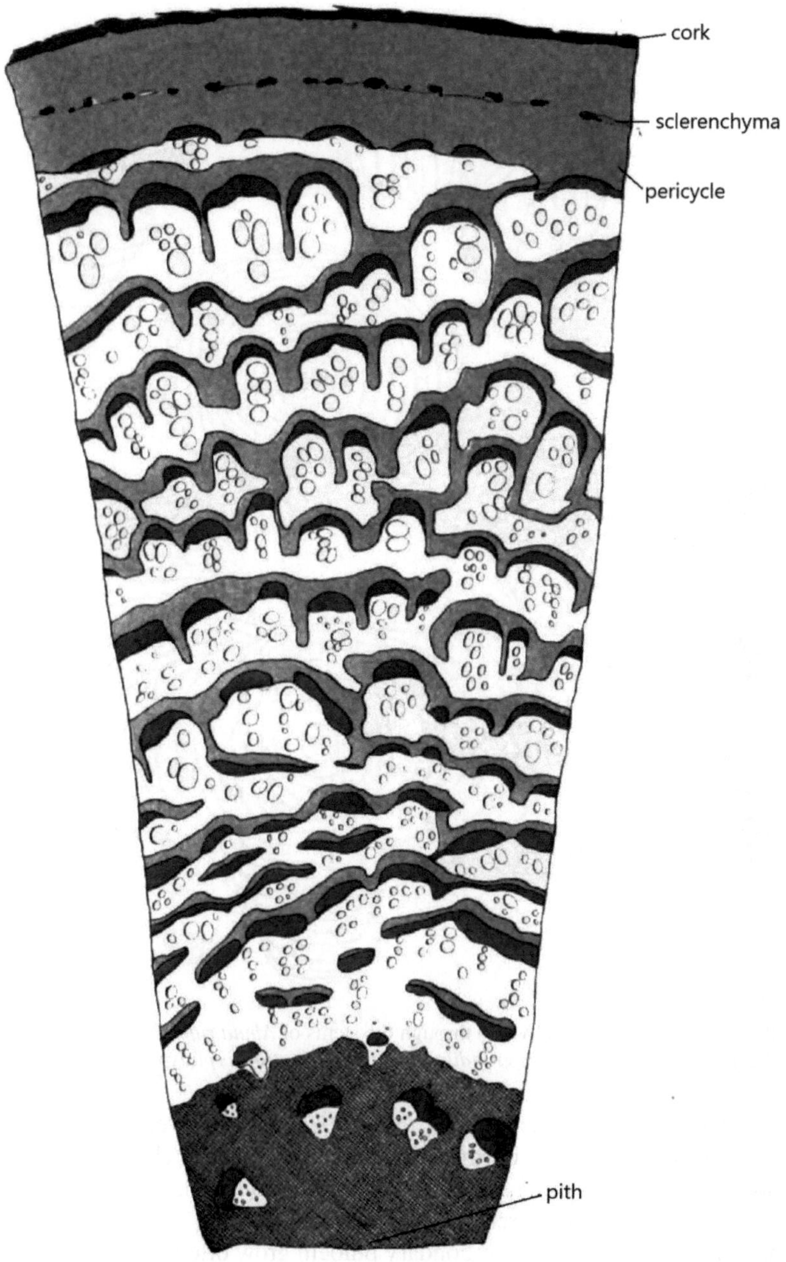

Fig. 49 Cross section through the stem of *Hebanthe pulverulenta* (Schenck 1893)

Fig. 50 Cross sections (general view) through the stems of *Abuta rufescens* (1), *Anomospermum schomburgkii* (2), *Anomospermum grandifolium* (3), and *Botryopsis platyphylla* (4) (Schenck 1893)

At certain bignoniaceous species, the grooves remain ever in number of four (Fig. 61); to others species, on the other part, additional ditches are continuously interpolated, as a result of a reductions in xylem production, in an increasing number of points on the circumference of cambium.

The embedded plates of the secondary phloem grow often in width to their outer ends, through a series of step-like growths (Fig. 62). In this way, the whole xylem cylinder is divided into many lobes and separate bands.

At certain species from the Malpighiaceae and Bignoniaceae families, splitting of the xylem is carried further (Fig. 62), because of secondary modifications of the xylem parenchyma and the pith. The xylem cylinder, initially continuous, but very

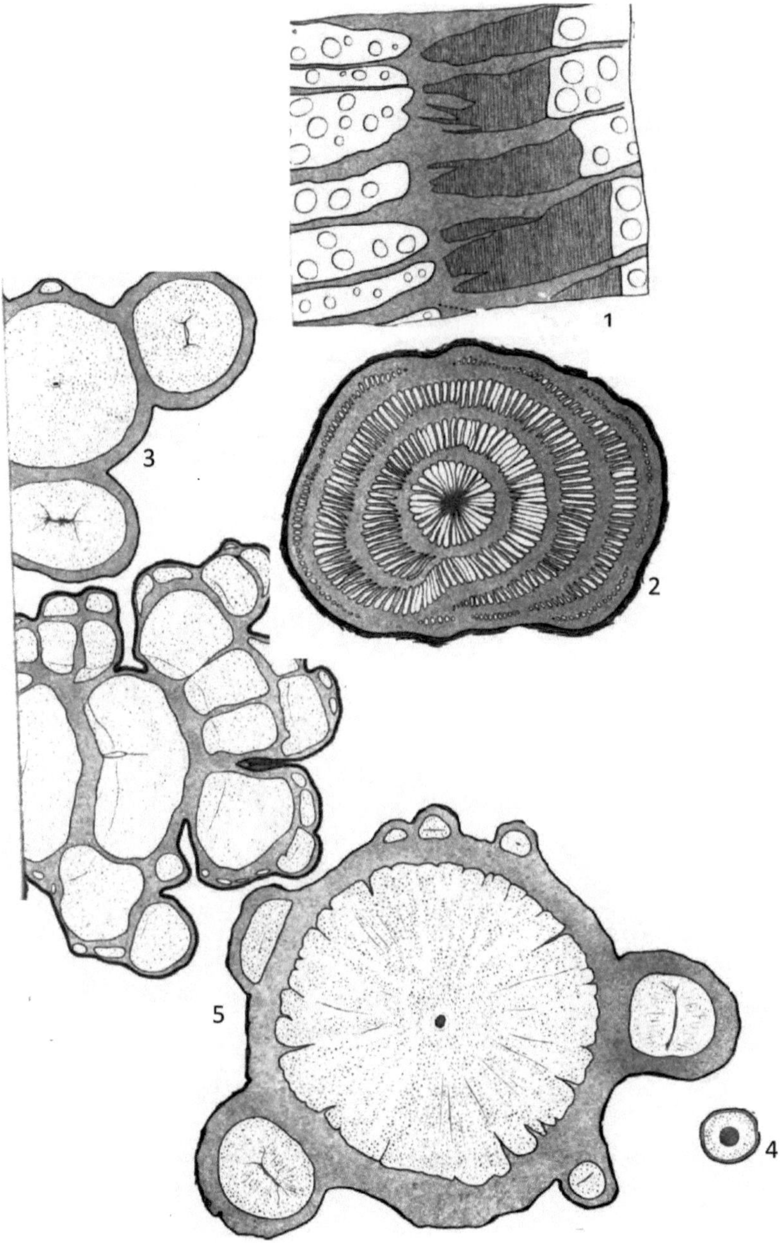

Fig. 51 Cross sections (general view) through the stems of *Doliocarpus rolandri* (1), a species from Dilleniaceae family (2), and *Thinouia scandens* (3–5) (Schenck 1893)

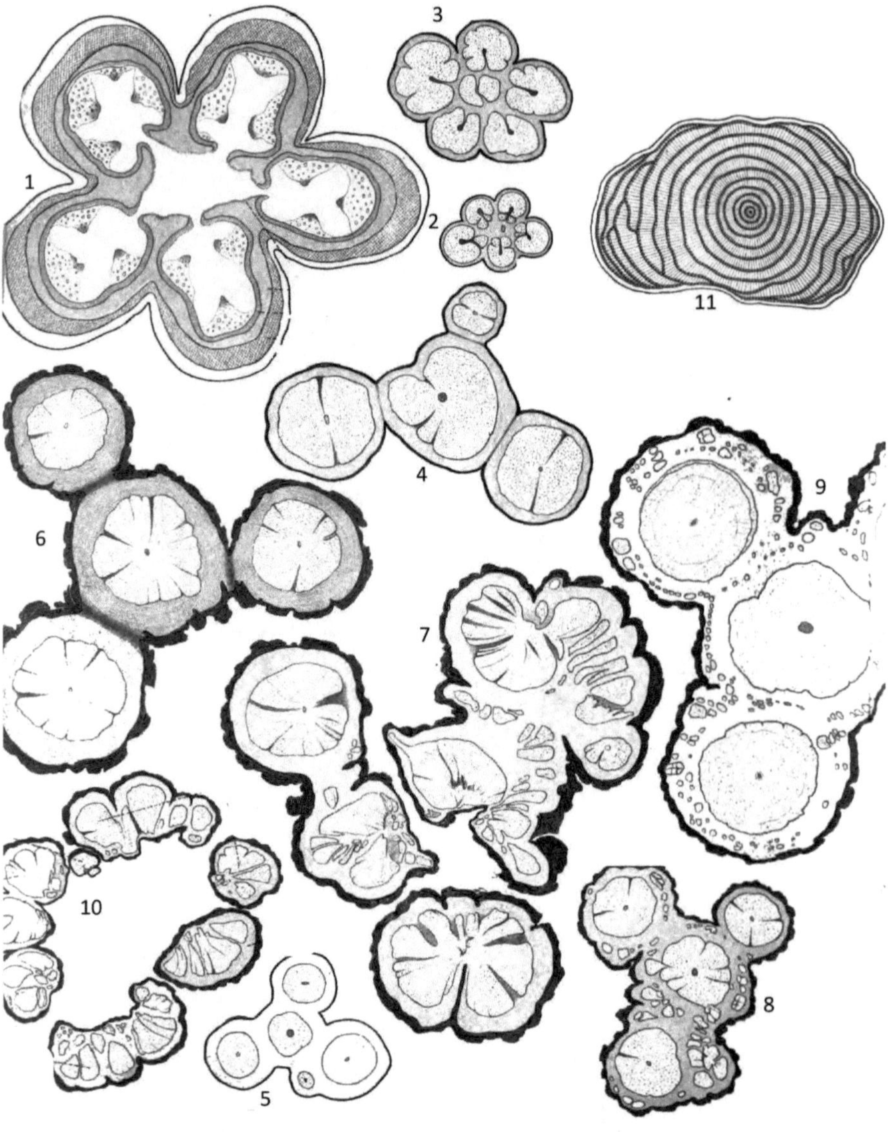

Fig. 52 Cross sections (general view) through the stems of *Serjania elegans* (1–3), *Serjania lamprophylla* (4), *Serjania lethalis* (5), *Serjania ichthyoctona* (6–10), and *Abuta selloana* (11) (Schenck 1893)

lobed, becomes completely broken, because of active intercalary growth of interpolated masses of the parenchyma. According to Schenck, this "dilatation parenchyma" appears, in the many cases, from the xylem parenchyma, having non-lignified thin walls. In some cases, however, the lignified elements associated with vessel or cell ray xylem are the ones that suffer this growth and secondary fragmentation. In

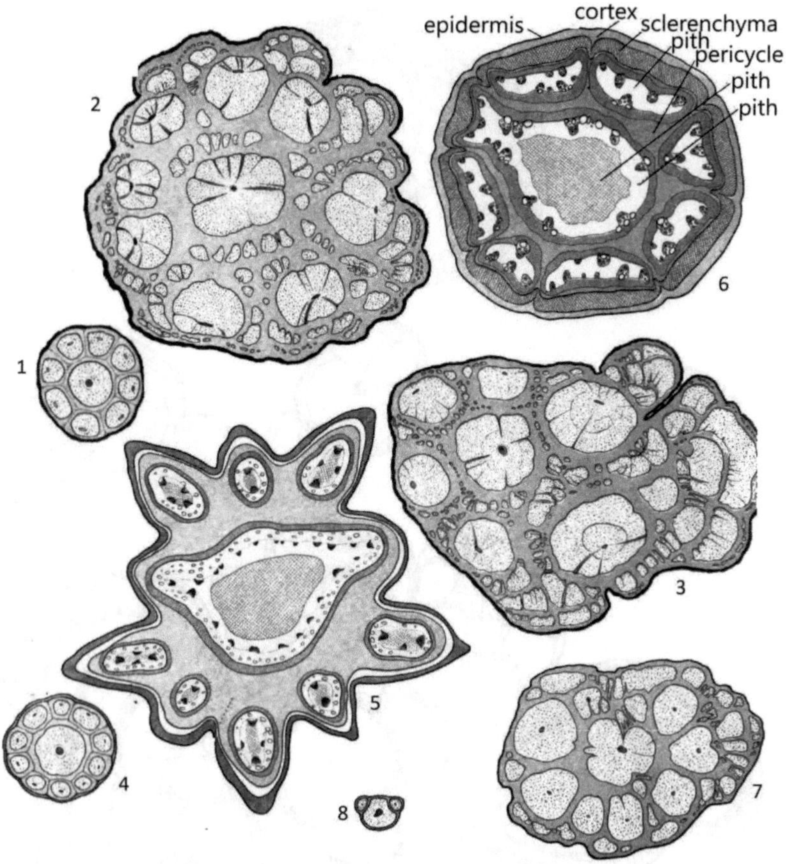

Fig. 53 Cross sections—general view through the stems of *Serjania multiflora* (1, 2, 3), *Serjania noxia* (4), *Serjania clematidifolia* (5), *Serjania grandiflora* (6, 7), and *Serjania erecta* (8) (Schenck 1893)

other *Bauhinia* species, the initials cells of the strips and wedges of dilatation parenchyma have their origin in the phloem and enter just secondarily in the xylem.

Once separate cords of the xylem were formed, each may grow in the thickness through intermediate secondary independent meristems; additional cords may also be interpolated between the first formed bundles. All stem becomes therefore formed from a mass of xylem cords interlaced and in consequence, presents often a very complicated aspect in cross section (Fig. 63). The comparison of such liana stem with a thick rope is therefore appropriate, where tissue plates interpolated between different xylem cords break and are partially replaced by a periderm. In such cases, simple macroscopic external inspection shows that the stem is divided into a number of longitudinal threads covered by cork and braid or partially fused in a complicated way.

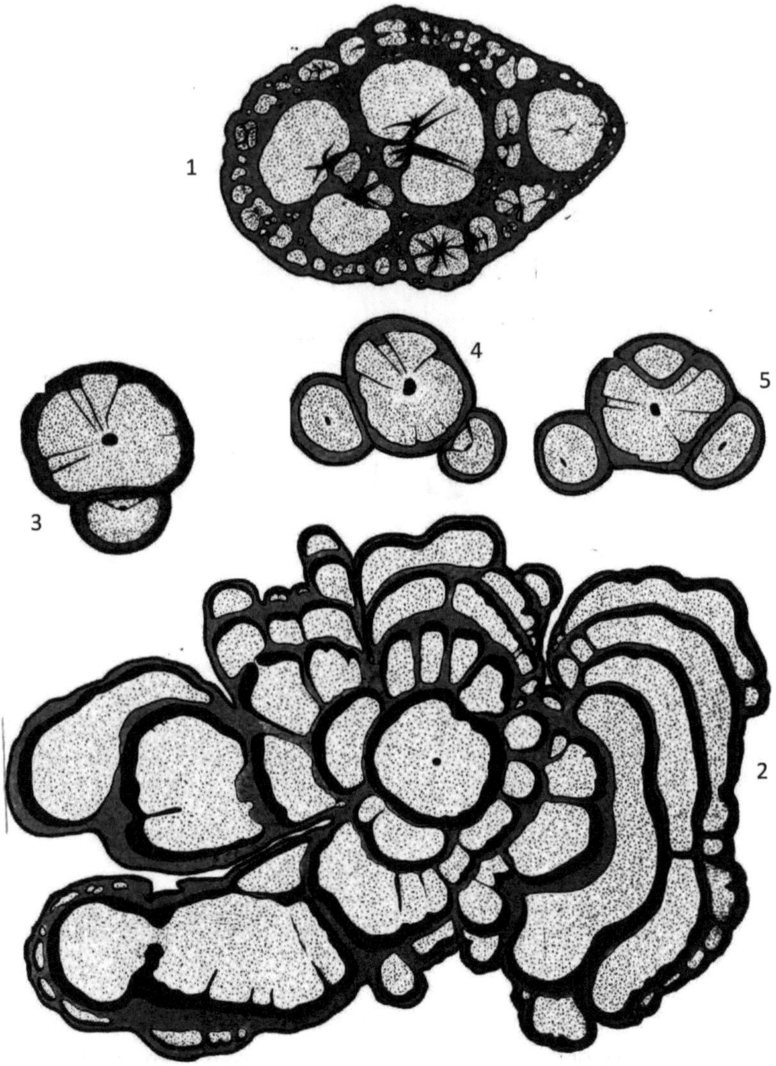

Fig. 54 Cross sections (general view) through the stems of *Serjania piscatoria* (1), *Paullinia pseudota* (2), and *Paullinia pinnata* (3–5) (Schenck 1893)

In all cases that were discussed until now (except for *Bauhinia*), secondary developments always start from one cambial layer having normal character. On the other hand, to the tendril-climbing plants of the family Sapindaceae (*Serjania*, *Paullinia*, *Thinouia*), many cambial layers are present, in addition. These can be arranged in different modes in the cross section of the stem. Sometimes, the center portion of the organ results from the activity of the main cambium cylinder, in around to which few peripheral small cylinders are grouped (Figs. 64, 65, and 66);

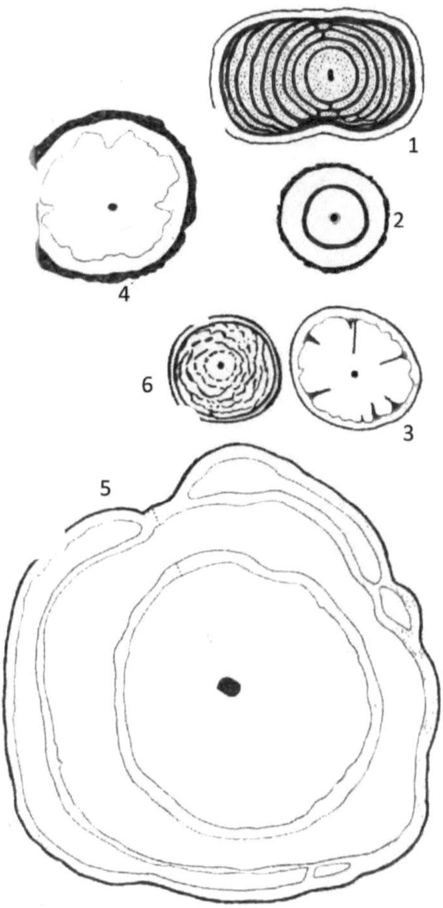

Fig. 55 Cross sections (general view) through the stems of *Securidaca sellowiana* (1), *Bredemeyera schenckiana* (2), a species from the Hippocrateaceae family (3), *Hippocratea ovata* (4), *Salacia* sp. (5), *Salacia serrata* (6) (Schenck 1893)

rarely, there is no central cambium, but only a circle of five until seven peripheral cambia.

Nageli (mentioned by Haberlandt 1914), who paid special attention to the ontogenetic development of individual cambial layers, found that a circle of primary bundles of leaf traces is more or less deeply divided in some points; in this way, it looks like the groups of individual bundles tend to be removed from the circle even and in this early stage. Actually, the first evidence of this particular trend is indicated by procambial cords; therefore, when these become linked to form a continuous cambial cylinder, the separate groups behave as independent circles of cords and, finally, form an equal number of separated cambial cylinders. Next, each individual cambial layer behaves normally separating the phloem secondary outside and the xylem secondary inside. Therefore, the stem is formed from one to many separate xylem cylinders; in other words, the cleavage of the xylem is not a secondary process (as in Bignoniaceae), but is determined by the behavior of the apical meristem. To specify that the secondary growth of some climbing stems in the family

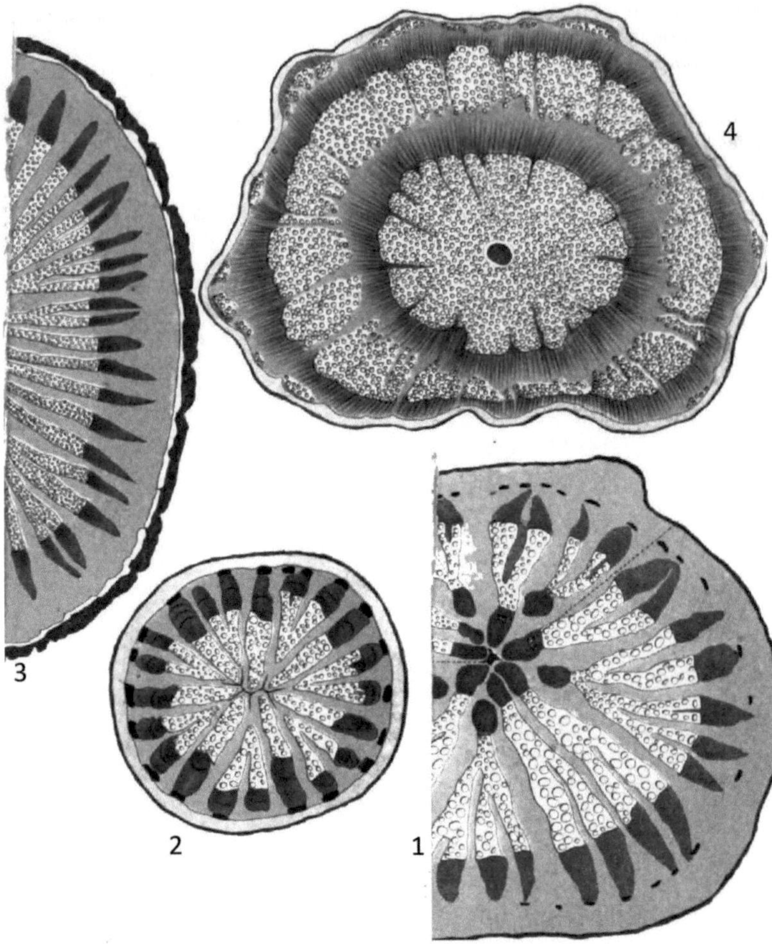

Fig. 56 Cross sections (general view) through the stems of *Wilbrandia verticillata* (1), *Anisosperma passiflora* (2), *Aristolochia triangularis* (3), and *Mucuna altissima* (4) (Schenck 1893)

Sapindaceae is closely related to the type of Bignoniaceae; the xylem cylinder is continuously, but deeply, lobed or equipped with prominent ridges. It is also worth noting the fact that at these non-climbing sapindaceous species, the stem structure is perfectly normal; therefore, the very particular anatomical structure of the climbing species seems indeed to be an adaptive characteristic (Haberlandt 1914).

A third type of abnormal thickening found in the liana stems is characterized by the occurrence of some additional cambial cylinders, complete or partial, external to the primary (initial) cambium. Every individual cambial layer produces a secondary xylem and phloem normally for a while; sooner or later, its activity ceases, after which a new secondary meristem layer appears further from the parenchyma. This process can be repeated many times. Two different subtypes can be distinguished, in

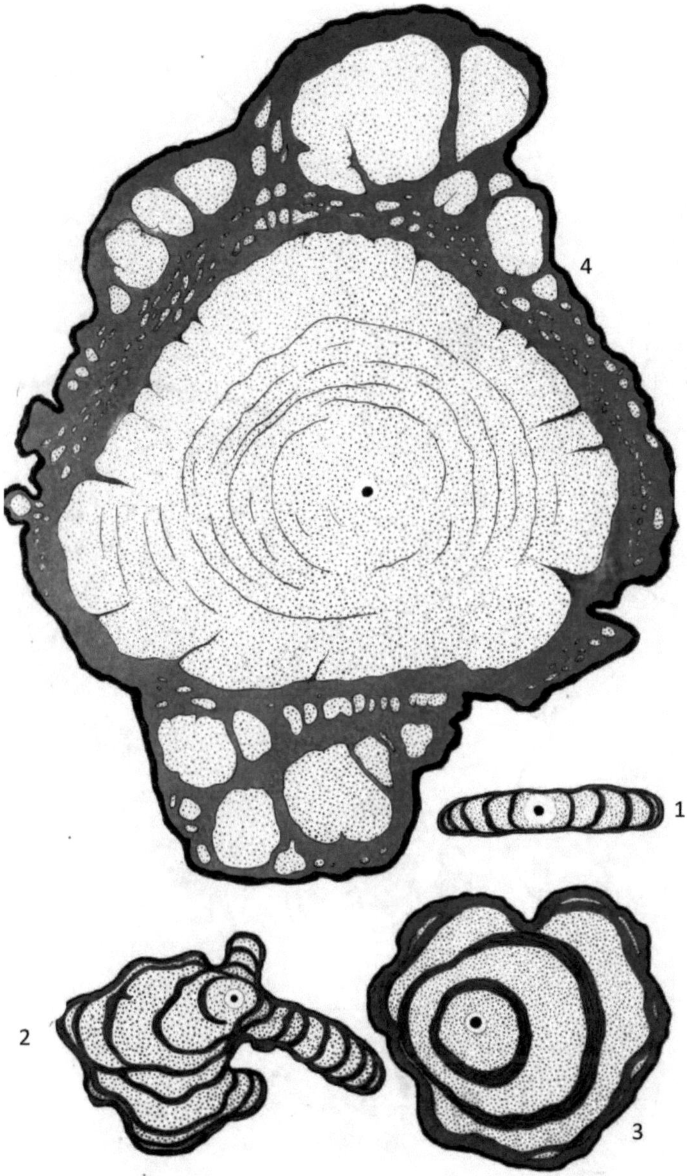

Fig. 57 Cross sections (general view) through the stems of *Machaerium aculeatum* (1–2), *Machaerium tounateifolium* (3), and old stem of an unspecified liana species (4) (Schenck 1893)

terms of the point of origin of these successive cambial layers. Additional meristems can appear either in the primary cortex, as in some species of the family Menispermaceae, in *Wisteria* and *Rhynchosia*, in the secondary phloem, at *Gnetum* (Fig. 67), *Bauhinia*, *Machaerium*, *Phytocrene*, and some species from Bignoniaceae

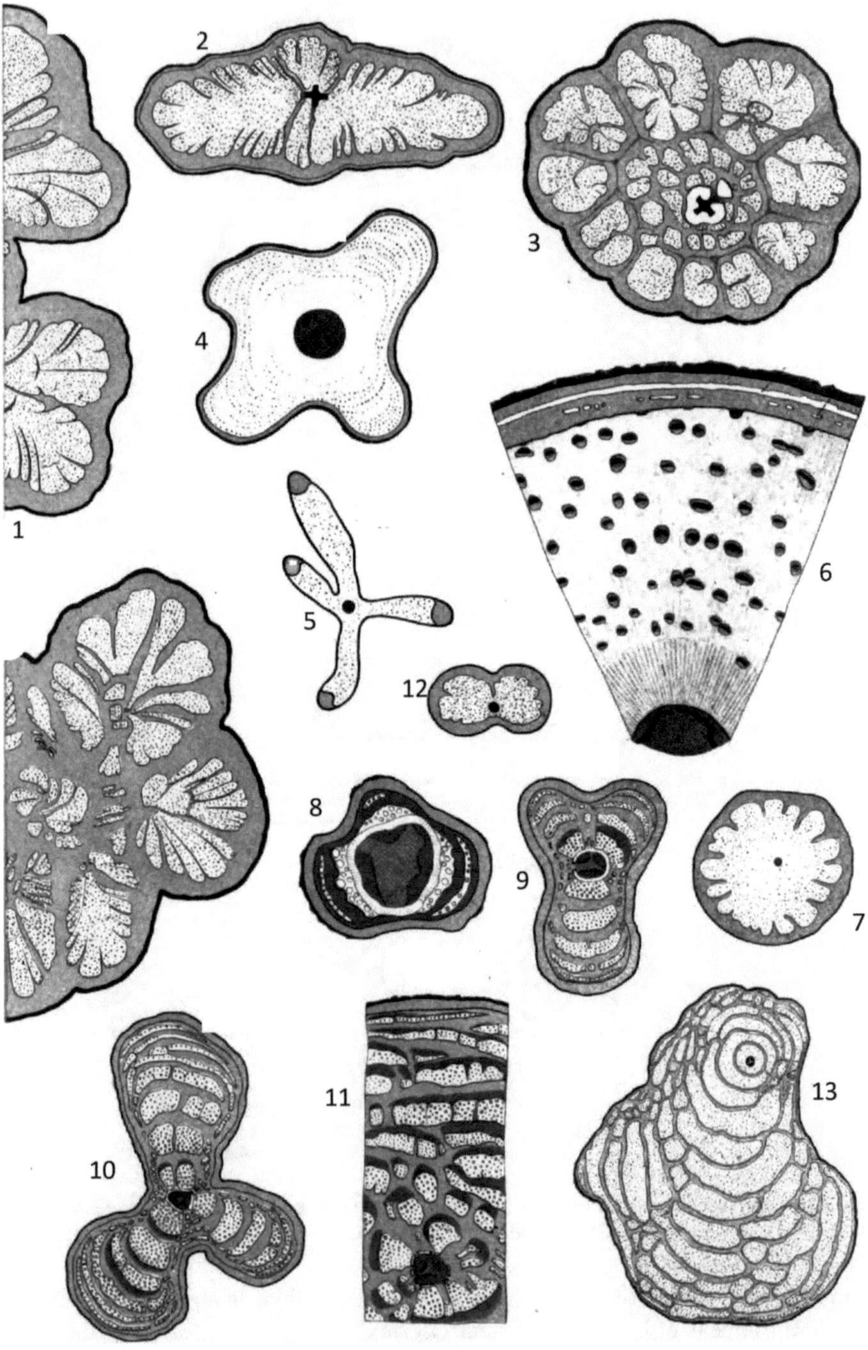

Fig. 58 Cross sections (general view) through the stems of *Bauhinia* ssp. (1–3), mimosa species (4–5), *Strychnos triplinervia* (6), *Condylocarpon rauwolfiae* (7), *Ipomoea umbellata* (8–11), *Ipomoea glabra* (12), and *Porana volubilis* (13) (Schenck 1893)

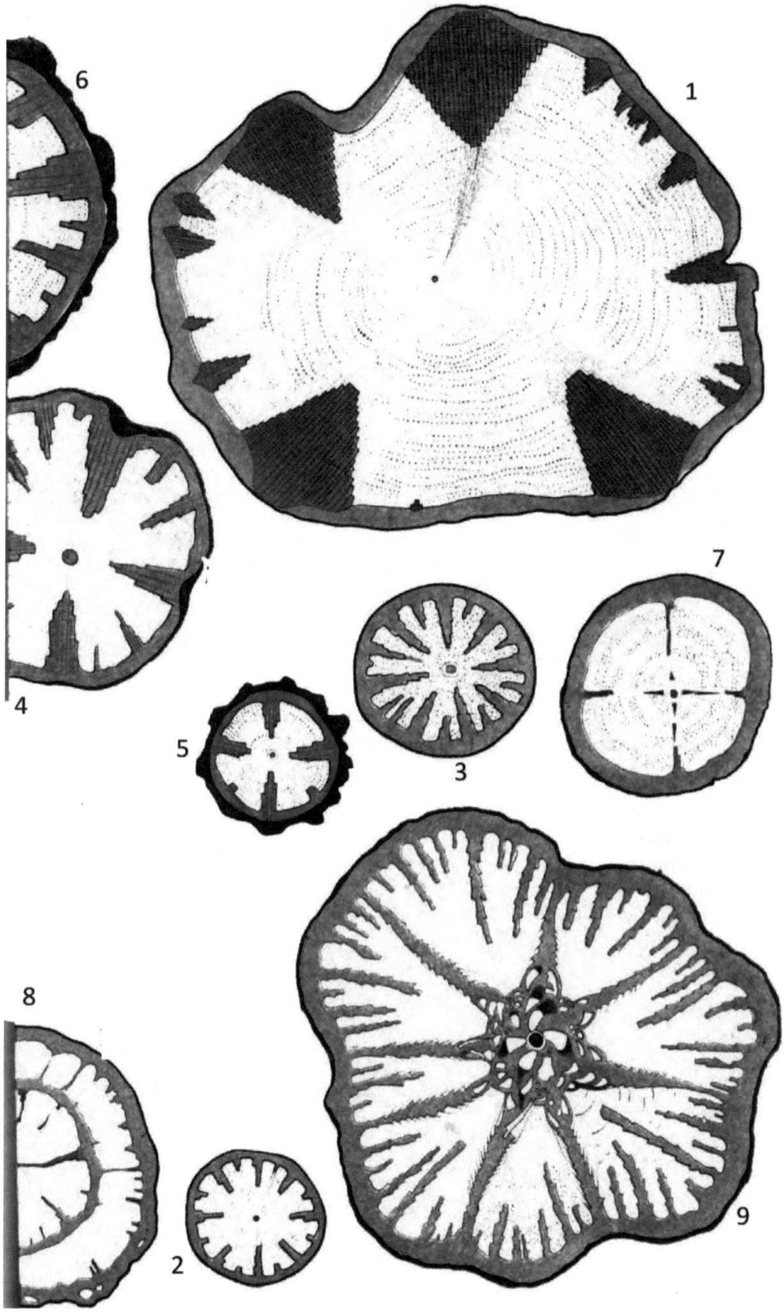

Fig. 59 Cross sections (general view) through the stems of *Tynanthus elegans* (1), *Clytostoma callistegioides* (2), Bignoniaceae species (3–6), *Pithecoctenium phaseoloides* (7), *Haplolophium* ssp. (8), and *Bignonia* sp. (9) (Schenck 1893)

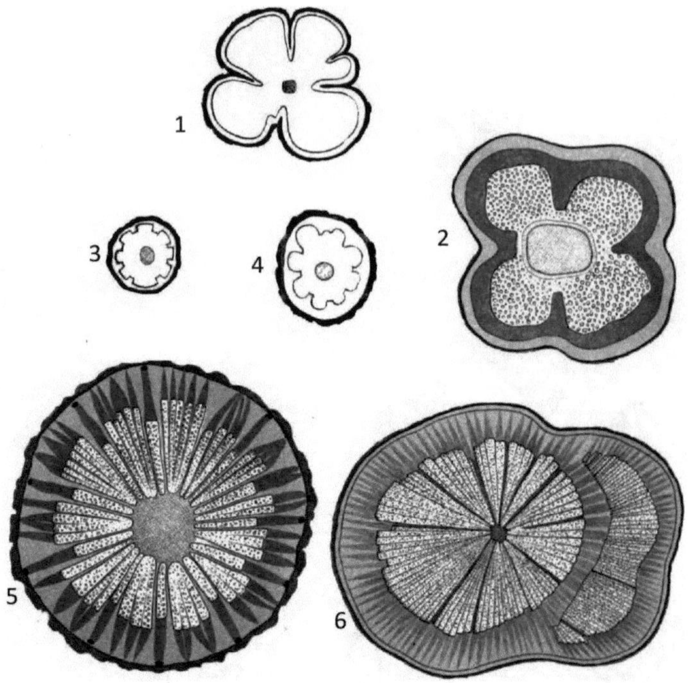

Fig. 60 Cross sections (general view) through the stems of *Lantana* ssp. (1), *Manettia luteorubra* (2), a species from Asteraceae (3–4), *Bidens rubifolius* (5), and *Mikania* (6) (Schenck 1893)

Fig. 61 Cross section through a 4-year-old stem of *Bignonia capreolata*; the xylem cylinder is traversed by four furrows or grooves occupied by the phloem (de Bary 1884)

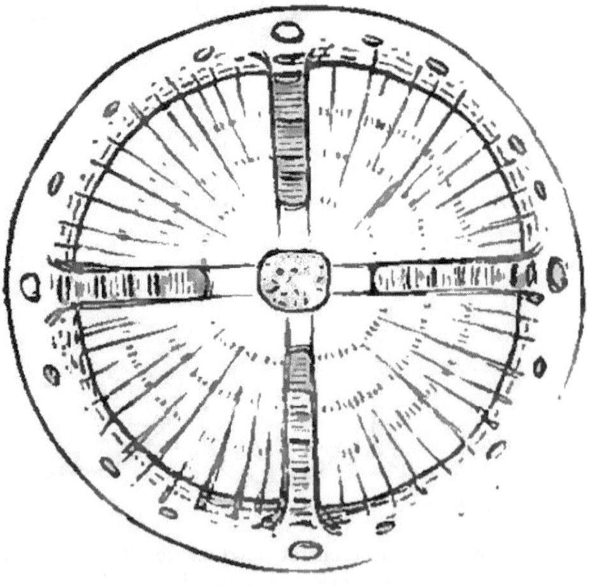

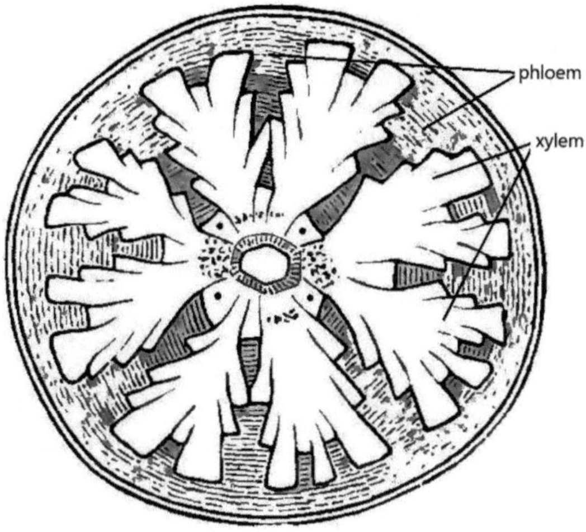

Fig. 62 Cross section through the stem of *Melloa populifolia* (by de Bary 1884)

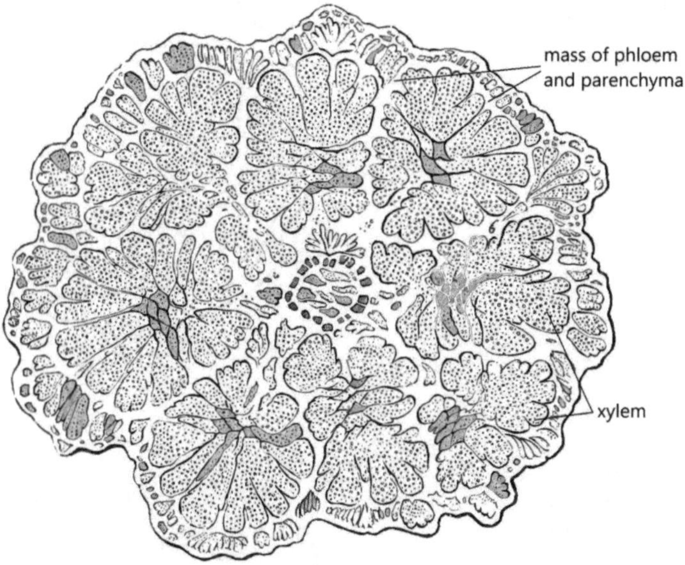

Fig. 63 Cross section through the stem of *Bauhinia* sp. (Haberlandt 1914)

(*Haplolophium*, *Glaziovia*, *Anisostichus*) and Convolvulaceae (*Ipomoea*, *Convolvulus*).

Another very common type of thickening (secondary growth) is found in those lianas to which a secondary cambium is formed on the inner face of the normal xylemic cylinder. At *Tecoma radicans* (*Campsis radicans*), the internal cambium generates both xylem and phloem, but in the reverse order. A somewhat similar situation can be found in *Periploca graeca*, but to which, unlike *Tecoma radicans* (in which the medullary phloem is mainly secondary), the perimedullary phloem is

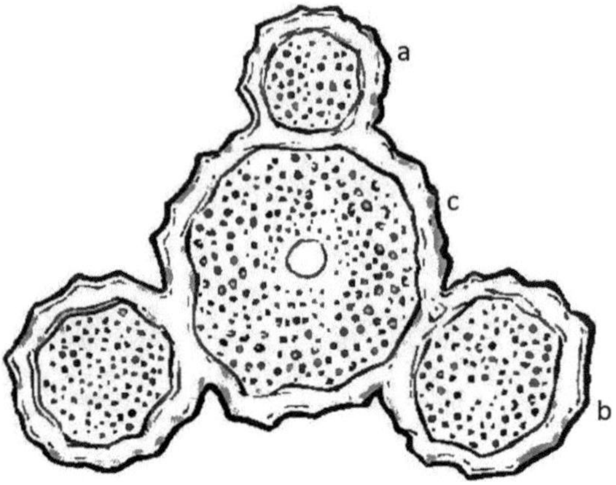

Fig. 64 Cross section through the stem of *Serjania* ssp. (or *Paullinia* ssp.); (**c**) main or central xylem cylinder; (**a, b**) peripheral xylem cylinders (de Bary 1884)

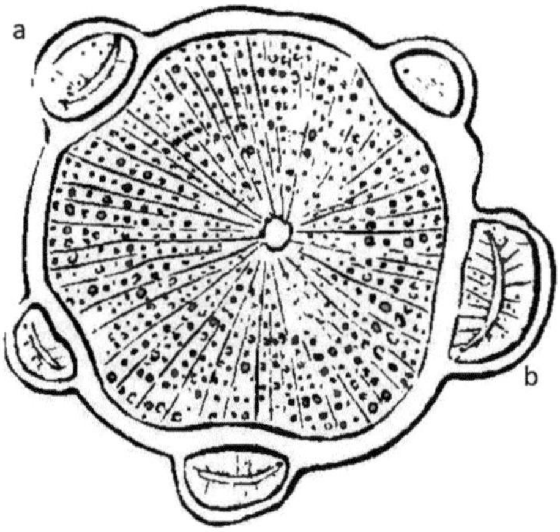

Fig. 65 Cross section through the stem of *Serjania* (or *Paullinia* sp.); (**a, b**) peripheral xylem cylinders (de Bary 1884)

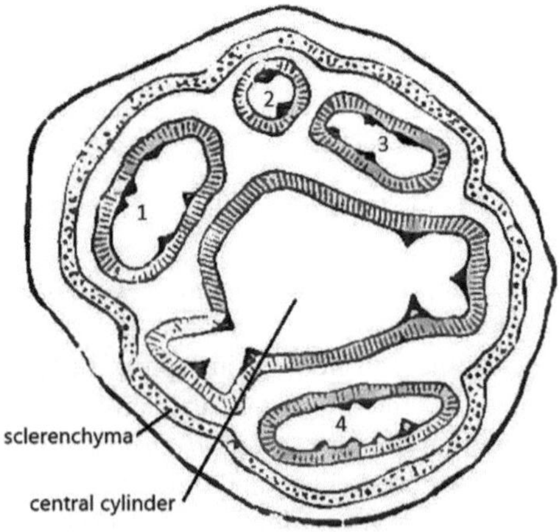

Fig. 66 Cross section through the stem of *Serjania caracasana*; 1–4, peripheral xylem cylinders (Haberlandt 1914)

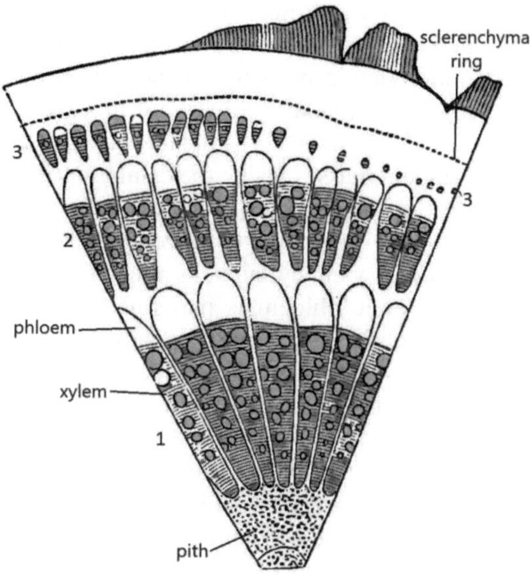

Fig. 67 Cross section through the stem of *Gnetum scandens*; 1–3: successive vascular cylinders (the third just beginning to develop) (de Bary 1884)

almost always of primary structure (Şerbănescu-Jitariu and Toma 1980). *Periploca graeca* (family Asclepiadaceae) is a liana species also present in the flora of Romania, with thin and long stems up to 12 m long, growing in forests, meadows, and rocky places in the southern part of the country (Dobrogea).

The most typical anatomical features of the liana stems are the ones that tend to produce a structure related to a rope or of a cable and thus make the organs more flexible. Tensile resistance is the most important in the young stem, where the first secondary or central xylem contains a very large proportion of mechanical elements (xylem fibers and tracheids fibrous), while the vessels are unusually narrow, and the xylem parenchyma is very poorly developed.

The portion of secondary xylem, formed later, contains, on the contrary, large vessels and abundant xylem parenchyma. According to Schenck (1893), this contrast between the secondary central xylem and the peripheral one of liana stems is, close invariably, very obvious.

Younger stems, are therefore, clearly inextensible through construction, after the pattern of the roots that have a cord of stereome in the center of eustele. However, as the stems grow and thicken, their inextensible character becomes less pronounced, as inflexible as trunks and the branches that do not keep peripheral the disposition of their stereome, after the secondary thickening (growth) is in the process for some time. In both cases, abundance of scattered mechanical elements is the one that compensates their inability to group in the center or close to the periphery, in accordance with the validated mechanical principles.

The abnormal ways of thickening, characteristic to climbing stems, in addition to the fact that lead to the construction pattern of a rope shape based on a split of xylem cylinder, implies another deviation from the arrangement of normal tissues; the secondary phloem, instead of taking a peripheral position, as it does in stems with normal secondary growth, moves inside—to a greater or lesser extent—and becomes included between different cords or xylem secondary cylinders. The formation of a such intraxylemic phloem depends often by the fact that the cambium, besides clearing the secondary phloem on its external side, generates in addition, on its internal side, isolated cords of sieve tubes, accompanied by parenchymatous tissue; these phloem masses inside the cambium (like those found in *Dicella*, *Combretum*, *Mucuna*, and *Entada*) become therefore embedded in the secondary xylem.

In *Strychnos* type, the cords of the phloem of extra-cambial origin become later closed in the xylem cylinder, as a result of some special growth processes.

Intraxylemic phloem is particularly well protected against lesion mechanical injuries, probably representing an advantageous condition from an adaptive point of view, because liana stems, as already stated, are particularly prone to be subjected to violent bending, in addition to radial compression pie produced by the expansion of living support.

In conclusion, in their later stages of development, the liana stems often differ from the usual vertical woody stems, even in their morphology (see Grigore 2022). The most remarkable example of this observation is provided by the very flat stems in the ribbon shape of certain species (*Rhynchosia*, *Dalbergia*, and *Machaerium*, *Bauhinia*), whose characteristic shape is due to an uneven distribution of cambial activity and, in some cases, to the occurrence of some additional cambial layers. Such ribbon shape stems are sometimes fold in an extraordinary way; often the two edges do not count for curling tendency.

3 Stem Anatomy in Some Xerophytic Species

In the case of xerophytes, which represent—as pointed out in the present work and in other works (Grigore 2022)—a very heterogeneous (supra) ecological group of plants, the identification of some anatomical features of the stem, with ecological significance, is somewhat more difficult.

The identification of such structural features can be done by isolation, by discussing particular aspects.

Some xerophytes show thin, switch-shaped stems (Grigore 2022), which are a good example of extreme reduction in leaf area, combined with the formation of assimilative green tissue in the stem cortex. This group of plants is characterized by slender-like rods, dry and hard, switch-shaped stems and branches (*sclerocauly*), while cactiform plants, on the contrary, always have sparsely branched axes and massive, thickened, fleshy, and rigid cauline structures that are not affected by the wind (*chylocauly*). The leaves of some species, such as *Genista tinctoria* and *Spartium junceum*, have relatively considerable dimensions, but fall early, once they have fulfilled their assimilation functions; but the leaves of other species are from the beginning very reduced in form and function.

In these species, the stem is circular in cross section, or deeply furrowed, with stomata and palisade tissue located in furrows and mechanical tissue in ridges.

The species of the genus *Spartium*, although they are not completely aphylous (because they form lance-shaped leaves, from place to place, on the long branches), do not carry out their photosynthesis at the expense of the leaves, except in a very small proportion. The leaves are therefore of secondary importance, because their green, assimilatory tissue can synthesize only a small part of the organic substances necessary for the subsequent growth of the plants. On the contrary, the photosynthetic function relies mainly on the cortex of the green branches (stems), which, moreover, present anatomical adaptations for photosynthesis (Fig. 68). This situation also applies for other desert (xerophytic) plants, such as *Retama raetam* (Fig. 69).

Under the epidermis, which presents strongly thickened walls, covered with wax, a green tissue (chlorenchyma) is located, consisting of five until seven rows of cells. This green tissue does not form a continuous layer around the stem, but it is divided into—from 10 to 15—thick cords, through sclerenchyma pillars (Fig. 68). Under this green cortex, which alternates with pillars of sclerenchyma, the phloem, cambium, xylem, and a very large pith are found. In the chlorenchyma cords found in the cortex of *Spartium* species, the agglomerated chlorophyll cells are adjacent tight with each other, leaving very small air spaces between them. At the level of the epidermis, stomata are located, with an obvious substomatal chamber; stomata are proportionally small, but their number is very big. Because cells' stomata are not so strongly thickened in their external walls, compared to other epidermal cells, they appear to be relatively sunken below the level of epidermal cells. Through this particular disposition, typical for xerophytes, through the wax layer from the level of the epidermal cells, the stomata are protected against excessive transpiration.

In *Casuarina* species and in *Cytisus radiatus* (Fig. 70), the chlorenchyma is distributed in the cortex of the stems, as in the case mentioned in *Spartium*; cords of green tissue (chlorenchyma), which crosses the stem, are interrupted deeply by longitudinal furrows. In other species of aphyllous shrubs, such as species in the genus

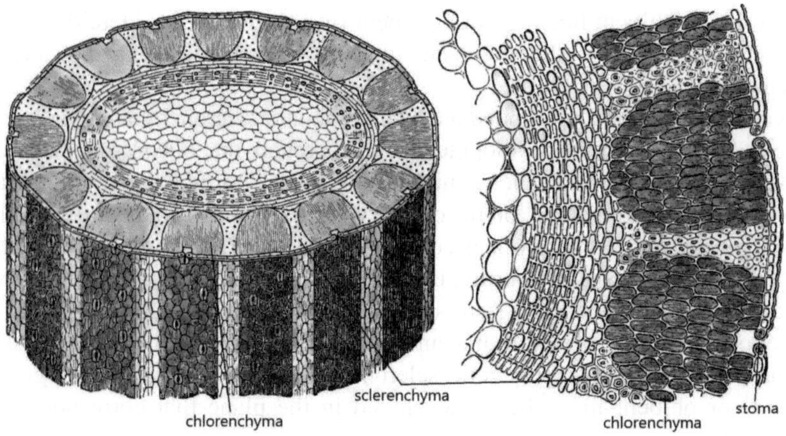

Fig. 68 Cross section through the stem of *Spartium scoparium*; left, overview; right, detail (von Marilaun 1902)

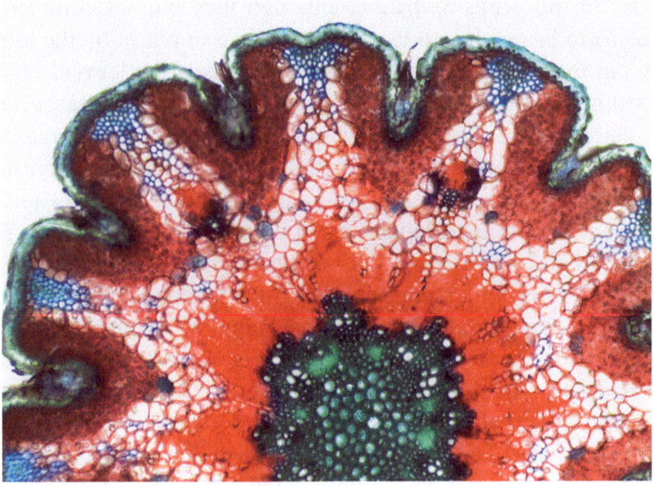

Fig. 69 Cross section through the stem of *Retama raetam* (Photo courtesy of Hafida Trabelsi)

Ephedra, chlorenchyma form a continuous and uniform layer around the stem, uninterrupted by sclerenchyma pillars. But, in this case, stomata are evenly distributed throughout the surface of thin branches, while in *Genista* species, *Casuarina*, and *Cytisus radiatus*, the stomata are absent from those portions of the epidermis that cover the sclerenchyma pillars.

One of the most extensive studies on the anatomy of some xerophytes (in this case, desert plants) is that of Sabnis (1919–1921). At the level of the organ that interests us here, the stem, Sabnis (1919–1921) identified some modifications of the pericycle, in the sense of its sclerenchymatization. Thus, there can be several situations: the pericycle is composed of either a composite and continuous ring of sclereids (cells—isolated mechanical elements) or of phloem fibers, or from large groups of sclereids or phloem fibers, or of a phloem ring consisting of sclereids or phloem fibers. In some cases, the pericycle consists of several isolated sclereids or phloem fibers.

According to Sabnis, the abundant development of sclerenchyma in the pericycle is due to ceasing from the development of the parenchyma, due to water scarcity. In the succulent species of the families Violaceae, Portulacaceae, Burseraceae, Rosaceae, and Gentianaceae, on the contrary, parenchyma is abundantly developed, and sclerenchymatized pericycle is completely absent. Apparently, a sclerenchymatized pericycle serves as supporting tissue against strong wind from the desert.

At some species of Asteraceae, Verbenaceae, Amaranthaceae, and Chenopodiaceae, the pericycle has an isobilateral symmetry, which is caused by unequal development of pericyclic sclerenchyma in two plans, larger groups of sclereids or phloem fibers being developed in the plane that corresponds to the

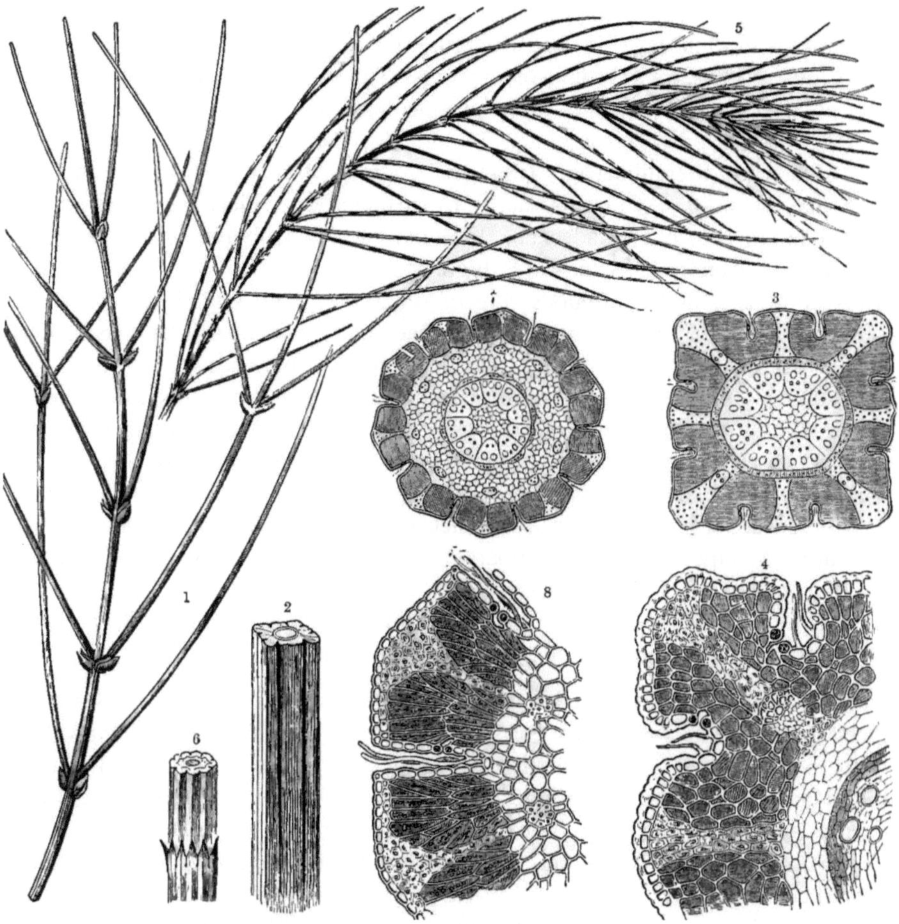

Fig. 70 Branch of *Cytisus radiatus* (1) and branch segment (2); 3, cross section through a branch (general view and detail—4); 5, branch of *Casuarina quadrivalvis*; 6, branch segment; 7, cross section through a branch (general view and detail—8) (von Marilaun 1902)

direction of dominant wind. In the inclined stems of some Zygophyllaceae species and Convolvulaceae, for example, the tissue consisting of sclereids or phloem fibers is more developed on a side that seems to be the superior side of the stem, while on the other side, the pericycle consists of several small, isolated groups of sclereids or phloem fibers. The abundant development of a mechanical tissue on the upper side is apparently necessary to protect the stem from the tensions induced by stretching on the upper side and in consequence, prevents it from bending.

There are several types of pericycle that were described by Sabnis (1919–1921) in the stems of desert plants:

(a) Isolated sclereids or small groups of sclereids, for example, in species of the families Elatinaceae (Fig. 71), Boraginaceae (Fig. 72), and Convolvulaceae.
(b) A loose ring of small groups of sclereids, as in some species of Elatinaceae.
(c) Large groups of sclereids, tight grouped, as in species of Menispermaceae (Fig. 73), Capparidaceae (Figs. 74 and 75), Caryophyllaceae, Tamaricaceae, Malvaceae, Sterculiaceae, Tiliaceae, Zygophyllaceae (Figs. 76 and 77), Simaroubaceae, Celastraceae, Cucurbitaceae, Ficaceae, Asteraceae, Salvadoraceae, Asclepiadaceae, Boraginaceae, Convolvulaceae, Verbenaceae, Labiatae, Amaranthaceae, and Polygonaceae.
(d) A composite and continuous ring of sclereids, as in Sapindaceae (Fig. 78), Caesalpiniaceae, Cucurbitaceae, Ficaceae (Figs. 79, 80, and 81), Asteraceae, Convolvulaceae, Verbenaceae, Liliaceae, and Commelinaceae.
(e) Isolated phloem fibers or small groups of phloem fibers, as in species of Cruciferae, Scrophulariaceae (Figs. 82 and 83), Acanthaceae (Fig. 84), Nyctaginaceae (*Boerhavia diffusa*—Fig. 85), and Euphorbiaceae (Fig. 86)
(f) A lax ring of small groups of phloem fibers, as in species of Lythraceae, Rubiaceae, Solanaceae, Scrophulariaceae, Nyctaginaceae, and Euphorbiaceae.
(g) Large groups of phloem fibers, as in Cruciferae, Sterculiaceae, and Asclepiadaceae (Figs. 87 and 88)
(h) A composite and continuum ring of phloem fibers, for example, at Polygalaceae (Figs. 89 and 90) and *Rhamnaceae* (Figs. 91 and 92).

In addition to the sclerenchymatized pericycle, small groups of sclereids and phloem fibers appear in the soft phloem of some species of Rhamnaceae and Malvaceae, respectively. In *Melhania denhamii*, the pericycle is formed of an external composite and continuous ring of sclereids, and of an internal groups of phloem fibers, with a parenchymatous pericycle interleaved between them.

The composite and continuous ring of phloem fibers from *Polygala erioptera* is layered (Fig. 90).

The conducting system of the stems of desert plants studied by Sabnis (1919–1921) consists of xylem that usually sets up a composite cylinder, empty inside. The vessels are less bit abundant in desert plants than in common mesophytes. This reality is explained by the fact that all the tissues of desert plants are adapted to control transpiration, due to water deficit. Vessels are embedded in a ground xylem parenchyma tissue, which is usually made up of cells with thickened and lignified walls, having small diameter. The occurrence of lignified tissues is, in the case of xylem, and of narrow medullary rays, when present, is characteristic to desert plants (Sabnis 1919–1921).

In the spirit of the discussions so far, from the root of xerophytes, we can observe, also in the case of the stem, that lignification is a phenomenon directly correlated with xeromorphic adaptations.

Coming back to the Hayden's (1919) study on prairie plants—mentioned first at the ecological anatomy of roots in eremophytes—she found out that, within the ten examined plants from alluvial basin and nine upland plants, of the subterranean stems (rhizomes), of the alluvial basin plants, 65% had aerenchyma and only 10%

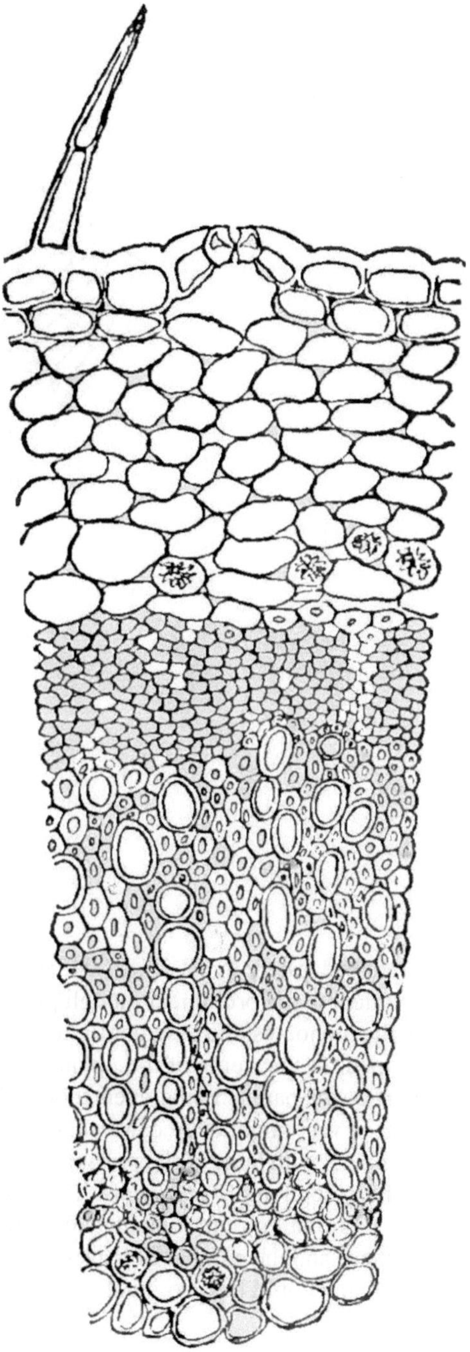

Fig. 71 Cross section through the stem of *Bergia ammannioides* (Sabnis 1919–1921)

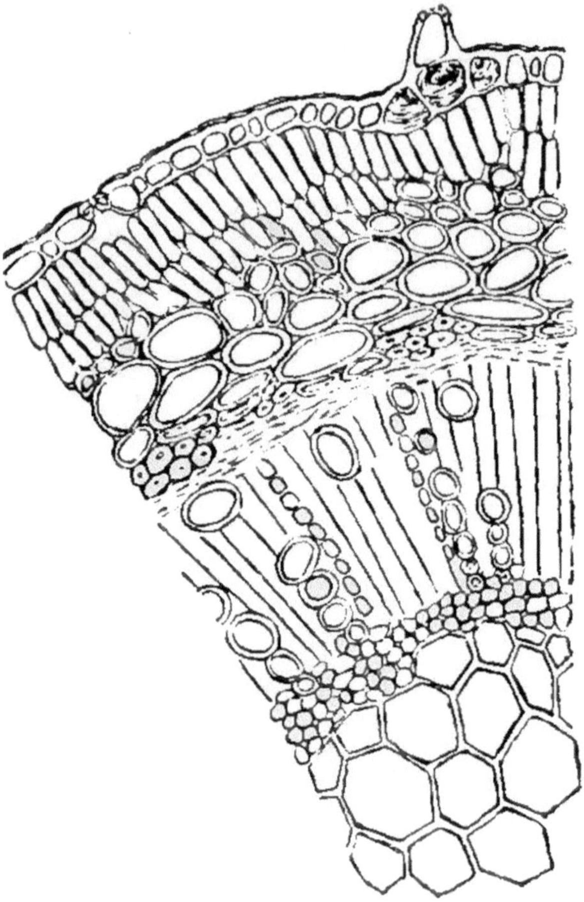

Fig. 72 Cross section through the stem of *Heliotropium undulatum* (Sabnis 1919–1921)

had reduced parenchyma (Figs. 94, 95, 96, 97, and 98). In only 10 percent was mechanical tissue prominent though present in the sheaths of monocots.

In the monocots, the vascular tissue was fairly prominent; in dicots, it was poorly developed. In the stems of the upland plants, there is high variation in the relative proportion of parenchyma, even though in the plants from xeric habitats, it was present in very slight degree. Mechanical tissue was found to be developed in most of the plants with large aerial portions, and usually absent in the low-growing plants with the exception of *Helianthus*. The tracheae were also variable in size and number. In this small group of plants, no correlation could be seen between the number of tracheae and the number of leaves. The cells with thick walls provide mechanical strength and could be also involved in protecting water-conducting tissues against water loss.

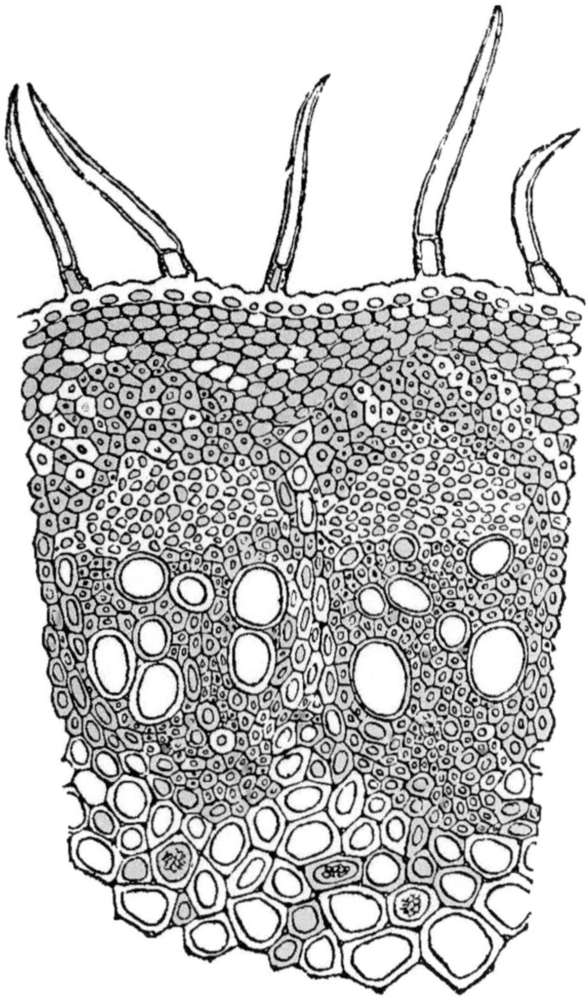

Fig. 73 Cross section through the stem of *Cocculus cebatha* (Sabnis 1919–1921)

In summary, Hayden (1919) emphasized that the subterranean stem is predominant as an equivalent of the primary root, especially in moist lowland regions; in addition, it has a pith that serves as a reservoir for water and hence increases its efficiency for radial distribution.

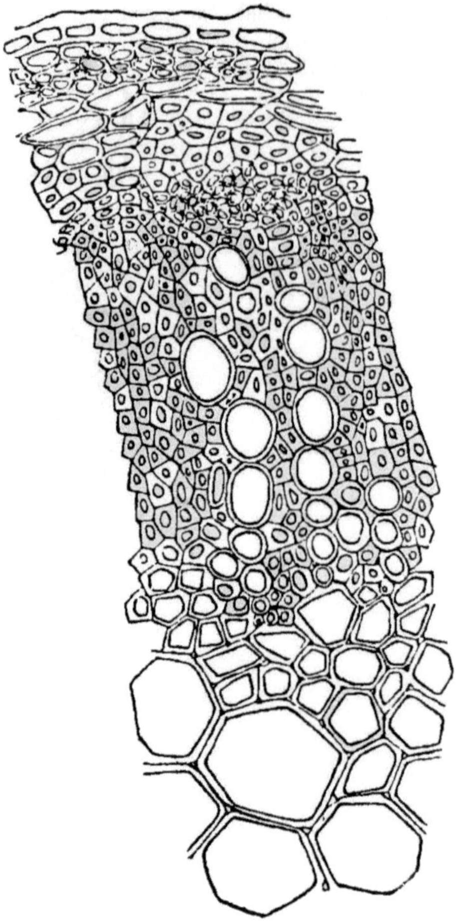

Fig. 74 Cross section through the stem of *Cleome papillosa* (Sabnis 1919–1921)

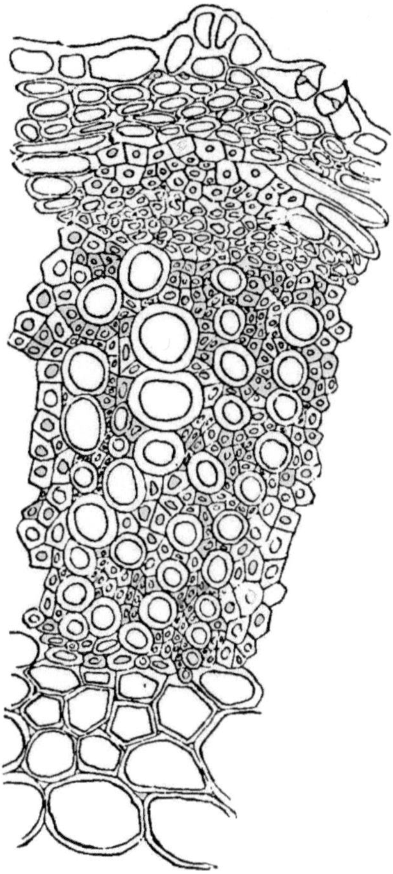

Fig. 75 Cross section through the stem of *Cleome brachycarpa* (Sabnis 1919–1921)

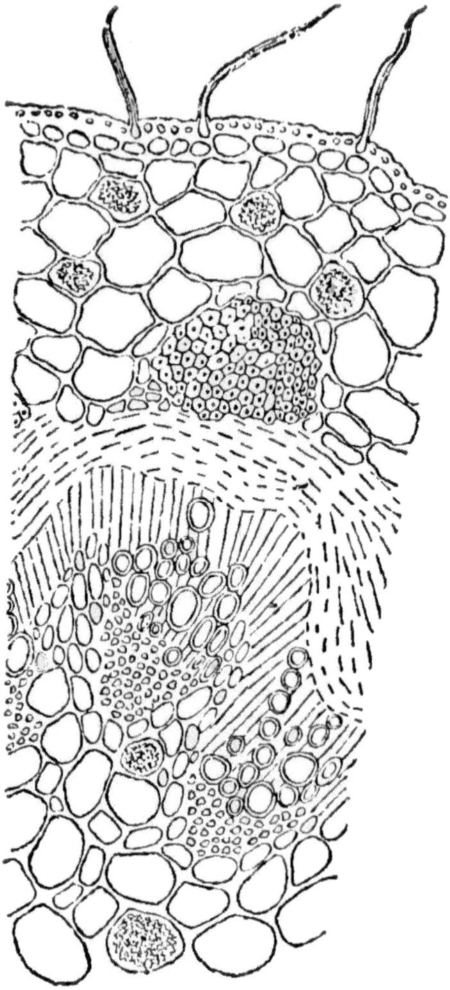

Fig. 76 Cross section through the stem of *Tribulus terrestris* (Sabnis 1919–1921)

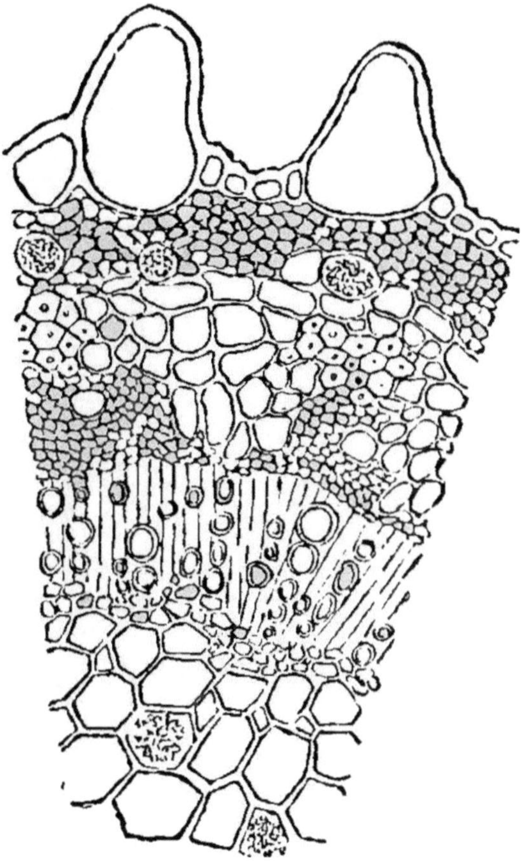

Fig. 77 Cross section through the stem of *Seetzenia orientalis* (Sabnis 1919–1921)

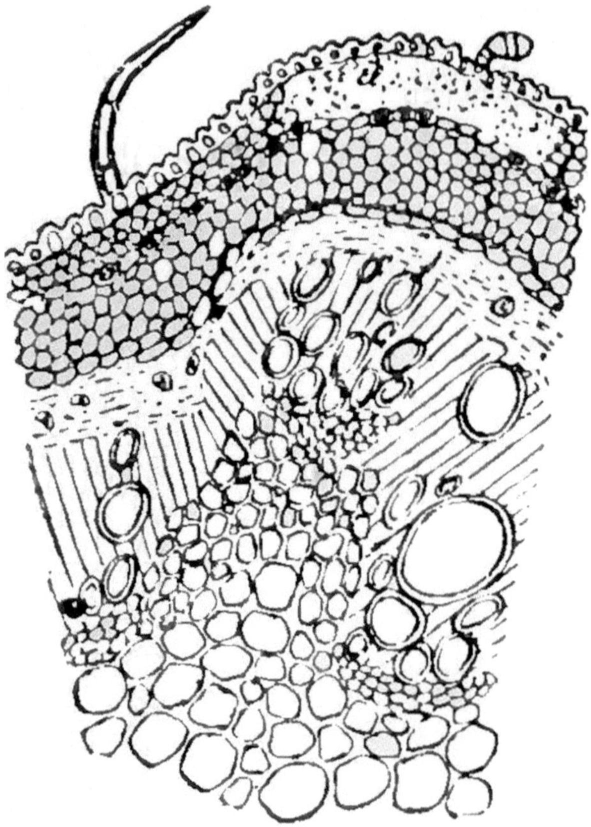

Fig. 78 Cross section through the stem of *Cardiospermum halicacabum* (Sabnis 1919–1921)

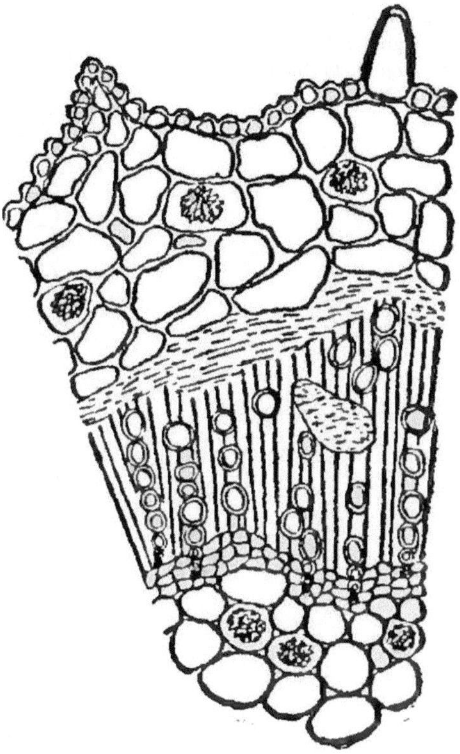

Fig. 79 Cross section through the stem of *Trianthema triquetra* (Sabnis 1919–1921)

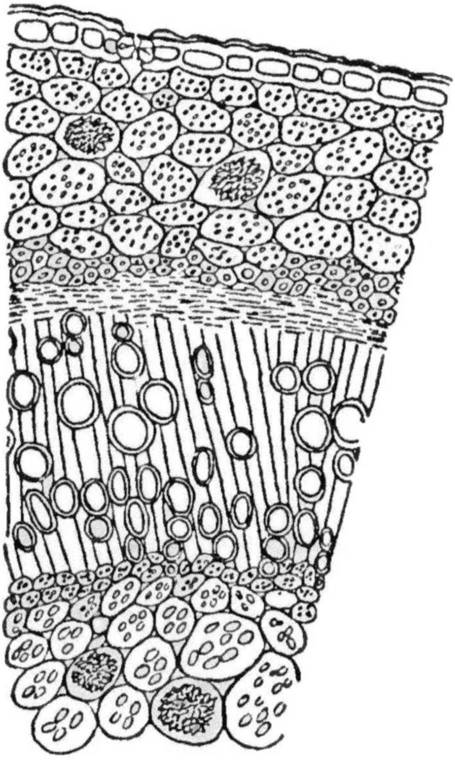

Fig. 80 Cross section through the stem of *Orygia decumbens* (Sabnis 1919–1921)

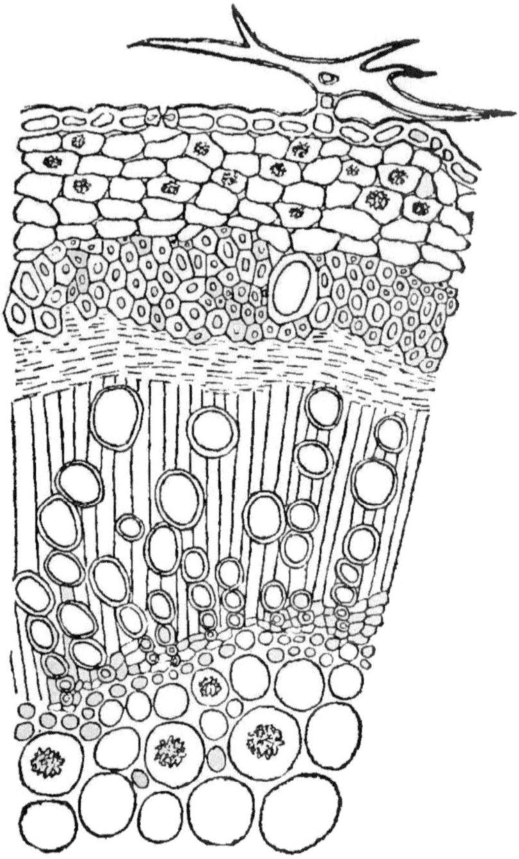

Fig. 81 Cross section through the stem of *Mollugo hirta* (Sabnis 1919–1921)

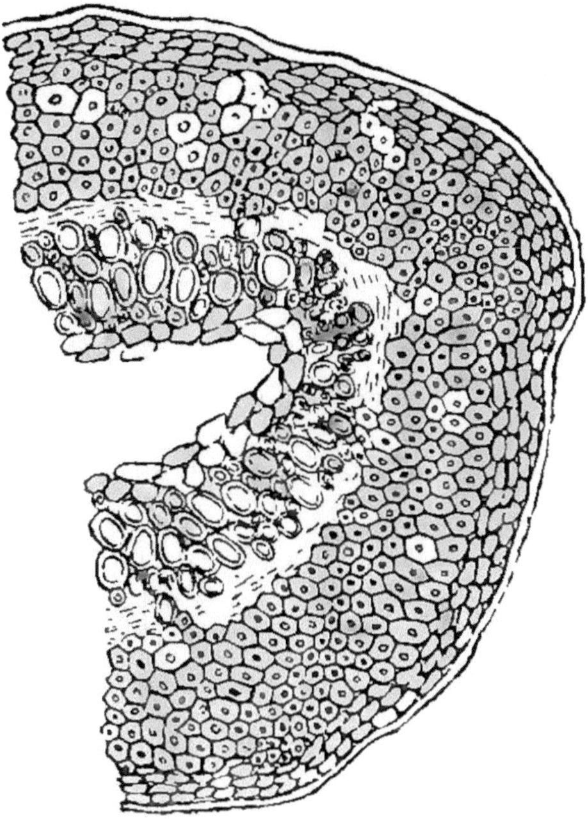

Fig. 82 Cross section through the stem of *Mollugo cerviana* (Sabnis 1919–1921)

Fig. 83 Cross section through the stem of *Anticharis linearis* (Sabnis 1919–1921)

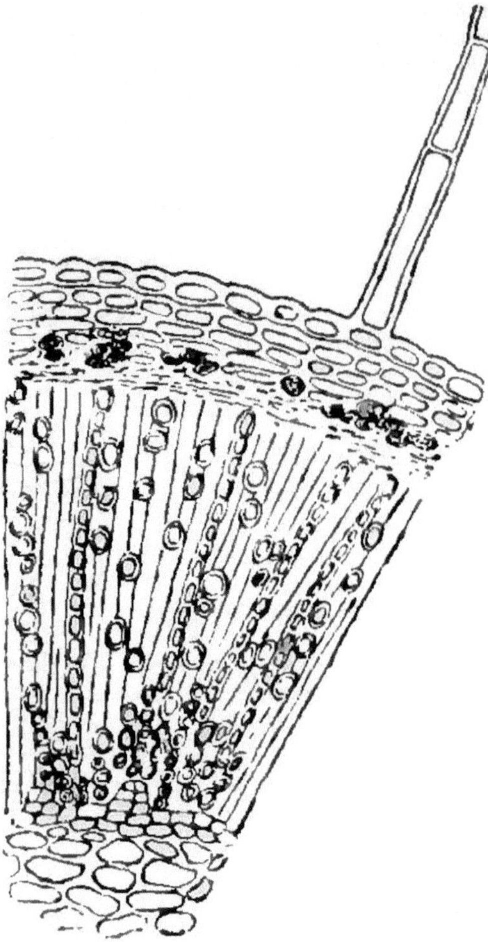

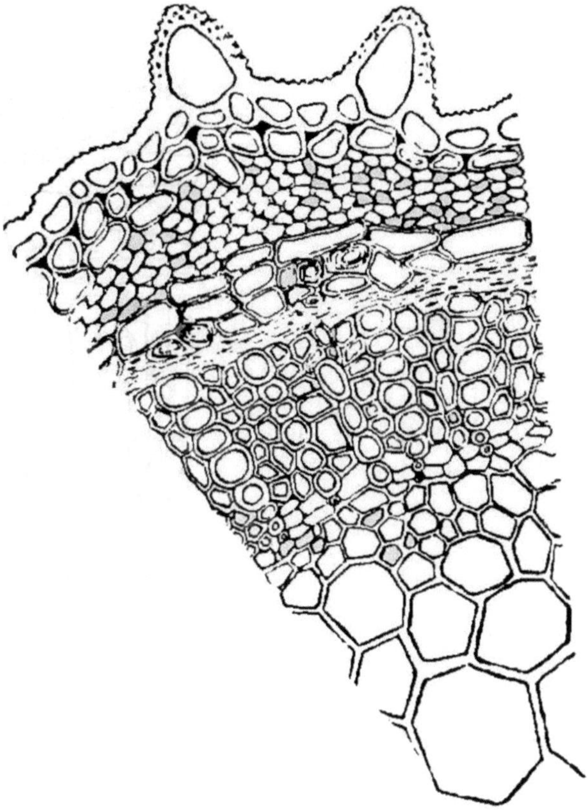

Fig. 84 Cross section through the stem of *Schweinfurthia sphaerocarpa* (Sabnis 1919–1921)

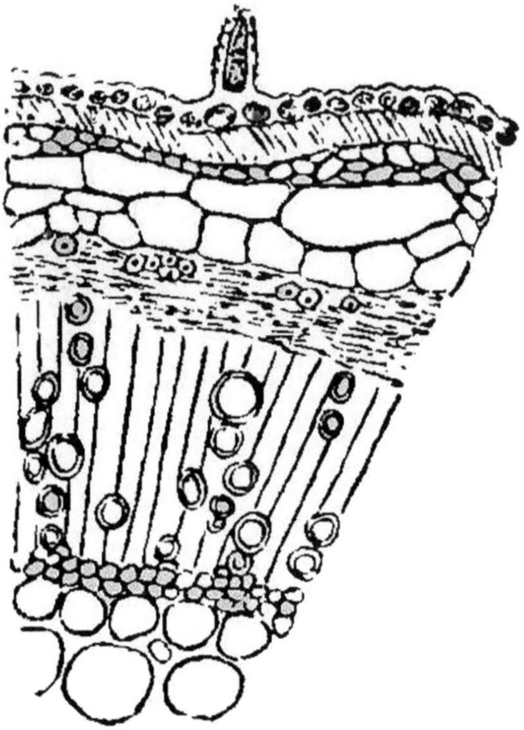

Fig. 85 Cross section through the stem of *Blepharis sindica* (Sabnis 1919–1921)

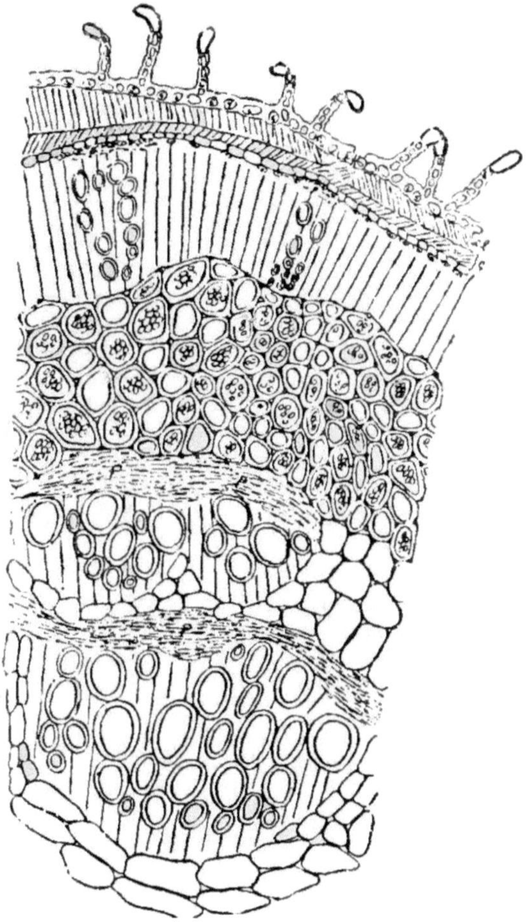

Fig. 86 Cross section through the stem of *Boerhavia diffusa* (Sabnis 1919–1921)

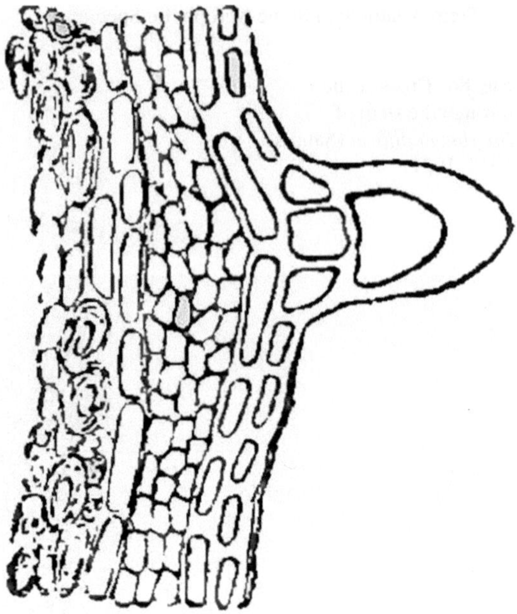

Fig. 87 Cross section through the stem of *Phyllanthus niruri* (Sabnis 1919–1921)

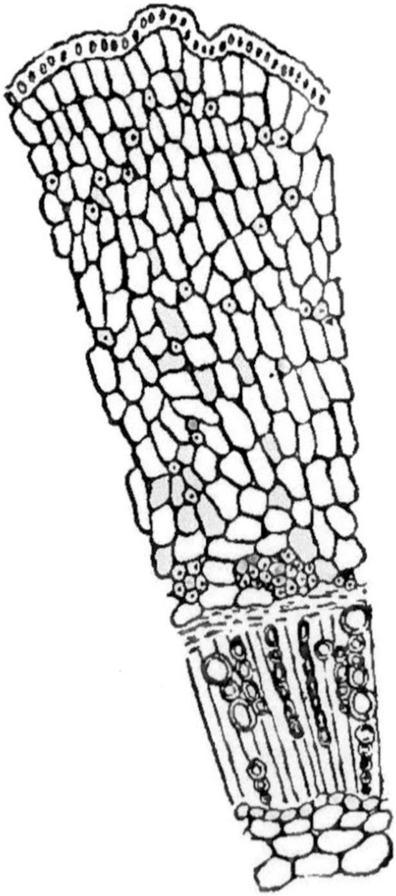

Fig. 88 Cross section through the stem of *Sarcostemma brevistigma* (Sabnis 1919–1921)

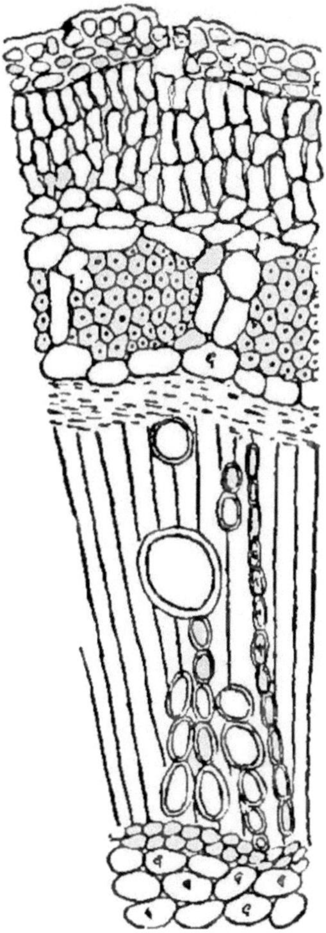

Fig. 89 Cross section through the stem of *Leptadenia spartium* (Sabnis 1919–1921)

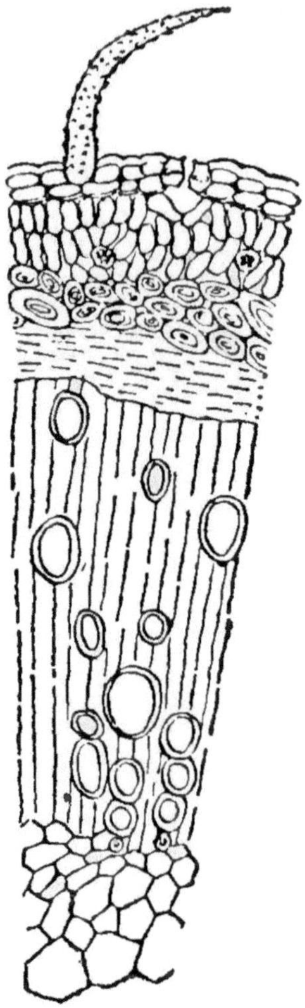

Fig. 90 Cross section through the stem of *Polygala erioptera* (Sabnis 1919–1921)

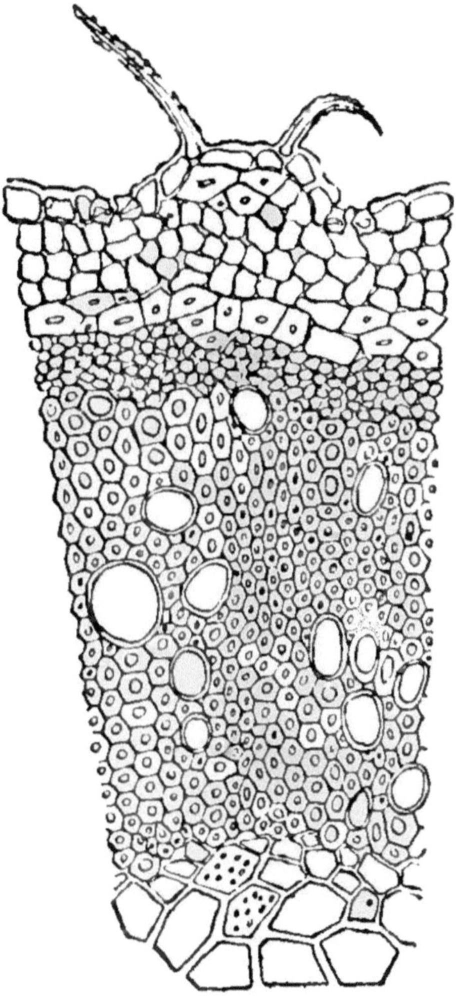

Fig. 91 Cross section through the stem of *Polygala irregularis* (Sabnis 1919–1921)

Fig. 92 Cross section through the stem of *Zizyphus rotundifolia* (Sabnis 1919–1921)

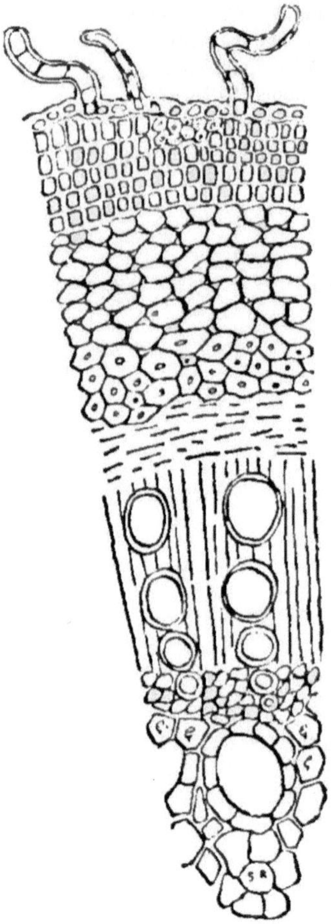

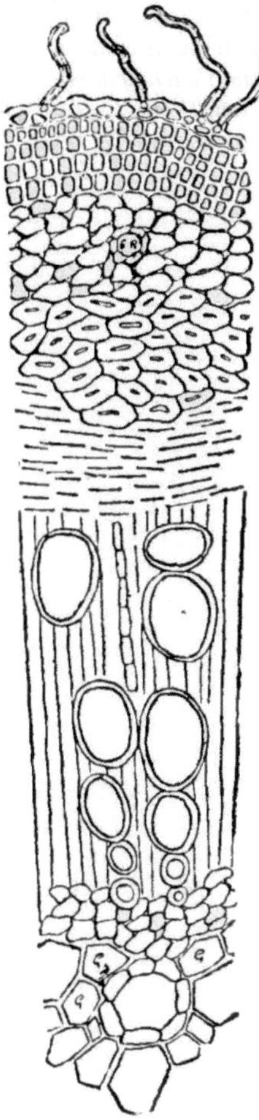

Fig. 93 Cross section through the stem of *Zizyphus truncata* (Sabnis 1919–1921)

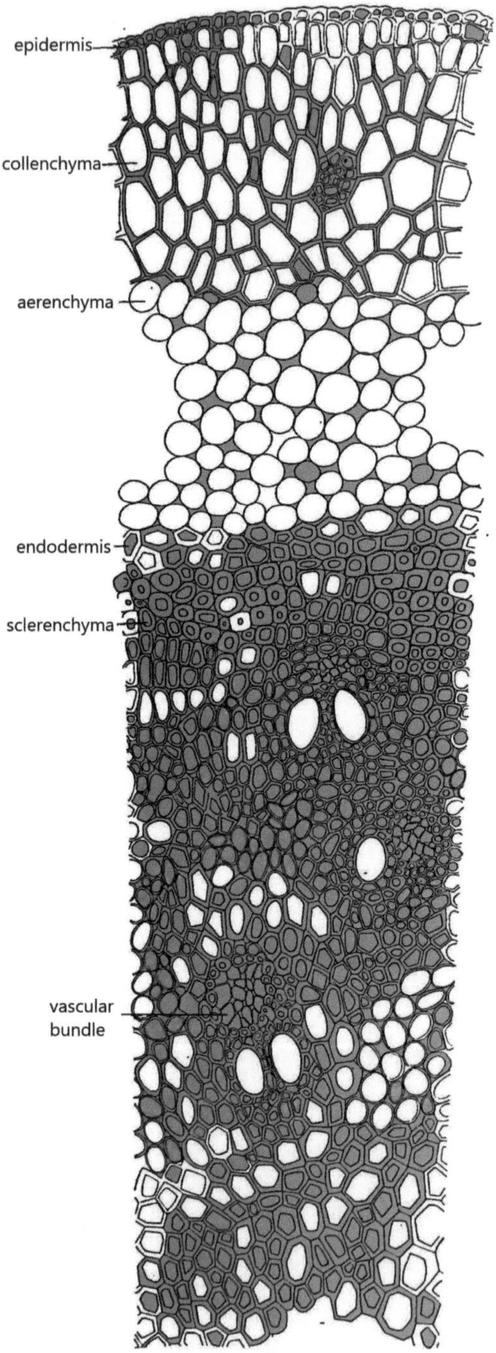

Fig. 94 Cross section through the rhizome of *Spartina michauxiana* (Hayden 1919)

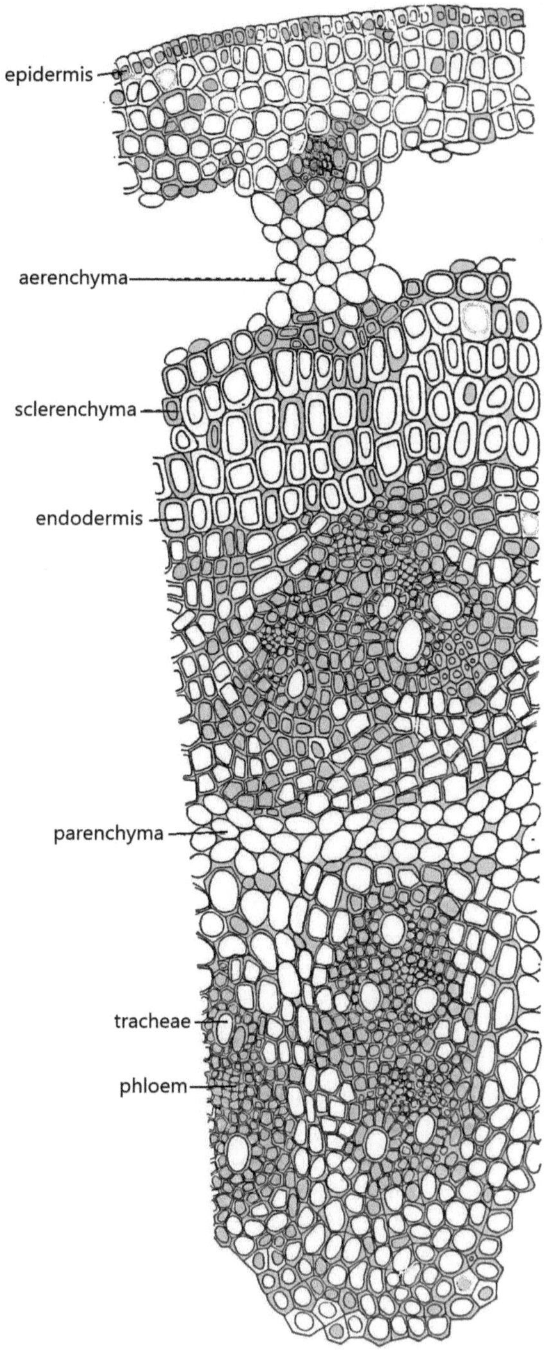

Fig. 95 Cross section through the rhizome of *Panicum virgatum* (Hayden 1919)

Fig. 96 Cross section through the rhizome of *Aster azureus* (Hayden 1919)

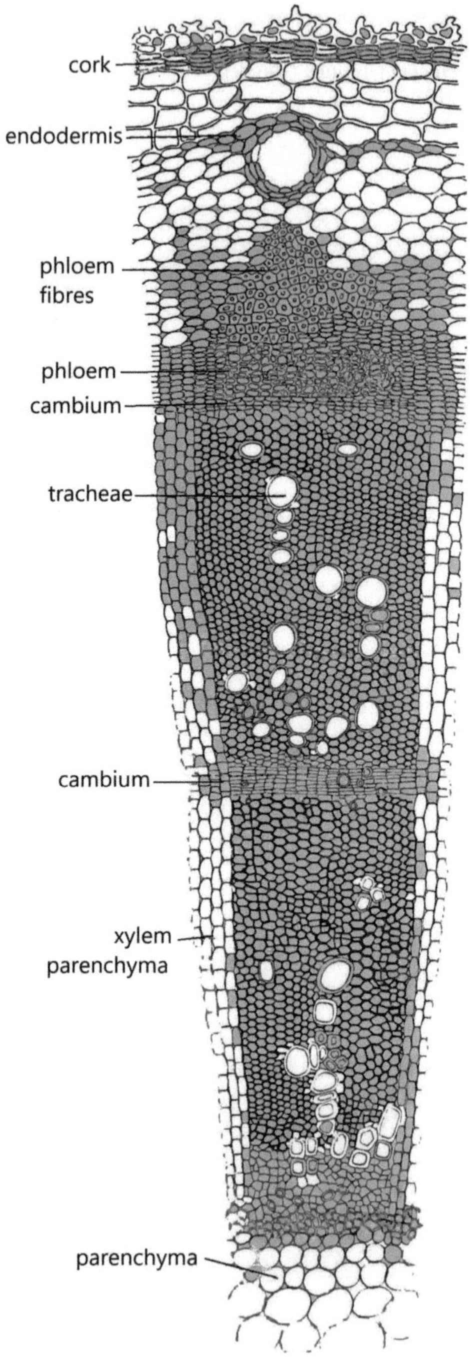

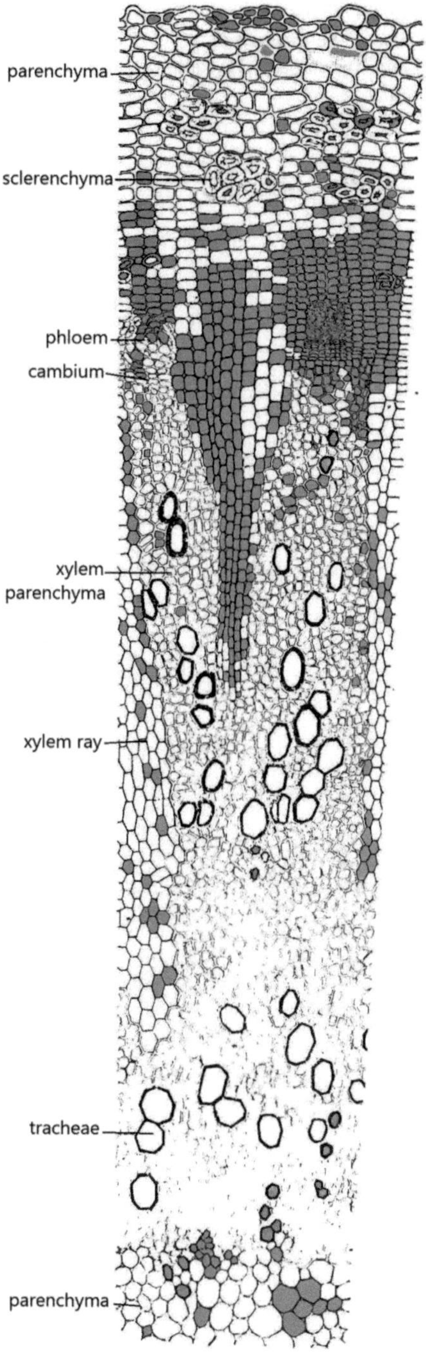

Fig. 97 Cross section through the rhizome of *Heliopsis scabra* (Hayden 1919)

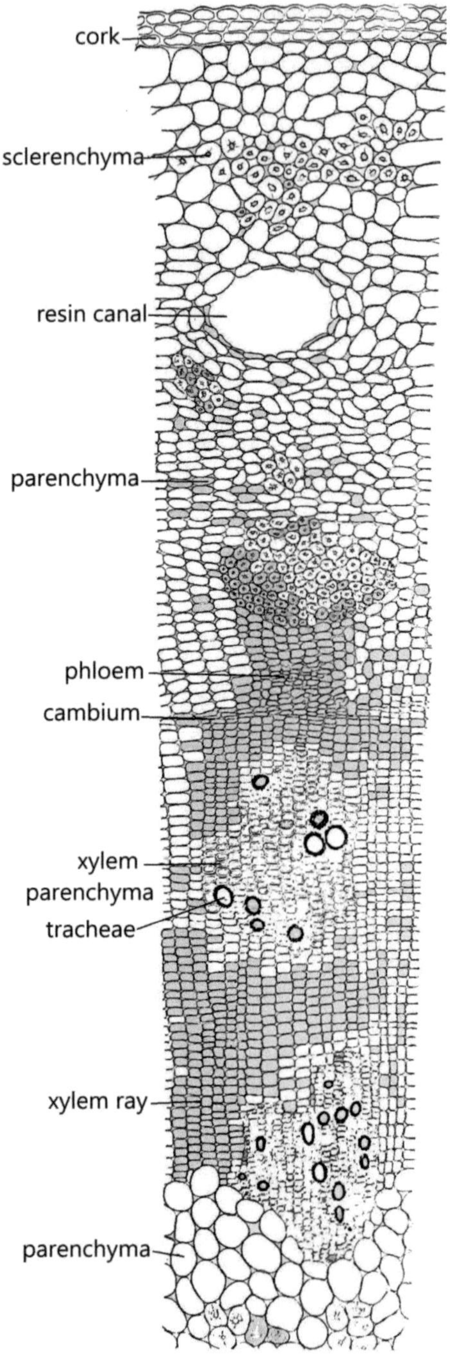

Fig. 98 Cross section through the rhizome of *Lepachys pinnata* (Hayden 1919)

Bibliography

Arber A., 1920 – *Water plants. A study of aquatic angiosperms*, Cambridge at the University Press

Arteca, R.N., 1997 – Flooding. In *Plant Ecophysiology*, M. N. V. Prasad (Ed.), pp. 151–171, New York. John Wiley & Sons

Barnes C. R., 1898 – *Plant life considered with special reference to form and function*, New York, Henry Colt & Company

Barnéoud F. M., 1848 – Mémoire sur l'anatomie et l'organogénie du *Trapa natans* (Linn.). *Ann. Sci. Nat., 3 ème sér., Bot.*, 9: 222–244

Barratt K., 1916 – The origin of the endodermis in the stem of *Hippuris*. *Ann. Bot.*, 30: 91–99

Batten L., 1918 – Observations on the ecology of *Epilobium hirsutum*. *J. of Ecol.*, 6: 162–177

Blanc M., 1912 – Sur les diaphragmes des canaux aérifères des plantes. *Revue gén. de Bot.*, 24: 233–243

Chatin G.-A., 1856 – *Anatomie comparée des végétaux. Plantes aquatiques*. Paris, J.-B. Baillière, Londres, New-York

Chrysler M. A., 1907 – The structure and relationships of the *Potamogetonaceae* and allied families. *Bot. Gaz.*, 44: 161–188

Coulter J. M., Barnes C. R., Cowles H. C., 1911 – *Textbook of botany for colleges and universities, vol. II. Ecology*, New York, Cincinnati, Chicago, American Book Company

Crang R., Lyons-Sobaski S., Wise R., 2018 – *Plant anatomy. A concept-based approach to the structure of seed plants*, Springer Nature Switzerland AG

Cronk J. K., Fennessy M.S., 2001 – *Wetland plants. Biology and ecology*, Lewis Publishers, Boca Raton, London, New York, Washington, D.C.

de Bary A., 1884 – *Comparative anatomy of the vegetative organs of the phanerogams and ferns*, Oxford: Clarendon

Ellenberg H., Weber H. E., Düll R., Wirt V., Werner W., Paulissen D., 1991 – Zeigerwerte von Pflanzen in Mitteleuropa. *Scr. Geobot.*, 18: 1–258

Glück H., 1911 – *Biologische und morphologische Untersuchungen über Wasser- und Sumpfgewächse*, Verlag von Gustav Fischer in Jena

Goebel K., 1891–1893 – *Pflanzenbiologische Schilderungen*, zweiter teil, erste lieferung, zweiter teil, zweite lieferung, Marburg N. G. Elwert'sche Verlagsbuchhandlung

Grace J. R., Wetzel R. G., 1998 – Long-term dynamics of *Typha* populations. *Aquatic Bot.*, 61: 137–146

Grigore M.-N., 2022 – *Morfologia plantelor. Aspecte ecologice*, Editura Universităţii „Ştefan cel Mare" Suceava

Grigore M.-N., Toma C., 2010 – *Halofitele. Aspecte de anatomie ecologică.* Editura Universităţii „Alexandru Ioan Cuza", Iaşi

Grigore M.-N., Ivănescu L., Toma C., 2014 – *Halophytes. An integrative anatomical study.* Springer, Cham, Heidelberg, New York, Dordrecht, London

Haberlandt G., 1914 – *Physiological plant anatomy*, Macmillan and Co, London

Hayden A., 1919 – The ecologic subterranean anatomy of some plants of a prairie province in central Iowa. *Am. J. Bot.,* 6: 87–105

Henslow G., 1895 – *The origin of plant-structures by self-adaptation to the environment*, London, Kegan Paul, Trench, Trübner, & Co, Ltd, Paternoster House, Charing Cross Road

Holzman B. A., 2008 – *Tropical forest biomes*, Greenxylem Press, Westport, Connecticut, London

Howard-Williams, C. 1993 – Processes of aquatic weed invasions: the New Zealand example. *J. of Aquatic Plant Management*, 31: 17–22

Jackson M. B., 1982 – Ethylene as a growth promoting hormone under flooded conditions. In *Plant Growth Substances*, P.E. Wareing (Ed.), pp. 291–301, London, Academic Press

Jackson M. B., 1990 – Hormones and developmental change in plants subjected to submergence or soil waterlogging. *Aquat. Bot.*, 38: 49–72

Jackson M. B. 1994 – Hormone action and plant adaptations to poor aeration. *Proc. of the Royal Soc. of Edinburgh*, 102B: 391–405

Kemp W. M., Lewis M. R., Jones T. W., 1986 – Comparison of methods for measuring production by the submersed macrophyte, *Potamogeton perfoliatus* L. *Limnology and Oceanography*, 31: 1322–1334

Laurentius A.C., Voesenek J., Banga M., Rijnders J. G. H. M., Visser E. J. W., Blom, C. W. P. M., 1996 – Hormone sensitivity and plant adaptations to flooding. *Folia Geobotanica Phytotaxonomica*, 31: 47–56

Lewakoffski N., 1873 – Ueber den Einfluss des Wassers auf das Wachsthum der Stengel und Wurzeln einiger Pflanzen (Gelehrte Schriften der k. Universitat in Kasan), Abstract in *Just's Bot. Jahresbericht, Jahrg.*, p. 594

Lüttge U., 2008 – *Physiological ecology of tropical plants*, second edition, Springer-Verlag Berlin Heidelberg

Ogden E. C., 1974 – *Anatomical patterns of some aquatic vascular plants of New York*, BULLETIN 424, New York State Museum and Science Service

Ridge I., 1987 – Ethylene and growth control in amphibious plants. In *Plant Life in Aquatic and Amphibious Habitats*, R. M. M. Crawford (Ed.), pp. 53–76, Oxford. Blackwell Scientific

Rosanoff S., 1871 – Ueber den Bau der Schwimmorgane von *Desmanthus natans* Willd. *Bot. Zeit.*, 29: 829–838

Sabnis T. A., 1919–1921 – The physiological anatomy of the plants of the Indian Desert. *J. of Indian Bot.*, 1(2): 33–43; 1(3): 65–83; 1(4): 97–113; 1(6–7): 183–205; 1(8): 237–251; 1(9–10): 277–295; 2(1–2): 1–20; 2(3): 61–79; 2(4–5): 93–115; 2(6–7): 157–173; 2(8–9): 217–235; 2(10): 271–299

Schenck H., 1886 – Vergleichende Anatomie der submersen Gewächse. *Bibliotheca Bot.*, 1: 1–67

Schenck H., 1893 – *Beiträge zur Biologie und Anatomie der Lianen, im Besonderen der in Brasilien einheimischen Arten.* Theil II. Beiträge zur Anatomie der Lianen, Heft V. Jena, Verlag von Gustav Fischer

Schenck H., 1889 – *Üeber das Aërenchym, ein dem Kork homologes Gewebe bei Sumpfpflanzen*, Berlin, Druck von G. Bernstein

Schimper A. F. W., 1903 – *Plant geography upon a physiological basis*, Oxford: Clarendon Press

Schulze E.-D., Beck E., Müller-Hohenstein K., 2005 – *Plant Ecology*, Springer Verlag, Berlin, Heidelberg

Schulze E.-D., Beck E., Buchmann N., Clemens S., Müller-Hohenstein K., Scherer-Lorenzen M., 2019 – *Plant ecology*, 2nd ed., Berlin: Springer Nature

Schrenk J., 1889 – On the Floating-tissue of *Nesaea verticillata* (L.) H.B.K. *Bull. Torr. Bot. Club*, 16: 315–323

Schweingruber F. H., Börner A., Schulze E.-D., 2011 – *Atlas of stem anatomy in herbs, shrubs and trees*, vol. 1, Springer Heidelberg Dordrecht London New York

Sculthorpe, C. D., 1967 – *The biology of aquatic vascular plants*, London, Edward Arnold Publishers

Schwarz F., 1883 – *Die Wurzelhaare der Pflanzen Ein Beitrag zur Biologie und Physiologie dieser Organe*, Breslau, Druck der Breslauer Genossenechafts-Buchdruckerei

Snell K., 1907 – Untersuchungen über die Nahrungsaufnahme der Wasserpflanzen. *Flora oder Allgem. Bot. Zeit.*, 98(2): 213–249

Solereder H., 1908 – *Systematic anatomy of the Dicotyledons. A handbook for laboratories of pure and applied Botany*, vol. 1, 2. Oxford: Clarendon.

Strasburger E., Noll F., Schenck H., Schimper A. F. W., 1894, 1923 – *Lehrbuch der Botanik für hochschulen*, Jena, Verlag von Gustav Fischer

Summers J. E., Ratcliffe R. G., Jackson M. B., 2000 – Anoxia tolerance in the aquatic monocot *Potamogeton pectinatus*: absence of oxygen stimulates elongation in association with an unusually large Pasteur effect. *J. of Exp. Bot.*, 51: 1413–1422

Şerbănescu-Jitariu G., Toma C., 1980 – *Morfologia şi anatomia plantelor*, Editura Didactică şi Pedagogică, Bucureşti

van Tieghem Ph., 1871 – Recherches sur la symétrie de structure des plantes vasculaires. *Ann. Sci. Nat. Bot.*, 13: 6–314

Vartapetian B. B., Jackson M. B., 1997 – Plant adaptations to anaerobic stress. *Ann. of Bot.*, 79: 3–20 von Marilaun A. K., 1902 – *The natural history of plants*, vol. 1, 2, London Blackie & Son, Limited, 50 Old Bailey, E.G., Glasgow and Dublin

Vöchting H., 1872 – *Zur Histologie und Entwickelungsgeschichte von Myriophyllum*. Dresden, Druck von E. Blochmann & Sohn

Walter H., 1974 – *Vegetația Pământului în perspectivă ecologică*, Editura Științifică, București

Warming E., 1909 – *Ecology of plants. An introduction to the study of plant communities* (English edition prepared by P. Groom, I. B. Balfour), Oxford: Clarendon Press

Yeo R. R., Falk R. H., Thurston J. R., 1984 – The morphology of *Hydrilla* (*Hydrilla verticillata* (L.f.) Royle). *J. of Aquat. Plant Management*, 22: 1–17

Leaf

Leaf

Among all the vegetative organs of the plant, the leaf is considered to have the highest morphological and anatomical plasticity; the influence of environmental factors on the anatomical structure of the leaf is much more obvious and relevant than in the case of the other vegetative organs of the plant (root, stem), mainly due to the multiple functions played by the leaf, leaf metamorphoses, the position of the leaf in relation to the stem, and the living environment (Bannister 1976; Grigore 2022).

During the process of ontogenesis, both sides of the leaf primordium can develop (resulting in bifacial leaves) or only the lower side (resulting in monofacial leaves) (Toma 1977).

Phylogenetically and ontogenetically, the structure of the leaf lamina is the result of the action of a complex of factors, internal and external, of which light plays a very important role—without minimizing the importance of others.

Light has a very important effect on the structure of the lamina, its distribution on the two sides being conditioned by the orientation of the leaf in space; from this point of view, two main categories of leaves can be distinguished:

- Euphotometric or bifacial dorsiventral (heterofacial): arranged more or less perpendicularly or obliquely to the stem, with the lamina receiving more light on the upper side, and less on the lower side; therefore, the lamina receives light differently on the two sides, which is why they are also called *diphotic* (Clements 1907—prominent representative of the American school of plant ecology, from the turn of the nineteenth and twentieth centuries).
- Panphotometric or bifacial isofacial (equifacial, isolateral): arranged more or less parallel to the stem, so somewhat vertically, with both sides more or less uniformly illuminated (Toma 1977; Toma et al. 2003); these leaves are also called *isophotic* (Clements 1907), meaning they receive light (approximately) equally on both sides.

It should be noted that the influence of light on the anatomical structure of the leaf is much more complex and subtle, and the discussions on this subject can be expanded throughout a separate chapter; in addition, the approach can be further complicated if we talk, separately, about "regulation" ("adjustment") and adaptation to stimuli (environmental factors) in the life of the plant, a valid discussion therefore also in the case of light.

M.-N. Grigore, *Plant Ecological Anatomy*,
https://doi.org/10.1007/978-3-032-01332-3_6

Bifacial leaves have two sides: the upper side (adaxial, ventral), corresponding to the upper epidermis, and the lower side (abaxial, dorsal), corresponding to the lower epidermis.

Broadly speaking, we can describe the general plan of structure of the leaf lamina as follows (Fig. 1): an upper side (adaxial, ventral); a lower side (abaxial, dorsal); upper epidermis; mesophyll ("the middle of the leaf"), with veins (conducting tissues) and mechanical tissues—the totality of the tissues located between the two epidermises; and lower epidermis (Şerbănescu-Jitariu and Toma 1980).

Leaves are called bifacial because they have two distinct sides, the upper (upper epidermis) and the lower (lower epidermis).

Depending on the organization of the mesophyll (between the two epidermises), bifacial leaves can be as follows:

- Heterofacial (dorsiventral): palisade tissue under the upper epidermis and spongy tissue under the lower epidermis; in this case, the leaf receives more light on the upper side and less light on the lower side.
- Inverse heterofacial (inverse dorsiventral): palisade tissue under the lower epidermis and spongy tissue under the upper epidermis, so the leaf receives more light on the lower side and less light on the upper side.
- Equifacial (isofacial), when light falls equally on both sides of the leaf.

In turn, equifacial leaves can be equifacial centrically homogeneous: the same type of tissue (either palisade or spongy) under both epidermises, and in the middle of the leaf, e.g., in *Pinus sylvestris* (pine) and *Vallisneria spiralis* (eelweed); or equifacial centric heterogeneous: the same type of tissue under both epidermises, but in the middle of the leaf, a different type of tissue is located, for example, in species of *Dianthus* (carnation).

In the spirit of the discussion above, regarding the influence of light on the anatomical structure of the leaf, we can say, in short, that all these structural variations represent, in fact, changes in the chlorenchyma in relation to the action of light; the uneven illumination, at the level of the leaf, is thus translated by the "conversion" of

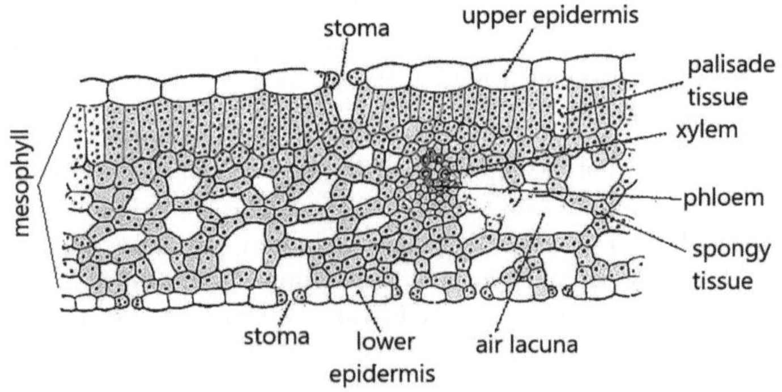

Fig. 1 The anatomical structure of the leaf limb *Ranunculus acris*—typical mesophyte leaf (plant that grows in conditions of sufficient humidity) (Bonnier and du Sablon 1905)

the chlorenchyma in the two major tissues of the lamina: the palisade tissue and the spongy tissue.

In this chapter, we will focus our attention only on the anatomy of the leaf lamina; the anatomy of the leaf petiole is described in the classic works on plant anatomy (Toma 1977; Şerbănescu-Jitariu and Toma 1980).

1 Leaf Structure in Gymnosperms

The epidermis consists of cells with all the walls (*Pinus*)—or only the external one—very thick, cutinized, and waxy. The stomata are located below the level of the epidermis, thus forming deep epistomatal cavity. All of these are considered xeromorph adaptations, which we will discuss in the chapter dedicated to xerophytes' leaf anatomy.

Usually, under the epidermis, there is a hypodermis (Fig. 2), which is sclerenchymatous, continuous, or discontinuous—always next to the substomatal chambers—and monolayered or multilayered (especially at the edges of the leaf).

The mesophyll (assimilatory parenchyma) can be as follows:

- Homogeneous: with cells with folds (*Pinus*) (Fig. 3) or with a tendency to acquire a palisade shape (*Picea*—Figs. 2 and 4); in this case, the leaves are bifacial equifacial.

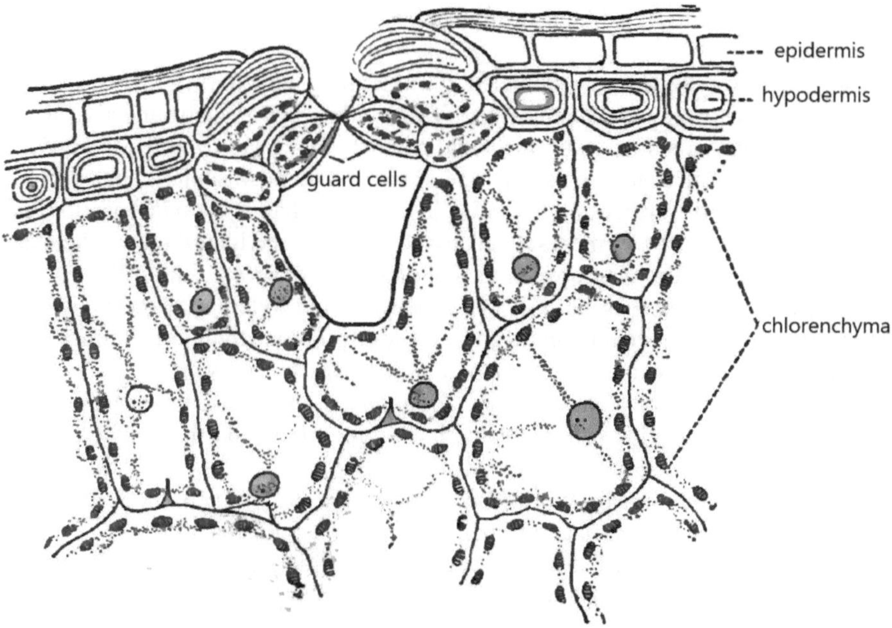

Fig. 2 Cross section through the leaf of *Picea excelsa*—details (Scott 1906)

Fig. 3 Cross section through the leaf of *Pinus* ssp. (original); black arrow indicates position of stomata

- Heterogeneous: with palisade tissue on the upper side and spongy tissue on the lower side (*Abies*) (Fig. 6)—the leaves having a heterofacial structure, with typical dorsiventrality—or with spongy tissue on the upper side and palisade tissue on the lower side, leaves with heterofacial structure, with reverse dorsiventrality (Toma 1977; Toma et al. 2003).

In the mesophyll, there are often one or more resin ducts (Fig. 4); however, there are also species of conifers without resin ducts, such as yew (*Taxus baccata*—Fig. 5) (Frunzete 2023).

The central cylinder is surrounded by a primary endodermis (cells showing Caspary thickenings in the lateral walls). Vascular bundles are surrounded by a special tissue, called transfusion or tracheidal parenchyma (Esau 1965; Beck 2005); in the central cylinder, there are usually one or two collateral vascular bundles, closer together (*Picea*) or further apart (*Pinus*). The phloem has an abaxial location, and the xylem, adaxial; sometimes, sclerenchyma fibers can be identified, on the external face of the phloem and between the bundles (where there are several).

Transfusion tissue (Fig. 7) is located between the xylem and mesophyll and apparently serves to recover solutes from the transpiration stream before they leave the (fibro)vascular bundle and return them to the adjacent phloem.

The phloem consists of sieve cells and phloem parenchyma cells, and the xylem of rows of dead tracheids with areolate pits (having torus) and living xylem parenchyma cells (Esau 1965).

The general contour of the cross section through the limb can be semicircular (*Pinus*), rhomboidal (*Picea*), and ribbon-shaped (*Abies*) (Toma 1977).

Many of the mentioned structures represent xeromorph features, thus ensuring protection against excessive transpiration, because plants suffer from physiological

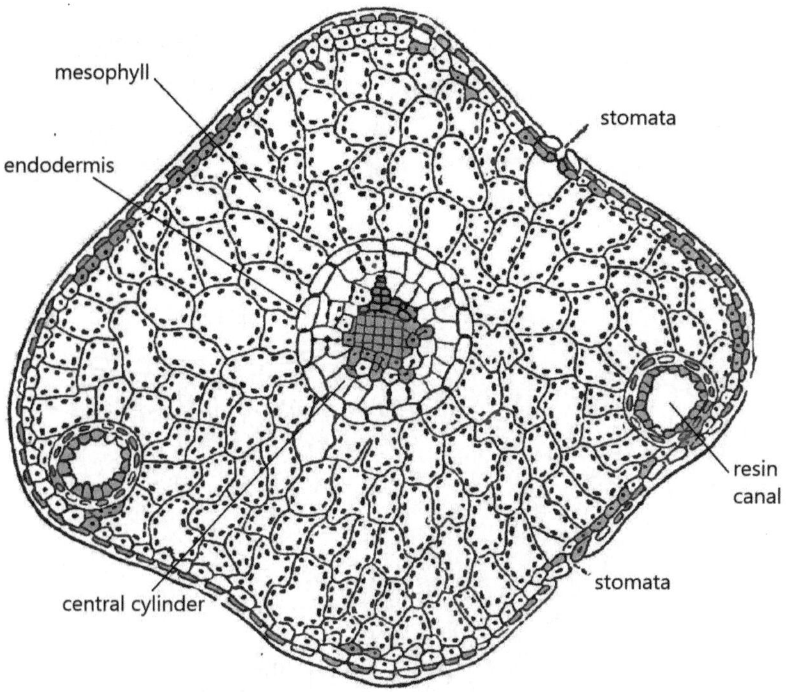

Fig. 4 Cross section through the leaf of *Picea excelsa* (Scott 1906)

drought (among other things, the substrate—the soil is rich in water, but it is cold and therefore difficult to absorb) (Grigore 2021, 2022).

2 Leaf Structure in Angiosperms

2.1 Bifacial Leaves Heterofacial (Dorsiventral) (Figs. 8 and 9)

These leaves have a perpendicular or oblique orientation on the stem, so the lamina is illuminated unevenly on the two sides. They are found in most woody and herbaceous plants, especially dicots.

The epidermis is single-layered, sometimes provided with hairs, glands, and often localized stomata: exclusively on the lower side of the lamina, hypostomatic leaves (as in most plants, especially woody ones), and on both sides, amphistomatic leaves (as in many labiates, crucifers, legumes, grasses, etc.). The leaves of submerged aquatic plants do not have stomata, and in the floating ones, the stomata are present only on the upper side of the lamina (water lilies)—epistomatic leaves (Grigore 2022). We will return to the anatomical structure of the leaf of aquatic plants.

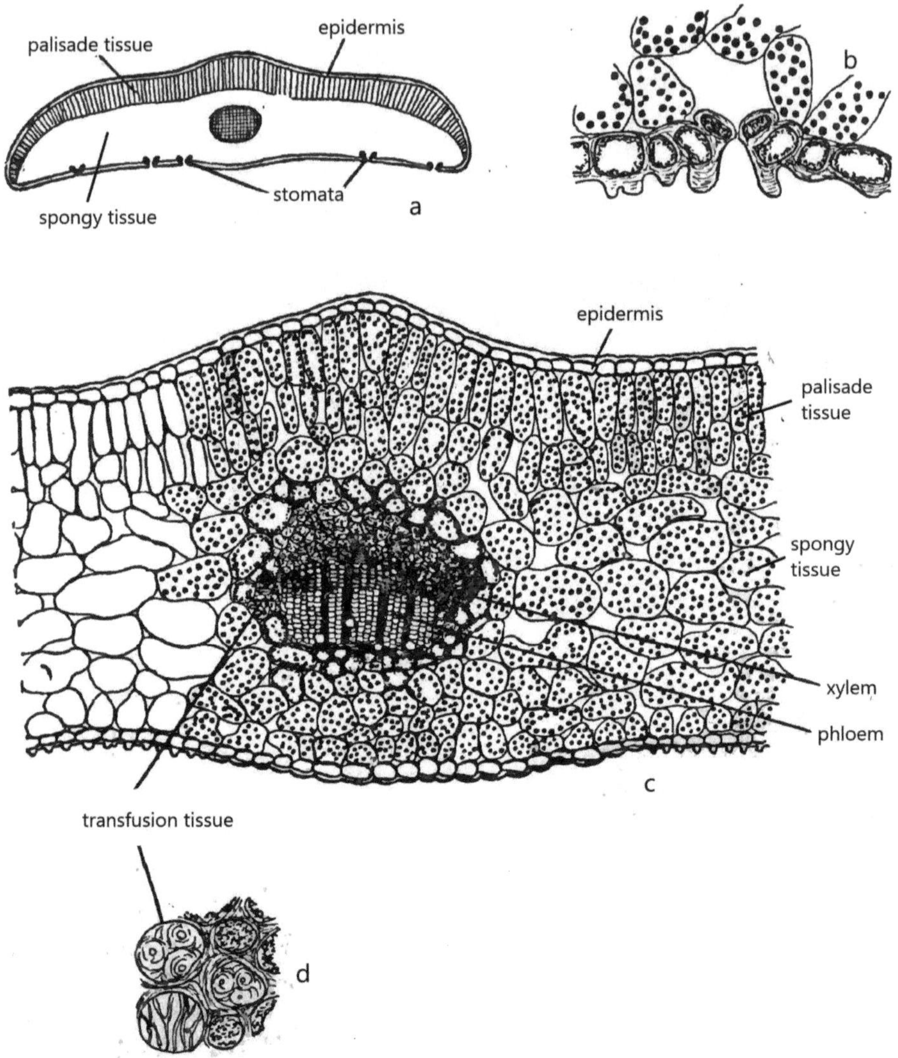

Fig. 5 Structure of the leaf of *Taxus baccata*; (**a**) cross section, general view; (**b**) stoma; (**c**) central part of cross section; (**d**) few cells of the transfusion tissue (Fritch and Salisbury 1920)

In broad leaves, the epidermal cells are more or less isodiametric, and the linear ones are noticeably longer than wide (Toma 1977).

The epidermis of the leaves of land plants is covered with a cuticle and often wax and is thicker on the upper side of the lamina, thus helping to regulate the plant's transpiration process. In grasses and sedges, the epidermis may be lignified and mineralized (silica impregnation of the outer cell wall).

As a rule, epidermal cells do not contain chloroplasts; however, there are exceptions—the leaves of submerged aquatic plants, many shade plants and few light

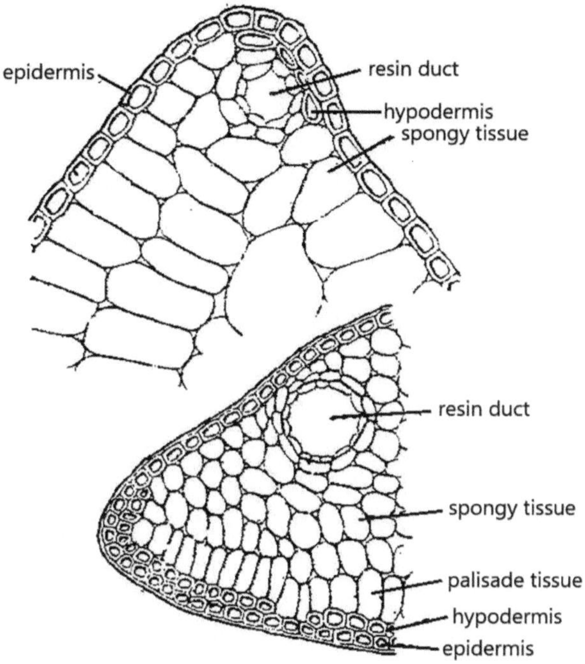

Fig. 6 Resin duct—cross section through the leaf of *Abies pectinata*; top, young leaf; mature leaf (Belzung 1900)

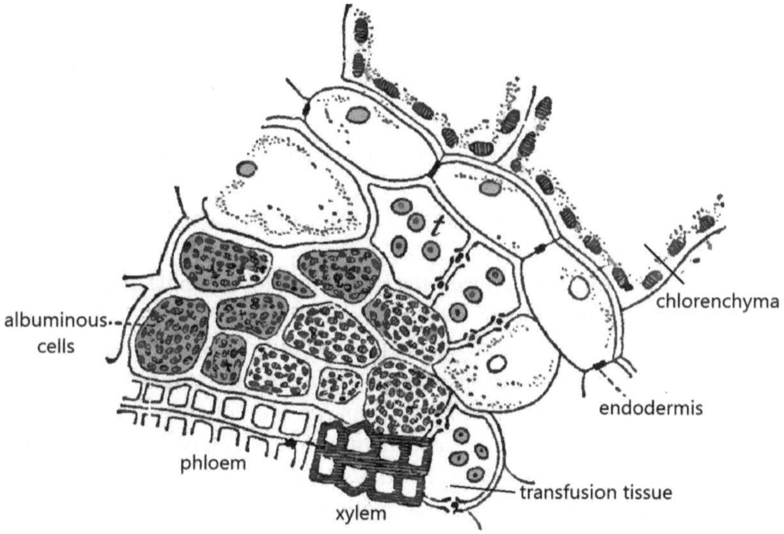

Fig. 7 Cross section through the leaf of *Picea excelsa*—details (Scott 1906)

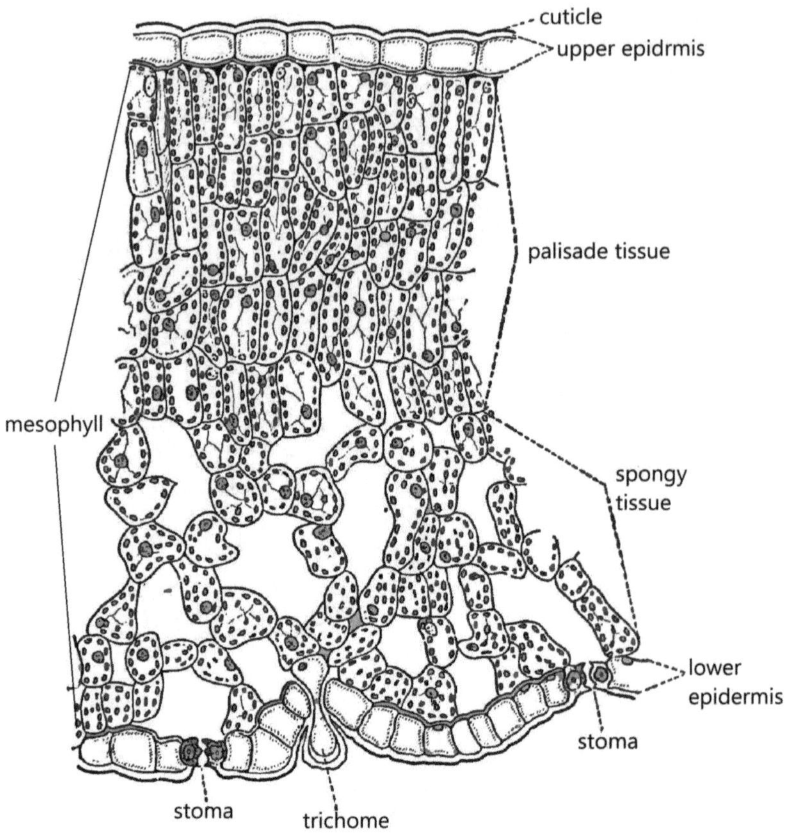

Fig. 8 Leaf with heterofacial bifacial structure (dorsiventral)—*Cheiranthus cheiri* (Scott 1906)

plants, and some psychrophytes (plants of cold habitats—see below, foliar ecological anatomy of psychrophytes).

The mesophyll is located between the upper and lower epidermis, in the middle of the leaf (from the Greek word *meso*, "middle," and *phyllos*, "leaf"; see Fig. 1). Mesophyll cells are parenchymal cells rich in chloroplasts and surrounded by a large volume of intercellular space, which facilitates gas exchange necessary for efficient photosynthesis. The intercellular spaces are connected with the external atmosphere through stomata. As a rule, the mesophyll is multilayered, but in some aquatic plants, for example, *Elodea canadensis* (Canadian waterweed), it is monolayered or absent, the lamina being practically represented only by the two epidermises (except for the midrib region) (Fig. 19) - Ecological Anatomy of the Leaf, with meatus and air gaps between the cells.

The mesophyll is often organized into two layers, a palisade layer (palisade tissue) and a spongy layer (spongy tissue). The cells of the palisade layer are columnar, while those of the spongy layer have an irregular shape (Fig. 1). Although the palisaded layer may appear very compact in sections through the leaf, in reality, the

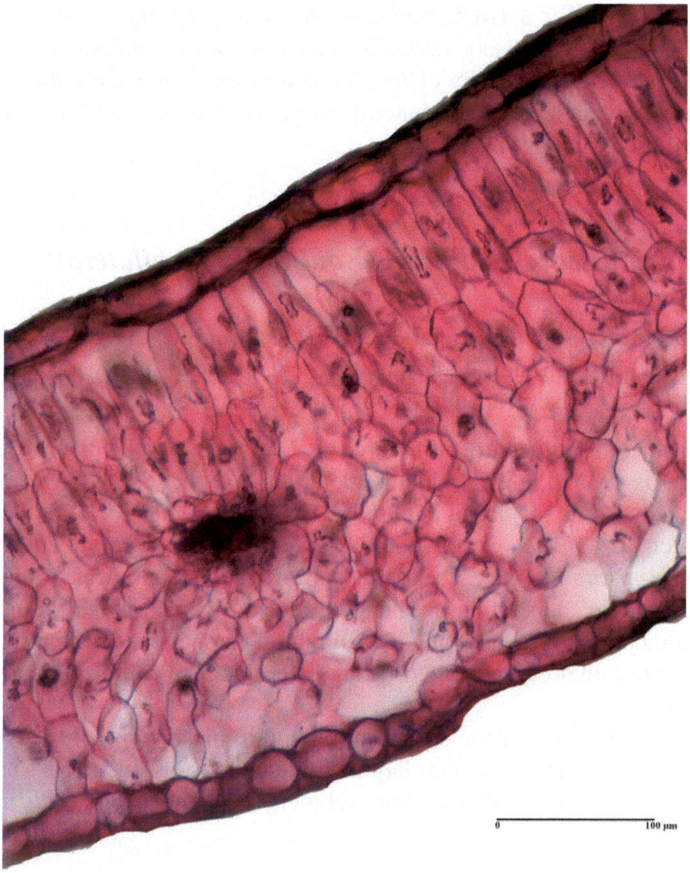

Fig. 9 Leaf with heterofacial bifacial structure (dorsiventral)—*Glaux maritima* (Grigore et al. 2014)

palisaded cells are actually surrounded by intercellular spaces, and most photosynthesis occurs within them rather than in the spongy tissue. The open arrangement of the cells in the spongy tissue favors rapid gas exchange, with carbon dioxide entering and oxygen and water vapor exiting through the stomata.

Hydrophytes (aquatic plants), in particular, have a large amount of air spaces (aerenchyma) in their mesophyll. This aerenchyma helps in floating of the leaves and conducting air throughout the plant and down to the roots.

The palisade tissue is usually located under the upper side of the leaf, and the spongy tissue, under the lower side. Leaves exposed to high light (sunny environments) may have two or more layers of palisade cells under the upper side, and xerophytes may have a palisade layer under both the upper and lower sides. The leaves of monocots, such as those of maize and other grasses, have a simpler mesophyll. In this case, the cells of the mesophyll are similar in shape and there is no proper differentiation into spongy and palisade tissue.

If the leaves suffer a twist, becoming involute (thus having their lower face towards the light), the palisade tissue lies below the lower epidermis, the leaf blade having a reverse dorsiventrality (*Thymelaea*, *Tafalla*); in this case, the palisade tissue, the most specialized for the function of photosynthesis, is always arranged on the side exposed to light (Toma 1977).

2.2 *Bifacial Leaves Isofacial (Equifacial, Isobilateral)*

They are arranged more or less parallel to the stem, so they have the lamina illuminated equally on both sides. They are present in some dicots, but especially in monocots. There are several subtypes of equifacial leaves:

1. Heterogeneous with centric structure (Fig. 10): under both epidermises, there is palisade tissue, and in the middle is spongy or compact tissue, formed by isodiametric cells (some species of *Dianthus*, carnation; *Medicago*, alfalfa; *Astragalus*, milkvetch; *Artemisia*, wormwood; *Salix*, willow; *Eucalyptus*, eucalyptus). At *Linaria genistifolia*—flax weed—depending on the ecological factors in which the plant grows, depending on the stem level at which the leaf is inserted, and depending on the level at which the leaf lamina is sectioned, different categories of structures are observed: heterofacial, heterofacial to equifacial, and equifacial (Toma 1977).
2. Homogeneous with centric structure: the entire mesophyll is homogeneous, of spongy type, with gaps and large air chambers, as in some shade plants (*Chlorophyllum*, *Jasione*) and many submerged aquatic plants (*Vallisneria*—see

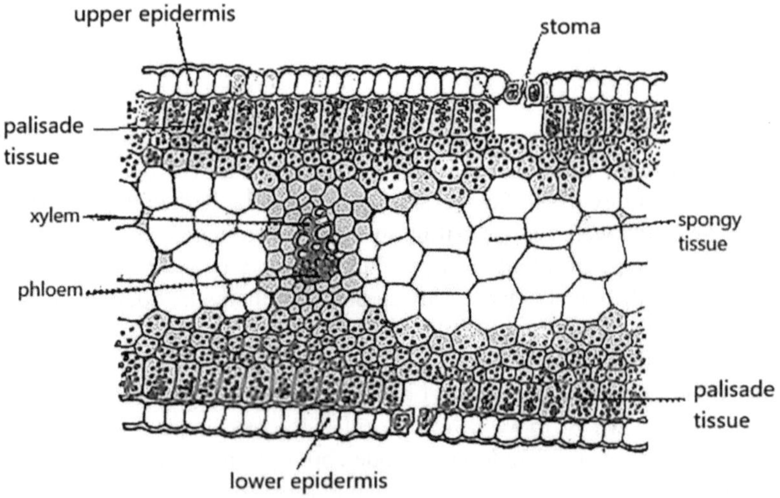

Fig. 10 The structure of the leaf lamina of *Narcissus tazetta*—cross section (Bonnier and du Sablon 1905)

Fig. 21) - Ecological Anatomy of the Leaf; in grasses, the mesophyll is homogeneous but compact, with small air spaces between the isodiametric cells (except for the hypodermic cells that are slightly tall and arranged in an orderly manner, perpendicular to the epidermis).

3. Intermediate structure: the mesophyll is entirely of the palisade type, but the cells are higher under the upper epidermis and lower under the lower epidermis, as in many littoral species (*Lathyrus maritima*, *Silene maritima*).

In all three situations, the two sides of the foliar lamina can only be specified by the position of the xylem, which is oriented, as we have seen, towards the upper epidermis (Toma 1977; Toma et al. 2003).

The mesophyll, in some species, can present various particularities, many of them having ecological significance. Thus, the mesophyll can present:

- Unilayered or multilayered hypodermis under the upper epidermis, considered as a water-storage tissue (Fig. 11) and a protective tissue of chlorophyll cells

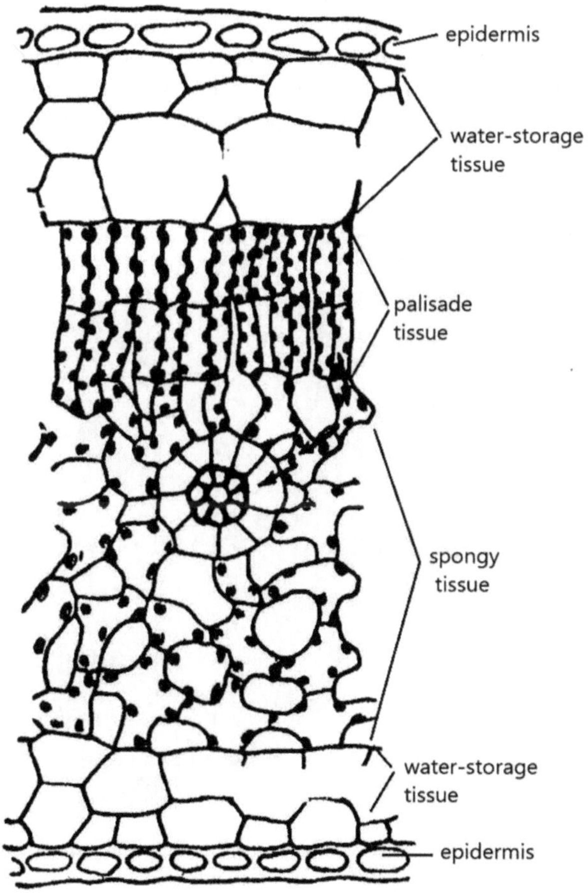

Fig. 11 Cross sections through the leaf lamina of *Ficus elastica* (Stevens 1924)

against a strong insolation (*Ficus carica*; *Nerium oleander*, oleander; *Musa paradisiaca*, banana tree).

- Mechanical elements: cords of sclerenchyma, collenchyma, or sclereids.
- Various secretory structures (isolated cells, laticiferous cells, secretory cavities, oxaliferous cells, cells with mucilage, cells with tannin, etc.)

In some grasses, many of them being xerophilous species (*Stipa*, sedge; *Arundo* sp., Italian cane), the upper epidermis shows groups of bulliform cells (motors) (Figs. 12, 13, 14, and 15); by reducing the volume during the drought period (through water loss), they cause the rolling of the leaf blade, protecting the upper side (the one with the stomata) against excessive transpiration (Grigore and Toma 2010, 2011, 2017, 2021).

Also in grasses, the stomata are dumbbell-shaped, the epidermal cell walls are silicified and lignified; the vascular bundles are surrounded (sometimes incompletely) by a double sheath: parenchymatous on the outside and sclerenchymatous on the inside (the cells having the appearance of those that make up tertiary endodermis). This structural peculiarity, also found in many species of dicots from arid (saline) habitats, is called Kranz type anatomy. In Kranz anatomy, the cells of the mesophyll form a ring around the cells surrounding the vascular bundles (Fig. 16). In cross section, this arrangement of cells resembles a crown, from which the German name (*Kranz*) derives. This type of anatomy is associated with a special photosynthetic pathway, called C_4 type photosynthesis (Grigore 2008; Grigore and Toma 2010, 2017, 2021; Grigore et al. 2012, 2014).

Fig. 12 Bulliform cells—cross section through leaf lamina of *Alopecurus arundinaceus* (Grigore et al. 2014)

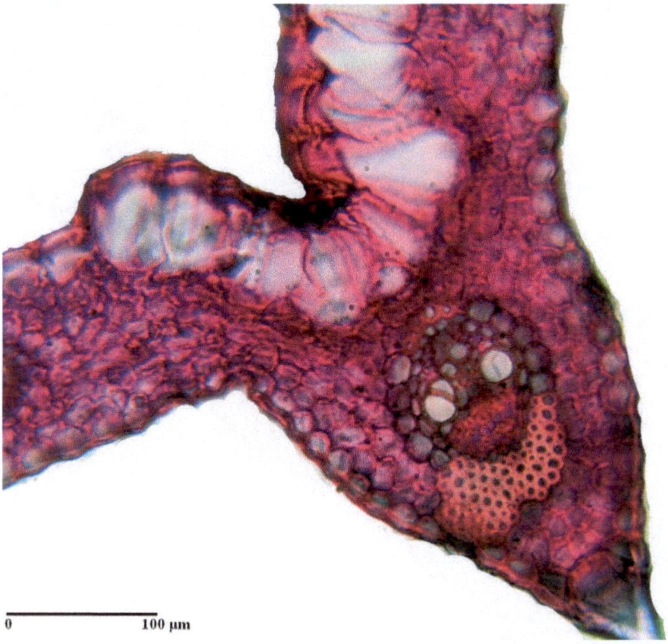

Fig. 13 Bulliform cells—cross section through the leaf lamina of *Carex distans* (Grigore et al. 2014)

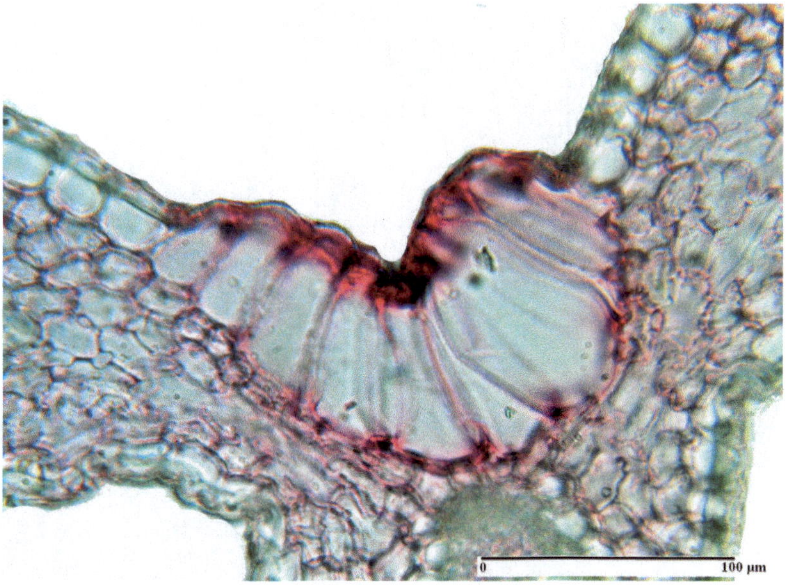

Fig. 14 Bulliform cells—cross section through the leaf lamina of *Carex vulpina* (Grigore et al. 2014)

Fig. 15 Bulliform cells—cross section through the leaf lamina of *Puccinellia distans* (Grigore et al. 2014)

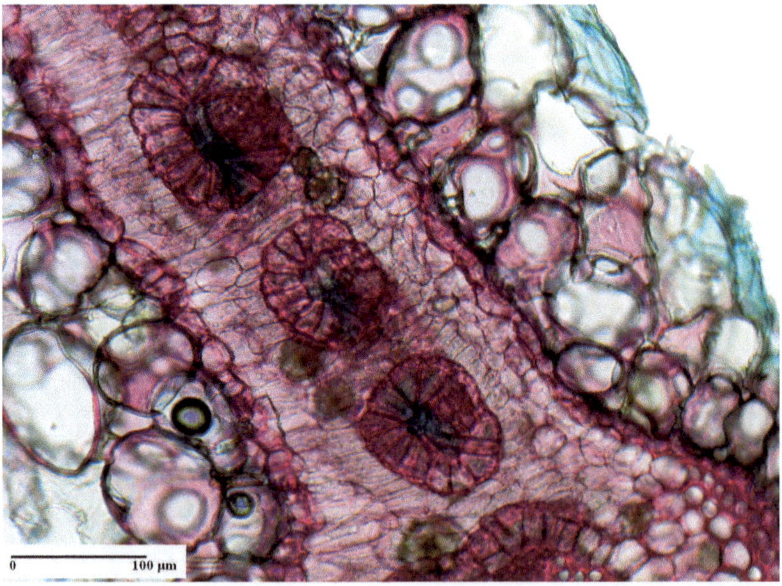

Fig. 16 Cross section through the leaf lamina of *Atriplex halimus*—salt plant (halophyte) with *Kranz* anatomy structure; secretory vesicular (water-storage) hairs are also observed at the level of the two epidermises (Grigore et al. 2014)

Also in grasses, sometimes a multilayered sclerenchymatous hypodermis is present on the lower side of the lamina; the mesophyll is compact and homogeneous.

2.3 Structure of Monofacial Leaves (= a Single Side)

Monofacial leaves have on most of the length of the lamina, only the lower epidermis; the mesophyll is usually homogeneous, consisting of isodiametric cells; in the fistulous (hollow inside) leaves of onion (*Allium cepa*), beneath the external (lower) epidermis, a multilayered palisade tissue lies all around, while spongy tissue lines the central air cavity.

Along the ensiform leaves of *Iris germanica* or *Gladiolus imbricatus*, the transition from bifacial to monofacial structure can be observed. Sections taken at different levels of the leaf sheath and lamina reveal the gradual narrowing of the adaxial groove in favor of the abaxial side. In the *Iris*, most of the lamina, compressed in the median plane, has a monofacial structure, the conducting bundles being arranged in two rows, with the xylem face to face, i.e., where the upper face should have been (Figs. 17 and 18) (Toma 1977; Toma et al. 2003).

In different *Allium* species, including *A. sativum* (garlic), the lamina is compressed in the transverse plane, having a strip shape similar to that of the bifacial leaf, but with a monofacial structure (Toma 1977; Şerbănescu-Jitariu and Toma 1980; Toma et al. 2003).

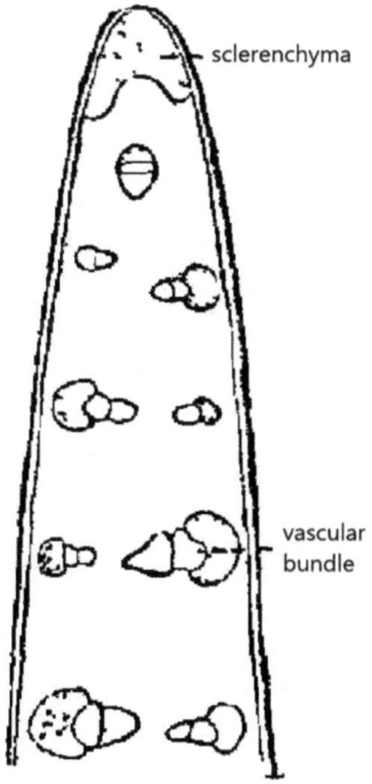

Fig. 17 Structure of the leaf lamina of *Iris* ssp.—general view (Balicka-Iwanowska 1892)

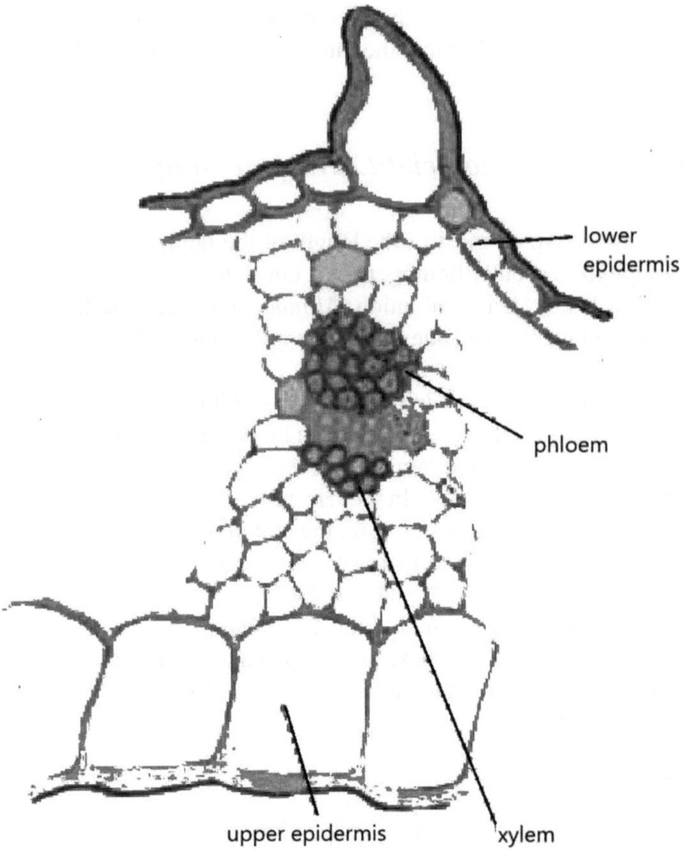

Fig. 18 Structure of the leaf lamina of *Iris* ssp.—general view (Balicka-Iwanowska 1892)

In many epiphytic orchids, the leaves are cylindrical, but retaining, on their upper side, a narrow groove; at *Sansevieria* and *Juncus*, the leaves are completely devoid of the adaxial face (Toma 1977; Toma et al. 2003).

Bibliography

Bannister P., 1976 – *Introduction to physiological plant ecology*, Blackwell Scientific Publications, Oxford, London, Edinburgh, Melbourne

Balicka-Iwanowska, G., 1892, 1893 – Contribution a l'étude anatomique et systématique du genre *Iris* et des genres voisins. *Archiv. des Sci. Phys. et Nat. de Genève*, 1892. 28: 413–435; 1893: 29: 185–200, 225–243

Beck C. B., 2005 – *An introduction to plant structure and development. Plant anatomy for the twenty-first century*, Cambridge University Press, Cambridge, New York

Belzung E., 1900 – *Anatomie et physiologie végétales*, Paris, Ancienne Librairie Germer Baillière et Cie Félix Alcan, Éditeur

Bonnier G., du Sablon L., 1905 – *Cours de Botanique. Phanérogames*, Paris, Librairie Générale de l'enseignement

Clements F. E., 1907 – *Plant physiology and ecology*, New York, Henry Holt and Company

Esau K., 1965 – *Anatomy of seed plants*, John Wiley & Sons, Inc., New York, London

Fritch F. E., Salisbury E. J., 1920 – *An introduction to the structure and reproduction of plants*, London, G. Bell and Sons Ltd.

Frunzete M.-E., 2023 – *Cercetări privind biologia unor taxoni de plante aparținând genului Taxus L. – Surse de extracte vegetale cu potențial de biocontrol evaluat în condiții experimentale*, Teză de doctorat, Universitatea „Alexandru Ioan Cuza" din Iași, Facultatea de Biologie

Grigore M.-N., 2008 – *Introducere în Halofitologie. Elemente de anatomie integrativă*, Editura PIM, Iași

Grigore M.-N., 2021 – Definition and classification of halophytes as an ecological group of plants. In: Grigore M.-N. (Ed.), *Handbook of halophytes. From molecules to ecosystems towards biosaline agriculture*, Springer, pp. 3–50

Grigore M.-N., 2022 – *Morfologia plantelor. Aspecte ecologice*, Editura Universității „Ștefan cel Mare" Suceava

Grigore M.-N., Toma C., 2010 – *Halofitele. Aspecte de anatomie ecologică.* Editura Universității „Alexandru Ioan Cuza", Iași

Grigore M.-N., Toma C., 2011 – Ecological implications of bulliform cells on halophytes, in salt and water stress natural conditions. *Studia Universitatis `Vasile Goldiș`, Ser. Șt. Vieții*, 21(4): 785–792

Grigore M.-N., Toma C., 2017 – *Anatomical adaptations of halophytes. A review of classic literature and recent findings*, Springer International Publishing

Grigore M.-N., Toma C., 2021 – Integrative anatomy of halophytes from Mediterranean climate. In: Grigore M.-N. (Ed.), *Handbook of halophytes. From molecules to ecosystems towards biosaline agriculture*, Springer, pp. 1293–1328

Grigore M.-N., Ivănescu L., Toma C., 2014 – *Halophytes. An integrative anatomical study.* Springer, Cham, Heidelberg, New York, Dordrecht, London

Grigore M.-N., Toma C., Zamfirache M.-M., Boscaiu M., Olteanu Z., Cojocaru D., 2012 – Ecological anatomy in halophytes with C_4 photosynthesis: discussing adaptative features in endangered ecosystems. *Carpathian J. of Earth and Environmental Sciences*, 7 (2): 13–21

Scott D. H., 1906 – *An introduction to structural botany. Part I. Flowering plants*, London, Adam and Charles Black

Stevens W. C., 1924 – *Plant anatomy from the standpoint of the development and functions of the tissues and handbook of micro-technic*, London J. & A. Churchill

Șerbănescu-Jitariu G., Toma C., 1980 – *Morfologia și anatomia plantelor*, Editura Didactică și Pedagogică, București

Toma C., 1977 – *Anatomia plantelor. II. Structura organelor vegetative și de reproducere*, Centrul de multiplicare al Universității „Al. I. Cuza" Iași

Toma C., Niță M., Rugină R., Ivănescu L., Costică N., 2003 – *Morfologia și anatomia plantelor. Manual de lucrări practice*, ediția a 2-a revizuită și îmbunătățită, Editura Universității „Alexandru Ioan Cuza" din Iași

Ecological Anatomy of the Leaf

1 Types of Leaves According to Light

Light is an essential factor in plant life (in relation to photosynthesis) (Oppenheimer 1959; Billings 1978), and, at the level of the leaf, as we have already seen, it induces numerous morphological (Grigore 2022), but, above all, significant anatomical changes. In relation to the morphology and anatomical structure of the leaf, one can talk about adjustments and adaptations of the leaf in relation to the action (influence) of light, although sometimes, it is more difficult to discriminate between adjustment and adaptation (Schimper 1903; Clements 1907; Warming 1909a; Polunin 1960; Daubenmire 1974; Billings 1978; Larcher 1975; Barbour et al. 1980; Collinson 1988).

Thus, regarding the regulation (adjustment) of the leaf in relation to the effects of light, it is of interest:

1. Chlorophyll production.
2. Breakdown of carbon dioxide and water to form carbohydrates.
3. Loss of water from the chloroplast.
4. Changes in the number and position of chloroplasts.
5. Daily opening and closing of stomata.
6. Movements of stems and leaves (see Grigore 2022).
7. The position of the leaves during the day and night.
8. Changes in the shape and structure of the leaf.

Regarding the adaptation of the leaf in relation to the action of light, it can be summarized as:

1. Influence on chloroplasts.
2. Changes in chlorenchyma.
3. The appearance of palisade tissue.
4. The appearance of spongy tissue.

M.-N. Grigore, *Plant Ecological Anatomy*,
https://doi.org/10.1007/978-3-032-01332-3_7

5. Changes in the epidermis.
6. The shape of the leaves.
7. Changes in leaf outline, shape, and thickness.
8. Types of leaves depending on the light, which we will refer to in the following lines.

Isophotic leaves are illuminated (almost) equally on both surfaces (sides) and therefore possess a more or less uniform chlorenchyma.

Diphotic leaves (Figs. 1 and 2) are unevenly illuminated and show an arrangement of chlorenchyma in palisade and spongy tissue, respectively; for this reason, as already discussed, common, horizontal, or dorsiventral leaves are usually diphotic.

Leaves of this type contain both palisade and spongy tissues, although the relative importance of these two tissues varies very widely in different species. Diphotic leaves are characteristic of plants from sunny marshes, meadows, and prairies and are common in xeromorphic habitats. Floating leaves, in which light is almost completely absent from the lower surface, are also diphotic. This type of structure is often found in the leaves of trees in shady forests.

Isophotic leaves are divided into three types, depending on light intensity: *the palisade*, or *staurophyll* leaf (language established by Clements 1907), is a sun leaf in which the chlorenchyma consists entirely of rows of palisade cells. It is the result of almost equal illumination of both surfaces, due to the vertical position of the leaf or the reflection of light by a light-colored soil. In a special form of this, the *diplophyll* leaf (double leaf), intense light does not penetrate to the middle of the leaf;

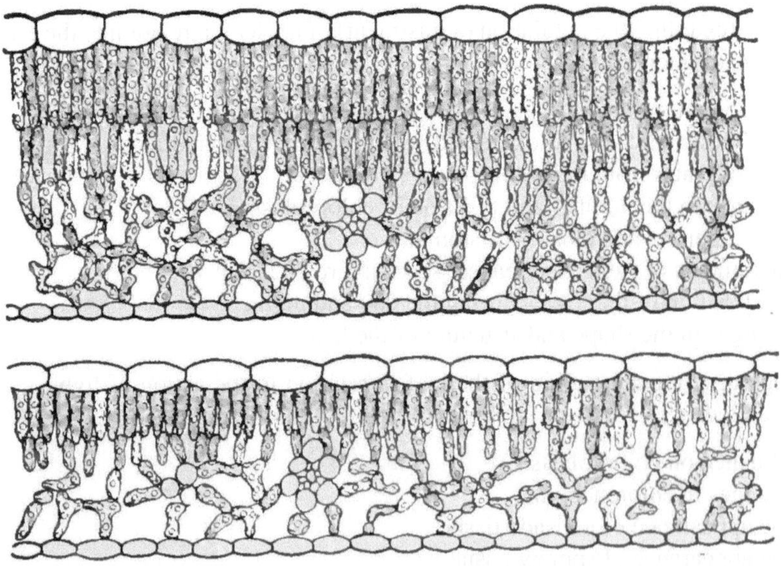

Fig. 1 Diphotic leaf in *Quercus novimexicana* (top); the shadow leaf (below) is of the same type, although a row of palisade tissue is lost and the leaf is thinner (Clements 1907)

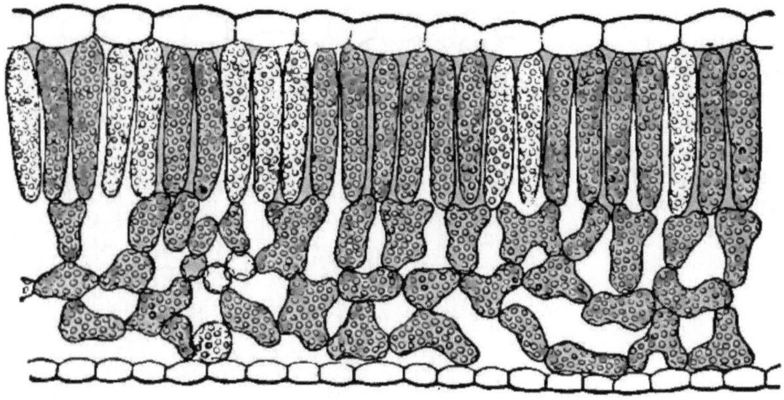

Fig. 2 Transverse section through a diphotic leaf of *Pedicularis procera* (mesophyte, heliophyte) (Clements 1907)

consequently, the upper and lower areas of palisade tissue are separated by a central, spongy, aquifer tissue (used for water storage).

The sponge or *spongophyll* (Clements 1907) leaf includes all shade leaves except those in which a portion of the palisade tissue persists from the ancestral sun form (sun-exposed) and submerged form. Its chlorenchyma consists only of spongy cells; certain monocots may also be included here, which grow in sunlight but lack palisade tissue.

2 Sun Plants (Heliophytes) and Shade Plants (Schiophytes)

Plants that grow exposed to sunlight are called heliophytes, and those that grow in the shade, sciophytes. The former comprise virtually all xerophytes, prairie and meadow mesophytes, and amphibious and floating hydrophytes. Shade plants include thickets and forest mesophytes and submerged hydrophytes. Most of the leaves of heliophytes are diphotic, as for example, *Pedicularis procera* (Fig. 2).

Plants with isophotic leaves are frequently found in xeromorphic habitats, although erect leaves of this type occur in most sunny habitats. Staurophyll leaves, in which the protection is due to the extreme development of the palisade tissue, are found in *Allionia linearis* (Fig. 3) and *Bahia dissecta* (Fig. 4). The diplophyll type, which is characterized by a central band of spongy tissue or storage cells, is found in *Mertensia linearis* (Fig. 5).

The *spongophyll* type is common among deep-shade plants, but, as cross sections through leaves of *Allionia* (Fig. 3) and *Quercus* (Fig. 1) show, the diphotic leaf is equally common among shade plants. The heliophilous spongophyll form is found in certain monocotyledons, such as *Gyrostachys striata* (Fig. 6).

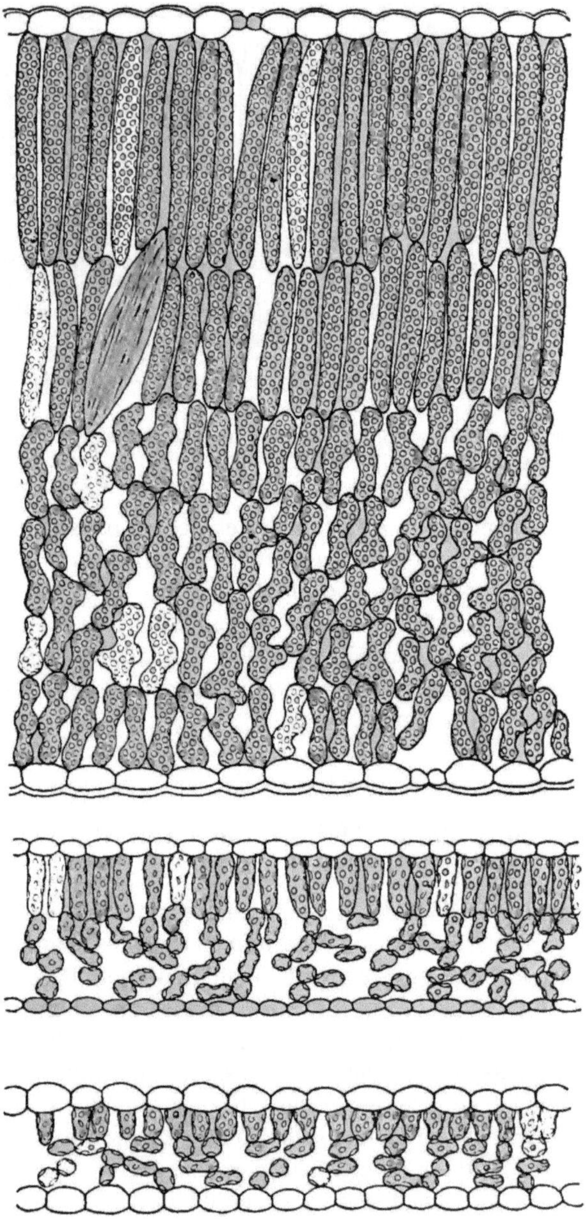

Fig. 3 Cross sections through the leaves of *Allionia linearis*, showing the changes chlorenchyma undergoes in light (top), when growing naturally in moderate shade (middle) and deep shade (bottom) (Clements 1907)

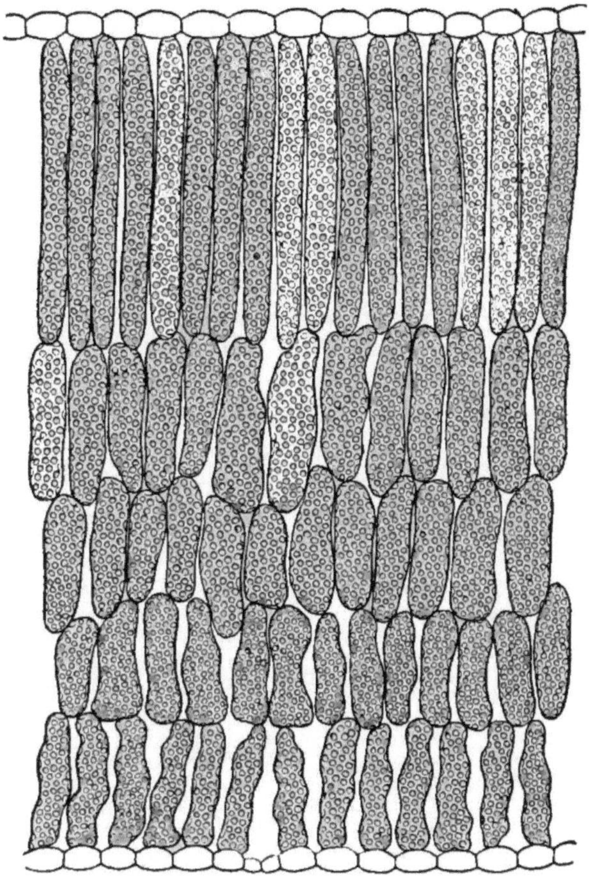

Fig. 4 Cross section through the leaf of a lowland species, *Bahia dissecta*, in which the chlorenchyma consists entirely of palisade tissue (isophotic, staurophyll leaf) (Clements 1907)

2.1 Aspects of Leaf Ecological Anatomy in Hydrophytes (Aquatic, Water Plants)

2.1.1 General Considerations

As pointed out in a previous work (Grigore 2022) and repeatedly emphasized in the present paper (when we dealt with the study of the ecological anatomy of the root and stem), hydrophytes (aquatic plants) represent a very complex ecological group, which provides examples of "extreme" anatomical adaptations in relation to the importance of water in the life of these plants. As explained (Grigore 2022), the ecological "extremes"—hydrophytes and xerophytes—are, in relation to water, some didactic concepts, first of all, and then, for specialists in plant ecology, problematic concepts difficult to manipulate when desired a more sophisticated approach to them. Very generally, operating with these "extremes" is useful and practical; to make the picture complete, in the "middle," the mesophytes (plants that grow in

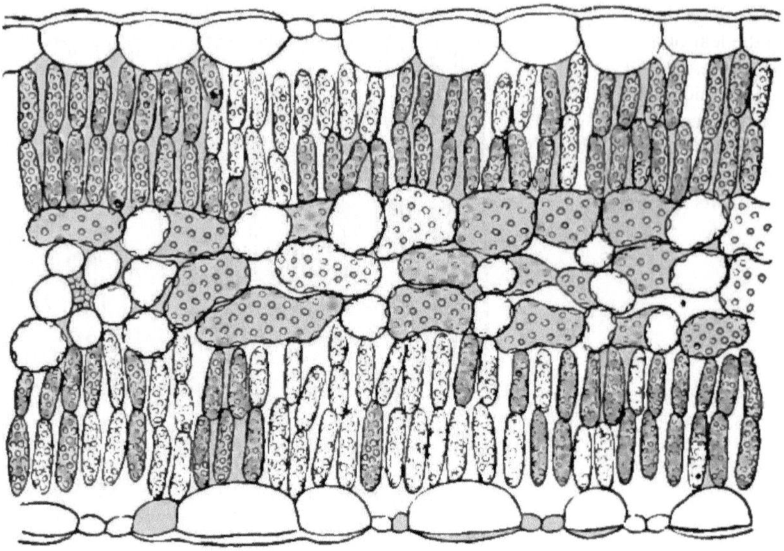

Fig. 5 Cross section through the leaf of *Mertensia linearis* (isophotic leaf, diplophyll) (Clements 1907)

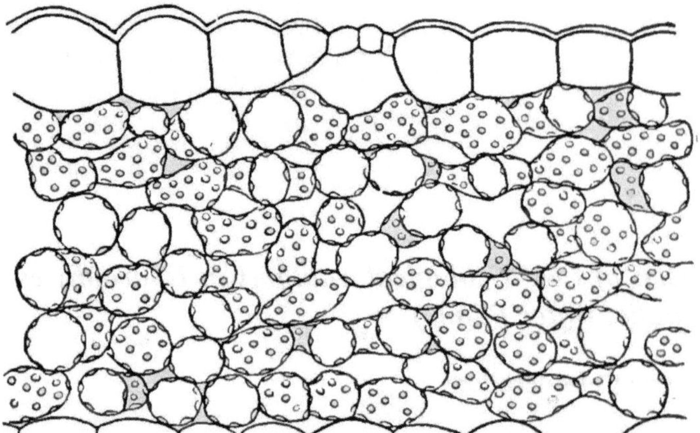

Fig. 6 Cross section through the leaf of *Gyrostachys striata* (isophotic, spongophyll leaf) (Clements 1907)

habitats with sufficient moisture/water) are usually placed. In addition, Schimper (1903) coined and used the term *tropophytes* (tropophilic plants) to denote those plants whose living conditions are, depending on the season of the year, alternatively those of hygrophytes and xerophytes, respectively.

Although the definition of hydrophytes is fragile, we can say that typical hydrophytes grow in water, in soil covered by water, or in soil saturated with water.

In terms of their relationship to water and air, they can be divided into three main groups, namely, amphibious, floating, and submerged plants.

Regarding the anatomy of the leaves of water plants, the discussion is limited to the leaves of amphibious plants (typical, usual structure), the floating leaves, and the submerged ones.

In amphibian forms, the leaves show the usual anatomical structure; they grow in the air, while the roots and stem are under water to a greater or lesser extent.

Floating plants have leaves whose upper face (surface) is in contact with air and the lower one with water.

In submerged forms, the leaves are usually below the surface of the water; that is, carbon dioxide and oxygen are obtained from the water, and not from the air. Both surfaces of the leaf—in amphibious plants—and the upper face of the floating leaf are of such a nature as to allow intense transpiration. Logically, this function is completely absent in the leaves of submerged plants.

Amphibious plants comprise species that are closely related to mesophytes; they are the least specialized of aquatic plants. As a rule, they grow in water-saturated soil or in shallow water.

Because of their frequent occurrence at the water's edge, many amphibious plants have a wide range of adaptability and can grow for a time as mesophytes or partially submerged. In most cases, the leaves are constantly above the water. The lower leaves of some species are covered with water, either normally or by an increase in the water level, taking the form and structure of submerged leaves (Grigore 2022).

At *Callitriche autumnalis* and *Hippuris vulgaris*, the submerged leaves show changes in size and structure, while in the species showing heterophily, such as *Ranunculus delphinifolius*, *Proserpinaca palustris*, and *Rorippa americana*, they differ from aerial leaves in that they are much divided (for the morphology of heterophily, including in aquatic plants, see Grigore 2022).

The leaves of amphibious plants, except those already mentioned, are usually large and entire, the stem well developed, and the roots numerous and branched.

The epidermis has a thin (or even absent) cuticle and is devoid of hairs (trichomes); when present, the trichomes are apparently involved in the secretion of mucilage or in the induction of transpiration (Warming 1909a). The functions of mucilage have not been fully elucidated; it could be involved in protection against injury due to rapid physical or chemical changes in the surrounding water, blocking, in this case, the passage of water into the plant's body; the mucilage could also have a role in defending against animal attack, or it could act as a lubricant, facilitating the movement of plants in the water mass.

In the leaves of amphibious plants, stomata are numerous and usually more abundant on the upper than on the lower surface of the leaf. The palisade tissue is well developed, being represented by one or more layers of cells, but this sector of the leaf is regularly thinner than the spongy tissue. It contains many air spaces, or numerous air chambers usually provided with thin plates (diaphragms) of cells.

In floating leaves, the stomata are located only on the upper side of the leaf (in contact with the atmosphere); the lower one, being in permanent contact with water, is devoid of stomata.

Wax coatings are absent in the epidermal cells of the leaves. The epidermis of the outer layer of cells often contains chlorophyll; there are species in which the entire structure of the leaf is practically reduced to the two epidermises of the leaf lamina.

Excretion of water is not excluded from submerged (submerged) hydrophytes, but when it occurs, it takes the form of guttation (excretion of liquid water) induced by internal activity. At the tips of the leaves of many aquatic species, aquiferous pores appear, or the tips are detached, and the ends of the vascular bundles are thus brought into direct contact with water; in this case, transpiration is absent, an aspect also correlated with the absence of stomata.

The chlorenchyma of aquatic plants is very little differentiated; as a rule, there is no distribution of it in palisade tissue and spongy tissue in submerged leaves of phanerogams. The leaves are, in this case, isolateral, and the floating ones are dorsiventral.

Regarding the general anatomical characteristics of the leaves of aquatic plants, the poor development of conductive and mechanical tissues should be mentioned (situation broadly valid for the root and stem as well).

2.1.2 Anatomy of Submerged Leaves

In submerged aquatic plants, the stem and root are greatly reduced (Grigore 2022), a fact explained by the fact that absorption is usually done by/through the entire surface of the stem and leaf, not just the root (when present). The greater density of water compared to air also accounts for less mechanical support being required, for which reason the stems are unusually long and thin, with poorly developed conducting bundles.

Submerged leaves are much reduced in size and thickness, having a ribbon shape, and they are linear, cylindrical, or finely divided. These changes seem to serve mainly to increase the surface area for absorption, especially of gases, and capture of light rays, which is diffused in water (Dring 1982). Stomata may sometimes be present, but in this case, they seem to be always nonfunctional.

Chlorenchyma has essentially the typical organization of a shade leaf. In the few cases in which palisade and spongy tissues are present, they are considered reminiscences of an earlier structure. Most submerged leaves have certain characteristics in common, the most obvious of which is the delicacy of their structure; this fragile aspect of submerged leaves is due to the thin mesophyll and the absence of its differentiation into palisade and spongy tissues, as repeatedly stated, as well as the (relative) lack of mechanical elements. The cuticle is very little developed. These epidermal characters, such as the reduction of the cuticle, are significant features that clearly distinguish submerged from aerial leaves.

As a very important parenthesis, it should be noted that a separate discussion can be developed about terrestrial forms (ecoforms) of aquatic plants and about the influence of water on terrestrial plants; it is known that most aquatic plants—except those exclusively adapted to the aquatic environment—can also develop terrestrial forms. The direction of study is very relevant, because it can clearly explain the influence of an environmental factor on the growth and development of plants (the same discussion can be held about the influence of salinity on plants—the case of maritime and continental forms of some halophytes—see Grigore 2021).

Going back, in general, the epidermal cells of dicot leaves tend to be round in cross section, while those of monocots are more rectangular. In the case of *Callitriche verna* (Fig. 7), which has both aerial and submerged leaves, it is found that, although the aerial leaves show epidermal cells of circular outline, typical of dicots, the epidermal cells of the corresponding elements in submerged leaves are straight-walled and, therefore, approach the characters of monocots.

The exact explanation for these histological differences is not known; an early hypothesis (Mer 1880) suggested that the epidermis is the tissue most directly affected by transpiration and that variations in this function could exert an influence on epidermal cell shape. According to his opinion, when the transpiration is scanty, as in the case of submerged plants, the epidermal cells are kept in a constant state of turgidity, and therefore their growth takes an even course, resulting in a regularity of form. But, on the other hand, when transpiration is active, as is the case for land

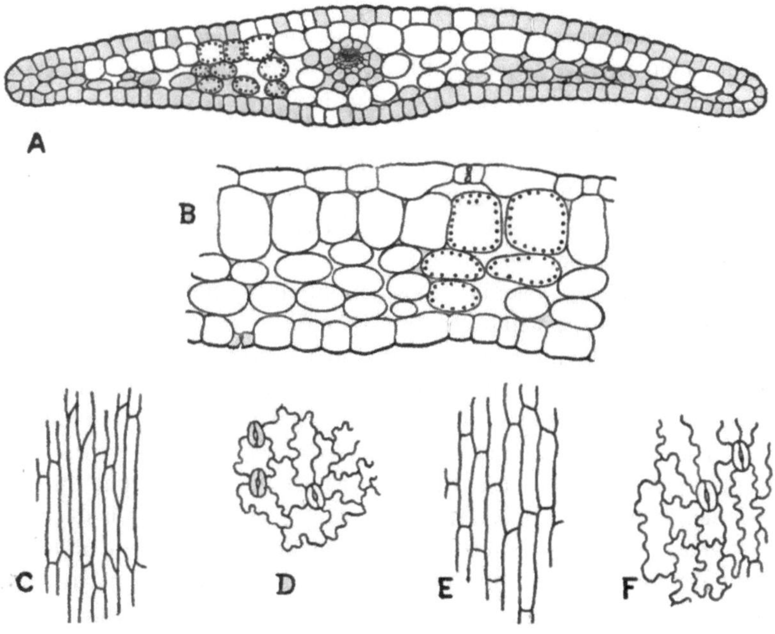

Fig. 7 *Callitriche verna*; (**a**) cross section through the submerged leaf; (**b**) leaf of the terrestrial form; (**c**) upper epidermis of the submerged leaf (front view); (**d**) upper epidermis of the leaf of the terrestrial form (front view); (**e**) lower epidermis of the leaf of the submerged form (front view); (**f**) lower epidermis of the leaf of the terrestrial form (front view) (Schenck 1886)

plants, it is subject to great variations that induce changes in the shape of the epidermal cells, producing sinuosity in the outline of the epidermal cells.

The epidermal cells of submerged leaves differ from those of aerial leaves not only in form but also in content. Chlorophyll, which is generally thought to be absent from the epidermal cells of land plants (with some exceptions), is often present in large quantities in the epidermal cells of submerged leaves (a character noted since the nineteenth century—see review—at Arber 1920). In some cases, for example, in *Zostera*, *Cymodocea*, and *Posidonia*, the epidermis is actually the leaf tissue richest in chlorophyll.

The presence of chloroplasts in the epidermal cells of aquatic plants need not, however, constitute such a defining difference as compared with terrestrial plants, and therefore this character should not be generalized to aquatic plants, nor excluded from terrestrial plants. In land plants, chloroplasts are usually observed only in the lower epidermis of the leaf, probably as an adaptation to the destructive action of light in the upper epidermis; in many land plants, in diffuse light, chlorophyll also occurs in the upper epidermis of leaves, whereas under normal conditions, chlorophyll exists only in the lower epidermis. Therefore, the presence of chloroplasts in the epidermal cells of the leaves of submerged plants can be considered as representing only the elaboration of a character already existing in terrestrial plants (aquatic plants are secondarily adapted to the aquatic environment); submerged leaves thus exploit favorable opportunities for development in the relatively poor illumination of the aquatic environment.

As a general statement—which we also made in the present paper—it is considered that the epidermis of submerged leaves is devoid of stomata. However, there are exceptions and nuances; it is, indeed, true that stomata are much less common in submerged leaves than in terrestrial ones; moreover, in certain aquatic plants, such as *Elodea* (Fig. 19), *Vallisneria*, *Thalassia*, and other species of the family Hydrocharitaceae, which always live exclusively submerged, stomata are completely absent. Of the cryptogams, *Isoetes lacustris* (fern species) is completely devoid of stomata (Goebel 1891–1893).

Submerged leaves are generally poor not only in stomata but also in hairs (trichomes), probably also related to low transpiration.

As exceptions, it should be remembered that stomata have been observed in a large number of submerged leaves, such as those of *Lobelia dortmanna*, *Villarsia ovata*, and *Pontederia cordata*, but also on the lower surfaces of certain floating leaves, such as *Limnocharis humboldtii* and *Hydrocharis morsus-ranae*.

It is not clear, even in this case, why the stomata were preserved in the epidermis of the submerged leaves of these species. Porsch (1903, 1905) was of the opinion that the stomatal apparatus must have evolved gradually over a long period of time, so that its characters became stable and efficient; in cases where the existence of the stomatal apparatus must have been not only superfluous, but accompanied by some danger to the plant, instead of the plant "giving up" it, the stomatal apparatus was/is often secondarily modified, in such a way that to be devoid of functionality.

Porsch (1903, 1905) showed that, in the case of submerged plants that retain their stomata, four different changes are observed, each of which must result in the prevention of the penetration of water into the tissues through the ostiole between the guard cells:

1. Guard cells can close on submergence, even under full illumination, e.g., in *Callitriche verna* and *Hippuris vulgaris*.
2. Ostiole may be permanently closed, as in *Potamogeton natans* (Fig. 8b), in which the entire stomatal apparatus remains covered by the cuticle.
3. The development of each stoma may effectively cease at an early stage. This situation is rare, but such "aborted" stomata are found in the submerged parts of a species of *Oenanthe*.
4. Stomata may develop normally, but the guard cells remain pressed together with their cuticular ridges interlocked, as, for example, in *Calla palustris* (Fig. 8a).

In addition to the usual (aerial) stomata, which, in the submerged (submerged) environment, are unable to exercise their normal function, submerged leaves are frequently provided with water stomata, which are probably important in maintaining the flow of "transpiration" by exudation (guttation) (Fig. 9); at *Potamogeton densus*, at the tip of the leaf, the tracheids communicate directly with the outside without the intervention of water stomata (Fig. 10).

2.1.3 Organization of Mesophyll in Submerged Leaves

As we have already seen, aerenchyma is a defining anatomical feature of aquatic plant organs (see above, discussions of hydrophytes root and stem aerenchyma). The aeration system of submerged leaves is a very pronounced feature.

The mesophyll of many water plants leaves, such as those of *Littorella* (Fig. 11), *Lobelia dortmanna* (Fig. 24), *Ceratophyllum demersum*, species of *Isoetes* (Fig. 12), *Alisma plantago* (Fig. 13), *A. natans* (Fig. 14), and *Callitriche vernalis* (Fig. 7), is traversed from one end to the other by air spaces, interrupted only by porous diaphragms.

The mesophyll of submerged leaves shows, as we have already pointed out, a poor differentiation into palisade tissue and spongy tissue. In many cases, the photosynthetic activity seems, to a large extent, limited only to the epidermis, the

Fig. 8 (**a**) *Calla palustris*—cross section—stomata in the petiole of a submerged leaf (thickening bands are tightly glued together); (**b**) *Potamogeton natans*—cross section—stomata in the petiole of a floating leaf (stoma completely covered with cuticle) (Porsch 1905)

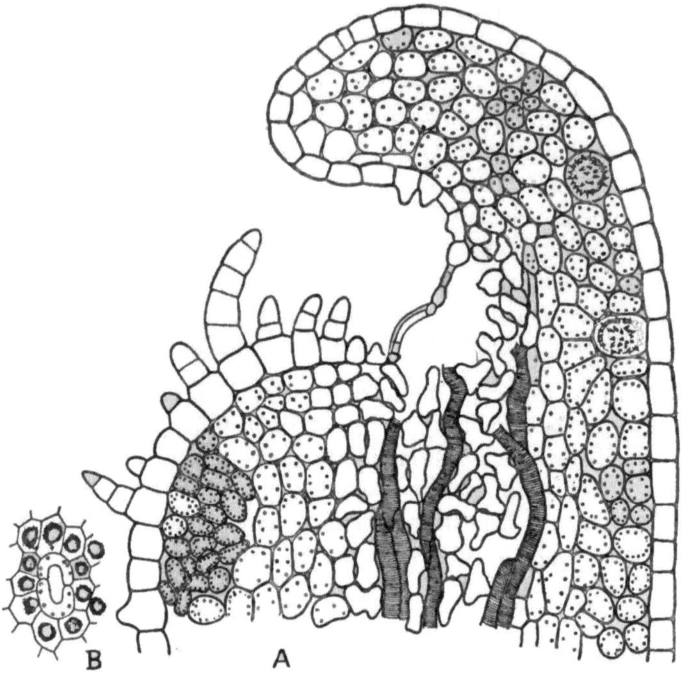

Fig. 9 *Pistia stratiotes*; (**a**) radial longitudinal section through the leaf apex showing the depression where the aquifer pores open and the space below them where the tracheids emerge; (**b**) front view of the aquifer pore (Minden von 1899)

Fig. 10 *Potamogeton densus,* longitudinal section through the tip of the leaf, passing through the midrib and showing the apical opening (Sauvageau 1891)

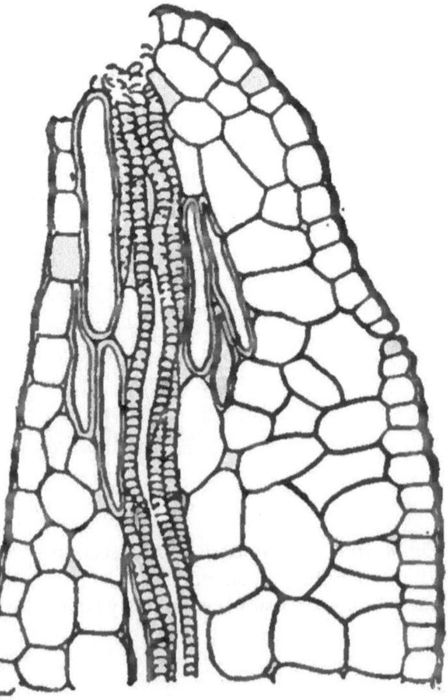

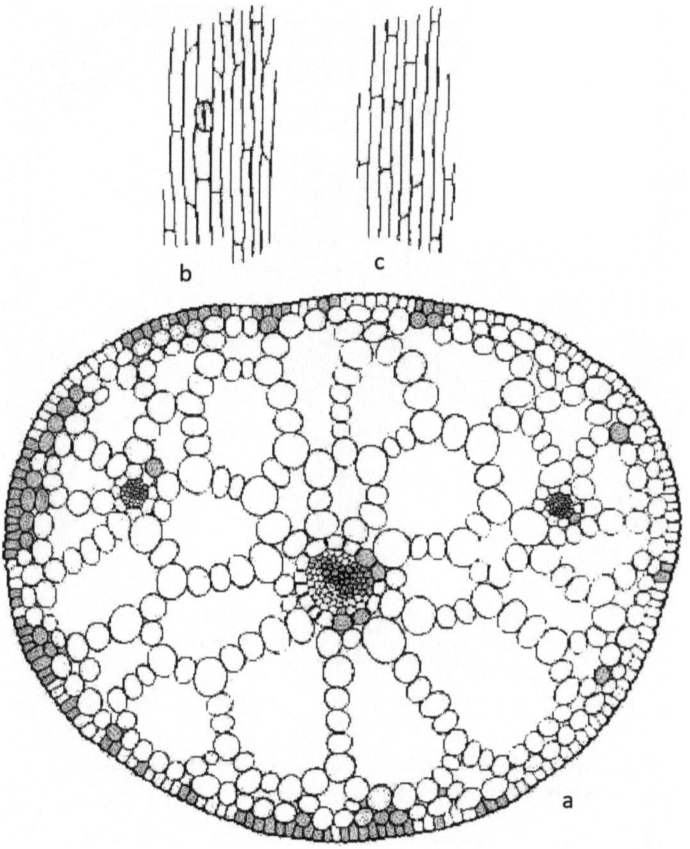

Fig. 11 *Littorella lacustris*; (**a**) cross section through the submerged leaf; (**b**) upper epidermis, front view; (**c**) lower epidermis, front view (Schenck 1886)

mesophyll having, rather, a storage function. This distinction is clearly observed in *Myriophyllum*: the epidermal cells are rich in chlorophyll, while the mesophyll accumulates starch (Fig. 15). This leaf is a good example of the subdivided, submerged type, showing a tendency to radial arrangement of tissues.

Littorella (Fig. 11), *Utricularia* (Figs. 16 and 17), and *Ceratophyllum* show roughly the same type of radial leaf anatomy. The effect of the environment on this type of leaf is illustrated by a comparison between the terrestrial and aquatic forms of the genus *Myriophyllum*.

In the case of *Myriophyllum alterniflorum* (Fig. 18), the terrestrial form, when growing in sunny conditions, has shorter and thicker leaf segments than the aquatic form; the segments are also dorsiventral and elliptical instead of radiating and cylindrical, while the xylem is more developed than in the aquatic form. The epidermis contains only a few small chloroplasts, and in addition, stomata appear. The epidermal cells also have the sinuous outline that is absent in the aquatic form. The absence

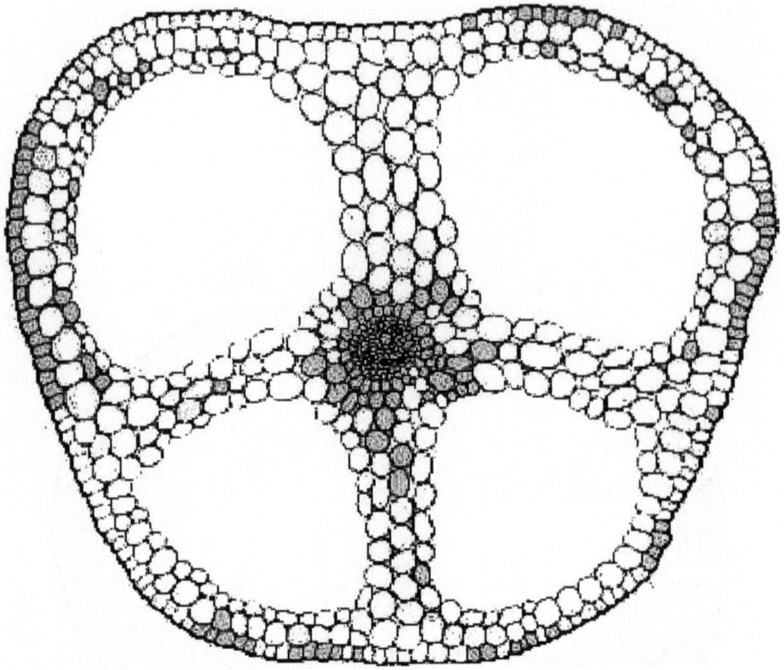

Fig. 12 *Isoetes lacustris*—cross section through the submerged leaf (Schenck 1886)

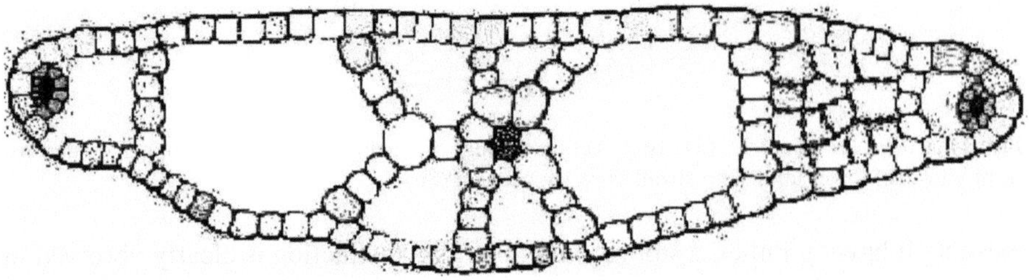

Fig. 13 *Alisma plantago*—cross section through the submerged leaf (Schenck 1886)

of clear dorsiventrality in the leaves of many submerged plants, such as *Myriophyllum*, can be attributed in part to the fact that they are constantly moved by water currents and thus maintain no constant position relative to incident light.

Callitriche (Fig. 7), *Elodea* (Fig. 19), and *Alisma* (Figs. 13 and 14) can be mentioned as examples of flattened submerged leaves, with a non-radial structure. In Fig.7, the difference between the aquatic and the aerial leaves in *Callitriche verna* is indicated. The aquatic leaf is thin, but still retains a mesophyte character; the outlines of the epidermal cells in the two forms show the distinctive characters to which we have already referred (see above). *Callitriches autumnalis*, which lives and flowers completely submerged, has a thinner leaf. The leaf of species of *Hottonia* (Fig. 20) resembles that of *Callitriche*.

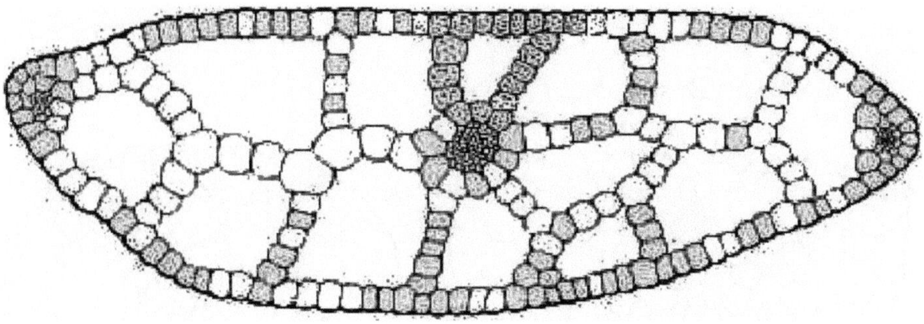

Fig. 14 *Alisma natans*—cross section through the submerged leaf (Schenck 1886)

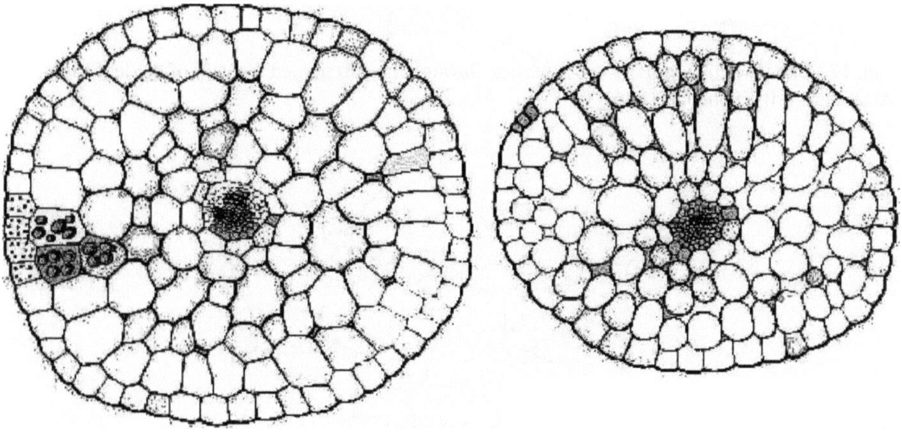

Fig. 15 *Myriophyllum spicatum*—cross section through the submerged leaf (left) and that of the terrestrial form (right) (Schenck 1886)

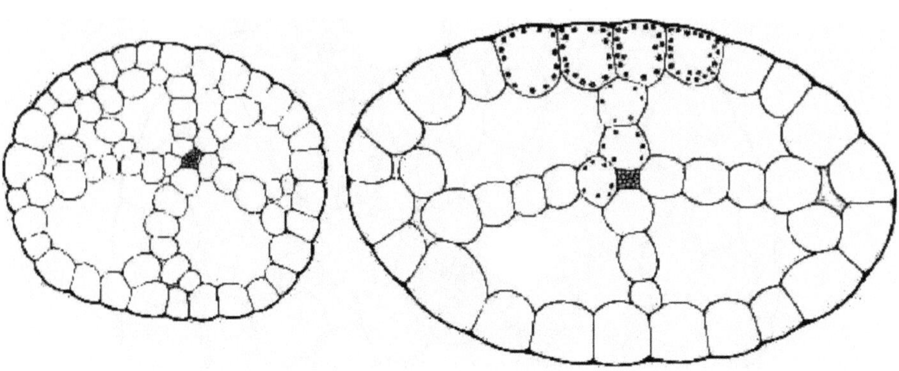

Fig. 16 *Utricularia minor*—cross section through the submerged leaf (tip of the leaf, left, and basal part, right) (Schenck 1886)

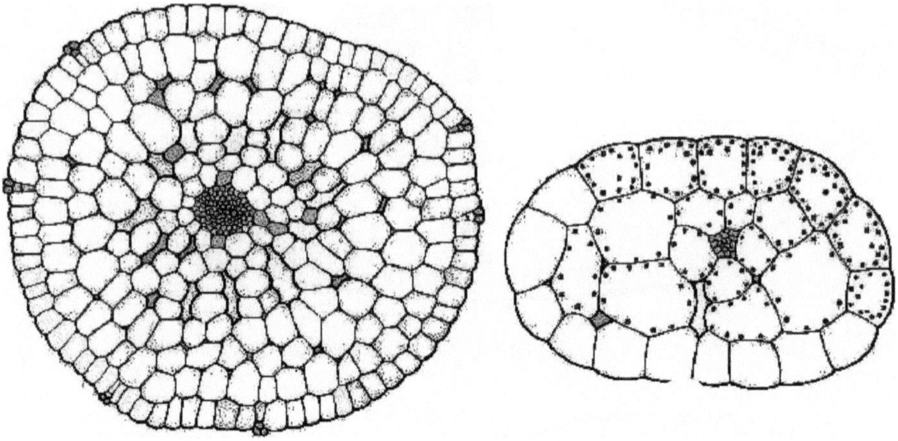

Fig. 17 *Utricularia vulgaris*—cross section through the submerged leaf (tip of the leaf, right, and basal part, left) (Schenck 1886)

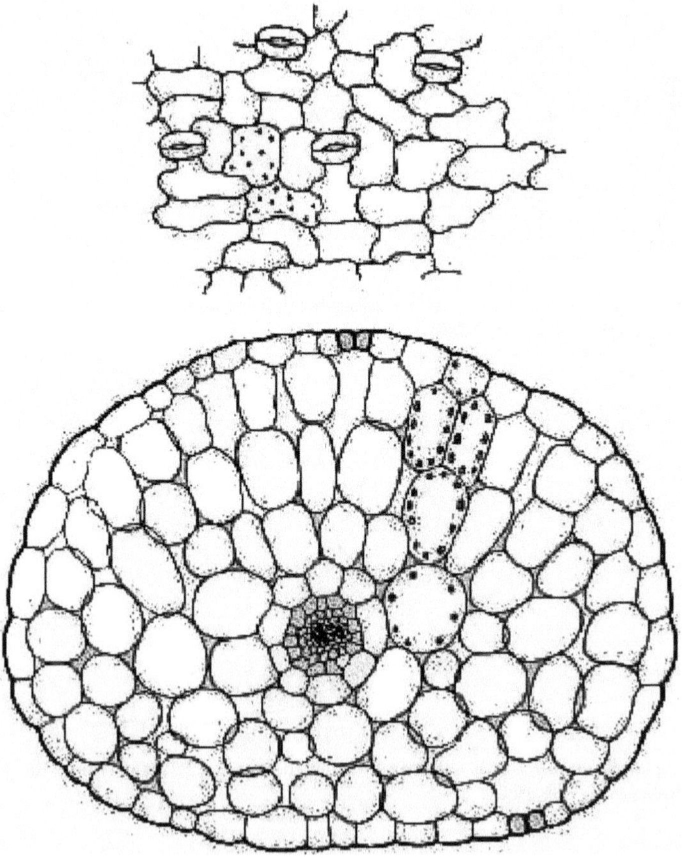

Fig. 18 *Myriophyllum alterniflorum*—cross section through the leaf of the terrestrial form (bottom); top, epidermis, front view (Schenck 1886)

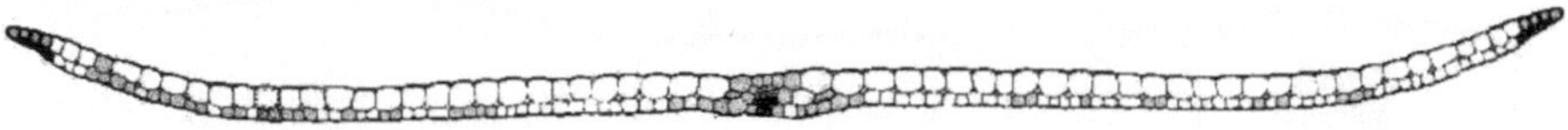

Fig. 19 *Elodea canadensis*—cross section through the leaf (Schenck 1886)

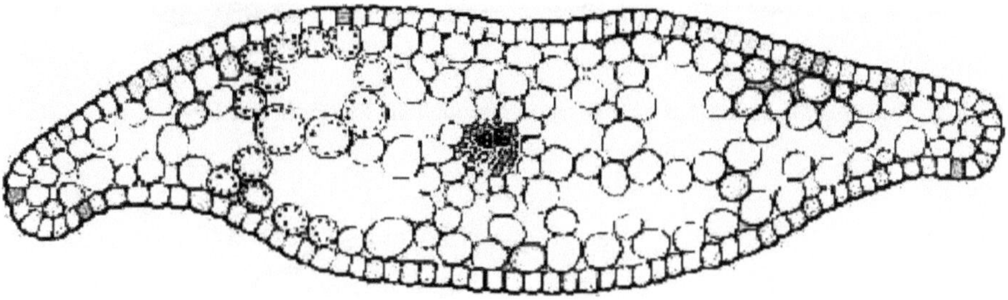

Fig. 20 *Hottonia palustris*—cross section through the leaf (Schenck 1886)

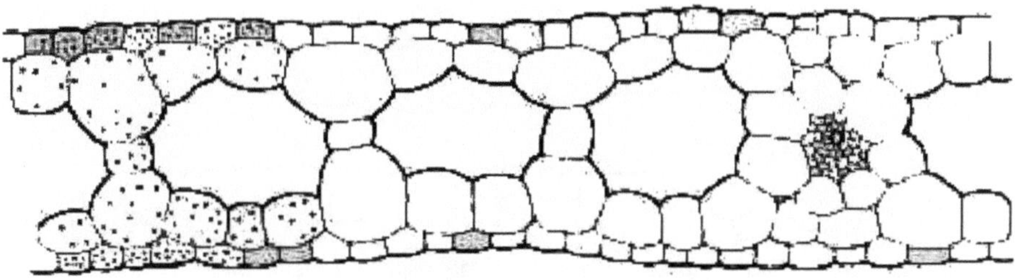

Fig. 21 *Vallisneria spiralis*—cross section through the leaf (Schenck 1886)

Ribbon-shaped leaf from *Alisma plantago* (Fig. 13) shows a slightly different type of structure. The chlorophyll-containing epidermis forms the main part of the leaf, and the large air ducts are bounded by it. There is a main leading bundle, accompanied by two smaller laterals placed near the edges.

Vallisneria also shows a ribbon-shaped leaf (Fig. 21).

At *Elodea canadensis* (Fig. 19), practically a process of complete reduction of the mesophyll is recorded, in this case the entire assimilating tissue being reduced to the two epidermal layers. The extremely delicate leaf is strengthened by some mechanical cells. A mesophyll practically reduced to the two epidermises, except perhaps in the midrib region, can also be seen in *Najas flexilis*.

Sclerenchyma is characteristic of a certain number of submerged leaves, such as those of some species of *Potamogeton* (Figs. 22 and 23).

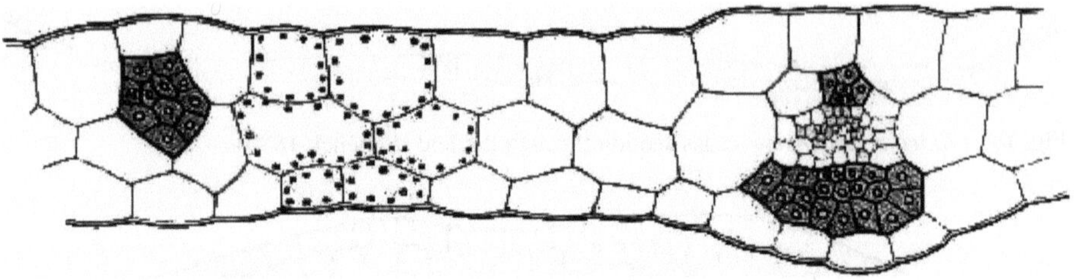

Fig. 22 *Potamogeton acutifolius*—cross section through the leaf (Schenck 1886)

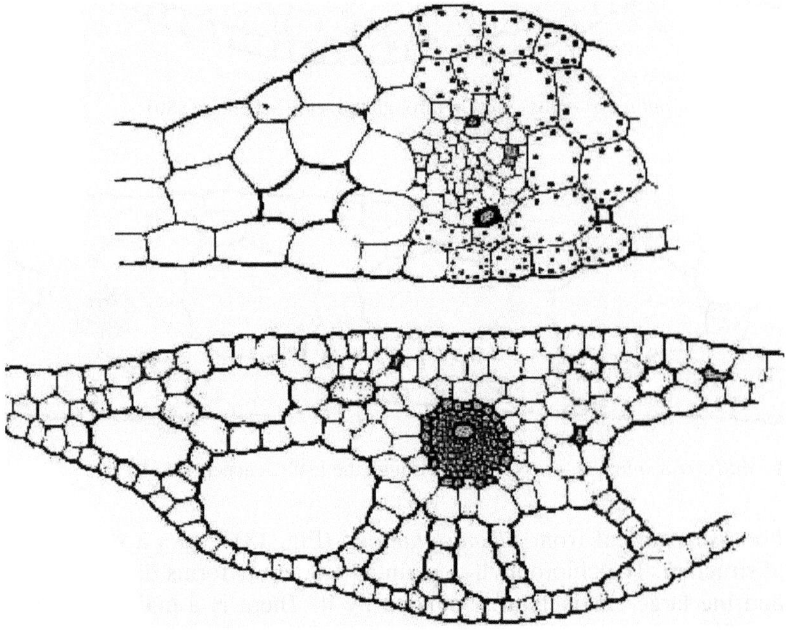

Fig. 23 *Potamogeton densus*—cross section through the leaf (top, lateral nervure; bottom, at the level of the midrib) (Schenck 1886)

There is a strong tendency in submerged leaves to reduce the tracheal system. Among species of Hydrocharitaceae, for example, although the typical spiral tracheids occur in the submerged leaves of *Stratiotes*, the leaves of other genera show either no tracheids or the elements are more or less transitory, with annular thickenings, as, for example, in *Elodea*, *Halophila*, *Vallisneria*, and *Thalassia*.There are also several species of marine aquatic phanerogams that show submerged leaves: *Cymodocea* (Figs. 25 and 26), *Zostera* (Fig. 27), *Halodule* (Fig. 28), and *Posidonia* (Fig. 29); some species of aquatic plants can also grow in salt water, such as *Ruppia* and *Zannichellia*, with representatives also in Romania's flora.

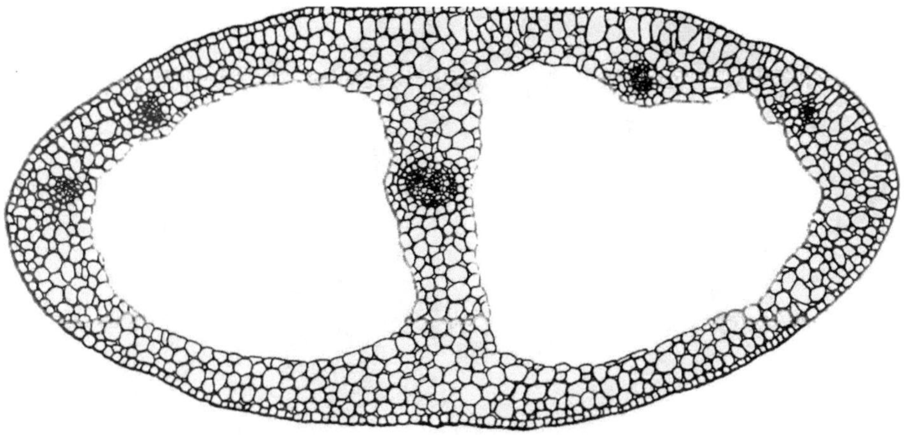

Fig. 24 *Lobelia dortmanna*—transversal section through the leaf (Ogden 1974)

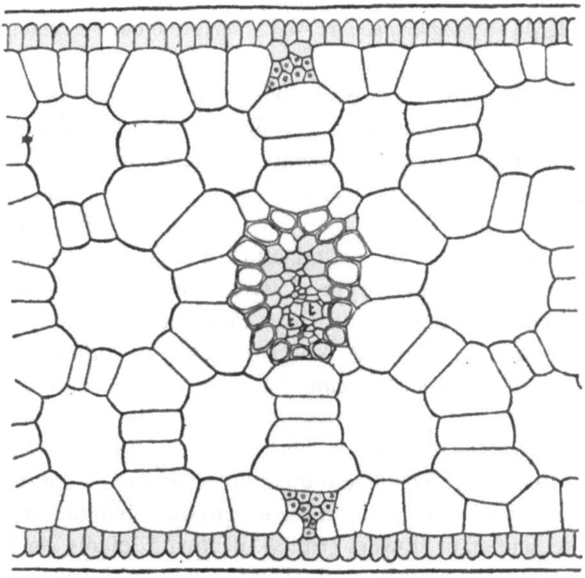

Fig. 25 *Cymodocea aequorea*—cross section through the base of the leaf blade, at the level of the midrib (Sauvageau 1891)

2.1.4 Anatomy of Floating Leaves

Among the examples that can be given here, we can mention the structure of the leaf lamina of *Hydrocharis morsus-ranae* (Fig. 30). The upper epidermis shows cells containing a few chloroplasts. The outer wall is sculpted on the inside and has a thin layer of cuticle on the outside. The stomata, which are located on the upper surface of the leaf, have external, slightly protruding cuticular ridges.

Haberlandt (1914) suggested that this form of stomata is adapted to lessen the risk of capillary occlusion of the opening by water.

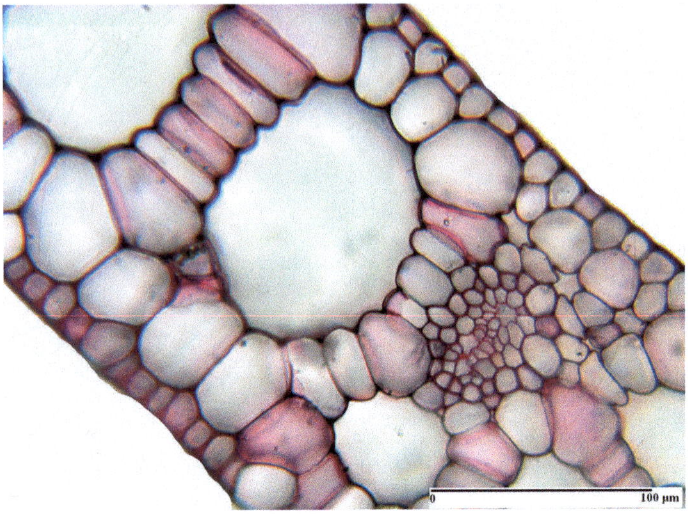

Fig. 26 *Cymodocea nodosa*—cross section through the leaf lamina (Grigore et al. 2014)

The palisade parenchyma, which is under the upper epidermis, is, in normal *Hydrocharis* leaves, very well differentiated. The spongy tissue is not distributed irregularly, as in land plants, but takes the form of plates of cells arranged in a polygonal network near the lower epidermis, which itself contains a small amount of chlorophyll.

Returning to the organization of the mesophyll and, in particular, to the presence of aerenchyma at this level, we can say that this characteristic is also strongly developed in the petioles of leaves (and peduncles) such as *Sagittaria* (Fig. 31), *Alisma plantago aquatica*, *Armoracia aquatica*, *Caltha palustris*, *Hydrocotyle umbellata* (Fig. 32), *Nuphar luteum*, *Nymphaea odorata*, *Nymphoides peltatum*, *Nymphoides cordatum* (Fig. 33), *Nuphar advena* (Fig. 34), *Peltandra virginica*, *Lobelia dortmanna* (Fig. 35), *Menyanthes trifoliata* (Fig. 36), *Potamogeton pulcher*, and *Trapa natans* (Fig. 37), where there are diaphragms that form support points for the secondary ribs that connect the longitudinal bundles; aerenchyma is also present in the peduncles of flowers (fruits) of aquatic plants, such as *Nuphar luteum*. In *Trapa natans*, the petioles of the leaves show float structures, adaptive morphological formations with a role in floating, also found in the stem.

In relation to the anatomical structure of the petiole and peduncle, in the examples mentioned above, we note a very interesting structural feature: in the species of the Nymphaeaceae family (Figs. 33 and 34), we note the presence of branched sclereids (mechanical cells) (hatched on the figures with black), called trichosclereids (Şerbănescu-Jitariu and Toma 1980) or, by older authors, "stellar hairs" (de Bary 1884), or sclereids, in general terms (Esau 1960) (Fig. 34—detail).

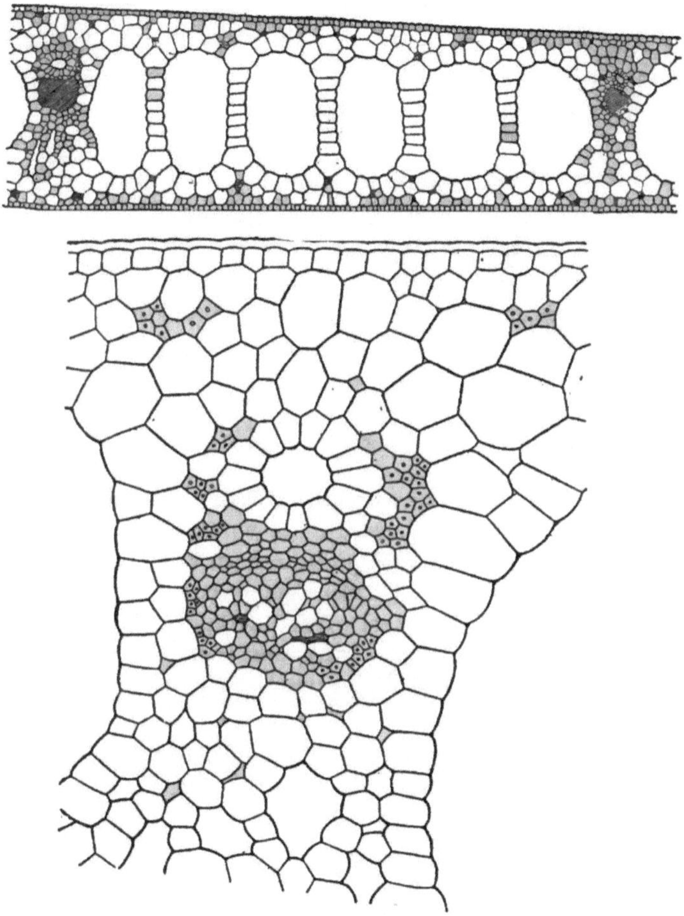

Fig. 27 *Zostera marina*—cross section through the leaf blade—between the midrib and a lateral nervure (top)—and at the level of the midrib (bottom) (Sauvageau 1891)

2.2 Ecological Anatomy of the Leaf in Xerophytes (General Considerations)

Xerophytes are probably the most studied plants from the point of view of the ecological anatomy of plants. Xerophytes (*xeros*, gr., *dry*; *phytos*, gr., *plant*) are plants that live (vegetate) in dry (xeric) places (habitats, environments); this is only a simplified, didactic definition, in reality the term xerophytes grouping several types of plants growing in very different climatic conditions and, in apparently, very heterogeneous habitats (Grigore and Toma 2010a, 2011a, 2017; Grigore et al. 2014; Grigore 2021, 2022); of these, desert plants are the best known and studied.

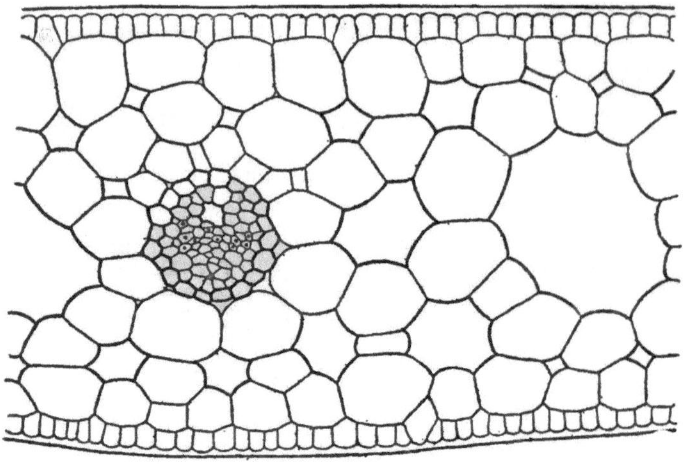

Fig. 28 *Halodule uninervis*—cross section through leaf blade (Sauvageau 1891)

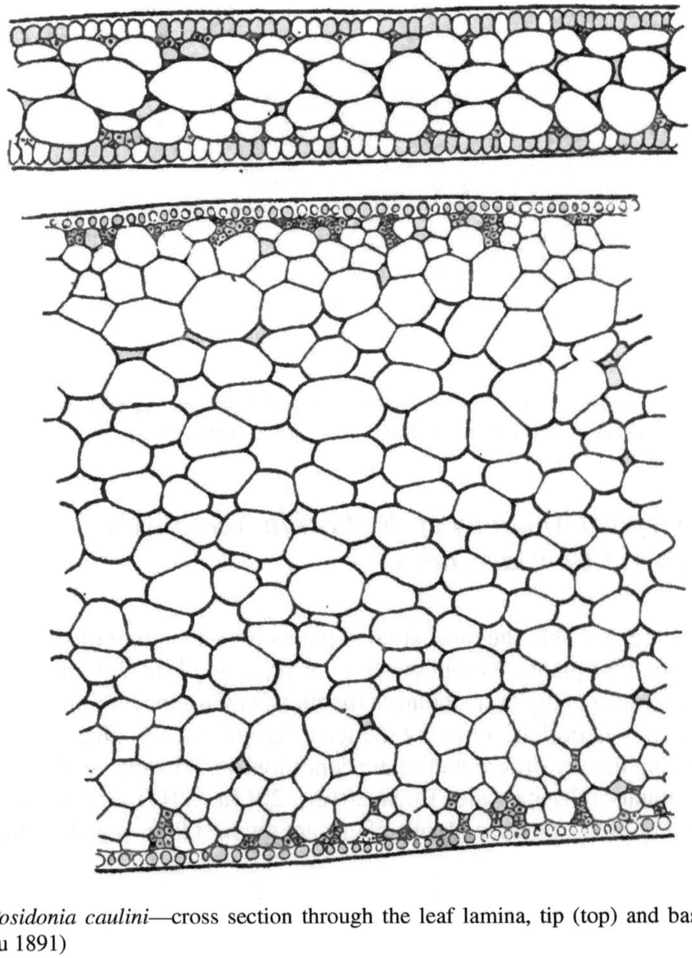

Fig. 29 *Posidonia caulini*—cross section through the leaf lamina, tip (top) and base (bottom) (Sauvageau 1891)

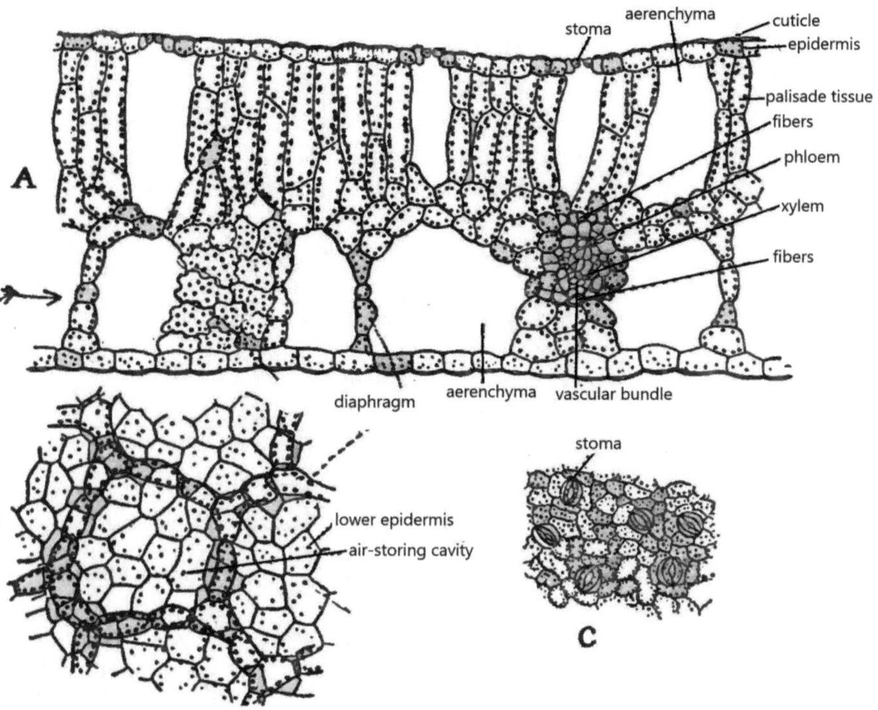

Fig. 30 *Hydrocharis morsus-ranae*; (**a**) cross section through floating leaf; (**b**) tangential section through the leaf at the level of the arrow indicated in **a**; (**c**) upper epidermis with open stomata (Arber 1920)

In this sense, we recall the importance of discussing general considerations regarding the definition and position of xerophytes within all ecological groups of plants (see the Introduction and discussion on pages 121–132, Ecological Anatomy of the Root).

As already stated, in xerophytes, physiological drought is caused by external factors that either reduce water uptake or accelerate transpiration, or, most frequently, there is a combination of these influences (Fig. 38). Factors that accelerate transpiration are dry atmosphere, high air temperature, air rarefaction, and light.

Broadly speaking, the adaptation of xerophytes to life in contact with air takes place following three main directions:

1. Controlling the exit of water from the plant body, i.e., regulating transpiration.
2. Increasing water intake, aimed at developing special mechanisms for its uptake.
3. Storing (depositing) water, i.e., the formation of water reservoirs in the form of water-storage tissues.

We will go through, one by one, the main anatomical adaptations of xerophytes, circumscribed by these three main directions.

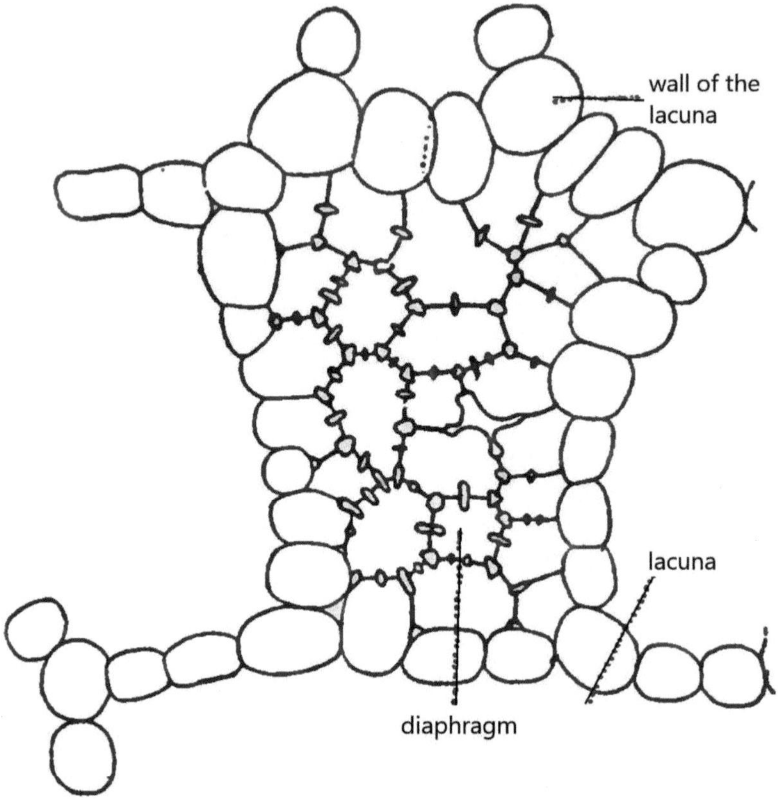

Fig. 31 Cross section through the petiole of the leaf of *Sagittaria sagittifolia* (Blanc 1912)

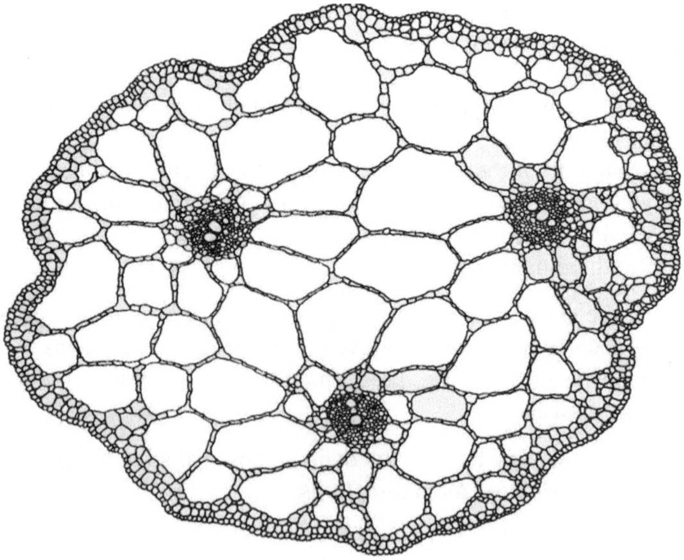

Fig. 32 Cross section through the petiole of the leaf from *Hydrocotyle umbellata* (Ogden 1974)

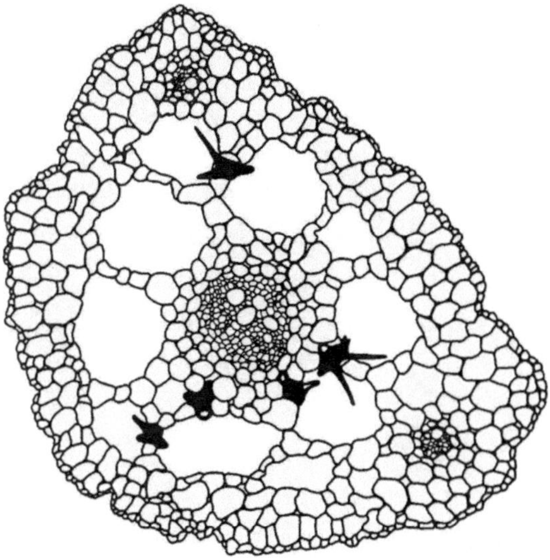

Fig. 33 Cross section through the petiole of the leaf from *Nymphoides cordatum* (Ogden 1974)

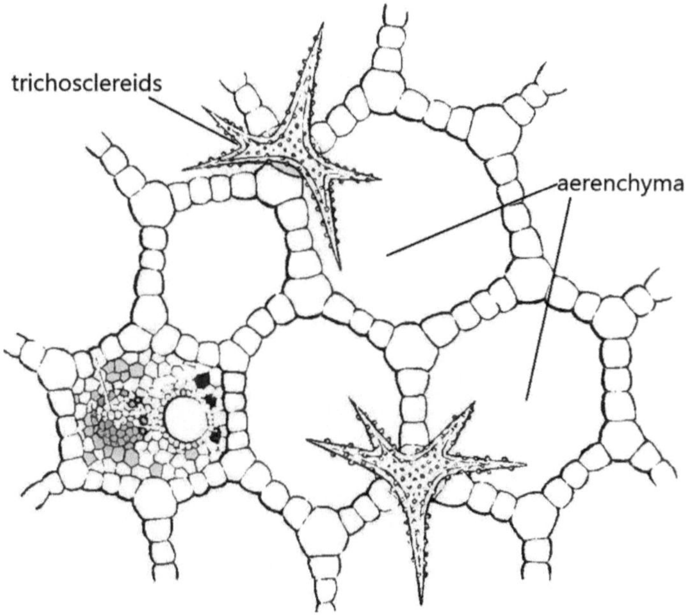

Fig. 34 *Nuphar advena*. Cross section through the petiole (de Bary 1884)

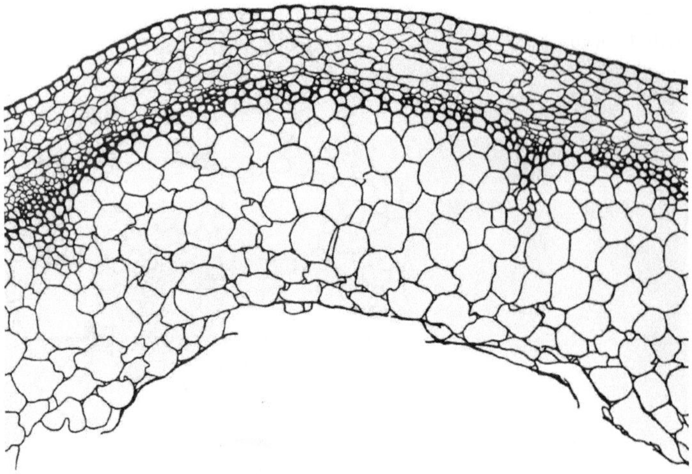

Fig. 35 Cross section through the peduncle of *Lobelia dortmanna* (Ogden 1974)

Fig. 36 Cross section through the leaf petiole of *Menyanthes trifoliata* (Ogden 1974)

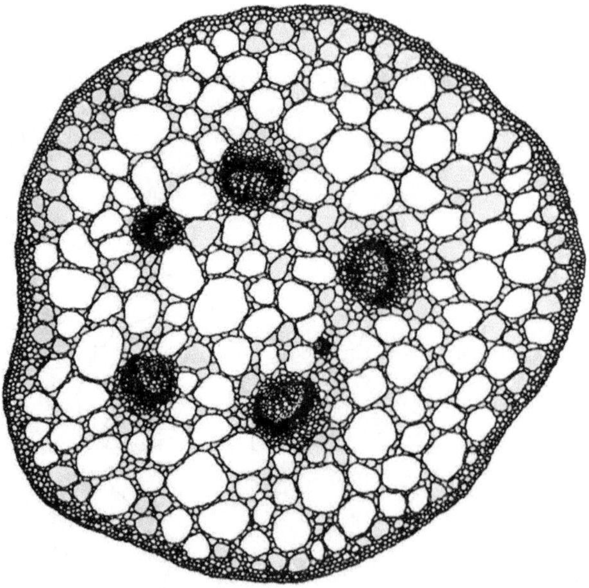

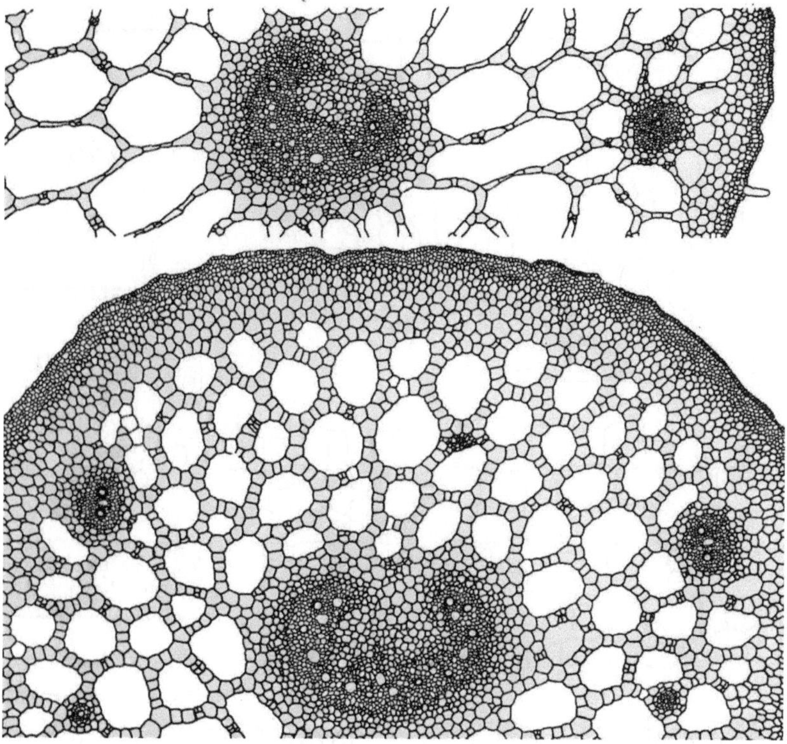

Fig. 37 Cross section through the leaf petiole of *Trapa natans*; top, petiole above the floater; bottom, petiole below the floater (Ogden 1974)

3 Regulation of Transpiration in Xerophytes

The regulation of transpiration and its control, at critical moments in the life of xerophytes, is done through the following strategies and components:

1. The anatomical structure that controls transpiration.
2. Decreasing the evaporation surface, either through characteristic movements or by reducing the surface of leaves or stems that adapt irreversibly and characteristically to the prevailing conditions (see Grigore 2022).
3. Adjusting the level of illumination at the level of the assimilative organs, either by temporarily adopting some photometric movements dependent on the light intensity or by a permanent, lie position (as in "compass" plants) (Grigore 2022).
4. Covering organs and structures, such as trichomes, and leaves that reduce light intensity and directly reduce transpiration.
5. Removal of rainwater by leaves (from leaf surface).

Of these listed strategies, not all have expression at the anatomical level; therefore, in the present work, we will only insist on anatomical adaptations;

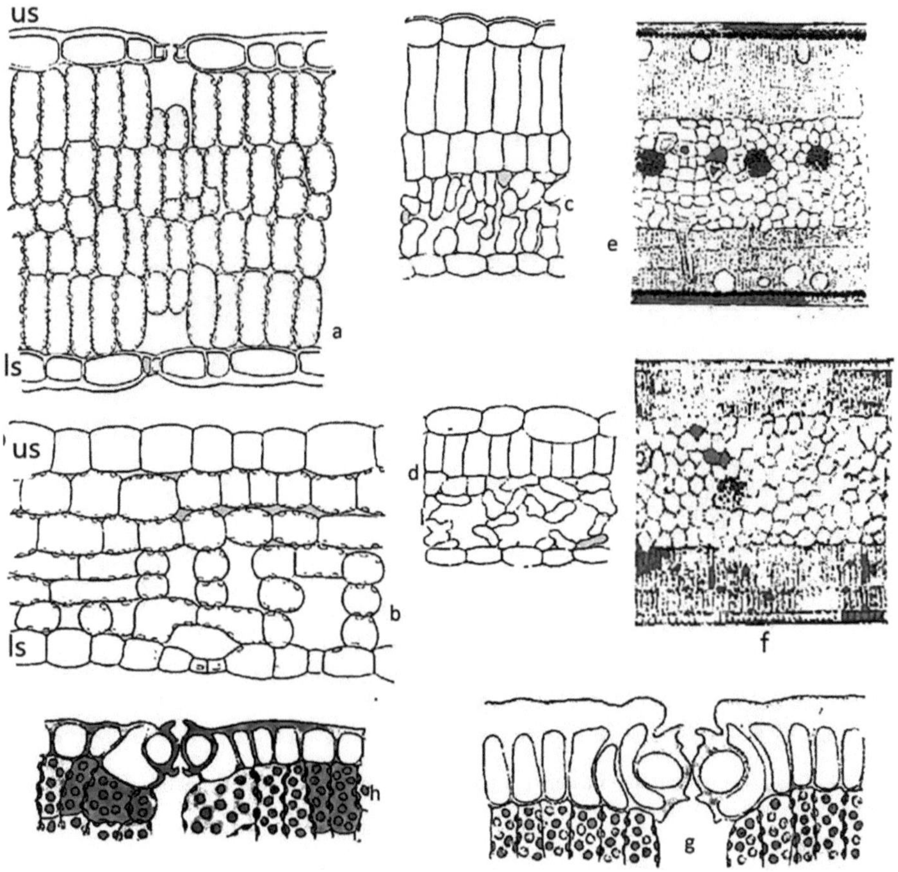

Fig. 38 The influence of transpiration on the differentiation of leaf tissues. Cross sections through leaves and stomata; (**a, b**): *Lactuca scariola*—(**a**) exposed to sunlight; (**b**) grown in the shade; us, upper face; ls, lower face; (**c, d**): *Robinia pseudoacacia*—(**c**) in normal air; (**d**) in air saturated with vapors; (**e–h**): *Sonneratia acida*—(**e**) on moist saline soil, (**f**) on normal soil in the Buitenzorg Botanical Garden; (**g**) on saline soil, (**h**) on normal soil (Modified after Schimper 1903)

morphological ones (such as permanent reduction of leaf shape, regulation of leaf light level, and removal of rainwater from the leaf surface) have been previously reviewed (Grigore 2022).

3.1 Anatomical Structure Involved in Regulation of Transpiration

In this sense, there is a fundamental distinction between land plants and especially xerophilous plants (xerophilic plants = *xerophytes*) and water plants. It is clear that there must be a great difference between the surface of a plant that is permanently immersed in water or moist air and one that is surrounded by dry air and prone to a permanent struggle with intense transpiration. The differences therefore concern, structurally, the covering structures of the aerial organs and the aeration system, including stomata and intercellular spaces, but also—of great importance—differences at the chlorenchyma level (photosynthetic tissues).

3.1.1 Cuticular Transpiration: The Cuticle

As is known, transpiration can be either cuticular or stomatal. Cuticular transpiration (through the cuticle) occurs through the plant's external cell walls, or, in most higher plants, through the epidermis. In relation to the regulation of cuticular transpiration, there are several structural devices involved in this process.

The cuticle is the main important regulator of transpiration (Stover 1951); relatively much has been written about the relevance of cuticle thickness in terms of transpiration limitation. It appears that the thickness of the cuticle (Figs. 39, 40, 41, 42, 43, 44, 45, 46, 47, 48, 49, 50, 51, 52, 53, 54, 55, and 56) is adjusted according to the plant's needs to limit transpiration. Of course, in the spirit of the vision pursued in this work, there are many nuances in this sense as well; for example, the cuticle of young leaves is more impermeable to water than that of old leaves. The cuticle of hydrophytes is usually very thin and permeable, but that of xerophytes is thicker and often completely impermeable.

In the leaf of xerophytes, the external walls of the epidermis can be strongly cutinized, a situation encountered especially in the case of some species that grow in the desert (Figs. 46, 47, 48, 49, and 50), but also in many littoral plants that group several types of xerophilous plants (psammophytes, sand plants; halophytes, salt-tolerant plants) (Chermezon 1910). In some cases, epidermal cell walls may even include calcium or silica oxalate crystals.

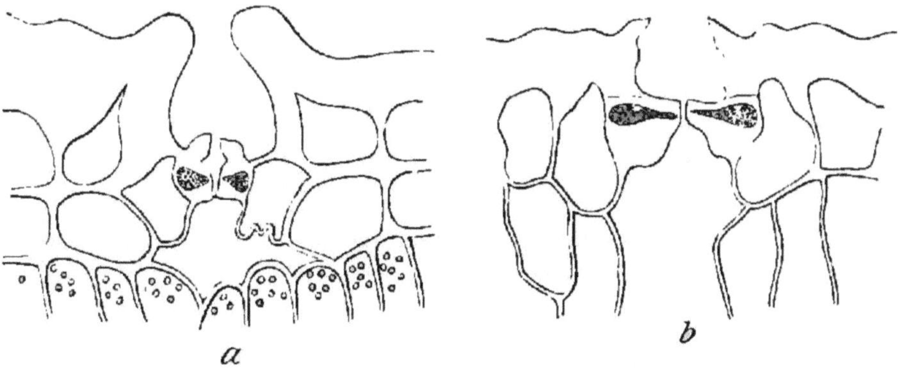

Fig. 39 Xerophilic structure—dry climate in temperate Australia; stomata at *Franklandia fucifolia* (**a**) and *Eucalyptus giganteus* (**b**) (Schimper 1903)

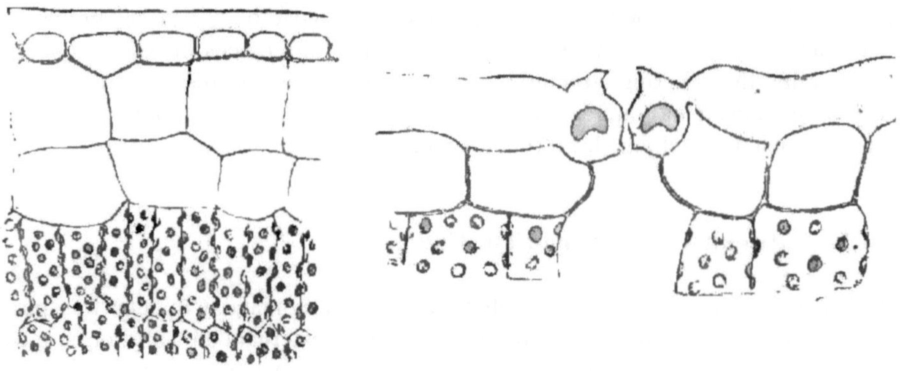

Fig. 40 Xeromorphic structure. Wet saline soil of Javanese mangrove swamps. Left, *Aegiceras majus*, the upper side of a leaf; right, *Rhizophora mucronata*, stomata and lower leaf epidermis (Schimper 1903)

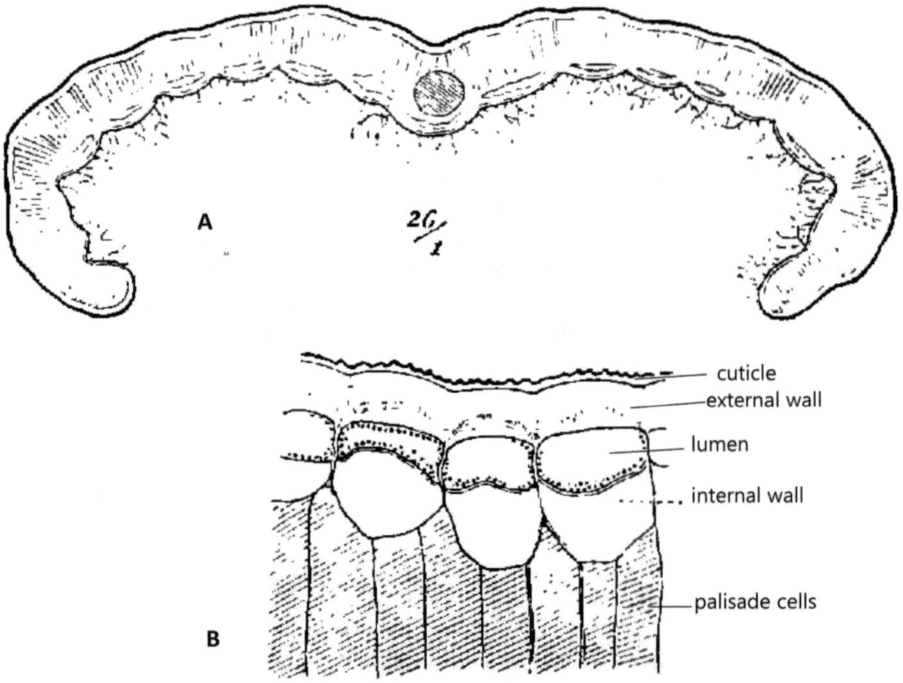

Fig. 41 Xeromorphic structure. The cold soil of Greenland. (**a**) *Dryas integrifolia*, cross section through the leaf; (**b**) *Loiseleuria procumbens*, portion of cross section through the leaf (Schimper 1903)

In addition to the very thick cuticle, in some desert xerophytes, the leaf can present a double epidermis (sometimes, even three-layered), most likely also related to adaptations to the xeric habitat (Figs. 51 and 52).

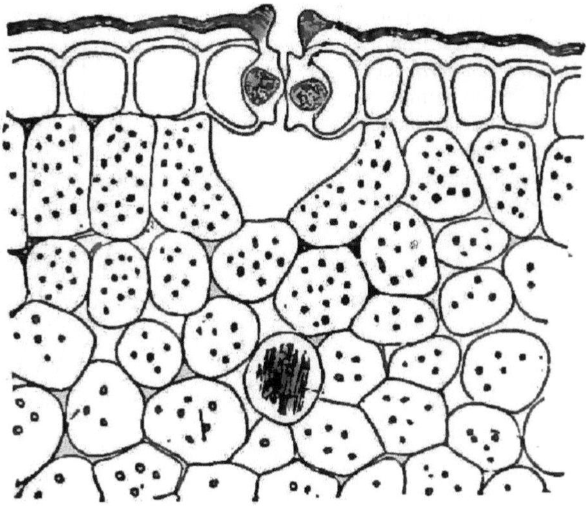

Fig. 42 Cross section through the leaf of *Aloe socotrina* (Barnes 1898)

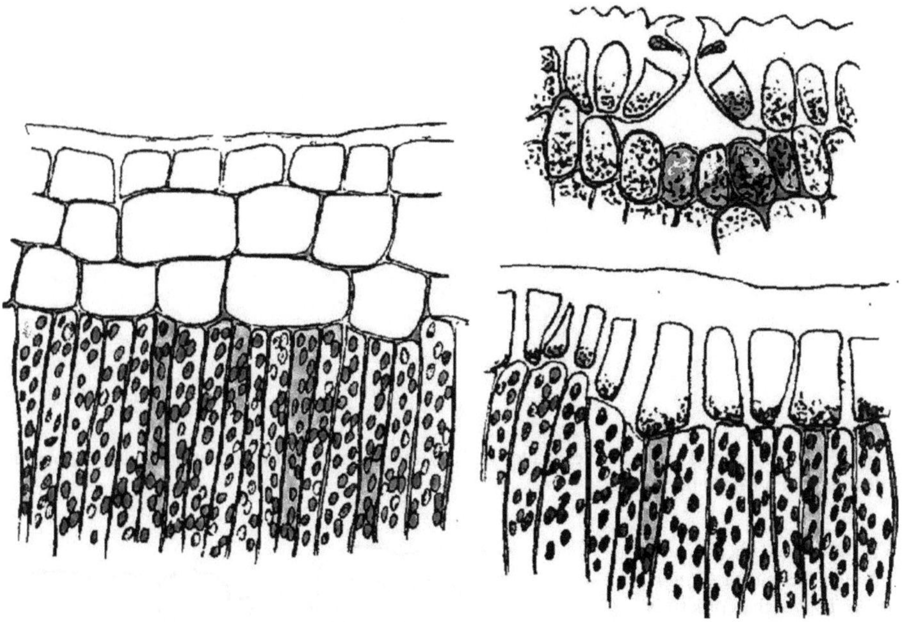

Fig. 43 Xeromorphic structure. Alpine climate. Left, *Myrica javanica*, a portion of the cross section through the leaf; right, *Photinia integrifolia*, cross section of the lower (top) and upper (bottom) side of the leaf (Schimper 1903)

The leaves, due to the particular nature of the epidermis, are often leathery and glossy, and this is a frequent and characteristic feature of tropical trees (sclerophyllous) (Grigore 2022), but it is also found in temperate climates, in the leaves of evergreens such as be *Ilex aquifolium*, some conifers, and *Vinca* species.

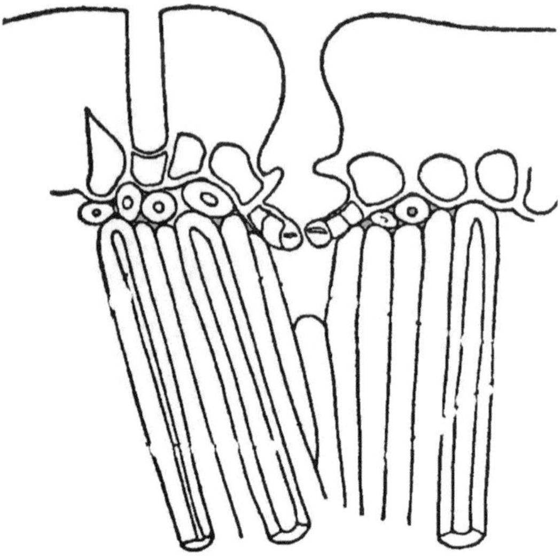

Fig. 44 Cross section through the leaf from *Hakea leucoptera*; very thickened epidermis (due to thick cuticle), sunken stomata, and sclerenchymatized hypodermis (Cannon 1921)

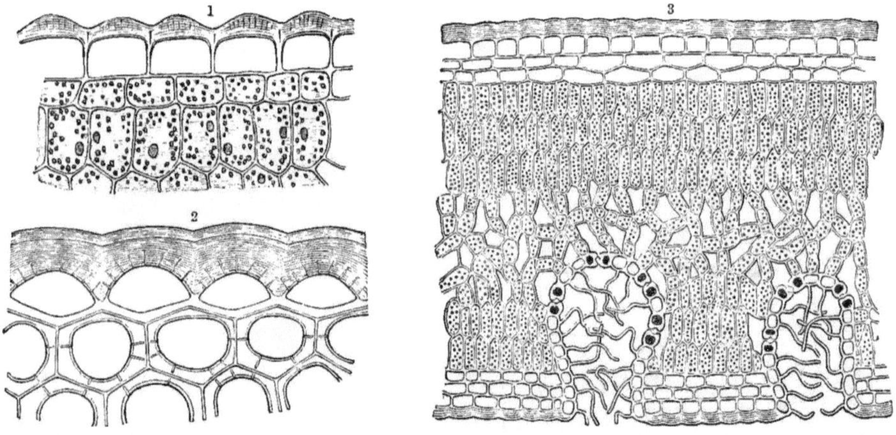

Fig. 45 Layered thickened cuticle: leaves of 1, *Viscum album*; 2, *Ilex aquifolium*; and 3, *Nerium oleander* (von Marilaun 1902)

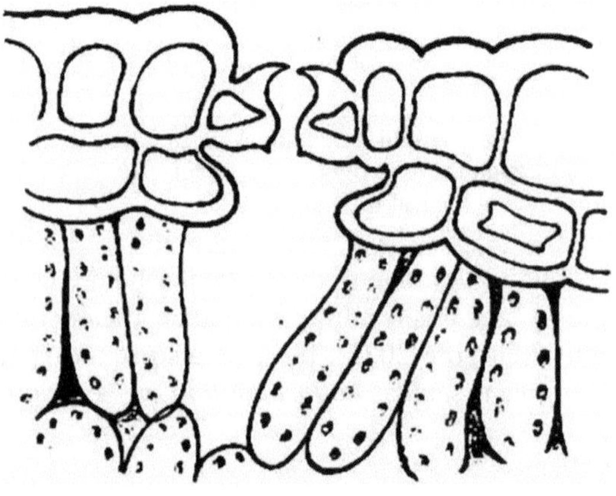

Fig. 46 Cross section through the leaf of *Alhagi manniferum*; the epidermis is very thickened, due to the thick cuticle, the hypodermis sclerenchymatized (Volkens 1887)

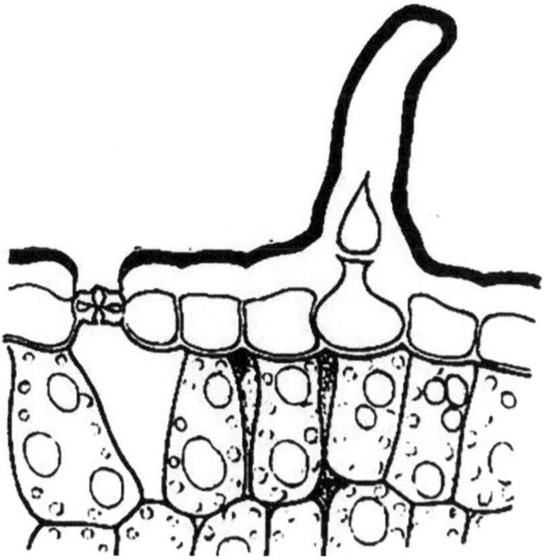

Fig. 47 Cross section through the leaf of *Lavandula coronopifolia*; the epidermis is very thickened, due to the thick cuticle (Volkens 1887)

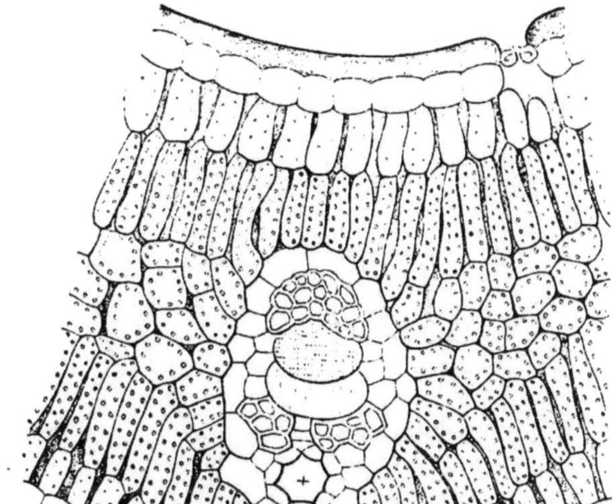

Fig. 48 Cross section through the leaf of *Carduncululus eriocephalus*; the epidermis is very thickened, due to the thick cuticle (Volkens 1887)

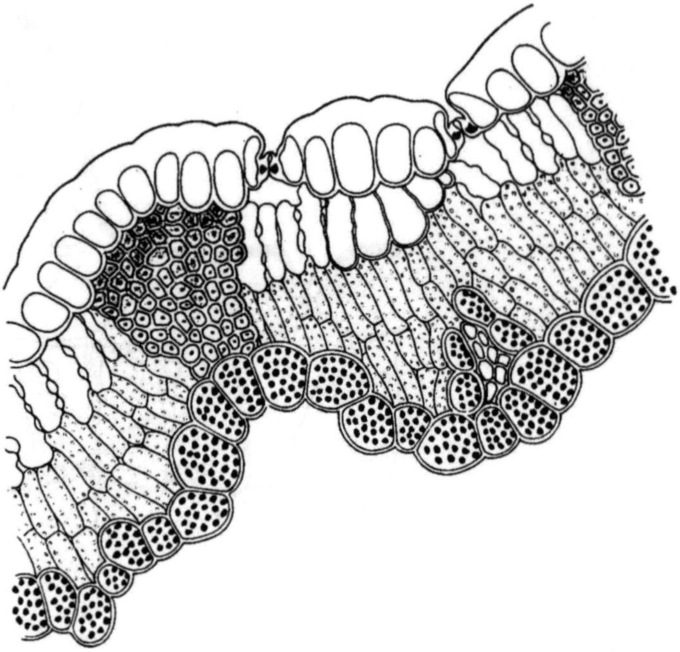

Fig. 49 Cross section through the leaf of *Panicum turgidum*; the epidermis is very thickened, due to the thick cuticle (Volkens 1887)

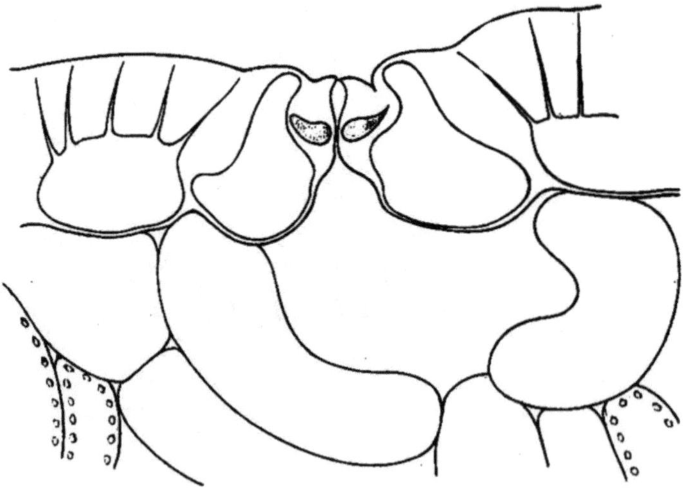

Fig. 50 Cross section through the leaf of *Sporobolus spicatus*; the epidermis is very thickened, due to the thick cuticle (Volkens 1887)

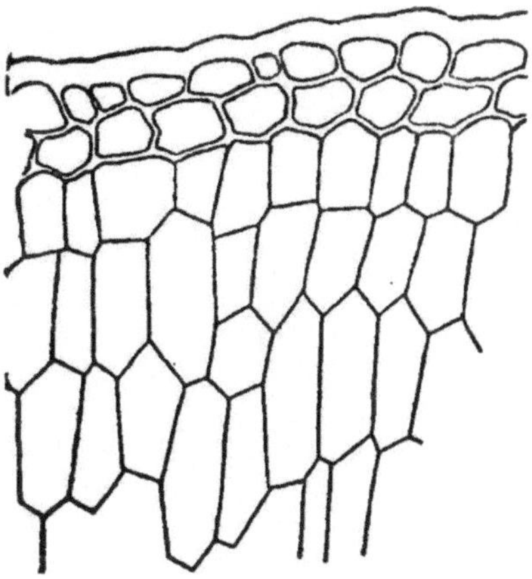

Fig. 51 Cross section through the leaf of *Fusanus acuminatum*; very thick; bilayered epidermis is observed, due to the thick cuticle (Cannon 1921)

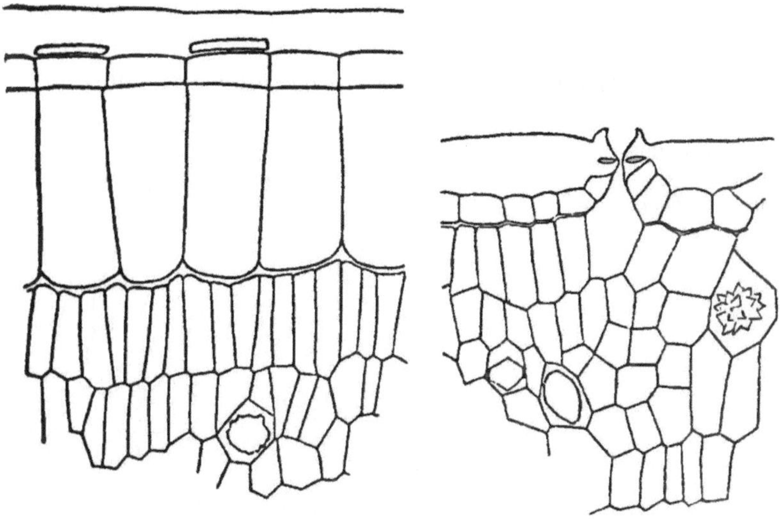

Fig. 52 Cross section through the leaf of *Pittosporum phillyreoides*; the bi- (tri)stratified epidermis is very thickened, due to the thick cuticle; left, dorsal side; right, ventral side (Cannon 1921)

The glossy surface reflects some of the incident light from the leaf and may therefore represent a useful adaptation.

Wax is excreted on the surface (Figs. 57, 58, 59, and 60) and can reduce transpiration (Henslow 1895; Mauseth 1988), an aspect established experimentally for a long time (Warming 1909a).

Usually, the wax layer produced is only a thin one; but, at *Capparis spinosa*, at the beginning of the dry season, in the Egyptian desert, a very thick layer of wax (Fig. 58) is produced on the entire upper surface of the leaf, which completely prevents transpiration. The wax layer can be very thick, more than a millimeter in South African *Sarcocaulon*, and up to 5 mm in some palm species. The wax also has

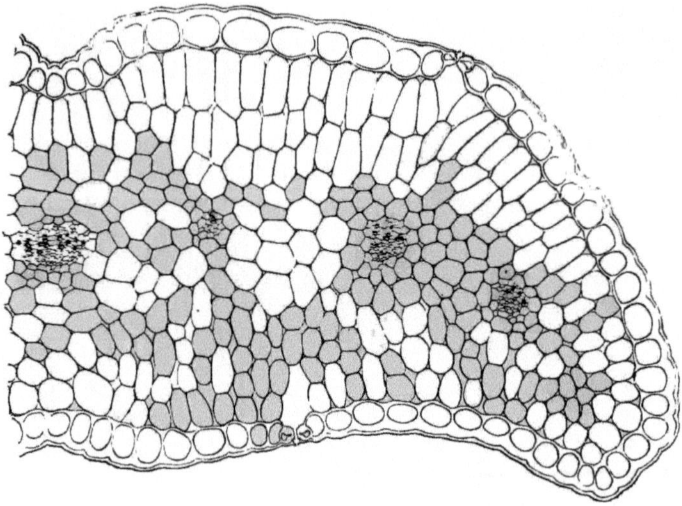

Fig. 53 Cross section through the leaf of *Galium arenarium* (Chermezon 1910)

Fig. 54 Cross section through the leaf of *Matricaria maritima* (Chermezon 1910)

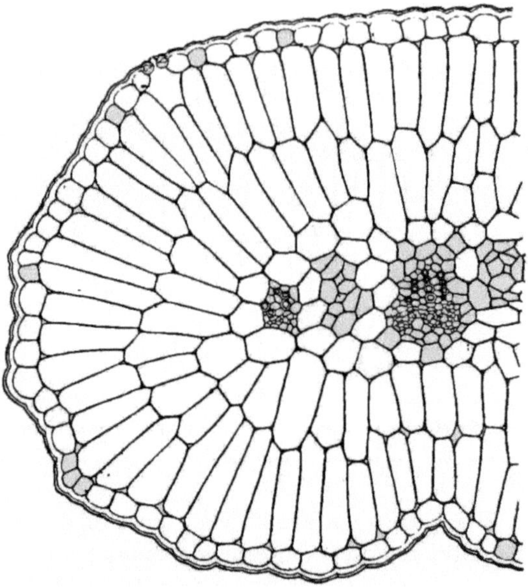

the role of preventing water from wetting the leaves, so as to protect the shade leaf from rain.

Other anatomical adaptations in relation to the leaf of xerophytes can refer to the following aspects:

Salt Incrustations on the Surface of the Leaf On the surface of the leaves (and stems, sometimes) of some desert plants, incrustations (films, deposits) of salt

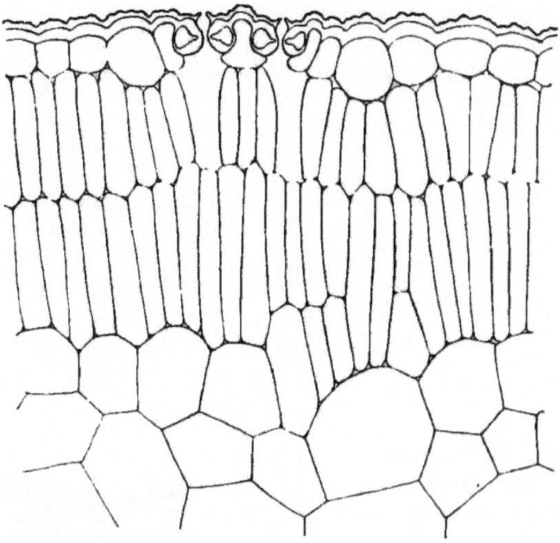

Fig. 55 Cross section through the leaf of *Echinophora spinosa* (Chermezon 1910)

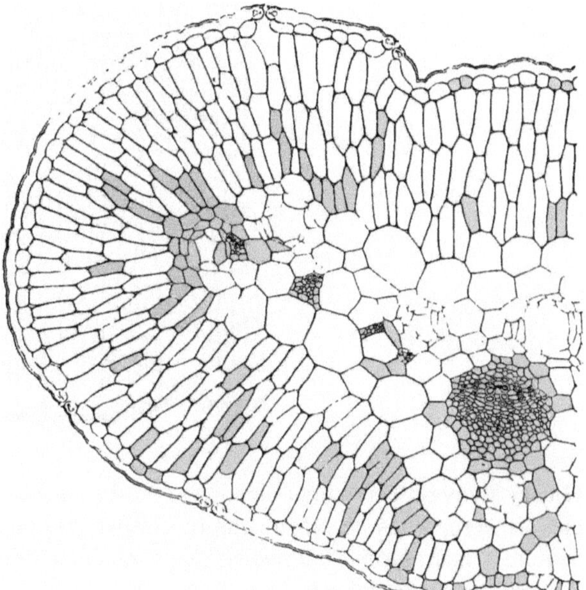

Fig. 56 Cross section through the leaf of *Artemisia crithmifolia* (Chermezon 1910)

appear; this way, leaves acquire a gray shade and seem to be thus protected against excessive transpiration; at night, the incrustations melt and become liquid as they absorb moisture from the atmosphere (Warming 1909a). Although these salt deposits are the result of the secretion of some salt glands, typical for Plumbaginaceae species (Grigore 2008, 2021; Grigore and Toma 2021a, b), the role of the salt deposit in regulating transpiration, on the surface of the leaf, requires further clarification.

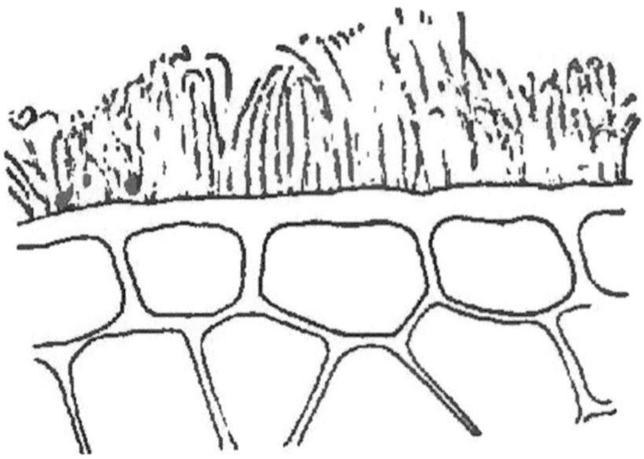

Fig. 57 Epicuticular wax—cross section through leaf of sugarcane (Warming 1918)

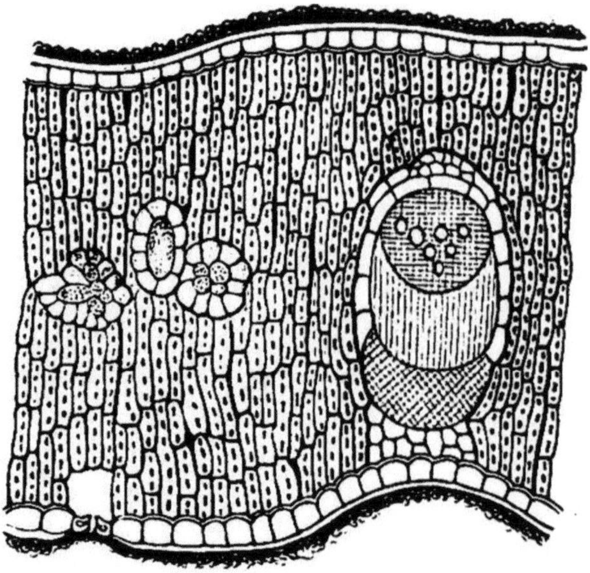

Fig. 58 Cross section through the leaf of *Capparis spinosa* (Volkens 1887)

Glossy Leaves The hairs (trichomes) on the surface of some leaves of many species of xerophytes secrete resins or similar compounds; in this way, the leaves acquire a macroscopic "varnished" appearance; in this case, the epidermal walls are thin and poorly cutinized.

Mucilages, or mixtures of gums, resins, and other compounds, are sometimes excreted by trichomes, in the buds of some species of Polygonaceae, for example; it appears that these mucilages can help absorb water and control transpiration during leaf water washing. In some species, the inner walls of the leaf epidermis are mucilaginous (Figs. 61 and 62).

Epidermal cell content may play a role in reducing transpiration; in addition, the epidermis itself can serve as a water reservoir when, in some xerophytic species, the epidermal cells are large, bulky, and colorless (we will return to this later when we

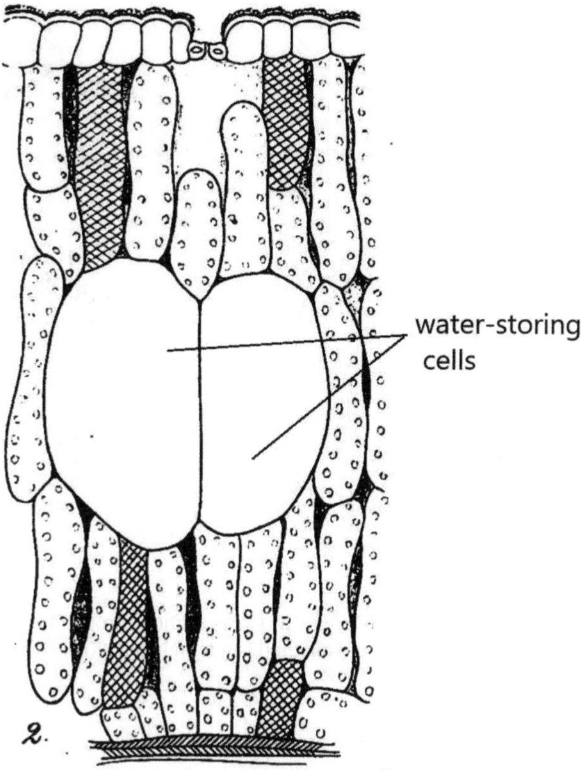

Fig. 59 Cross section through the leaf of *Nitraria retusa*; massive water-storage cells are also observed in the mesophyll (Volkens 1887)

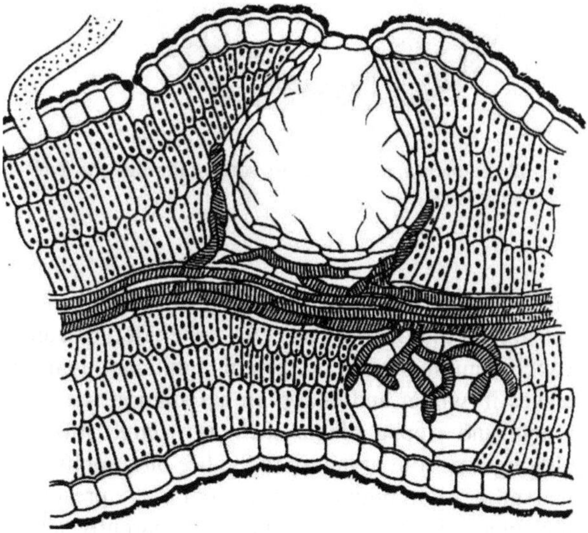

Fig. 60 Cross section through the leaf of *Haplophyllum tuberculatum* (Volkens 1887)

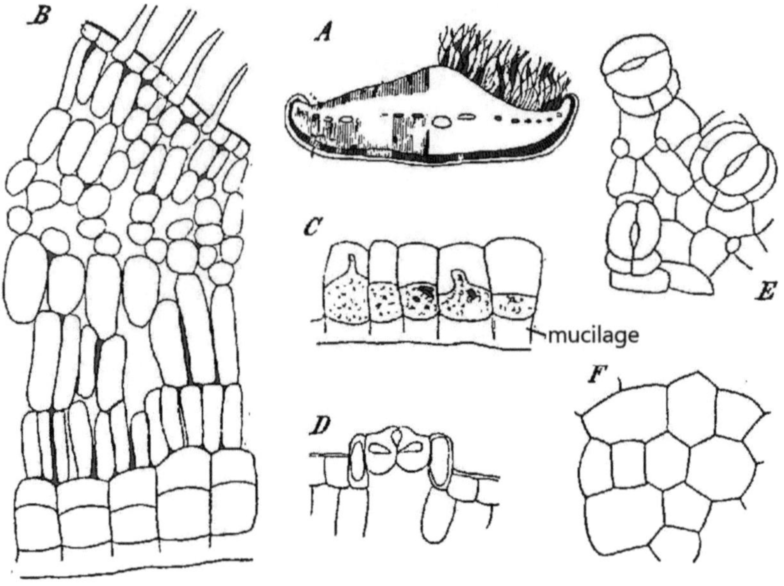

Fig. 61 *Thymelaea hirsuta*. (**a**) Cross section through the leaf—general view and detail (**b**); (**c**) lower epidermis, with mucilaginous internal walls; (**d**) stomata; (**e**) upper and lower epidermis (**f**), front view (Warming 1918)

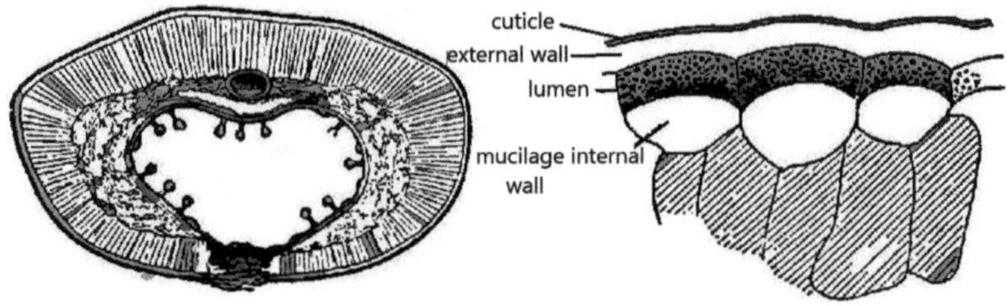

Fig. 62 *Empetrum nigrum*: cross section through the leaf—general view (left) and detail (right) (Warming 1918)

discuss the water storage structures of xerophytes at length). Through the various substances dissolved in its cells, the epidermis can become less permeable to water vapor. Tannin is often present in significant quantity in the epidermis of evergreen leaves in winter and occurs in association with the water-storage tissue of desert plants and steppe plants such as *Alhagi*, *Monsonia*, *Astragalus*, and *Tamarix*, although the role of tannin in the life of xerophytes is not fully understood.

Anthocyanin pigments, red in color, are present in many plants and especially in the epidermis; their role is unclear, apparently acting as a heat-absorbing substance and thus promoting transpiration; but according to other opinions, it does exactly the opposite (see Warming 1918).

3.1.2 Stomatal Transpiration and the Aeration System in Xerophyte Leaves

In addition to transpiration at the cuticle level, the intercellular spaces are also involved in transpiration; the transpiration surface of a plant is constituted not only by the surface exposed to the external atmosphere but also by the cell walls that delimit all the intercellular spaces. Therefore, a very simple but logical hypothesis takes into account the fact that the air-containing intercellular spaces of terrestrial plants, and especially of xerophytes, are much different and narrower than those of aquatic plants, in which the aerenchyma, in general, is highly developed, as discussed in the previous pages.

But perhaps the biggest and most relevant differences are related to the stomata, structures highly specialized in gas exchange between the plant and the external environment.

3.1.2.1 The Stomata

There is a lot of data about stomata—gathered over time, especially in xerophytes—mainly interested in the number of stomata, the distribution of stomata in the epidermis of the leaf, and, last but not the least, the position of stomata in relation to the general level of epidermal cells. The physiology of stomata, respectively the closing and opening movements of stomata (Öpik et al. 2005; Hopkins and Hüner 2009; Graham et al. 2014; Schulze et al. 2019), is not the subject of our discussion in these pages.

Stomata are adapted, by their structure and physiology, to regulate transpiration. As a rule, they close when transpiration becomes excessive and when the leaves wither due to lack of soil moisture, in the leaves of many dormant plants in winter, and reopen when there is, hypothetically, no more danger.

Guard cells of certain desert plants are motile only in young leaves; in old leaves, the cells become immobile due to strong thickening of their walls, or they may become blocked with wax or resin. The floating leaves of aquatic plants have stomata on the surface of the leaf exposed to the atmosphere (upper side), but these take, at a given moment, a particular shape and lose their ability to move.

The number of stomata, also a very important aspect in relation to xerophytes, depends on the nature of the environment. As a general (but not absolute) rule, the drier a habitat, the less numerous stomata are, as can best be seen when comparing closely related species (e.g., with aquatic and xerophilic representatives).

The distribution of stomata is closely related to transpiration and humidity conditions. Species of meadow grasses and other mesophilic terrestrial plants usually have stomata on both sides of the leaf (amphistomatic leaves); steppe grasses have stomata, as a rule, only on the furrowed upper face; other xerophytes have stomata usually only on the lower face, where they are often hidden, so as to hinder transpiration, or, on the contrary, the stomata are completely unprotected, but located on some small projections (Fig. 63). The stomata of xerophytes are often sunk below

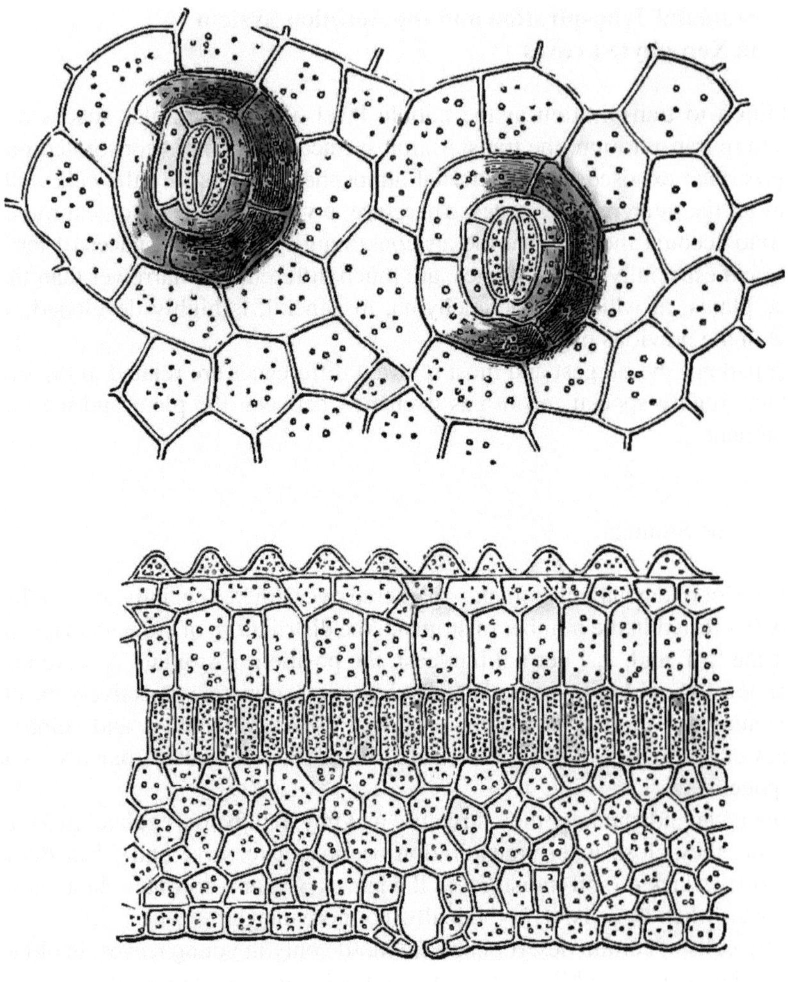

Fig. 63 *Peperomia arifolia*. Cross section through the leaf (bottom) and epidermis, front view (top) (von Marilaun 1902)

the general surface of the epidermis, in depressions, furrows, and other configurations, which are often lined with hairs (trichomes) (Fahn and Cutler 1992).

A space containing air is thus formed, which is lost with difficulty; the space is protected from the wind and charged with water vapor; the result is that transpiration is delayed. A few such arrangements are described below.

The simplest device is the production, next to each stomata, of an epistomatal cavity, which is a saucer, urn, or funnel-shaped cavity, which is formed either by cuticular processes that give rise to this external epistomatal cavity or by epidermal cells adjacent ones projecting above the stoma, which is thus immersed in an external respiratory chamber, as in *Pinus sylvestris* (Fig. 3, Leaf – general), species of

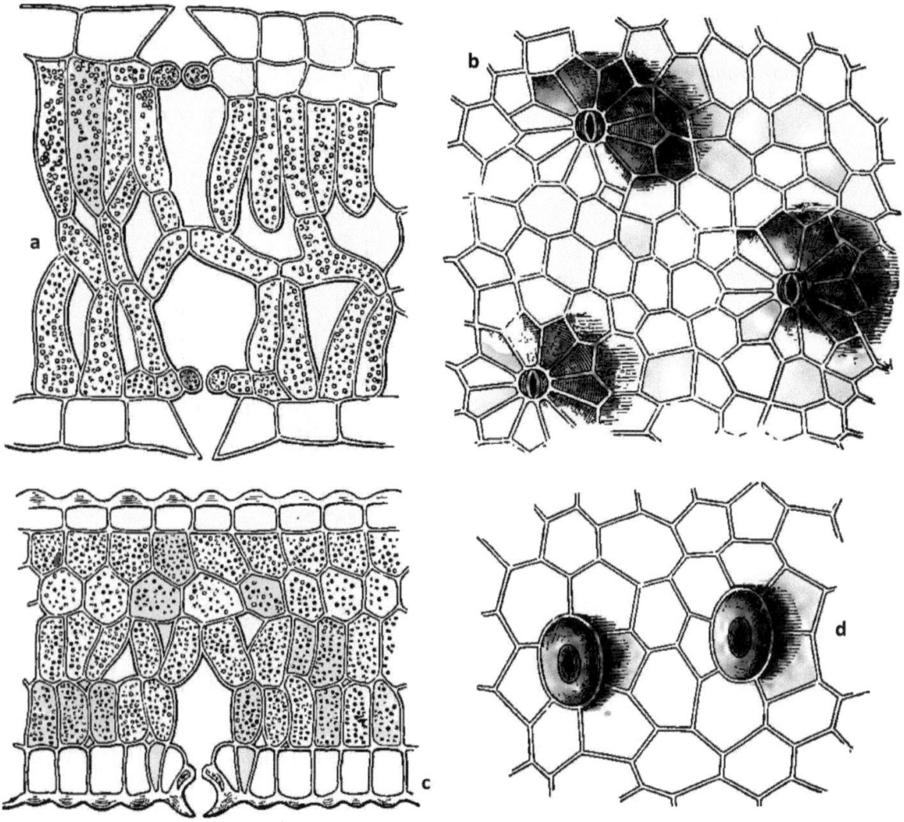

Fig. 64 Sunken stomata, thick cuticle in *Hakea florida* (**a** cross section through the leaf; **b** epidermis, front view) and *Protea mellifera* (**c** cross section through the leaf; **d** epidermis, front view) (von Marilaun 1902)

Proteaceae (Fig. 64, in which, in addition, the stomata are protected by an epidermal cuticular "armor"), and many other xerophilous species.

Sunken stomata are also found in many littoral species (Figs. 65, 66, 67, 68, 69, and 70), which, in reality, comprise several sub-groups of xerophytes (halophytes, salt marsh plants; psammophytes, sand plants; etc.), these being studied, among others, by Chermezon (1910) and Denis (1919). For the examples mentioned here, very thick cuticle is also observed.

At *Euphorbia paralias* (Fig. 67), and also in various grasses and sedges (Juncaceae), the stomata are surrounded by small papillae (Fig. 71).

Stomata located and grouped in small depressions ("crypts"). On the lower side of the leaves of *Nerium* (Figs. 45 and 73), *Banksia*, and other xerophytes (Fig. 72) groups of stomata arranged in small depressions ("crypts") occur, the stomata having the ostioles practically covered with hairs (trichomes).

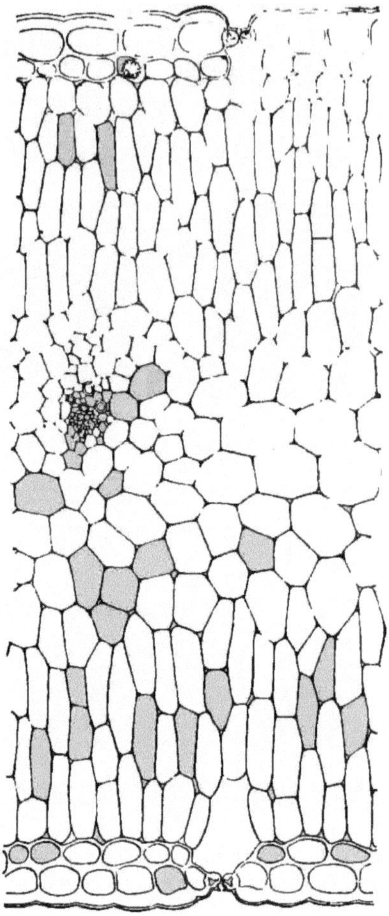

Fig. 65 Cross section through the leaf of *Eryngium maritimum* (Chermezon 1910)

Stomata Arranged in Longitudinal Grooves (Furrows) In many plants, stomata are deposited in longitudinal furrows whose margins are often more or less provided with trichomes. Furrows occur on the upper side of the leaf in many steppe grasses, such as *Weingaertneria canescens*, *Festuca ovina*, *Psamma* (*Ammophila*) *arenaria*, *Aristida*, *Stipa*, *Sporobolus spicatus*, and *Cynodon dactylon*; in such cases, the furrows may be narrowed above, and the stomata may be closed (covered) almost completely by rolling up the leaf.

In many species such as *Empetrum*, *Phyllodoce caerulea*, *Calluna*, *Erica*, *Loiseleuria procumbens*, *Ledum palustre*, and *Cassiope tetragona*, the leaves are revolute (with the edges rolled down) and are undivided, with a small area, generally linear, but sometimes linear-ovate, elliptic, or even circular in outline; they are always rigid and usually evergreen (many are alpine, mountain, or peatland plants) (Grigore 2022); in these, the lower part, which is oriented towards the ground, is

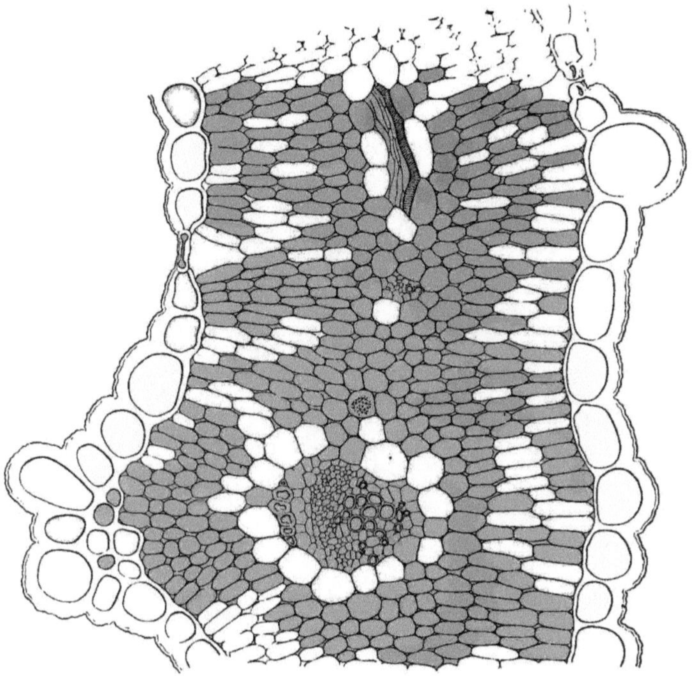

Fig. 66 Cross section through the leaf of *Polygonum maritimum* (Chermezon 1910)

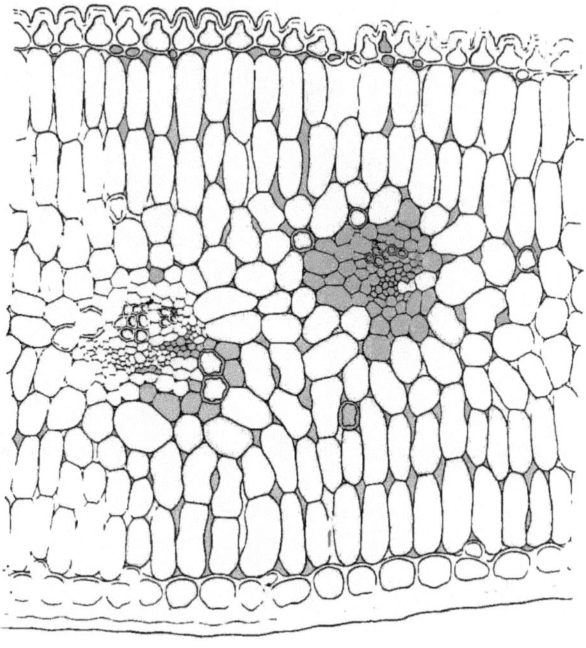

Fig. 67 Cross section through the leaf of *Euphorbia paralias* (Chermezon 1910)

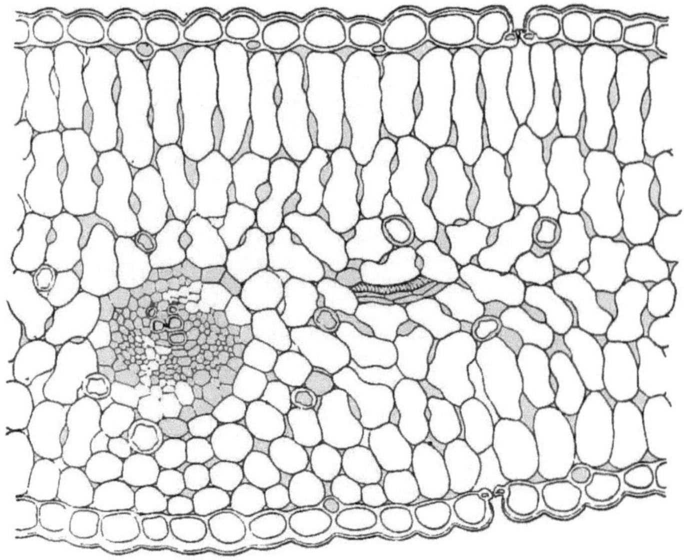

Fig. 68 Cross section through the leaf of *Euphorbia portlandica* (Chermezon 1910)

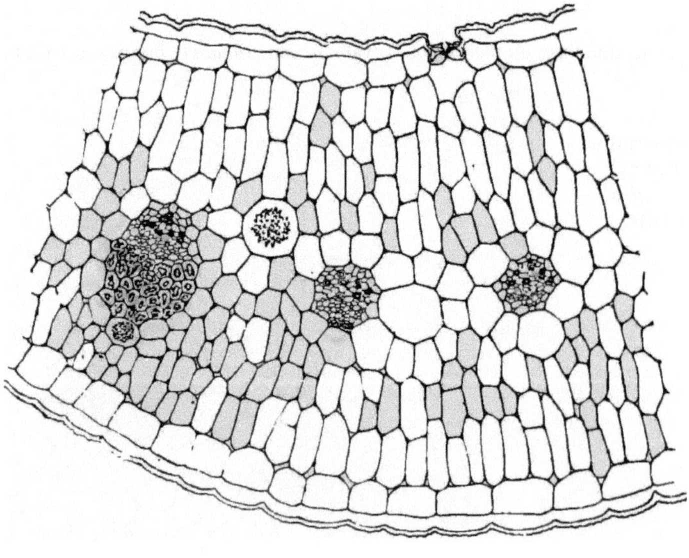

Fig. 69 Cross section through the leaf of *Dianthus gallica* (Chermezon 1910)

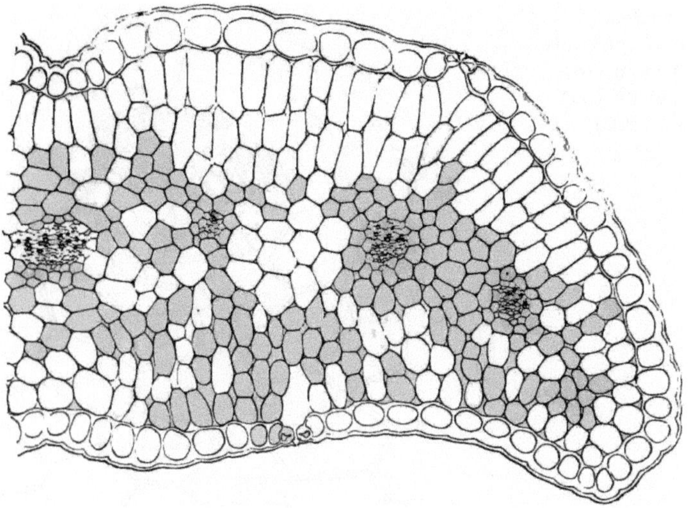

Fig. 70 Cross section through the leaf of *Gallium arenarium* (Chermezon 1910)

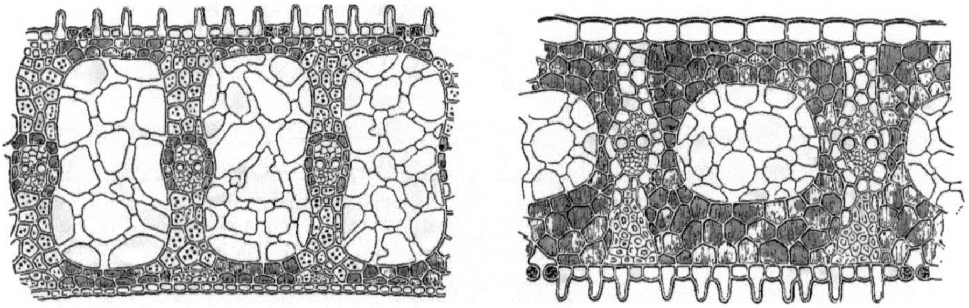

Fig. 71 *Glyceria spectabilis* (left) and *Carex paludosa* (right)—cross sections through the leaf (von Marilaun 1902)

hollow to a greater or lesser extent, while the upper part, which is oriented towards the outside (atmosphere), is arched. Often, the leaf thus conformed is twisted, in such a way as to close inside a specific cavity, which communicates with the outside only through a very narrow opening. The twisted edges of the leaves in this category almost touch, and in the lower epidermis of the leaf appear wider furrows or channels provided with trichomes and stomata (Figs. 74 and 75). To this category may be added leaves such as those from *Dryas octopetala*, with revolute edges, to a lesser extent, and with stomata located on the lower side of the leaf, equipped with trichomes. This is the situation of the twisted leaves of dicots xerophytes, while in the leaves of many xerophytic monocots, as we shall see, there is another type of adaptive (temporary) rolling of the leaves, due to the bulliform cells (see again—Figs. 12, 13, and 14 - Leaf - general aspects), in plants exposed to conditions of prolonged drought.

Fig. 72 *Dryandra floribunda* (top, epidermis, front view; bottom, cross section through the leaf) (von Marilaun 1902)

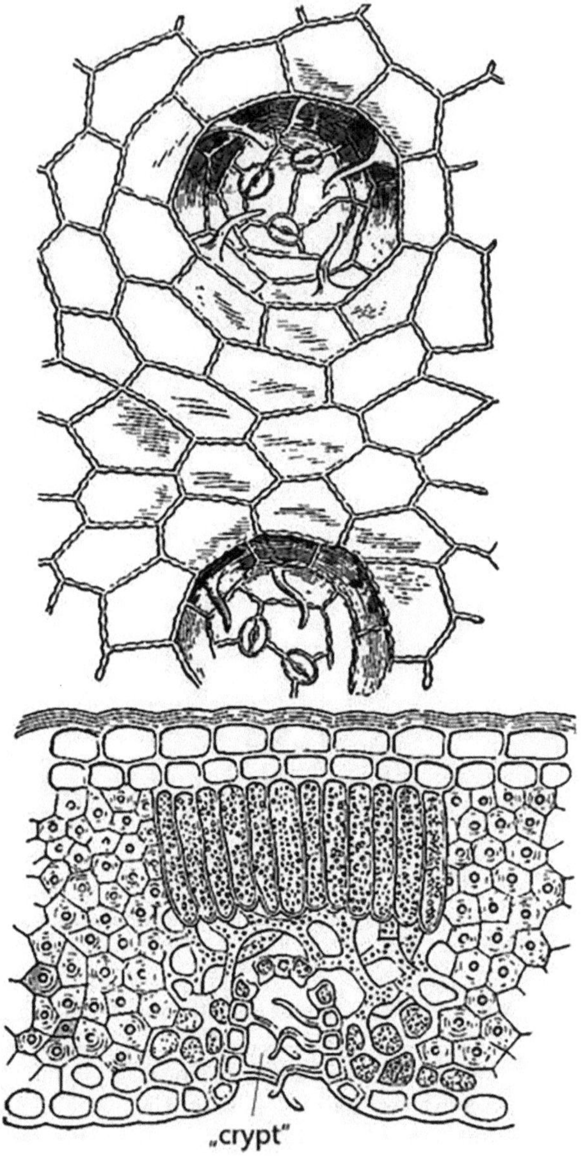

If the leaves are permanently and abruptly directed upwards, so that the lower side is more strongly illuminated, it may be differentiated, like the upper side, forming palisade tissue; in such cases, the stomata-furrow appears on the upper side, as in *Passerina filiformis*, *Ozothamnus*, and *Lepidophyllum*.

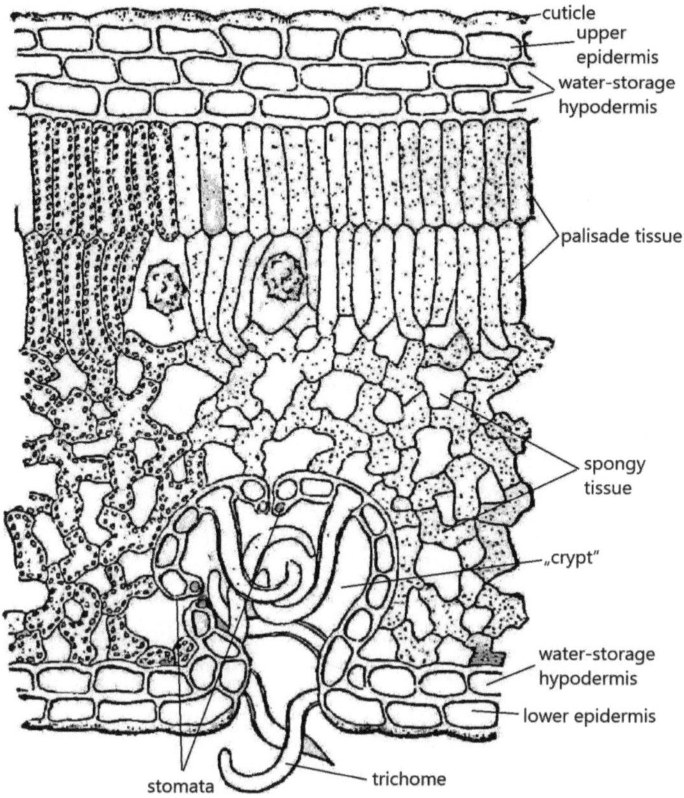

Fig. 73 Cross section through the leaf of *Nerium oleander* (Barnes 1898)

3.1.2.2 Intercellular Spaces

The size of the intercellular spaces varies according to external conditions; as a rule, intercellular spaces are larger in shade leaves (sciophyllous) than in sun leaves (heliophyllous), better developed in moist air than dry.

In xerophytes, the intercellular spaces containing air are very narrow (or even completely absent), as in many species of salinized deserts or maritime dunes (Chermezon 1910).

3.1.3 Chlorenchyma

As we anticipated in the discussion about the leaves of aquatic plants, the leaves of xerophytes, unlike those of hydrophytes (especially submerged ones), show dorsiventrality, translated, especially in the development of palisade tissue.

The palisade tissue is highly developed in xerophytes, by increasing either the number of palisade cell layers or the height of the cells, or by both means. Within

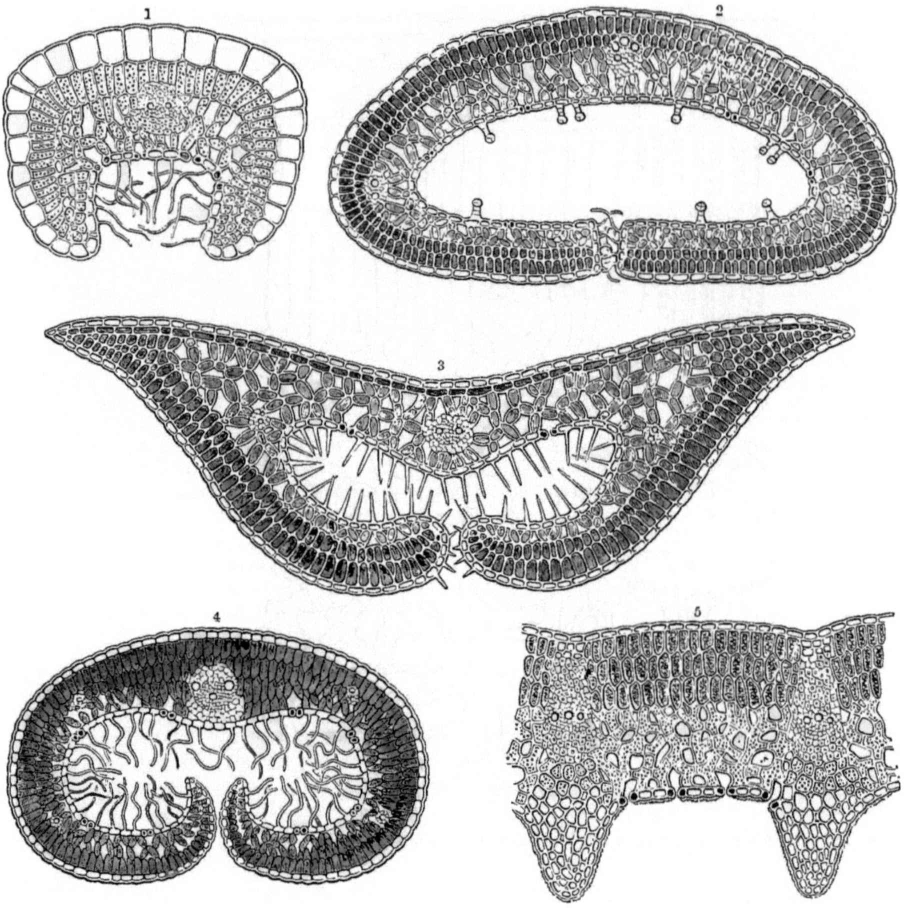

Fig. 74 Cross sections through rolled (twisted, revolute) leaves in *Erica caffra* (1), *Empetrum nigrum* (2), *Andromeda tetragona* (3), *Tylanthus ericoides* (4), and *Salix reticulata* (5) (von Marilaun 1902)

the general xerophyte-environment interrelationship, it is not very clear which environmental component induces this structural change; probably the development of the palisade tissue is most closely correlated with the dryness of the atmosphere and with transpiration. And light undoubtedly plays a role in this structural change, since the oblique orientation of the palisade cells must be due to illumination.

In some xerophytes, such as terrestrial halophytes, the height of the palisade tissue increases depending on the salts contained in the soil, as it was demonstrated in some classical experiments (Lésage 1890, commented by Grigore and Toma 2021a).

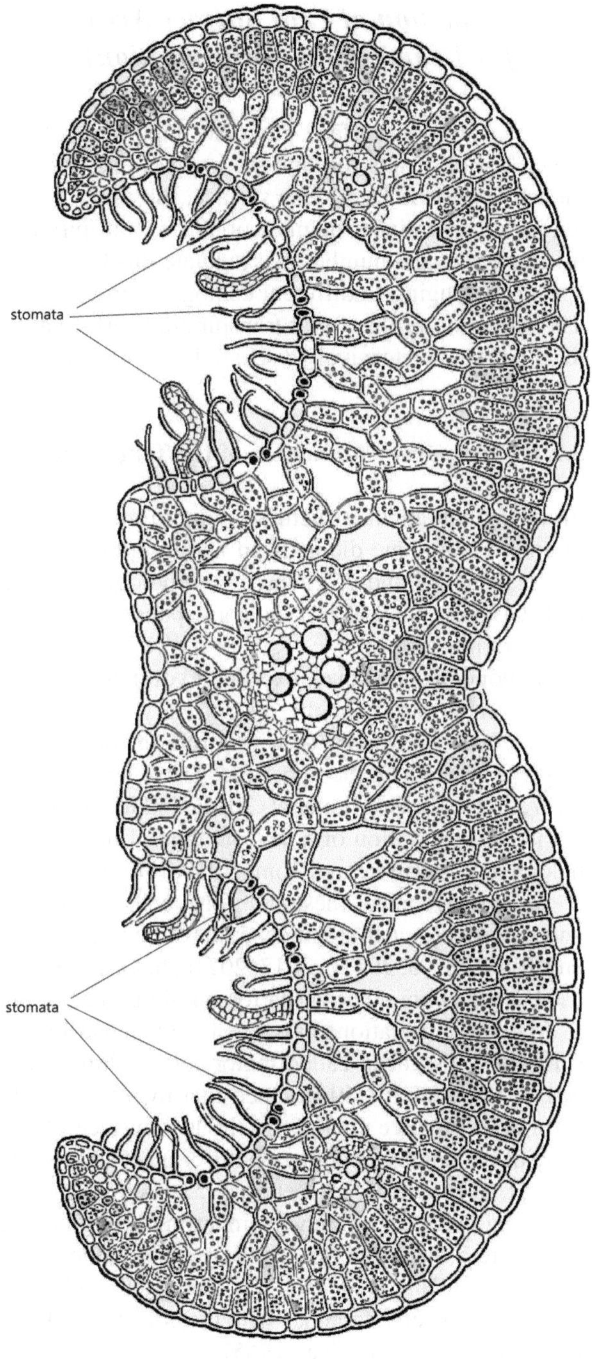

Fig. 75 Cross section through a rolled (twisted, revolute) leaf of *Azalea procumbens* (von Marilaun 1902)

3.2 *Reduction of Leaf Surface Area for Evaporation (Transpiration)*

The transpiration surface plays an important role in determining the amount of water lost by plants through the process of transpiration. In a very simplistic and didactic way, the larger this surface, the more intense the transpiration. Since the leaves are essentially the main organs of transpiration (but not always exclusively), it is their size and number that regulate this function and therefore vary in different species belonging to different ecological groups, according to climatic conditions. The reduction of this foliar transpiration surface at the organ at the morphological level has been extensively discussed in a previous work (Grigore 2022).

3.2.1 Temporary Reduction of the Transpiration Surface

From an anatomical (structural) point of view, we have already discussed the revolute leaves of some dicots from xeromorphic habitats (alpine, mountainous, peatlands) (Figs. 74 and 75), which morphologically correspond to the ericoid type of leaf (Grigore 2022). This type corresponds, rather, to a permanent reduction of the transpiration surface.

In monocots, however, exposed to dryness (water stress), at the level of the leaf lamina, there are, near the midrib, very voluminous cells, called bulliform cells (Figs. 76, 77, 78, and 79), with a role in rolling the leaves and protecting, in this way, the stomata on the upper face of the lamina (Grigore and Toma 2010a, 2011b. 2017, 2021a, b; Grigore et al. 2014); in halophytes, the presence of bulliform cells led to the description of a new subtype of halophytic plants, namely, "amphibious halophytes" (Grigore and Toma 2010c) (Figs. 12, 13, 14, and 15 - Leaf - general).

As already stated, these bulliform cells are widespread in grasses, and the role and their ecological significance have been previously widely discussed (Grigore and Toma 2010a, 2011b, 2017, 2021a, b; Grigore et al. 2012, 2014). They were also used as a taxonomic criterion for certain groups of plants (Metcalfe 1960).

Many interpretations have been made regarding the role of these bulliform cells (they have also been called motor cells or hinge cells) (Grigore and Toma 2011b, 2017). Haberlandt (1914) stated that hygroscopic changes in the growth of bulliform cells produce changes in the leaf surface of xerophyte leaves. Other authors were of the opinion that these bulliform cells have a water-storage role (Mader et al. 2020; Luo et al. 2021; Ali et al. 2022), being able to participate in the "expansion" of young leaves. Their involvement in curling or folding of mature leaves, due to water stress, has been discussed by other authors (Grigore and Toma 2011b; Yang et al. 2022; Wang et al. 2023). It seems that under conditions of intense water loss, the bulliform cells, together with the colorless cells, lose their turgor and allow the leaves to curl or fold (Esau 1965).

Clayton and Renvoize (1986) stated that the bulliform cells favor the penetration of light into the mesophyll cells. In certain situations, bulliform cells are not active or necessarily correlated with rolling, and the hygroscopic movements of the leaves,

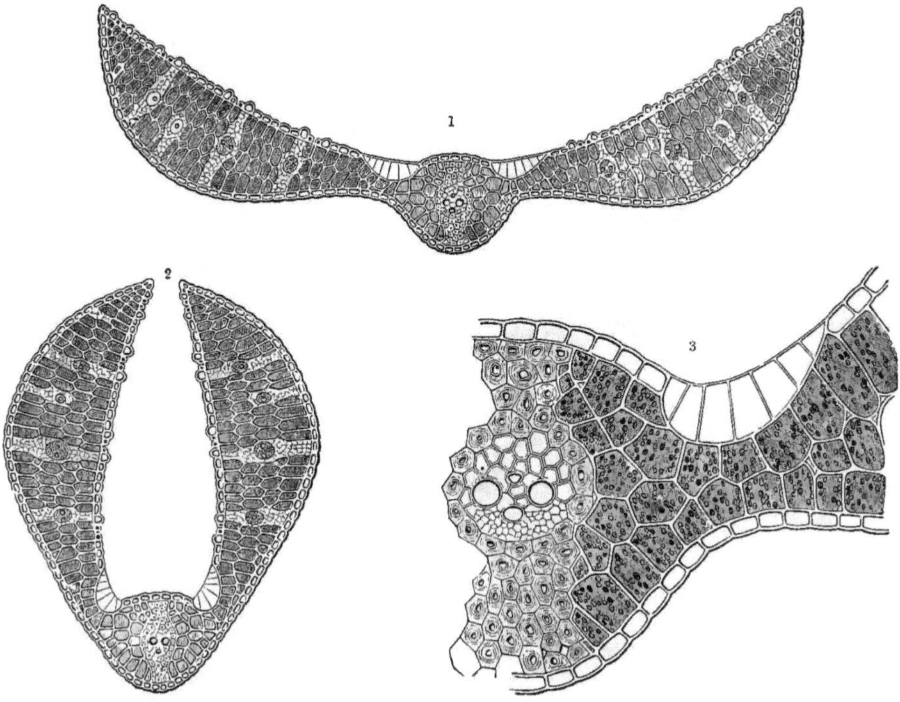

Fig. 76 Cross sections through the leaves (bulliform cells) in *Sesleria tenuifolia* (1, expanded leaf; 2, rolled leaf; 3, detail near the midrib) (von Marilaun 1902)

since they accumulate large amounts of silicon, and their external walls can be thickened and cutinized, thus becoming rigid (Nedukha 2022).

In conditions of water stress, some species such as *Loudetiopsis chrysothrix* and *Tristachya leiostachya* showed leaf curling (Alvarez et al. 2008). Merrium et al. (2022) were of the opinion that leaf rolling has adaptive value, reducing light interception and transpiration and protecting leaves from dehydration and overheating.

It would be a mechanism designed to reduce exposure to sunlight, which can lead to shrinkage transpiration, by maintaining the stomata in a microclimate with high humidity, preventing dryness (Silva et al. 2001).

And other species, in conditions of water stress, present, among other adaptations, large bulliform cells, including *Carex ligerica* (Toma and Dumitru 1973), corn (Ristic and Cass 1991), beans (Silva et al. 1999), and tomatoes (Sam et al. 2000).

Nawazish et al. (2006) showed that bulliform cells from a species collected from xeric and saline habitats, *Cenchrus ciliaris*, were more developed under severe drought conditions, and it was assumed that these structures play a crucial role in low humidity environments and are responsible for leaf curling and control of water loss at the leaf surface (Albernethy et al. 1998; Alvarez et al. 2003).

Most of ecological interpretation converge towards the xeric character of bulliform cells, towards the connection between water stress and the function of

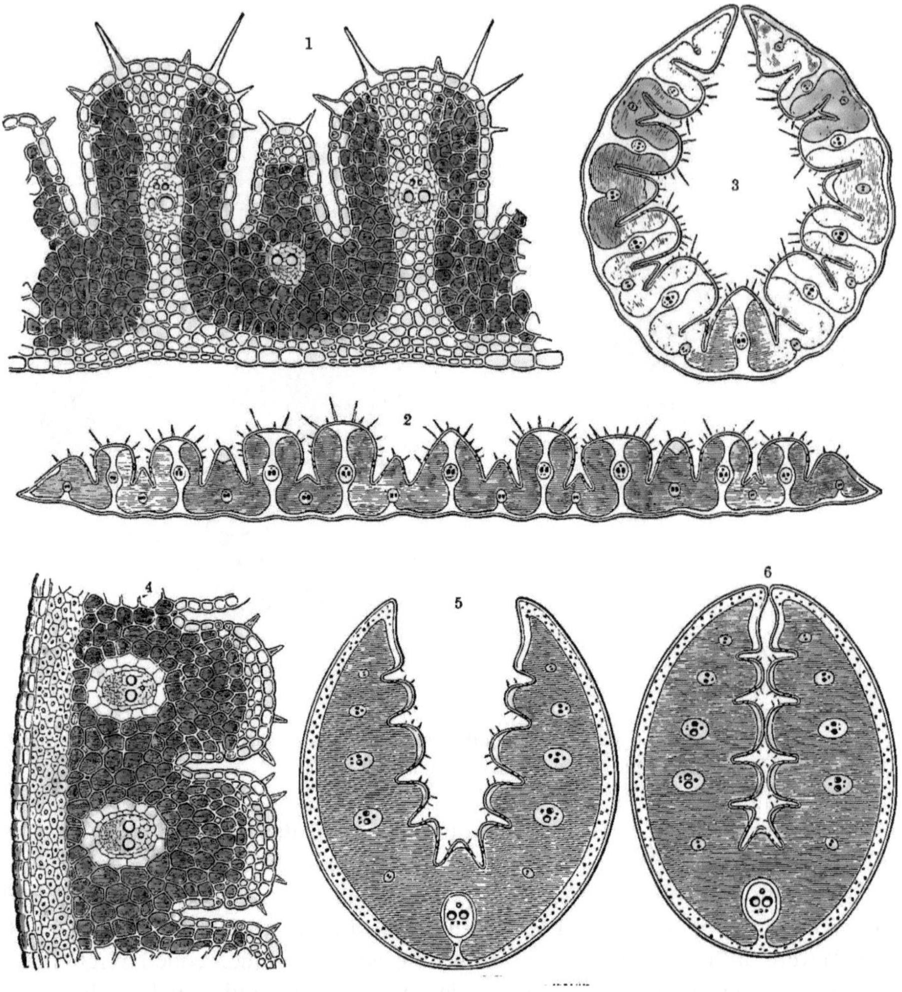

Fig. 77 Cross sections through the leaves of *Stipa capillata* (1, 2, expanded, detail, and general view; 3, closed) and *Festuca alpestris* (4, detail; 5, expanded; 6, closed) (von Marilaun 1902)

bulliform cells. The interpretations made by Grigore and Toma (2010a, 2010c) and the ecological characterization of "amphibious" halophyte species converge in another direction: all investigated halophyte species (in addition, *Puccinellia distans*, *Juncus gerardii*, and *Bolboschoenus maritimus*, in which bulliform cells are present) are plants of humid habitats, so hygrohalophytes.

Some authors (Sculthorpe 1967; Font-Quer 1970) use the term "amphibious" plants to denote those species that can live on dry soils, but flooded for a certain period of the year (see above, the definition of amphibian aquatic plants, but also Grigore 2022). These species show, in a structural and ecological sense, transitional features between terrestrial and aquatic plants. The plants investigated by Grigore

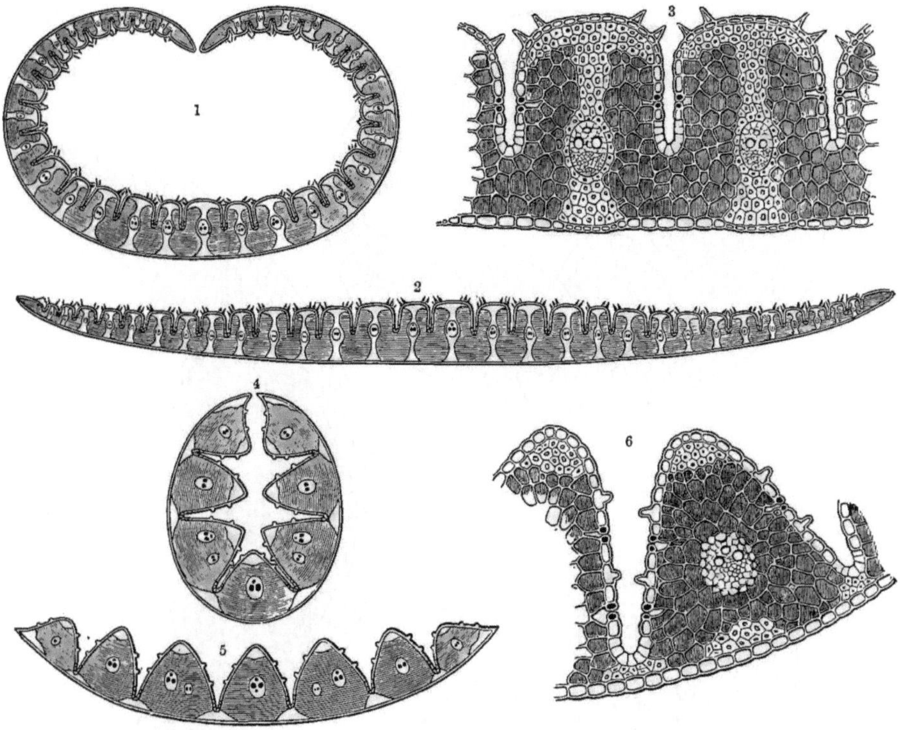

Fig. 78 Cross sections through the leaves of *Lasiagrostis calamagrostis* (1, closed; 2, 3, expanded, general view, and detail) and *Festuca porcii* (4, closed; 5, 6, expanded, general view, and detail) (von Marilaun 1902)

and Toma (2010b) could be included in the same category: they are hygro-halophyte species, but the wet character of the habitats is relative and not constant; these species were also observed in dry conditions, through water evaporation, due to periods of prolonged drought. This is the only way to explain the presence of these xeric characters in plants classified as hygrophilous; they are, in fact, if we take this definition, amphibian halophytes. Some of the above listed species also have air cavities, which is a characteristic of hygrophytes' life (Grigore and Toma 2010a).

3.3 *Organs and Covering Structures at the Level of the Leaf*

In general, transpiration is greatly reduced in quantity when the transpiration surface is covered by bodies containing air, within and between which the air is retained, in such a way that its circulation is obstructed.

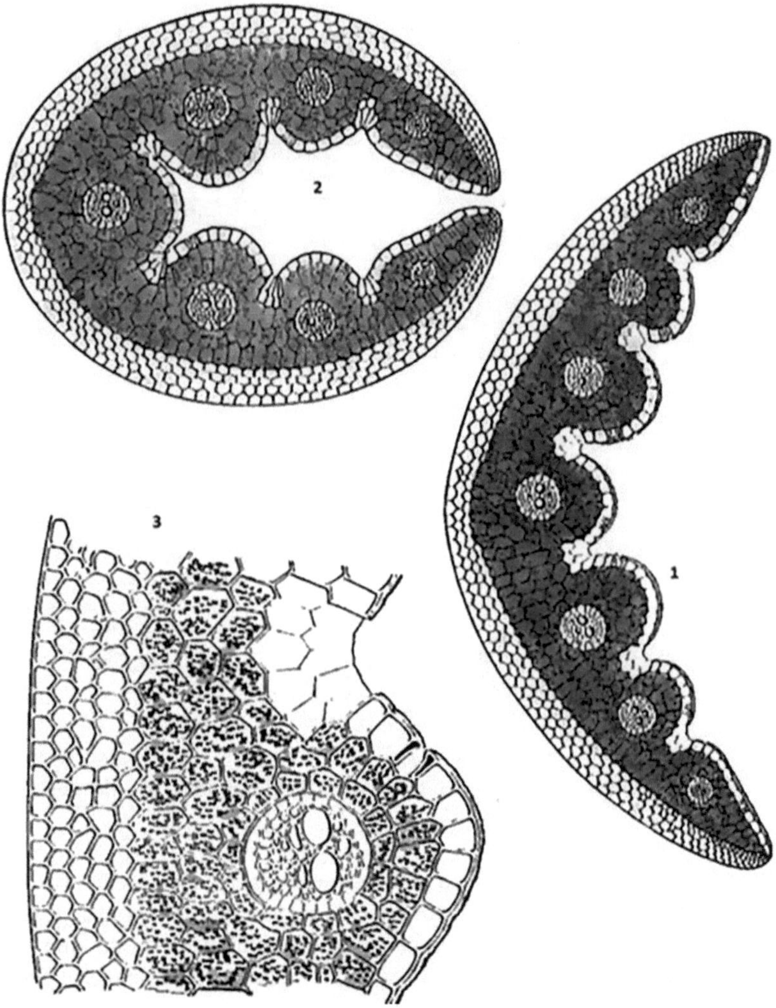

Fig. 79 Cross sections through the leaves of *Festuca punctoria* (1, expanded; 2, 3, closed, general view, and detail) (von Marilaun 1902)

3.3.1 Hairs (Trichomes)

The ecological relevance of the presence of trichomes at the level of the xerophytes' leaf epidermis—in the sense of regulating (controlling) transpiration—is best observed by (didactic) comparison with hydrophytes (aquatic plants); these are practically glabrous (hairless), and xerophytes are often covered with gray or white hairs. Basically, only the trichomes that are dead at maturity and contain air are specialized to perform this function.

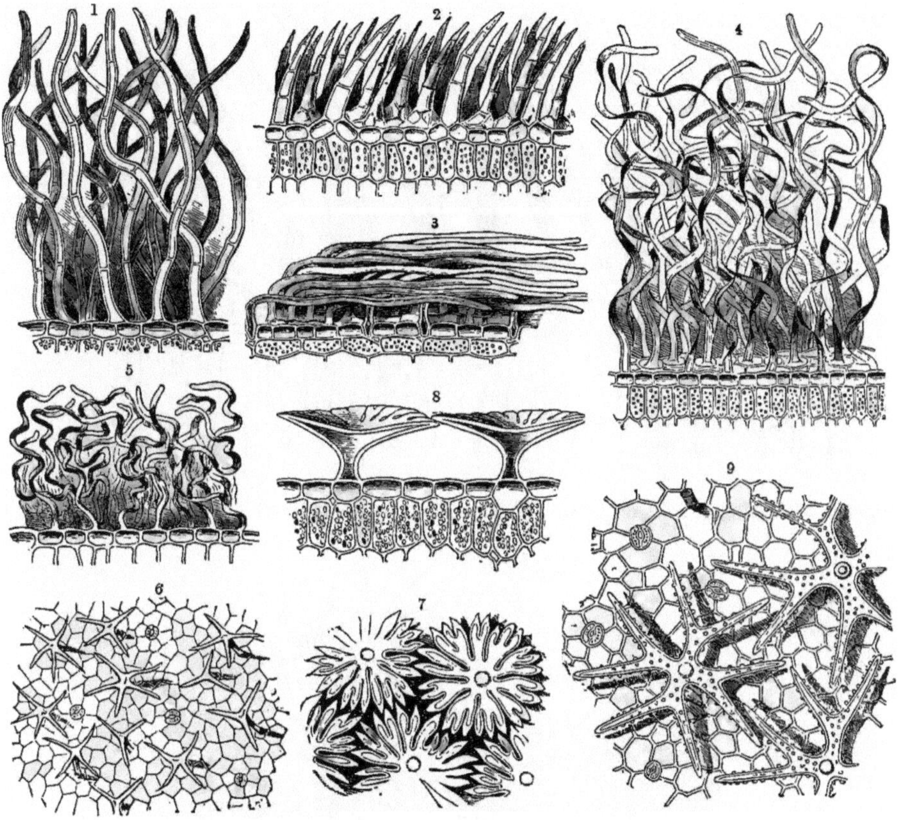

Fig. 80 Protecting hairs (trichomes) of *Gnaphalium leontopodium* (1), *Gloxinia speciosa* (2), *Convolvulus cneorum* (3), *Gnaphalium tomentosum* (4), *Centaurea ragusina* (5), *Alyssum wierzbicki* (6), *Koniga spinosa* (7, front view; 8, cross section), and *Draba thomasii* (9) (von Marilaun 1902)

In xerophilic plants, in form, these hairs are extremely diverse from the point of view of shape, size, and position in relation to the epidermis (Figs. 80, 81, 82, and 83).

In relation to the formation and presence of hairs, it is known that species that are glabrous in certain habitats become hairy in dry places and that hirsute (hairy) species become hairier in dry places than in wet places, as exemplified by *Ranunculus bulbosus*, *Polygonum persicaria*, *Mentha arvensis*, and *Stachys palustris*. Hair-covered plants grow in many arid habitats, such as rocks in Mediterranean countries, in dry scrub, and in the desert, steppe, or alpine area. This layer of hairs (with a woolly appearance) acts as a sunshade and is involved in controlling temperature changes to reduce transpiration. Normally, the covering of hairs is almost invariably denser on the underside of the leaf where the stomata occur.

Young stems and leaves are often covered with dense trichomes, more so than older leaves, because in the first stage, they need more protection from intense

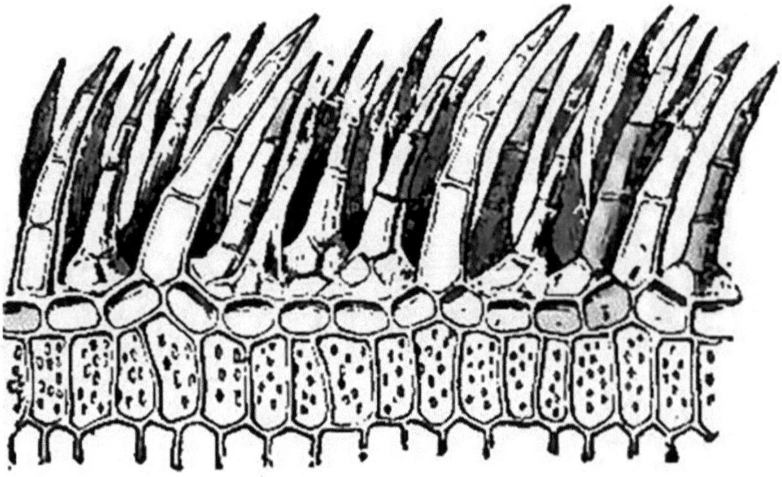

Fig. 81 Multicellular hairs on *Leontopodium alpinum* (Barnes 1898)

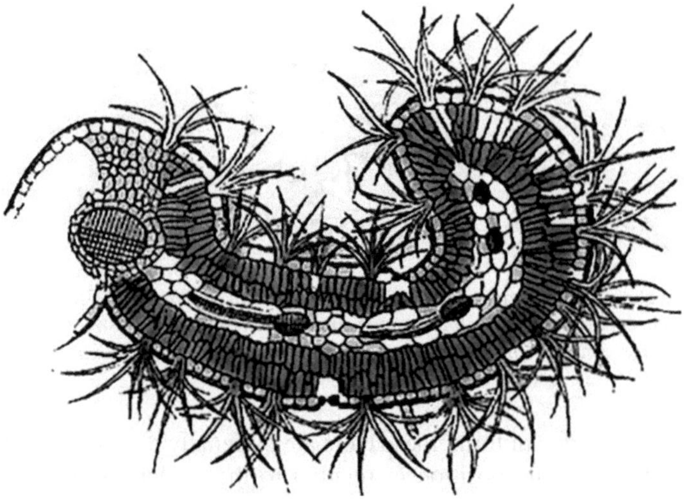

Fig. 82 Cross section through the leaf of *Helianthemum kahiricum*—dry climate of the Egyptian desert (Schimper 1903)

transpiration. Sometimes, in dry tropical areas, the leaves produced immediately after the dry season and those developed later differ greatly in appearance, the former being hairier and the latter larger and greener. When xerophilic plants lack protective hairs, they usually possess other structural protection mechanisms against excessive transpiration, such as succulent species (cacti), *Aloe* and *Agave* species, which usually have smooth leaves, of whose epidermal cells are provided with cuticle, and provided with water-storage tissue.

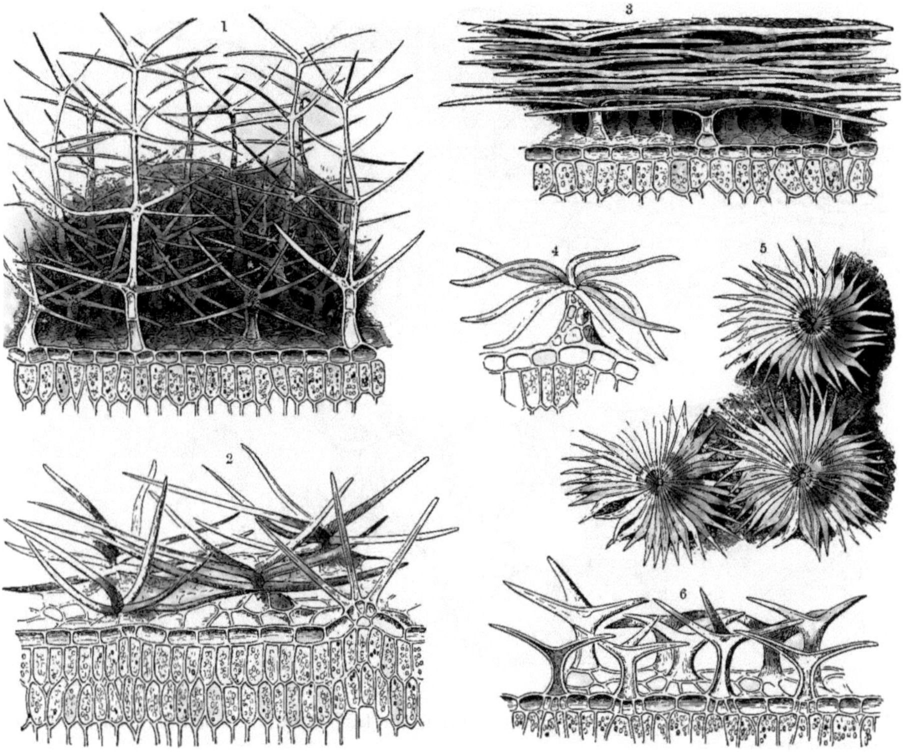

Fig. 83 Protecting hairs (trichomes) of *Verbascum thapsiforme* (1), *Potentilla cinerea* (2), *Artemisia nutellina* (3), *Correa speciosa* (4), *Elaeagnus angustifolia* (5), and *Aubretia deltoidea* (6) (von Marilaun 1902)

A very interesting and atypical situation occurs in the species *Rochea* (*Crassula*) *falcata*, in which, at the level of the epidermal cells and among them, some very large vesicular cells develop, which protrude far above the epidermis (Fig. 84); these cells, which form practically an epidermal "armor," appear not to be air-filled, as in the case of ordinary dead-at-mature, air-filled guard hairs, but are vesicular, with silicified cell walls.

4 Water Absorption by Xerophytes: Foliar Scales (Trichomed Leaf) of Epiphytes and Desert Species

In a previous work (Grigore 2022), we dealt with the morphology of the hypogeous and epigeous absorption organs (the case of some epiphytic orchids) in xerophyte species and, especially, those of the root, in relation to the adaptations for the uptake of water with mineral salts in plants from arid (xeric) habitats.

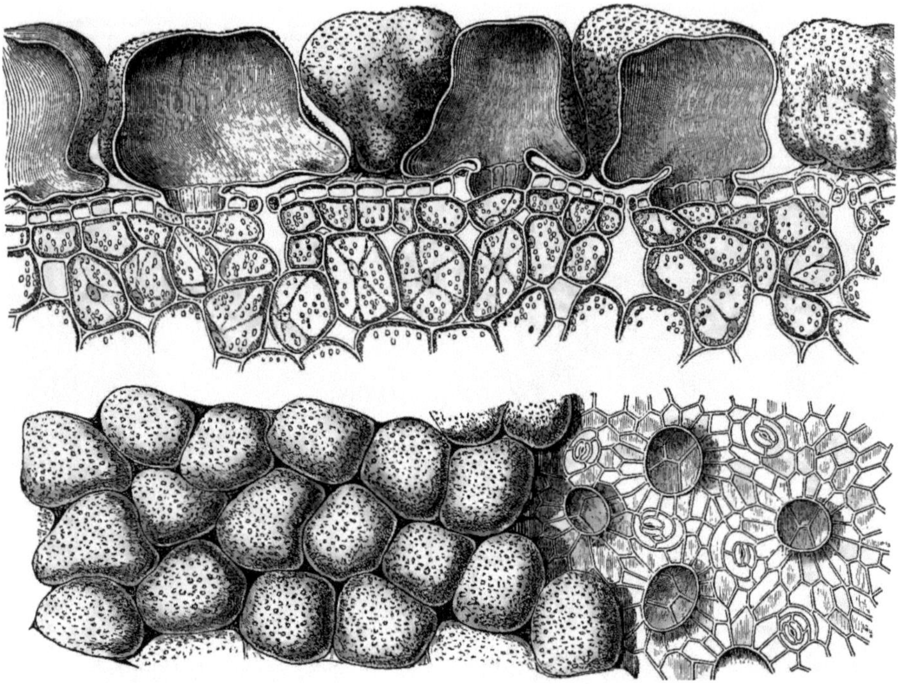

Fig. 84 *Rochea falcata*; top, cross section through the leaf; bottom, epidermis, front view (von Marilaun 1902)

In this chapter, dedicated to the leaves of xerophytes, we will refer to foliar hairs (scales) with a role in the absorption of atmospheric water, found in some tropical epiphytes. As we discussed before (Grigore 2022), but also in the present work, tropical epiphytes, by the nature of their adaptations, especially those regarding the absorption of water and its preservation in their body, are regarded to be xerophilous species (Ingrouille 1992). This observation may seem contradictory at first sight, since it is known that tropical areas receive quite a lot of rain; we remind that the epiphytes, by the fact that they are not rooted in the ground, must ensure an absorption of water at the aerial level.

The two major means by which tropical epiphytes obtain their (atmospheric) moisture are aerial roots—via the root velamen (which we have already dealt with – Ecological Anatomy of the Root)—and these foliar hairs (scales) found in some epiphytes, especially in the Bromeliaceae family.

These scales (Figs. 85, 86, and 87), studied mainly through the classic works of de Bary (1884), Schimper (1884, 1903), and Haberlandt (1914) and more recently of Benzing (1990), Papini et al. (2010), Ballego-Campos et al. (2020), and Raux et al. (2020), are found in many epiphytic ferns, in *Astelia*, probably in some orchids; however, the hair (trichome) from species of Bromeliaceae, especially in the subfamily Tillandsioideae, is the best known and shows the most elaborate structure.

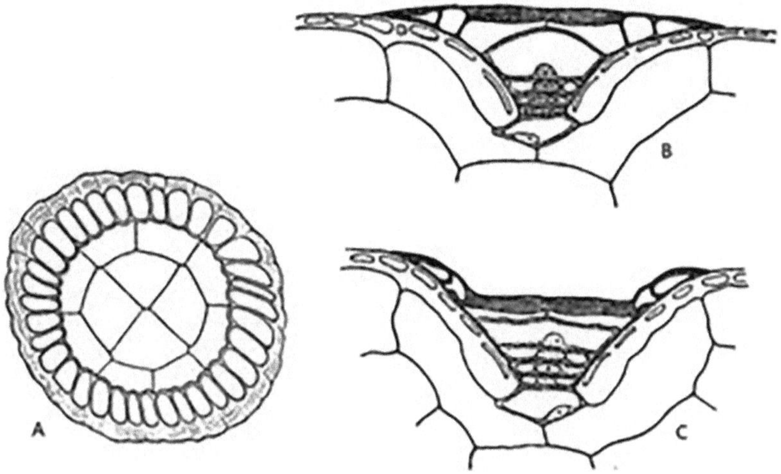

Fig. 85 Absorptive leaf scales in *Vriesea psittacina*; (**a**) top view of the disk of a scale; (**b**) cross section of a scale in a turgescent state; (**c**) cross section of a scale in a collapsed state (Haberlandt 1914)

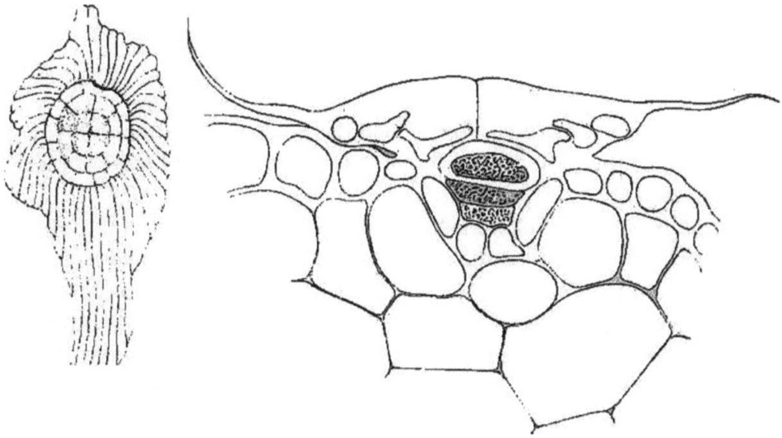

Fig. 86 Absorptive leaf scales in *Tillandsia usneoides*; left, front view; right, cross section (Schimper 1884)

Scales (another term for these peltate hairs) consist of shields formed of hollow cells arranged in an orderly fashion (Fig. 87). The most central shield, just above the point where the dome-shaped cell supports the shield, consists of four hollow, equal-sized, thick-walled cells. Several additional rings of dead cells may be present, each consisting of—twice—as many cells as the previous one. The outermost portion of the shield forms the wings, containing cells more than twice as many and more elongated than the preceding ring. Ring cells have thin areas in the outer walls

Fig. 87 Absorptive leaf scales in *Tillandsia* sp.; top, front view; bottom, cross section (Modified after Coulter et al. 1911)

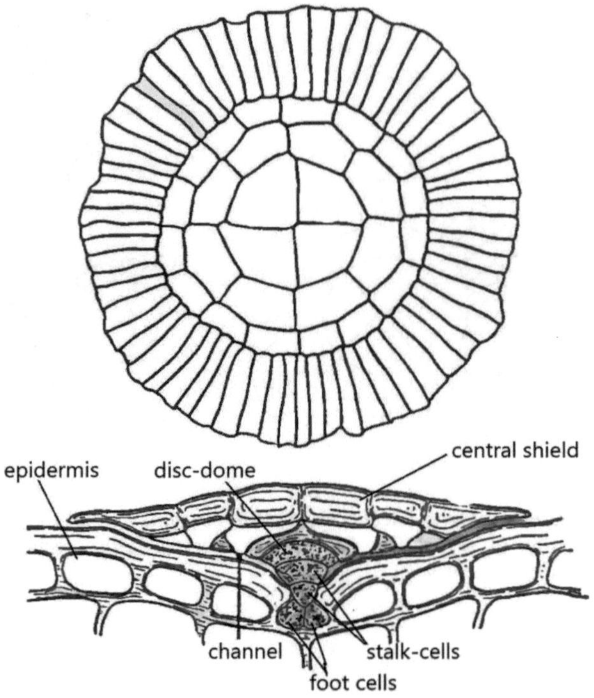

that allow wing extension. Thus, a bromeliad leaf scale is, like the velamentous root, bounded by an absorbent, nonliving tissue that traps moisture for slower absorption into the living tissue.

Most *Tillandsia* scale stalks consist of three to five transfer cells supported by two foot cells. The dome cell is the largest of the cell row and contains a dense protoplast with a prominent nucleus. Some studies have shown that this cell is provided with an elaborate plasmalemma, plus other membrane systems and numerous mitochondria believed to be involved in the absorption process (Dolzmann 1964, 1965). The cell walls of the shield, as well as those of the hair stalk, are modified in a way that suits the unfolded function. All transverse connections between cells: disk—dome cell, dome cell—stalk, stalk cells, and stalk—foot cells are uncutinized and crossed by numerous plasmodesmata. Shields have different shapes and orientations. Those on the leaf surfaces of shade species are small, immobile, and scattered; if they are wide and numerous, they are permanently pressed against the epidermis to reduce reflectance (e.g., in *Tillandsia bulbosa*) (Benzing et al. 1978). The foliage of species found in regular contact with clouds, fog, or high humidity tends to be thin and elongated.

Although some explanations have been proposed regarding the way water is absorbed by these scales (Coulter et al. 1911; Haberlandt 1914; Dolzmann 1964, 1965), absorption can be plausibly described only in terms of osmotic and mechanical forces (Benzing 1990).

Early authors (Haberlandt 1914; Mez 1904) correctly observed that plasmolysis first occurred in the mesophyll cells around the bases of the hair pedicels when hypertonic solutions were applied to intact leaf surfaces. Instead of multiplying, moisture droplets falling on a leaf spread rapidly through capillarity to form a uniform thin film between the trichome shields and the leaf epidermis. Shield cells quickly soak up water; as filling continues, the upper walls of the central disk are forced upwards, while the wing flattens against the surface of the leaf.

Until the end of the water supply, the inner space of the shield cells is flooded with water. A thick plug is then placed between the dome cell and the air mass, thus preventing the water from leaving the plant by the route it entered. In fact, the absorbent bromeliad scale serves as both a one-way valve and an energy sink, alternately hydrating the plant and insulating it against water loss and intense insolation (Benzing 1990). The functional analogy between this system and the velamen (or exodermis) is obvious, although the latter is less dynamic.

Although the absorptive function of these scales has been elucidated, ecological and ecophysiological studies are still needed to further explain how these structures actually contribute to atmospheric water supply.

Water-absorbing hairs are also found in other xerophytic species that grow in the desert (*Diplotaxis* species, *Heliotropium*, *Centaurea*, *Convolvulus*), which are not epiphytes (Fig. 88). Each hair consists of a sunken basal cell, one to three thin-walled, slightly elongated absorptive cells; a broader, somewhat thick-walled subterminal cell; and a very elongated, air-containing terminal cell that is separated from the subterminal cell through a thick transverse wall. A wilted leaf, which was immersed in water, except for the cut surface of the petiole, was found to resume its

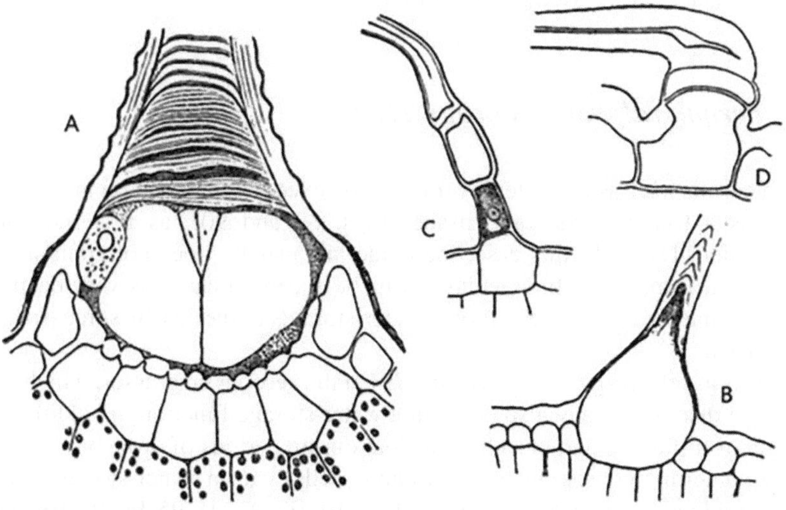

Fig. 88 Water-absorbing trichomes in xerophyte species; (**a**) *Diplotaxis harra*; (**b**) *Heliotropium luteum*; (**c**) *Centaurea argentea*; (**d**) *Convolvulus cneorum* (Haberlandt 1914)

normal turgor within 24 h, during which time it had absorbed 13% of its weight of water (Haberlandt 1914).

5 Storage (Deposit) of Water: Water-Storage Tissues

The formation and development of water-storage tissues is probably one of the most characteristic and defining morpho-anatomical features of xerophytes. Through this strategy, a land plant in xeric habitats can withstand air and soil dryness, thanks to the reserve of water in the aquifer tissues, which it consumes when necessary (Etherington 1975). Morphologically, many such organs possessing water-storage tissues have a succulent appearance. Succulence in xerophilous plants has several roles, in addition to strictly storing water, such as diluting salts in halophytes (Ranwell 1972; Baron 1979; Grigore 2008, 2021; Grigore and Toma 2010a, 2021b) and ensuring an erect port of the organs, in plants with poorly developed mechanical tissues.

Water storage is defining for xerophytes, while it is completely absent in hydrophytes (remember all the "dichotomous" discussions about the anatomical adaptations of xerophytes versus those of aquatic plants).

In general terms, a true water-storage tissue consists of thin-walled cells that contain water, but do not contain chlorophyll and lacks intercellular spaces (since no gas exchange takes place in it), and its cells are usually very large. It can shrink when it loses water and expands when cells absorb it from the environment.

There are several possibilities for classifying water-storage tissues in plants, the simplest being their classification into peripheral (epidermal or hypodermal) and internal water tissues (Warming 1909a).

5.1 Peripheral Water-Storage Tissues

The epidermis itself is the outer layer that acts (except in plants growing in water or shade) as a true water-storage tissue (Figs. 89 and 90), as first indicated by Westermaier (1884). In this case, the epidermis usually does not contain chlorophyll, and it forms a continuous layer which, in certain cases, is very deep and is directly connected with the internal water-storage tissue, as in some species of Velloziaceae.

Bulliform cells (Figs. 76, 77, 78, and 79) of the xerophytic grasses, which already have been discussed, may also perform a water-storage function, in addition to the function of rolling up and unrolling the leaf during periods of drought.

The epidermis consisting of voluminous cells, with a water-storage role, was also highlighted in some desert species (Figs. 91, 92, 93, 94, 95, 96, 97, 98, 99, 100, and 101) (Schimper 1884), among which the vesicular epidermal cells typical for

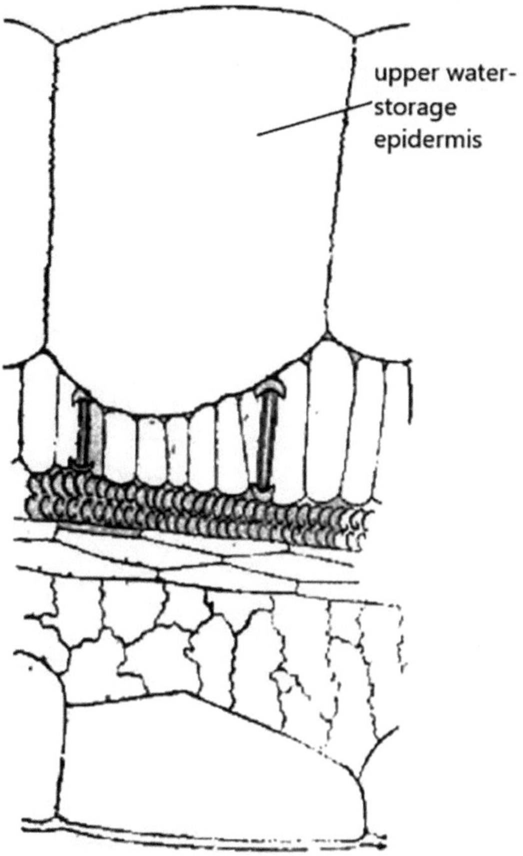

Fig. 89 Cross section through the leaf of *Dolichos axillaris* (Denis 1919)

species of the Aizoaceae (Mesembryanthemaceae) family stand out (Figs. 102 and 103) (discussed at length by Grigore and Toma 2010b).

Probably also the vesicular hairs, typical of some halophyte species of the Chenopodiaceae (Amaranthaceae) family (Fig. 102 and 103), play a role in water storage, in addition to the function of secreting excess salts (discussed in detail by Grigore 2008; Grigore and Toma 2010b, 2021a, b).

Bulky peripheral water-storage tissue can arise either by tangential division of the epidermis or by the formation of a hypodermis. The tissue is located mostly on the upper side of the leaf, and if present on the lower side, it is less developed. It controls the penetration of heat rather than light rays and therefore delays transpiration, in addition to acting as a water reservoir.

A multilayered epidermis with a role in water storage occurs frequently in xerophytes, lithophytes, and epiphytes; a voluminous tissue can thus be formed, the thickness of which far exceeds that of the chlorenchyma, such as in species of *Peperomia* (Fig. 104), *Begonia*, and *Ficus*.

Hypodermal water tissue is found in some species of xerophytes. It consists of one layer of cells in some species of Velloziaceae and Orchidaceae, and of two to

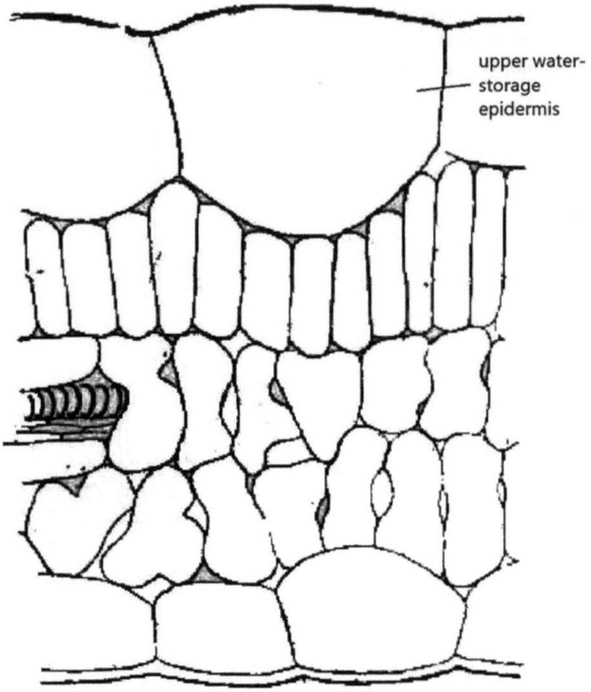

Fig. 90 Cross section through the leaf of *Ionidium buxifolium* (Denis 1919)

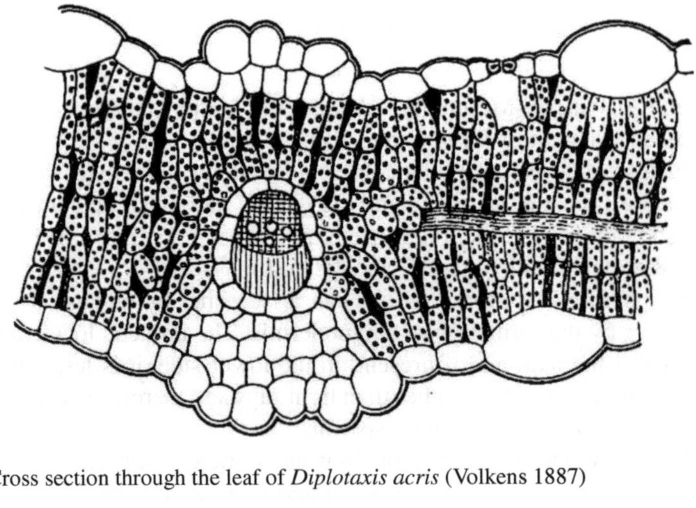

Fig. 91 Cross section through the leaf of *Diplotaxis acris* (Volkens 1887)

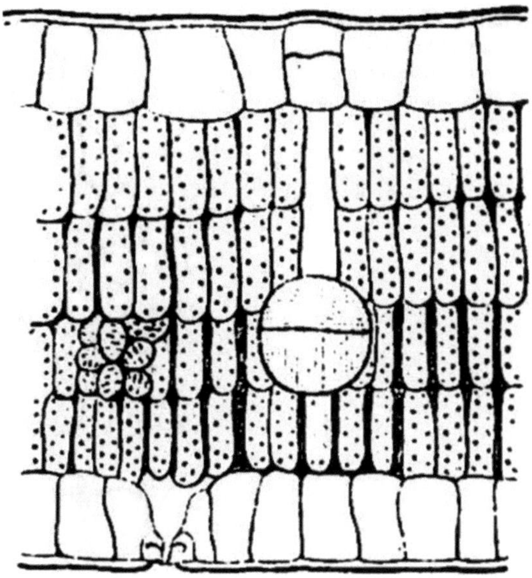

Fig. 92 Cross section through the leaf of *Ziziphus spina christi* (Volkens 1887)

three layers in *Nerium*, and is very voluminous in certain species of the families Commelinaceae, Bromeliaceae, Gesneriaceae (Fig. 105), and Rhizophoraceae

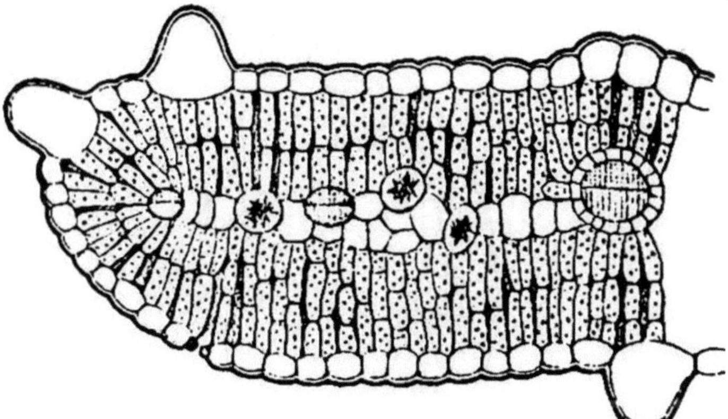

Fig. 93 Cross section through the leaf of *Pteranthus dichotomus* (Volkens 1887)

(mangrove species—Fig. 106). A collenchymatized hypodermis that functions as water-storage tissue is found in some cacti, being crossed by narrow intercellular spaces that connect the chlorenchyma to the stomata.

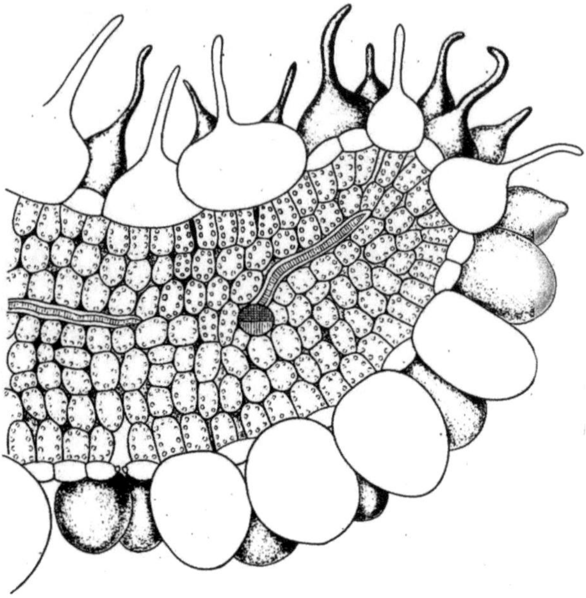

Fig. 94 Cross section through the leaf of *Mesembryanthemum crystallinum* (Volkens 1887)

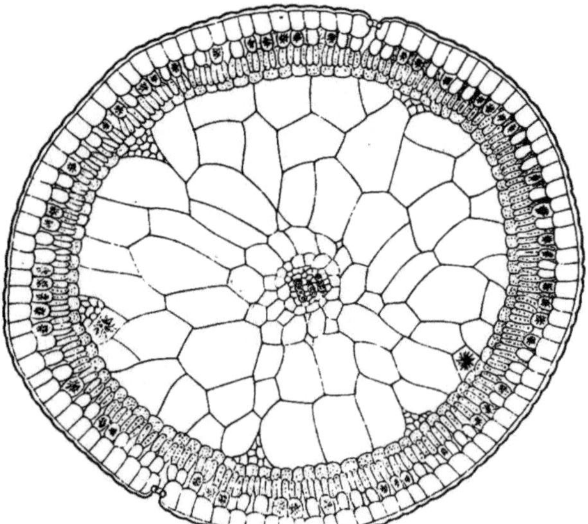

Fig. 95 Cross section through the leaf of *Halogeton alopecuroides* (Volkens 1887)

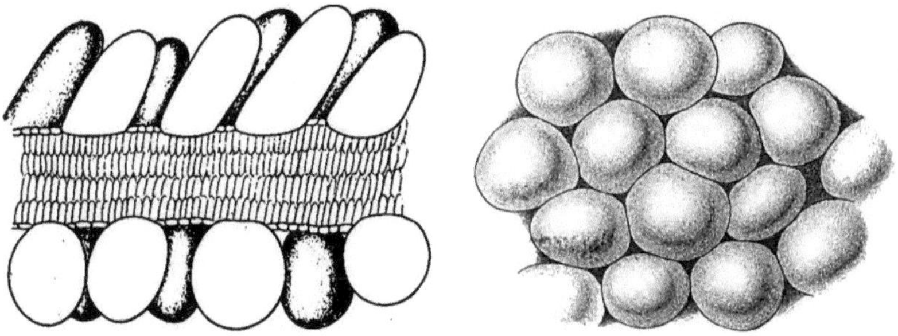

Fig. 96 Cross section through the leaf of *Aizoon canariense* (left) and front view of epidermis (right) (Volkens 1887)

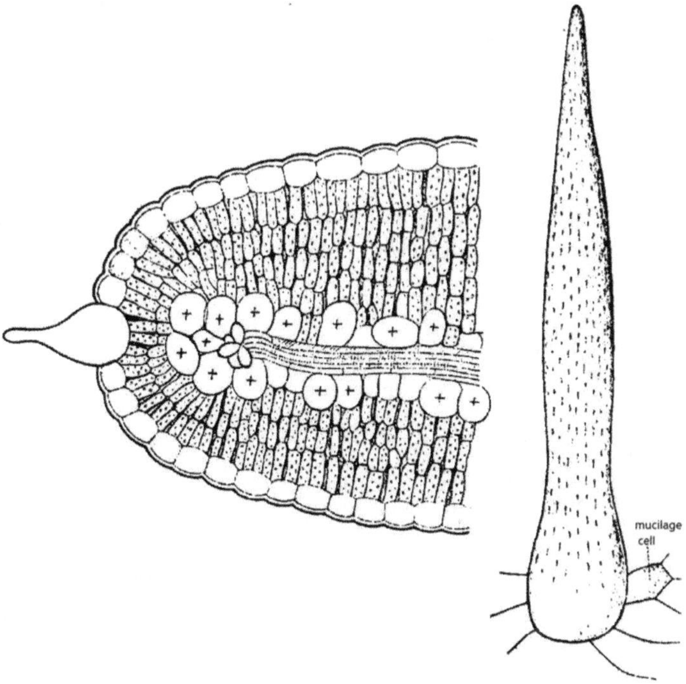

Fig. 97 Cross section through the leaf of *Caylusea canescens* (left) and a water-storage hair (right) (Volkens 1887)

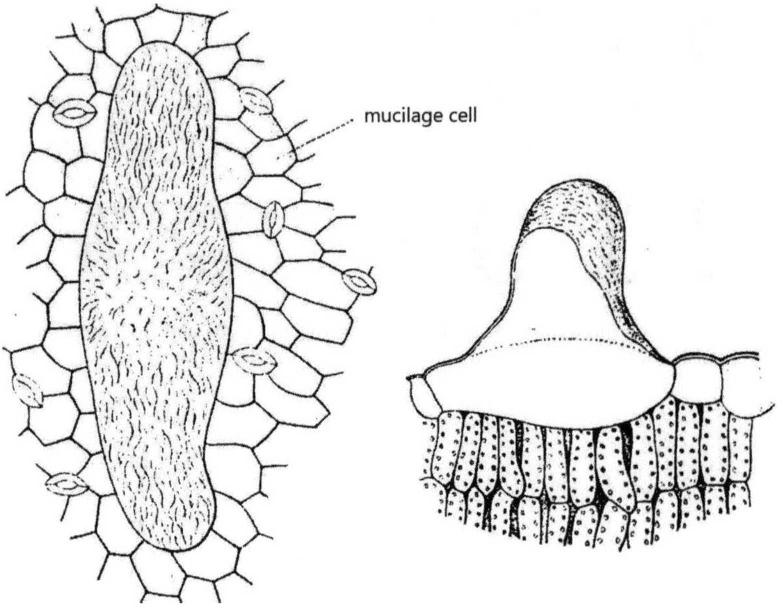

Fig. 98 Cross section through the leaf of *Reseda pruinosa* (right) and epidermis, front view (left) with large water-storage cell (Volkens 1887)

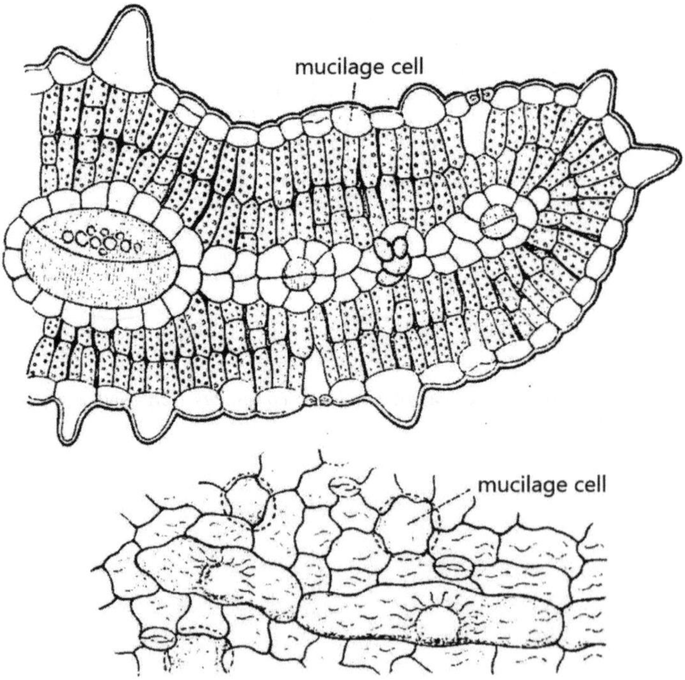

Fig. 99 Cross section through the leaf of *Oligomeris subulata* (top) and epidermis, front view (bottom) with large water-storage cells (Volkens 1887)

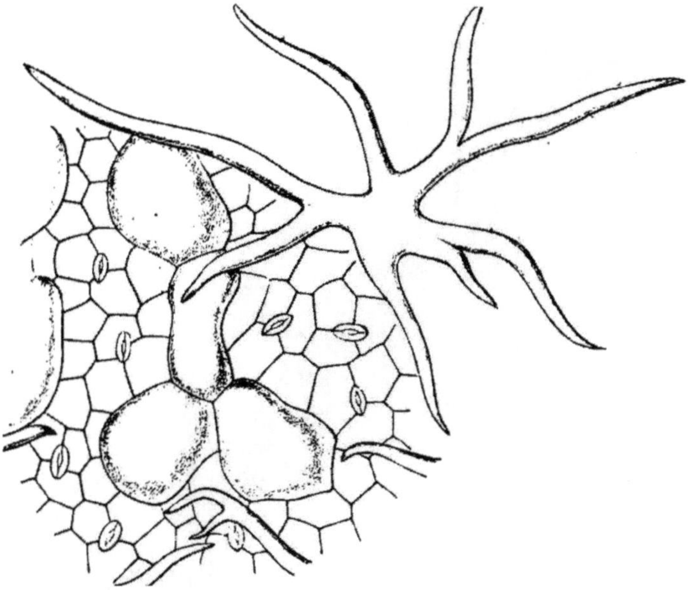

Fig. 100 *Malcolmia aegyptiaca*—leaf epidermis, front view—notice the large, water-storage cells and a branched trichome (Volkens 1887)

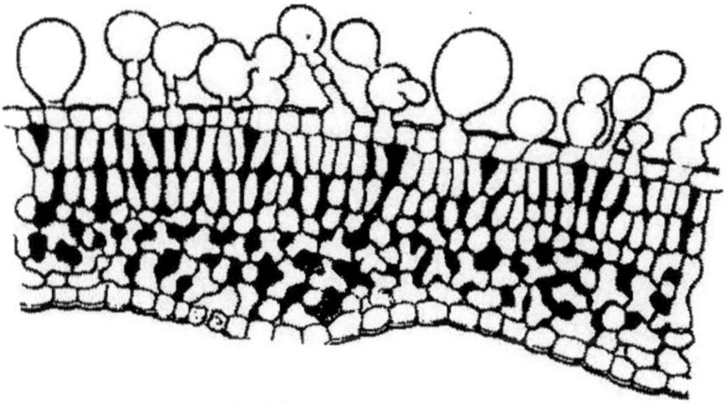

Fig. 101 Cross section through the leaf of *Zollikoferia nudicaulis*—large water-storage cells (Volkens 1887)

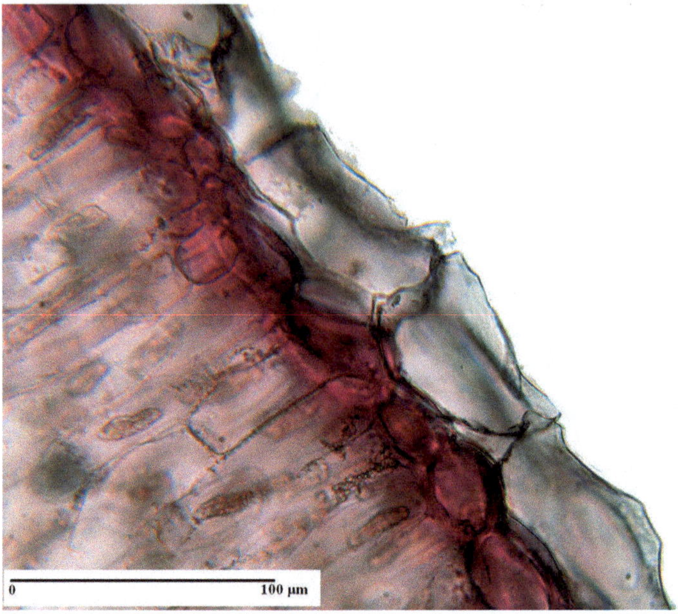

Fig. 102 Cross section through the leaf of *Halimione verrucifera* (Grigore and Toma 2010a)

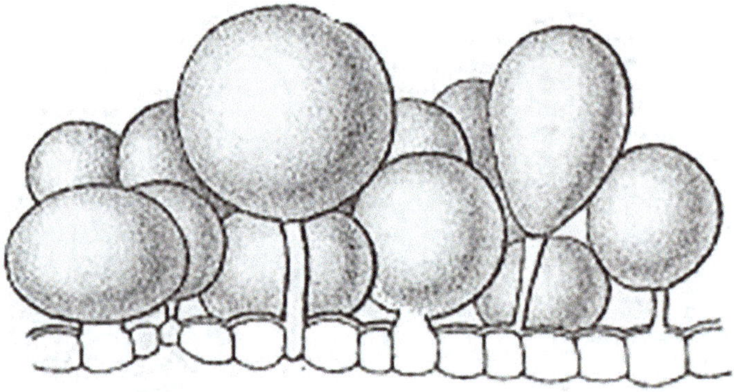

Fig. 103 Vesicular hairs at the level of the epidermis of the leaf of *Atriplex leucoclada* (Volkens 1887)

5.2 Internal Water-Storage Tissues

Internal water tissues are found in xerophytes in a variety of forms, as follows:

(a) Longitudinal bands of water-storage tissues, as in some desert grasses and *Phormium tenax*, bands that extend across the entire thickness of the leaf, from the upper face to the lower one. Bands of chlorenchyma, in which the veins are embedded, alternate, in this case, with bands of water tissue. In species of

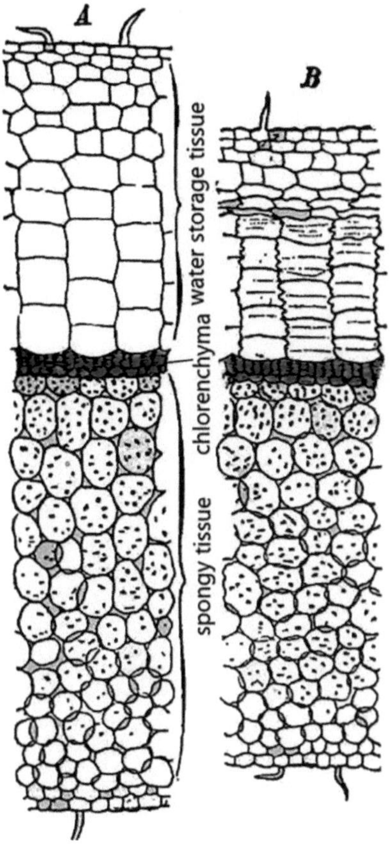

Fig. 104 Fragment of a cross section through a leaf of *Peperomia trichocarpa*; (**a**) from a fresh leaf; (**b**) leaf after 4 days of transpiration at 18–20°, with water-storage and spongy tissues that shrunk (Haberlandt 1914)

Velloziaceae, such longitudinal bands connect the epidermis on the upper face of the leaf with the water cells that form a sheath around the vascular bundles.

(b) Central water-storage tissue, occupying the center of the leaf and surrounded by a thin layer of chlorenchyma, is found in many xerophytes (Figs. 107 and 108) including *Aloe*, *Agave*, *Salsola*, *Atriplex*, *Petrosimonia*, *Camphorosma*, *Halogeton*, and *Zygophyllum*. In aphyllous stems, the water tissue may be distributed in the same way, as is the case in the halophyte species of *Salicornia*, *Arthrocnemum*, *Halocnemum*, *Sarcocornia* (Grigore et al. 2014; Grigore and Toma 2017, 2021b), and *Haloxylon*. The water tissue and the chlorenchyma can either be well demarcated from each other, entering, in many species, into characteristic structural architectures, of the type of "Kranz" anatomy (in plants with type C_4 photosynthesis), or they can gradually fuse, due to the cells inside the leaves that contain little chlorophyll, as in many species of Crassulaceae and Cactaceae.

(c) Water-storing tracheids are found in the chlorenchyma of some desert and halophyte species, including mangroves (discussed, in this last group, in detail, by Grigore and Toma 2017). In xerophytes, plants that grow in dry, sunny environ-

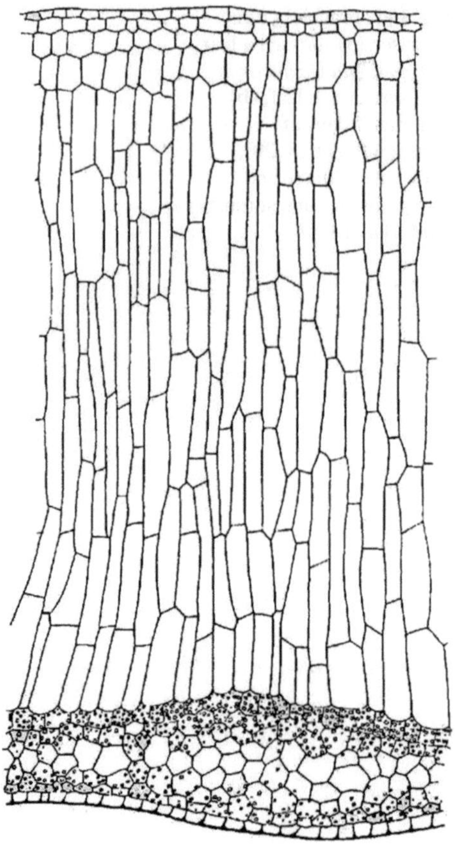

Fig. 105 Cross section through the leaf of *Codonanthe* ssp.—voluminous water-storage tissue (Schimper 1903)

ments are provided with storage tracheids that have no direct connection with the vascular system; the cells can be isolated, but in some cases form continuous layers, interpolated between the photosynthetic tissue and the epidermis. The last mentioned situation is found in *Physosiphon landsbergii* (Fig. 109), where, under the adaxial epidermis, there is a water-storage tissue (consisting of two layers of cells); under this tissue, there is a single layer of large storage tracheids, having the shape of palisade, with typical spiral thickenings that alternate on the two sides of each radial wall. Between these tracheids there are typical water-storage cells; under this water-storage tissue, there is a palisade tissue, continued below with a spongy tissue, where these water-tracheids may be also found; under the abaxial epidermis, another layer of water-tracheids is also found.

Similar water-tracheids are found in several species of *Capparis* (Figs. 110, 111, and 112), as evidenced by Vesque (1882) and named by him "réservoirs vasiformes." According to Vesque, these elements, which are only enormously enlarged vessels, occupy in groups in the free ends of the bundles in the leaf blade; they are

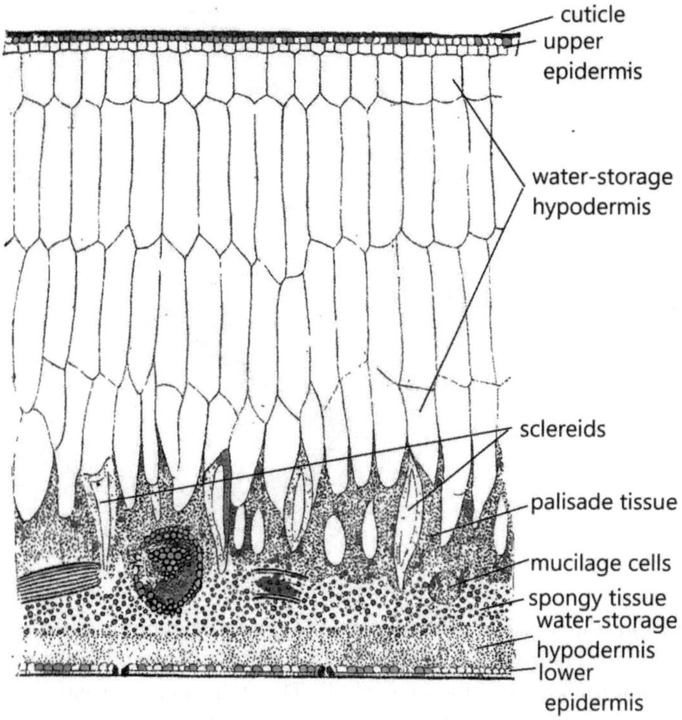

Fig. 106 Cross section through the leaf of *Rhizophora mucronata* (mangrove species) (Schimper 1903)

Fig. 107 Cross section through the leaf of *Salsola longifolia* (Volkens 1887)

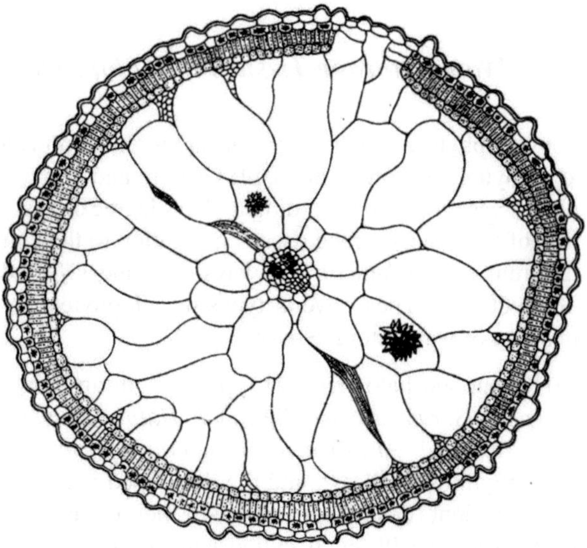

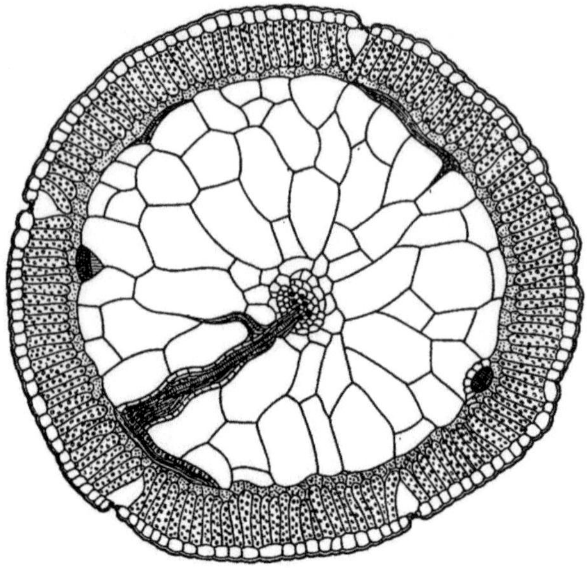

Fig. 108 Cross section through the leaf of *Zygophyllum simplex* (Volkens 1887)

punctuated like the vessels, and their walls generally remain quite thin although they are lignified. Their volume, extremely variable, can go so far as to occur a large part of the total thickness of the mesophyll. These "réservoirs vasiformes" are considered anatomical adaptations by which xerophytic plants store water in their organs.

d) In the mass of the mesophyll, isolated, very voluminous water-storing cells can be found, as in *Nitraria retusa* (Fig. 59).

5.3 *Water-Storage Tissues in Halophytes*

For halophytes, which represent, as we have repeatedly stated, a special case of xerophytes (Grigore et al. 2014; Grigore and Toma 2017), a classification of these plants was proposed according to their degree of succulence, depending on the position of the water-storage tissues in relation to the chlorenchyma (which also define different photosynthetic pathways—C_3 and C_4) and the epidermis (Grigore and Toma 2021b). Thus, several types of halophytes were described, according to the position of water-storage tissue.

(a) Chenopod halophytes of type C_3 with articulated segments: *Salicornia europaea*, *S. ramosissima* (Fig. 113), *Halocnemum strobilaceum*, *Sarcocornia fruticosa*, and *Arthrocnemum macrostachyum*. These species are apparently aphyllous (leafless), and the developed water-storage tissues contribute to the succulent (fleshy) appearance of their organs.

(b) C_3 species with small, cylindrical, or flattened leaves, in which the water-storage tissue occupies an important volume in the mesophyll: *Suaeda maritima*

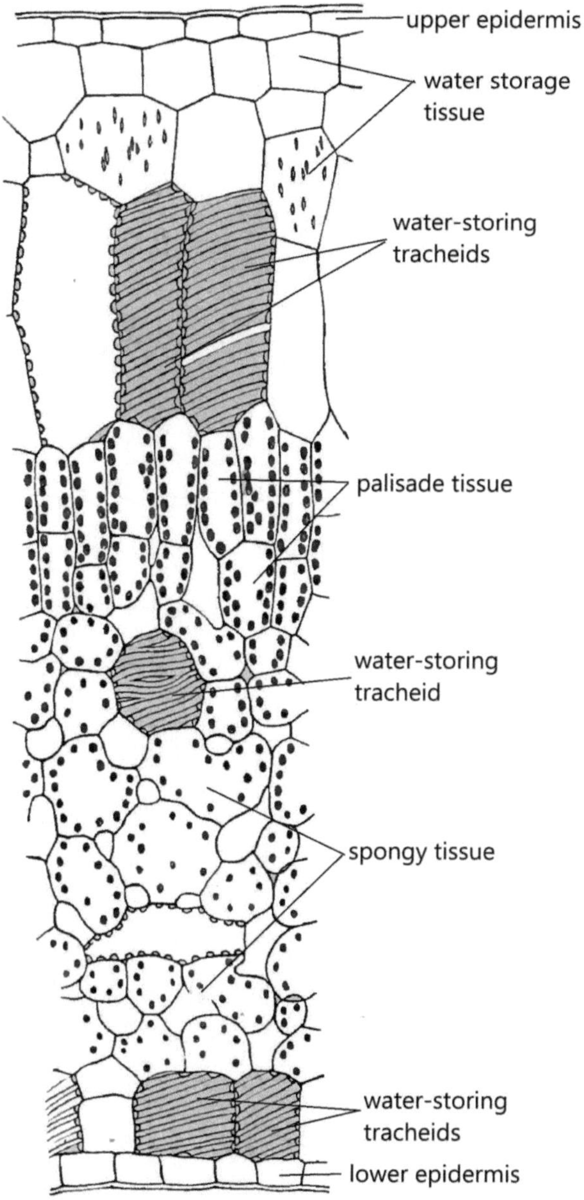

Fig. 109 Cross section through the leaf of *Physosiphon landsbergii* (Haberlandt 1914)

(Fig. 114), *S. spicata*, *Bassia hirsuta*, *B. sedoides*, and *Halimione portulacoides*. Succulency was also highlighted in *Plantago crassifolia* and *P. tenuiflora* and *Plantago maritima* and at *Inula crithmoides*, *Spergularia media*, and *Crithmum maritimum* (Fig. 115).

(c) C_4 chenopod halophytes, in which the water-storage tissue is positioned in the center of the leaf, being clearly and functionally delimited to the outside by two

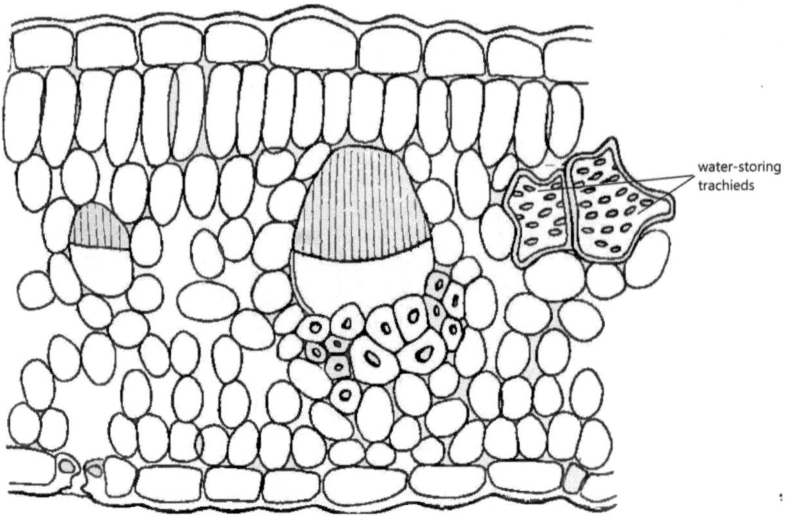

Fig. 110 Cross section through the leaf of *Capparis ericoclada* (Vesque 1882)

Fig. 111 Cross section through the leaf of *Capparis galeata* (Vesque 1882)

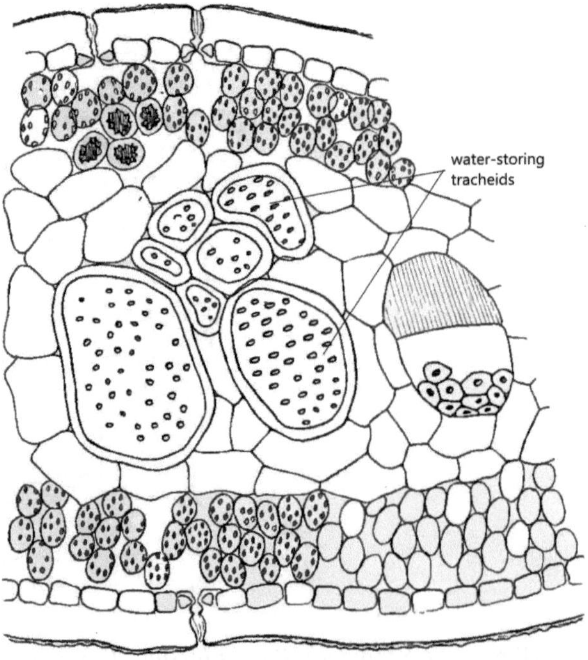

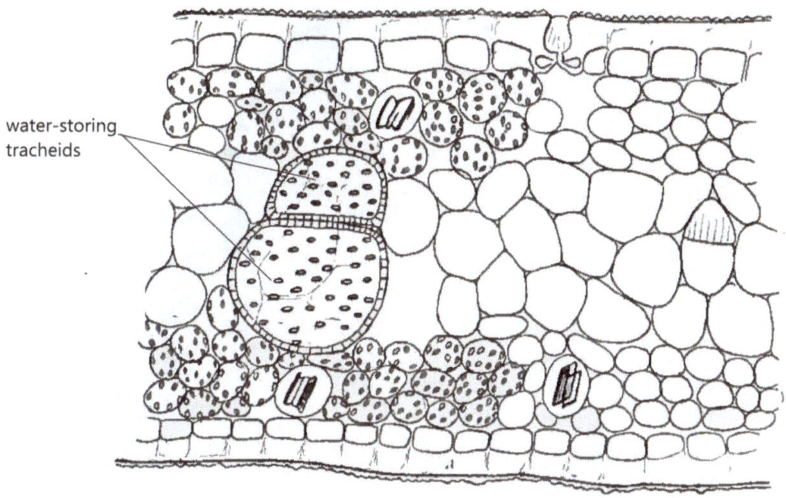

Fig. 112 Cross section through the leaf of *Capparis rupestris* (Vesque 1882)

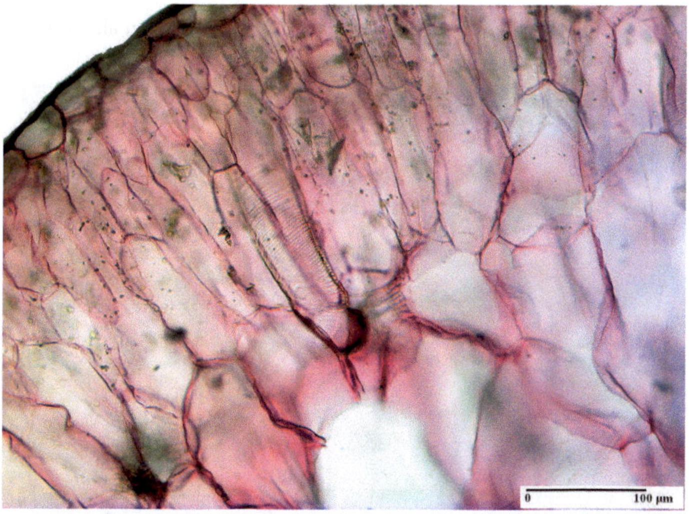

Fig. 113 Cross section through an articulated succulent segment from *Salicornia ramosissima* (Grigore et al. 2014)

chlorenchyma tissues typical for C_4 type species: *Petrosimonia oppositifolia*, *P. triandra* (Fig. 116), *Camphorosma annua*, *Salsola oppositifolia*, *S. kali* (Fig. 117), *Suaeda splendens*, and some Iranian species of *Bassia*: *B. turkestanica*, *B. pilosa*, and *B. stellaris* (Safiallah et al. 2017).

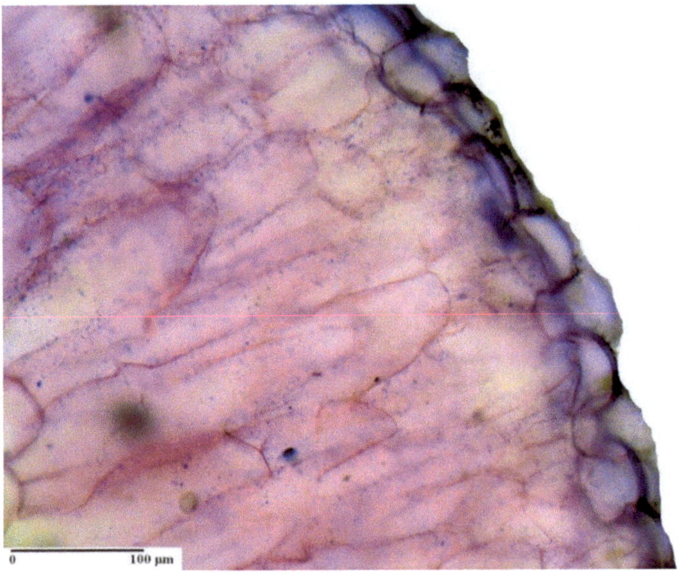

Fig. 114 Cross section through the leaf of *Suaeda maritima* (Grigore et al. 2014)

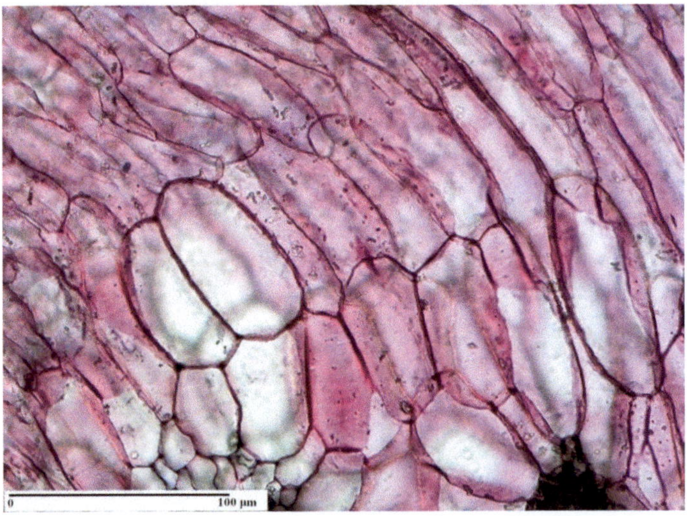

Fig. 115 Cross section through the leaf from *Crithmum maritimum* (Grigore et al. 2014)

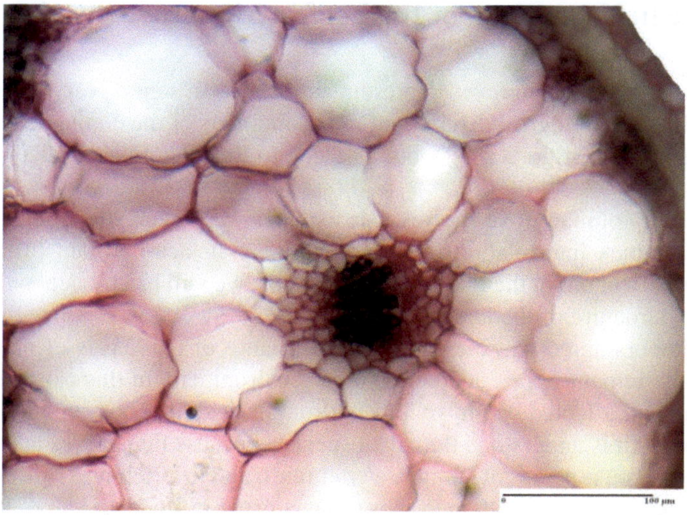

Fig. 116 Cross section through the leaf of *Petrosimonia triandra* (Grigore et al. 2014)

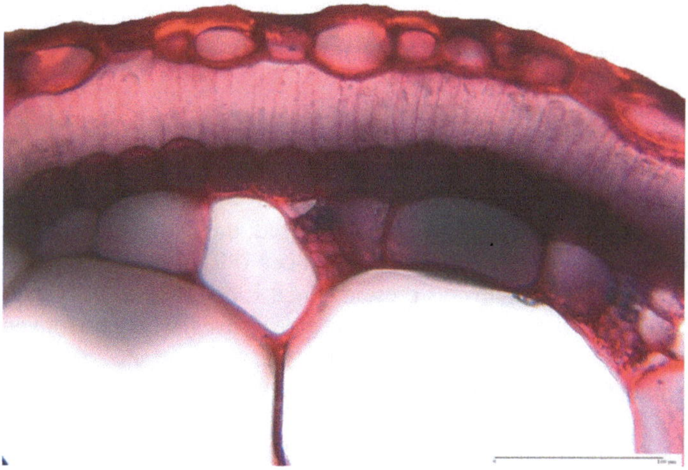

Fig. 117 Cross section through the leaf of *Salsola kali* (Grigore et al. 2014)

6 Ecological Anatomy of the Leaf in Mangrove Species

As already shown, mangroves represent unique ecosystems, very interesting in terms of the morphological diversity of the adaptations of the component plant species (reviewed by Grigore 2022), but also due to the anatomical structure of these species.

The biology of mangrove species, as a whole, is very interesting, being related to the very complex interrelationship of environmental factors, plant (Turner 2001; Tomlinson 2016), but in the following paragraphs, we will briefly deal only with the anatomical structure of the leaf.

Most authors agree that the anatomical structure of the leaves of mangroves is xeromorphic in nature, i.e., adapted to a dry environment, in this case being rather an environment affected by physiological drought. It should be reminded that mangrove species (it is understood that plant species, because mangroves are also populated by numerous species of animals, vertebrates, and invertebrates) are represented by shrubs and trees. Even though these plants grow in flooded, salty marshy soil, their aerial parts show several structural features that occur in plants adapted to withstand drought. In other words, as already emphasized in this work, their habitat is affected by physiological drought conditions (Warming 1909a; Schimper 1891, 1903; Grigore 2008; Grigore and Toma 2010a; Grigore et al. 2014; Grigore and Toma 2017).

Briefly, the xeromorphic adaptations of leaves of mangroves can be found in the following structural features:

1. The leaves are thick, leathery, or succulent to some extent: *Sonneratia*, *Lumnitzera* (Fig. 118), *Carapa* (Fig. 119), *Rhizophora* (water-storage hypodermis—see above, Fig. 106), and *Avicennia*.
2. The epidermis is thick and strongly cutinized; the leaves are often very glossy: *Rhizophora mucronata* (Fig. 106) (Schimper 1891, 1903).
3. Water tissue is almost always present (Figs. 106, 118, and 119) and sometimes massive. In *Rhizophora mucronata*, the older leaves, which no longer photosynthetically assimilate, become thicker than the younger ones; this is caused by a development of the waters-storage tissue and therefore the leaf changes its function.
4. Salt glands are anatomical structures involved in the secretion (elimination) of excess salts from the leaves (organs) of halophytes (Baylis 1940–1941; Grigore 2008; Grigore and Toma 2010b, 2017; Grigore and Toma 2021a; Grigore and Toma 2021b; Grigore et al. 2012) (Figs. 120 and 121).
5. Stomata are often sunk below the general surface of the epidermis.
6. The mesophyll has few intercellular spaces, but the palisade tissue is well developed, as the main or even the only chlorenchyma: *Sonneratia* (Fig. 122) and *Lumnitzera* (Fig. 118).
7. The ends of the veins dilate into aquifer tracheids: *Sonneratia* (Fig. 123), *Bruguiera*, *Avicennia*, and *Ceriops*.

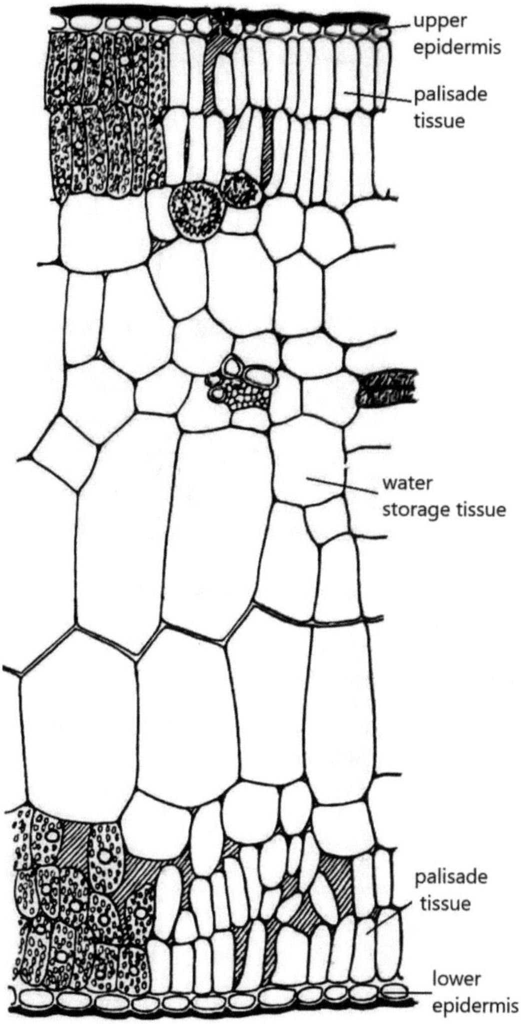

Fig. 118 Cross section through the leaf of *Lumnitzera racemosa* (Walter and Steiner 1936)

8. Characteristic mechanical elements (sclereids/trichosclereids—similar to those found in aquatic plants—see Figs. 33 and 34) arranged between the palisade cells of *Rhizophora* (Fig. 106), *Sonneratia* (Fig. 122), and *Carapa*.
9. Mucilaginous cells appear in the species of *Sonneratia* and *Rhizophora* (Fig. 106).
10. The leaves of some species are covered with many protective hairs (*Avicennia*).

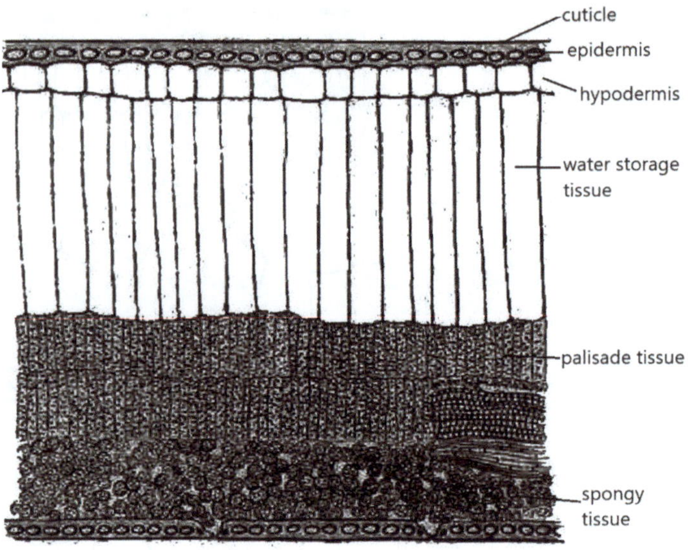

Fig. 119 Cross section through the leaf of *Carapa moluccensis* (Schimper 1891)

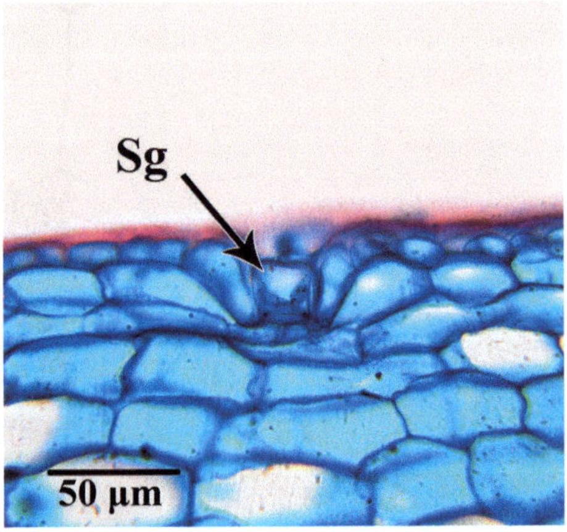

Fig. 120 Salt gland (Sg) of *Avicennia marina*, on the upper side of lamina (photo courtesy of Saikat Naskar)

7 Ecological Anatomy of the Leaf in Eremophytes (Steppe Species)

In plant ecology, the term "eremophytes" is used by a limited number of specialists, meaning plants "of desert and steppe formations" (Warming 1909a); the etymology of the term is most likely from the Greek word *eremos*, meaning "desert, wild,

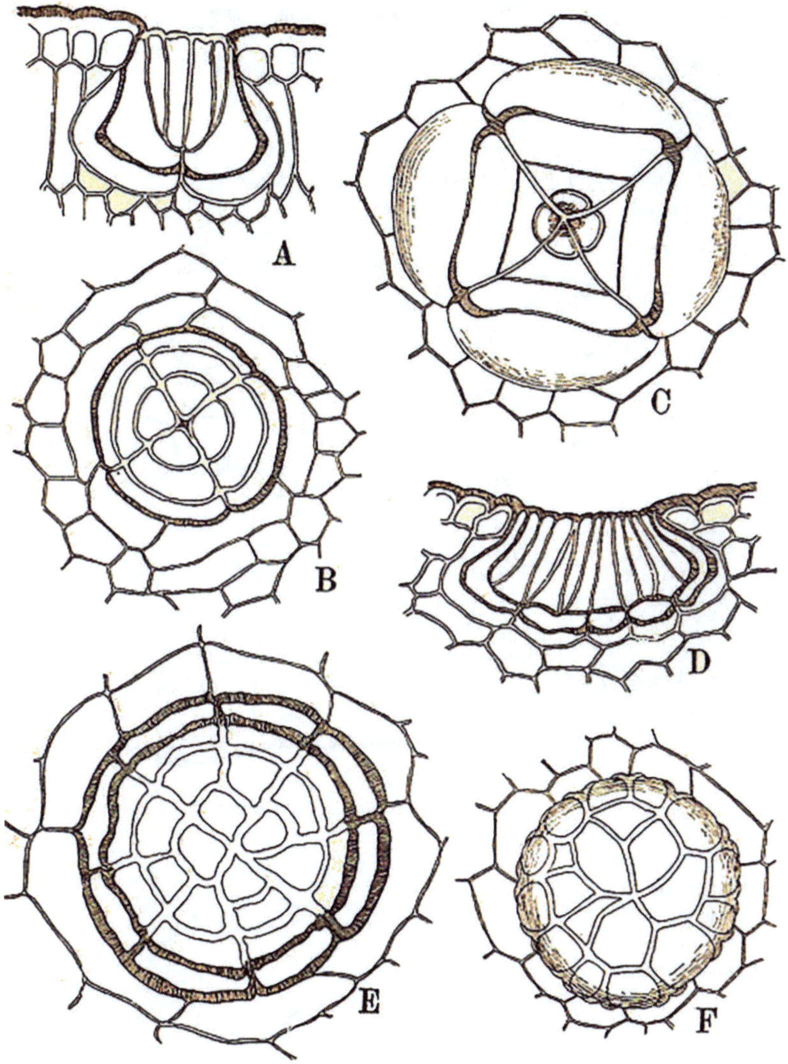

Fig. 121 Structure of the glands of *Aegialitis annulata*. (**a**–**c**) Chalk glands, (**a**) cross section, (**b**) surface view from outside, (**c**) surface view, from within, (**d**–**f**) mucilage glands, (**d**) cross section, (**e**) surface view from outside, (**f**) surface view from inside (Solereder 1908)

lifeless, uncultivated place." In the present work, we reserve the term only for steppe plants, xerophytic by nature of their adaptations, which also reflect the nature of environmental factors. The anatomical characters of desert plants, in a broad sense, were reviewed in the general characters of xerophytes.

It must be said that the term "steppe" is very pretentious and has become problematic in the language of plant ecology, due to the inconsistency with which it has

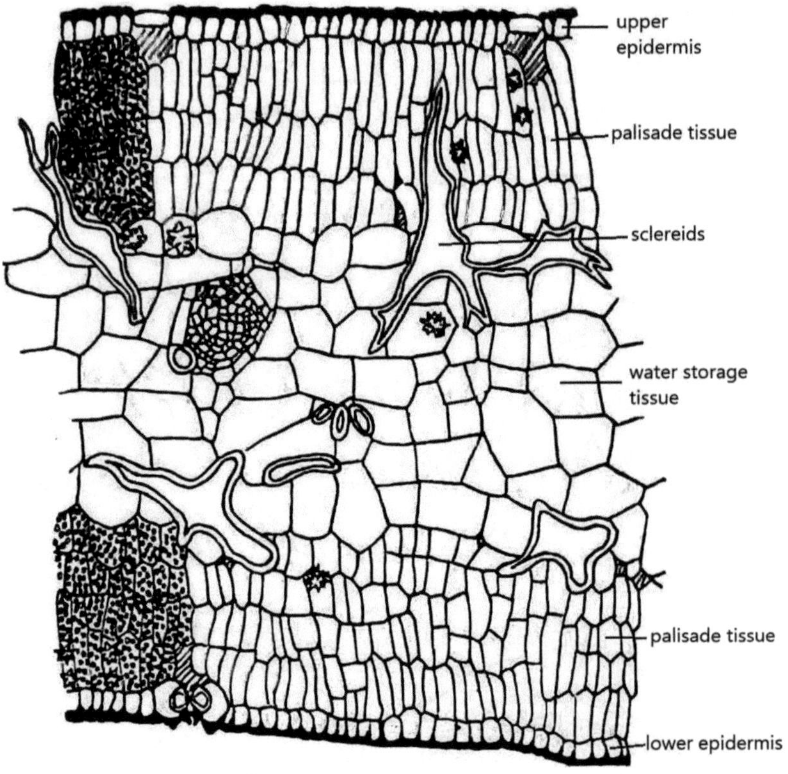

Fig. 122 Cross section through the leaf of *Sonneratia acida* (Walter and Steiner 1936)

Fig. 123 Water-storing tracheids (tr) in the leaf mesophyll of *Sonneratia acida* (Schimper 1903)

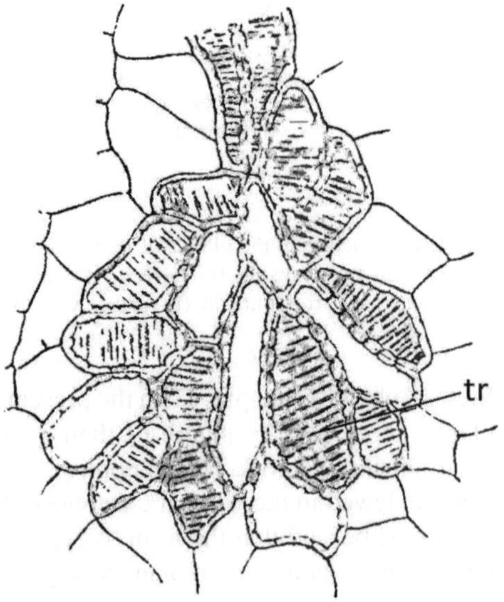

been used, by specialists, over time, an aspect that betrays, in reality, precisely the nuance heterogeneity of climatic and ecological conditions in different geographical areas of the globe (Schimper 1903; Riley and Young 1966; Collinson 1988; Archibold 1995; Pfadenhauer and Klötzli 2020).

Broadly speaking, steppes would correspond to "temperate grasslands" (Collinson 1988; Archibold 1995) or, more specifically, to dry, heather (cold-temperate) vegetation zones (Pfadenhauer and Klötzli 2020). Sometimes, "steppe" (the Eurasian variant) is used equivalently to "prairie" (for the North American steppes), although in some respects, there are differences between them. Moreover, for certain geographical areas and particular climatic conditions, "steppe" is also used with "local" terms, such as those of "puna" (high mountain steppes and semi-deserts, which include part of the vegetation of the tropical belt alpine, from the Andes mountain area of Peru, northern Chile, and Argentina) and "páramo" (high plateau steppes, where the montane forest cover is replaced by grasses and evergreens, of the northern Andes, which are cool, wet, and foggy; páramo extends from western Venezuela to northern Peru, at altitudes of 3000–4200 m, further south in Chile, merging with the slightly drier "puna," earlier mentioned) (Grigore 2022).

Put simply, in appearance, steppes (or prairies) are usually dominated by more or less xeromorphic grasses with intensive root systems mixed with other life forms, including annual and perennial herbaceous plants, bulbous plants, and woody shrubs; as a rule, it is characterized by the complete absence of trees. They occur under a variety of climatic regimes, at mid-latitudes, all of which are characterized by a dry season.

The steppes of Romania were very well described, from an ecological and floristic point of view, by Prodan (1939), who emphasizes the general characteristics of the steppes, but also some particularities of the "more pronounced" steppes, keeping the expression used by Prodan. Thus, the Romanian steppes, "in general" (Prodan's words), are characterized by drier climate, strong winds, lack of forests, low liquid precipitation, and low level of groundwater; the most pronounced also present some common botanical characters, such as the absence of beech, woody plants represented by small trees (dwarf cherry, *Prunus spinosa*, *Crataegus*, rose hip, *Paliurus spina-christi*—the latter in Dobrogea area), the small number of ferns, the genus *Primula* with few representatives, and large number of species of grasses, legumes, labiates and composites, with the great development of species of grasses. For a complete description of the steppes in Romania, and beyond, it is worth reading Prodan (1939). Of all the climatic elements, temperature has, in this case, the most pronounced influence; the vegetation must endure in the steppes' excessive heat during the day and dry air at night, often without dew.

Returning to studies of ecological anatomy strictly regarding eremophytes (steppe plants), they are very few. We will mention the extensive study of Urcelay (1923) regarding some anatomical adaptations of plants in the Spanish steppes, in order to reduce the intensity of transpiration. Thus, the results of Urcelay (1923) showed that the investigated plants possess clearly xeromorphic anatomical characters, characterized by:

(a) An abundant covering of trichomes at the level of the aerial organs (Figs. 126, 131, 132, 136, 137, and 138).
(b) Succulence and reduction in leaf size.
(c) The presence of stomata.
(d) Thick cuticle and cutinization of the outer wall of epidermal cells (Figs. 127, 130, 132, 134, and 137).
(e) Staurophyll or centric structure of the leaf mesophyll (see the details of these structures in the general anatomy of the leaf) (Fig. 130—bifacial structure with centric tendency; Figs. 132, 134, and 135).
(f) The presence of water-storage tissues (Figs. 124, 128, and 129).

Therefore, several types of adaptive strategies can be described:

(a) Narrow succulent leaves, with a generally centric structure.
(b) The leaf parenchyma is staurophyll or centric, leaves most often with a covering of trichomes on the surface.
(c) Dense covering with trichomes and glands (including salt glands, as in *Frankenia*—Fig. 133).
(d) Strongly developed sclerenchyma, forming a complete arch under the epidermis (Figs. 125 and 126) and the presence of stomata.

Looking at all these anatomical characters and structural types, we can see that they fall into the group of typical xeromorphic anatomical characters, which we

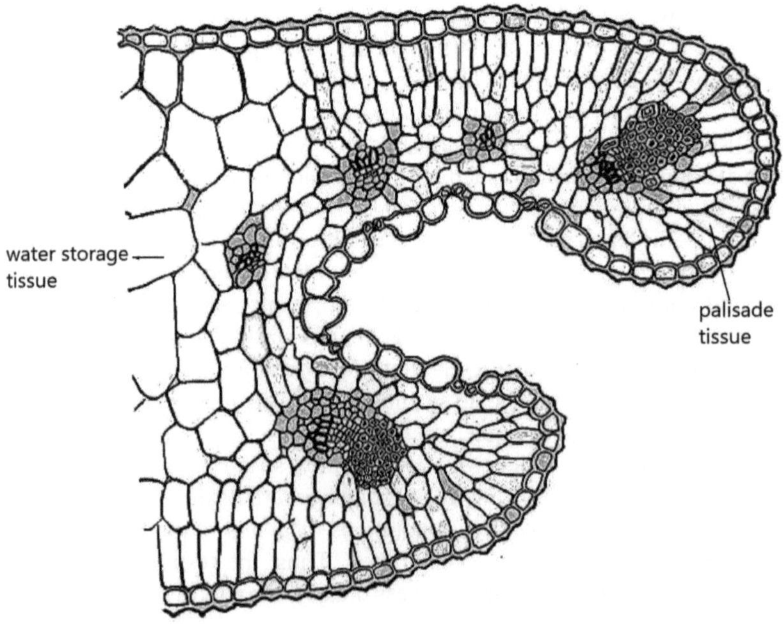

Fig. 124 *Crocus serotinus*: cross section through the leaf (Urcelay 1923)

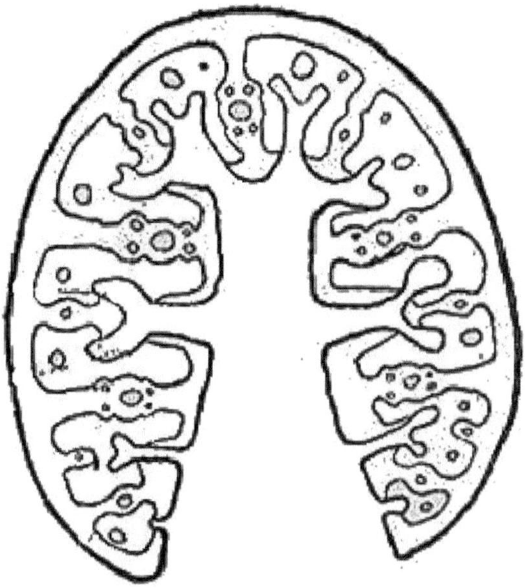

Fig. 125 *Macrochloa tenacissima*: cross section through the leaf (general view—next figure, detail) (Urcelay 1923)

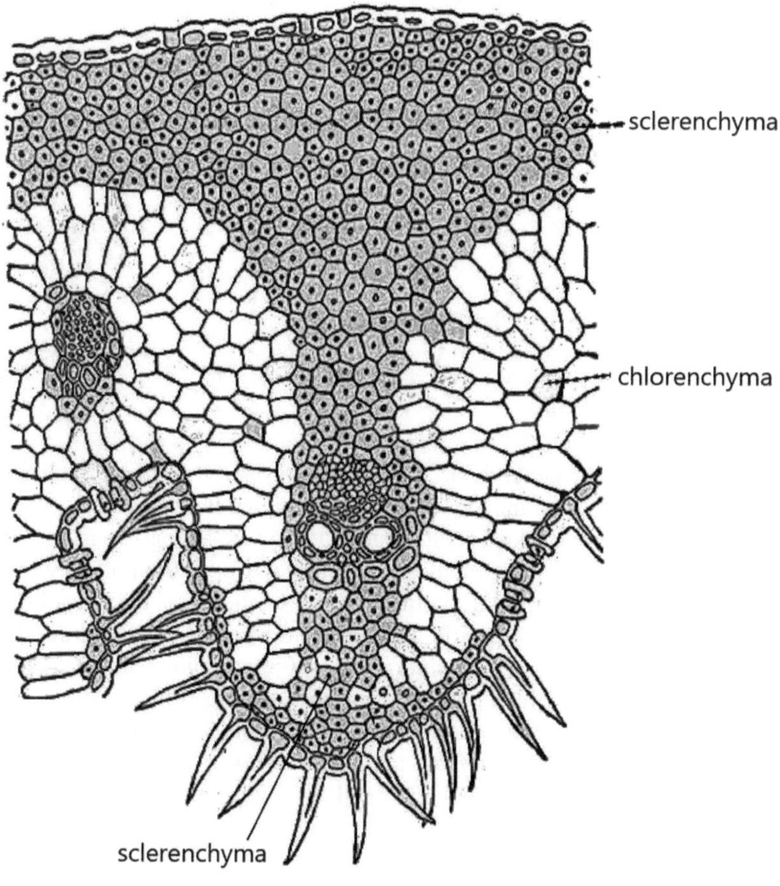

Fig. 126 *Macrochloa tenacissima*: cross section through the leaf (detail) (Urcelay 1923)

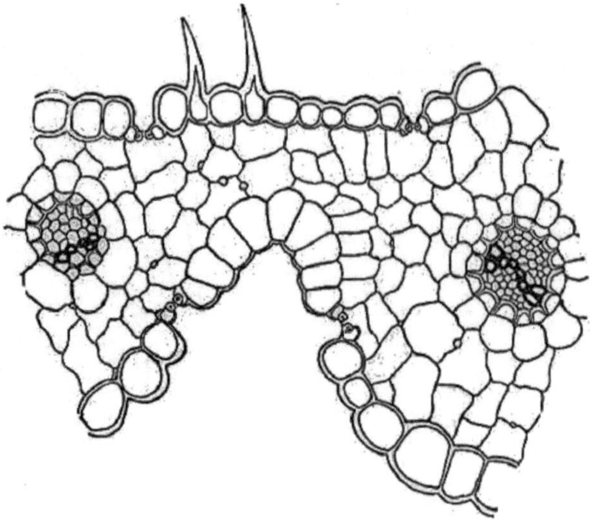

Fig. 127 *Echinaria capitata*: cross section through the leaf (Urcelay 1923); sunken stomata in the epidermis and bulliform (motor) cells are also observed

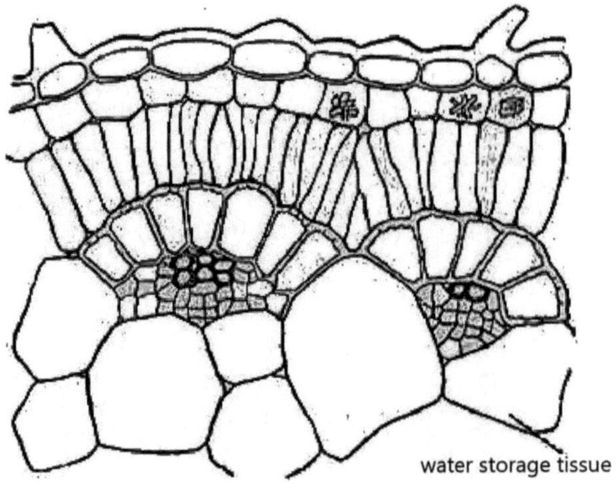

Fig. 128 *Kochia prostrata*: cross section through the leaf (Urcelay 1923); hypodermis and Kranz anatomy are also observed (plant type C_4)

discussed at length in the general part of leaf anatomy. All the characters listed here, minus the succulence, have, indeed, the role of reducing transpiration.

Another iconic study on leaf ecological anatomy on steppe (prairie species) is that of Hayden (1919), referring on plants vegetating in a prairie province in central Iowa. She found out that, by comparing the leaf anatomy of upland and alluvial basin plants, the epidermis of prairie hill crest and slope plants (*Andropogon scoparius*, Fig. 139; *Bouteloua curtipendula*, Fig. 140; *Anemone cylindrica*, Fig. 141; *Amorpha canescens*, Fig. 142; *Ceanothus americanus*, Fig. 143; *Eryngium yuccifolium*, Fig. 144; *Artemisia canadensis* and *Solidago canadensis*, Fig. 145; *Solidago rigida*, Fig. 146; *Coreopsis palmata* and *Helianthus grosseserratus*, Fig. 147; *Helianthus tuberosus*, Fig. 148), is thick, in all cases, or, if the wall is not thickened,

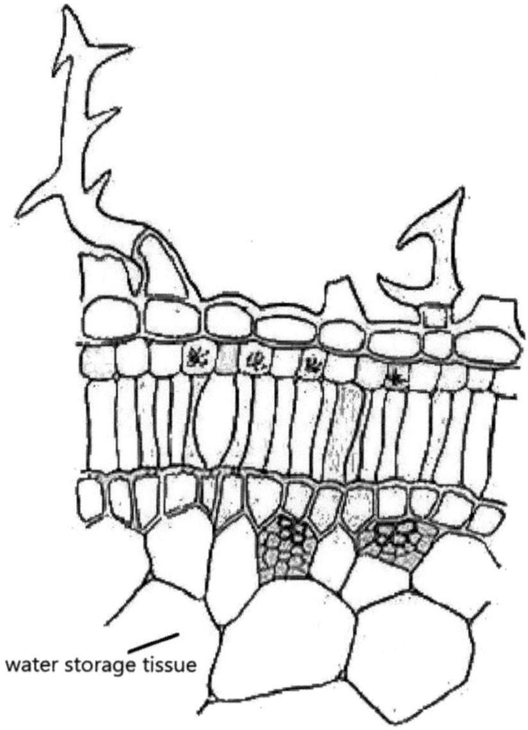

Fig. 129 *Salsola vermiculata*: cross section through the leaf (Urcelay 1923); hypodermis and Kranz-type anatomy are also observed (plant type C_4)

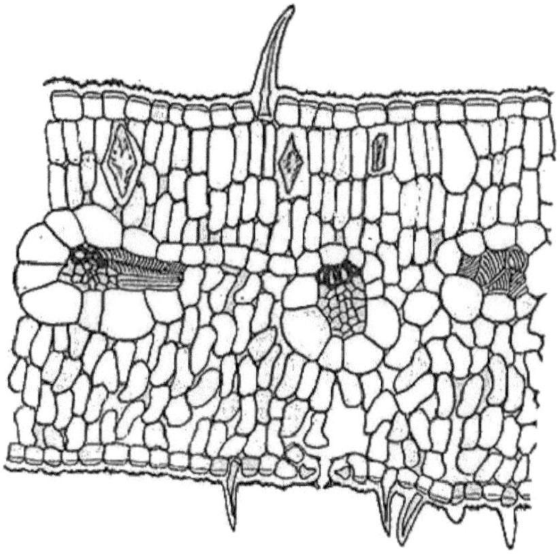

Fig. 130 *Rhamnus lycioides*: cross section through the leaf (Urcelay 1923)

Fig. 131 *Helianthemum squamatum*: cross section through the leaf (Urcelay 1923)

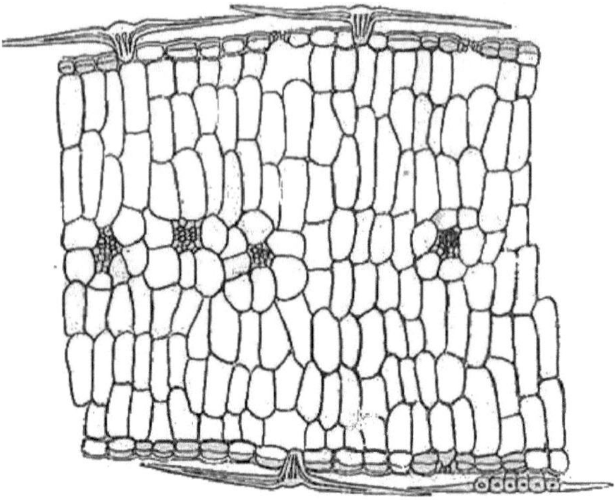

Fig. 132 *Helianthemum cinereum*: cross section through the leaf (Urcelay 1923)

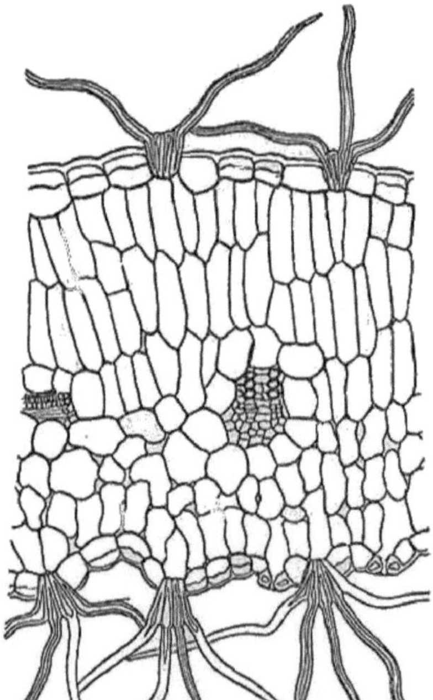

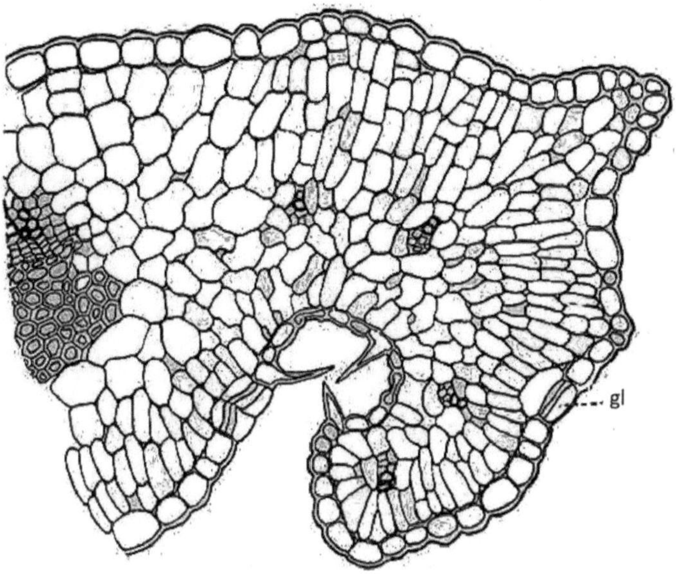

Fig. 133 *Frankenia reuteri*: cross section through the leaf; gl, salt gland (Urcelay 1923)

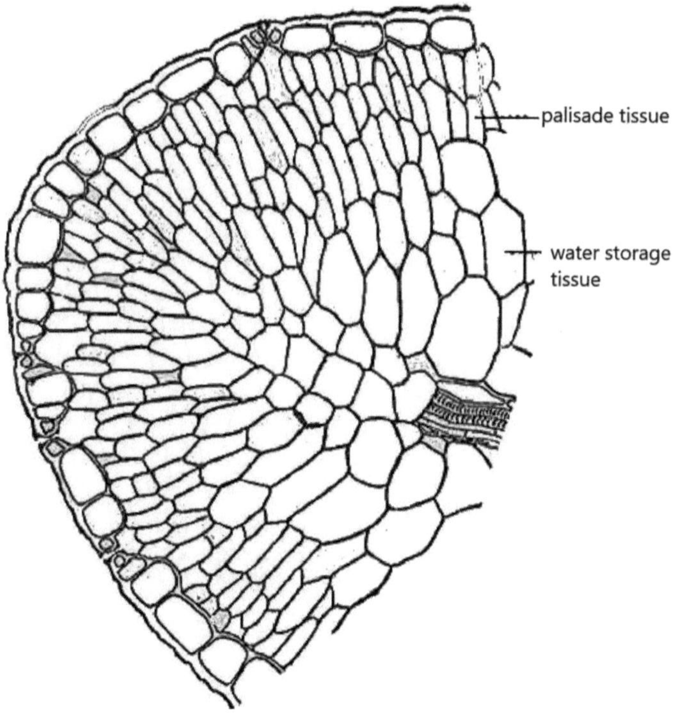

Fig. 134 *Spergularia dillenii*: cross section through the leaf (Urcelay 1923)

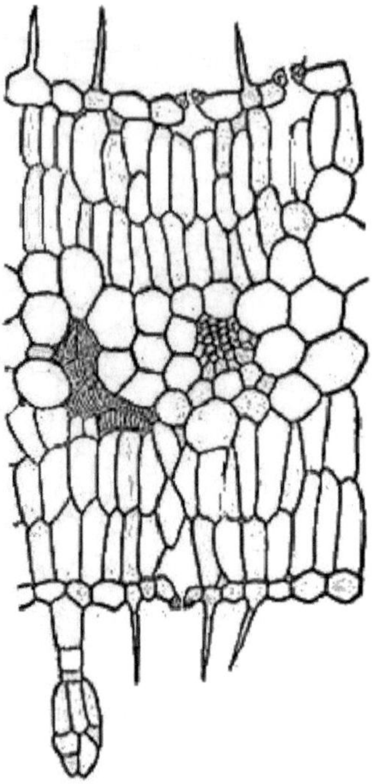

Fig. 135 *Odontites longiflora*: cross section through the leaf (Urcelay 1923)

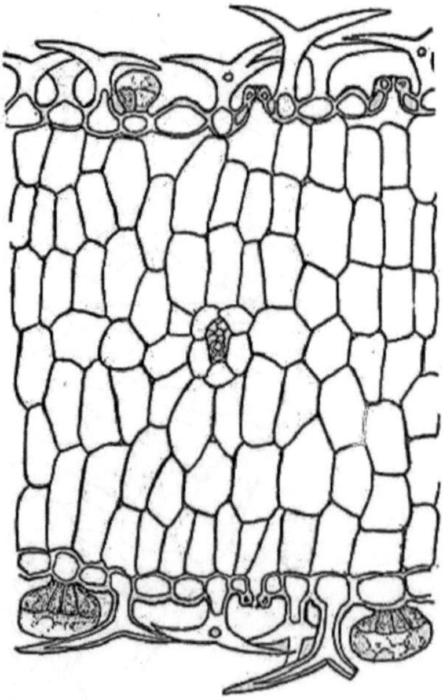

Fig. 136 *Lavandula latifolia*: cross section through the leaf (Urcelay 1923)

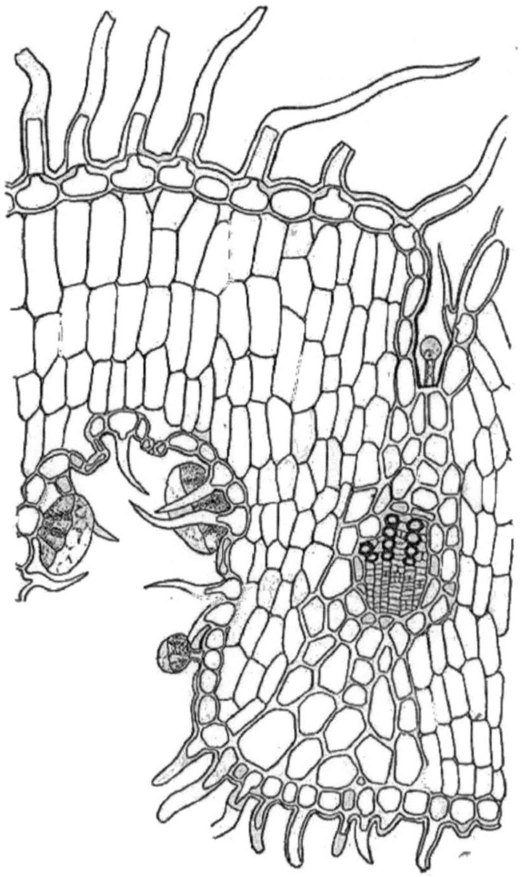

Fig. 137 *Salvia lavandulaefolia*: cross section through the leaf (Urcelay 1923)

there are prominent trichomes; in this case, the presence of trichomes is regarded as an equivalent of thick-walled epidermis; within studied species, two have bulliform cells.

Regarding the epidermis of alluvial basin species (*Leersia oryzoides*, Fig. 149; *Muhlenbergia mexicana* and *Polygonum muhlenbergii*, Fig. 150; *Baptisia leucantha*, Fig. 151; *Steironema lanceolata*, Fig. 152; *Asclepias verticillata*, Fig. 153; *Apocynum cannabinum*, Fig. 154; *Mimulus ringens*, Fig. 155; *Lycopus virginicus*, Fig. 156; *Mentha canadensis*, Fig. 157; *Physostegia virginiana*, Fig. 158; *Solidago serotina* and *Solidago graminifolia*, Fig. 159; *Vernonia noveboracensis*, Fig. 160; *Silphium laciniatum*; *Aster salicifolius*), nine have a lower epidermal wall relatively thinner than that on the upper side; of the 16 alluvial basin plants examined, ten are smooth.

In summary, Hayden (1919) revealed that the leaves of prairie plants show a xerophytic tendency in their leaf structure, indicated by the specialized palisade

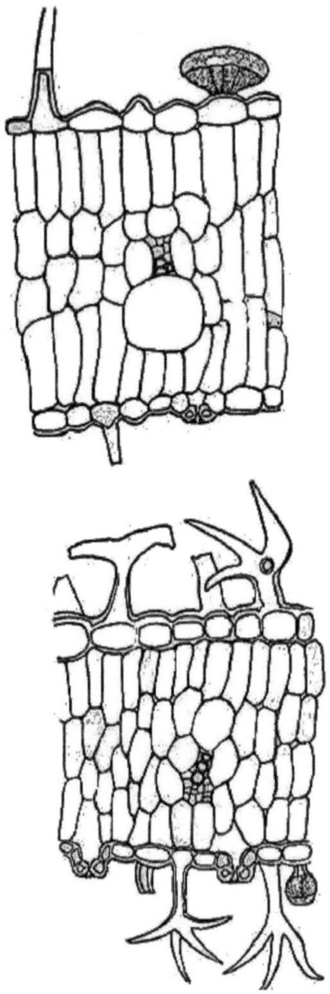

Fig. 138 *Teucrium sponosum* (top), *T. capitatum* (bottom): cross section through the leaf (Urcelay 1923)

tissue, the thick-walled and trichomeless epidermis, the presence of water-storing tissue, and sometimes of trichomes. The mere presence of these characters is not of primary significance as an indication of xerophytism, but their relative development correlated with other morphological features of the plant such as the extensiveness of the root system.

The upland plants have a thinner epidermis than those of the lowland, and 70% of those studied are without trichomes, while 75% of the upland species have trichomes. Of the alluvial basin leaves studied, 50% were bifacial, while 33% of the upland plants have bifacial leaves; 12% of the alluvial basin plants were centric to subcentric; 50% of the upland plants were centric to subcentric.

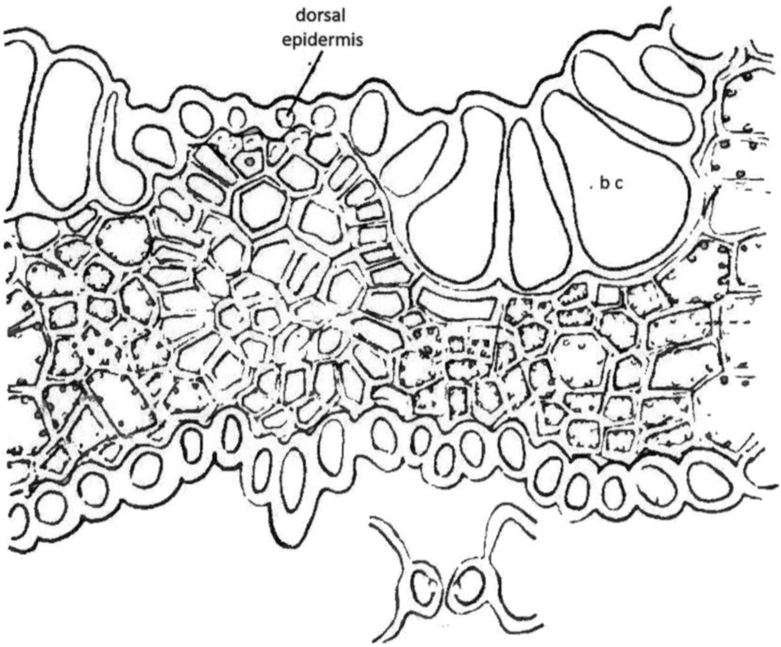

Fig. 139 Cross section through the leaf of *Andropogon scoparius* (Hayden 1919); b c – bulliform cells

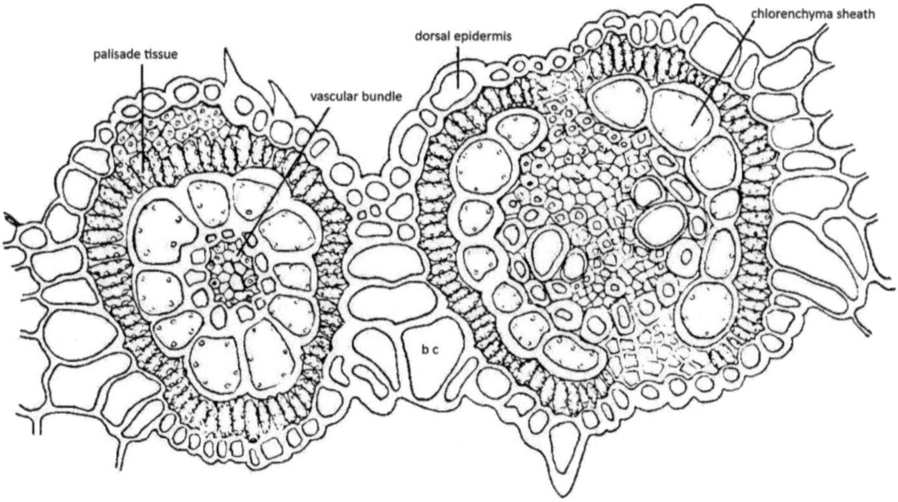

Fig. 140 Cross section through the leaf of *Bouteloua curtipendula* (Hayden 1919); b c, bulliform cells

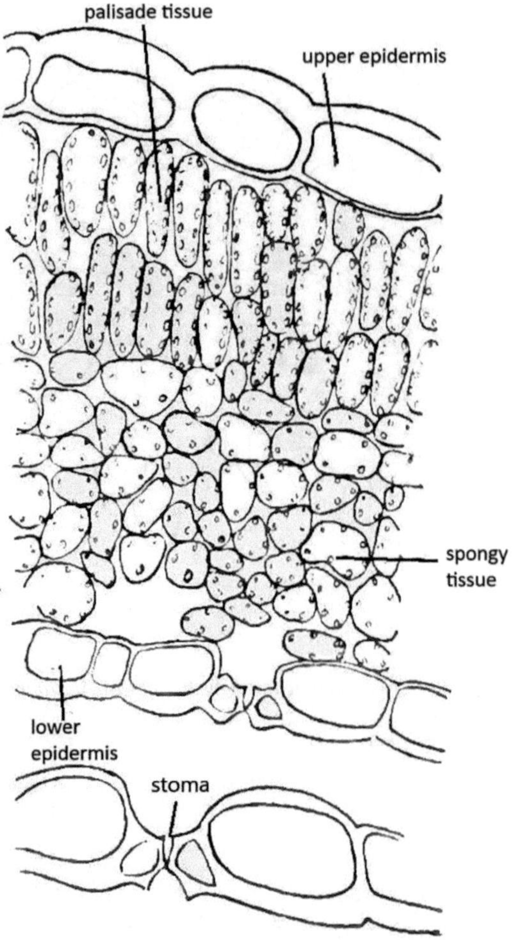

Fig. 141 Cross section through the leaf of *Anemone cylindrica* (Hayden 1919)

Shimek (1915) conducted an anatomical study of plants growing in Iowa prairies, demonstrating that prairie plants are essentially xerophytic, differing from native desert plants chiefly in the degree of modifications for protective purposes; prairie plants typically exhibit reduced leaf surface area, extensive root systems, hairy or pubescent surfaces, or smooth but leathery textures. Additionally, many of these plants form rosettes for enhanced protection. In summary, these plants display various structural adaptations characteristic of xerophytes. The microscopic structure, particularly of the leaves, exhibits xerophytic adaptations to varying extents. Often, the cuticle layer is notably thick (*Dalea alopecuroides*, Fig. 161, and *Oxytropis lambertii*, Fig. 162); in some cases, the stomatal guard cells are sunken into pits (*Dalea alopecuroides*, Fig. 161; *Amorpha canescens*, Fig. 163). The leaf parenchyma is typically very dense, with minimal intercellular spaces; additionally, specialized water-storage cells are present.

Fig. 142 Cross section through the leaf of *Amorpha canescens* (Hayden 1919)

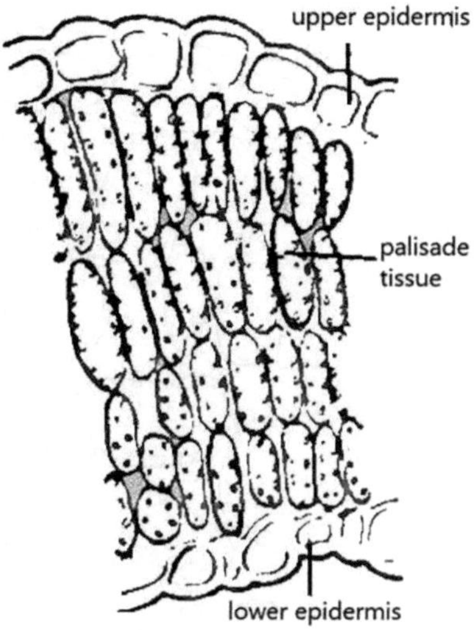

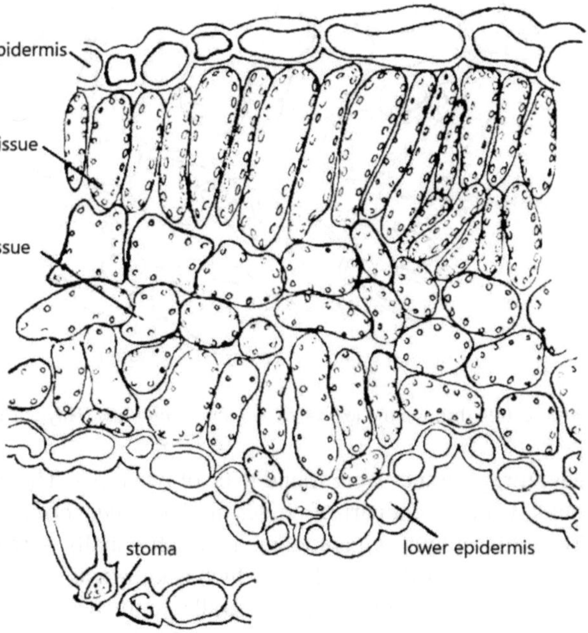

Fig. 143 Cross section through the leaf of *Ceanothus americanus* (Hayden 1919)

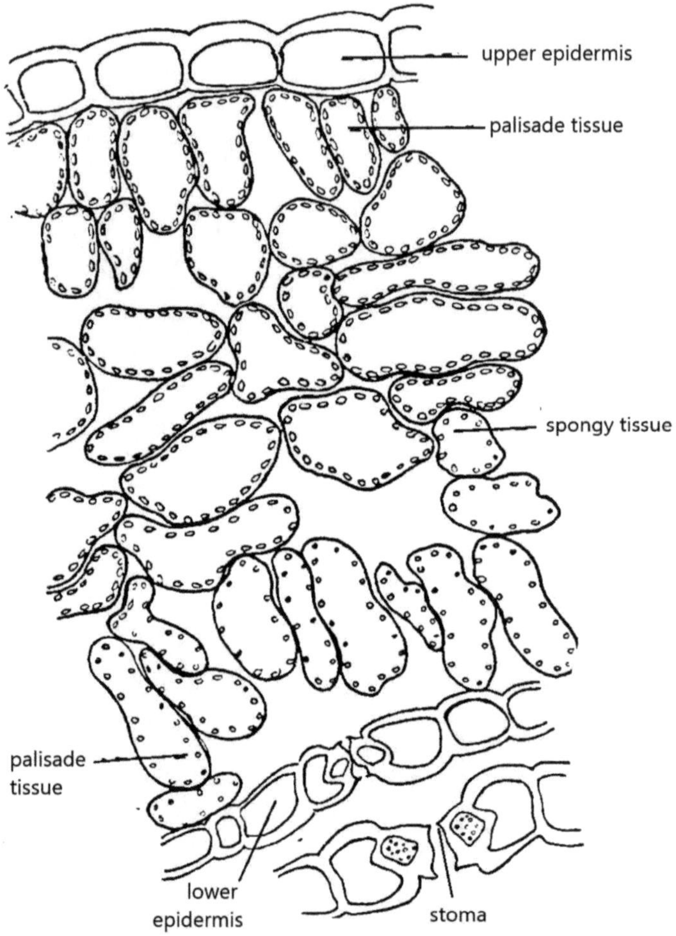

Fig. 144 Cross section through the leaf of *Eryngium yuccifolium* (Hayden 1919)

The water-storage tissues were found to be of two types. The first type comprises bulliform cells, which are a part of epidermis (*Panicum scribnerianum*, Fig. 164); the other is a chlorophylless parenchyma, which usually surrounds the vascular bundles (*Solidago nemoralis*, Fig. 165), which is known as border parenchyma; this sometimes extends in the form of a plate parallel to the two leaf surfaces (*Oxytropis lambertii*, Fig. 162, and *Solidago speciosa*, Fig. 166), or it forms vertical plates or columns (*Lactuca ludoviciana*, Fig. 167, and *Panicum scribnerianum*, Fig. 164), or both may appear (*Grindelia squarrosa*, Fig. 168).

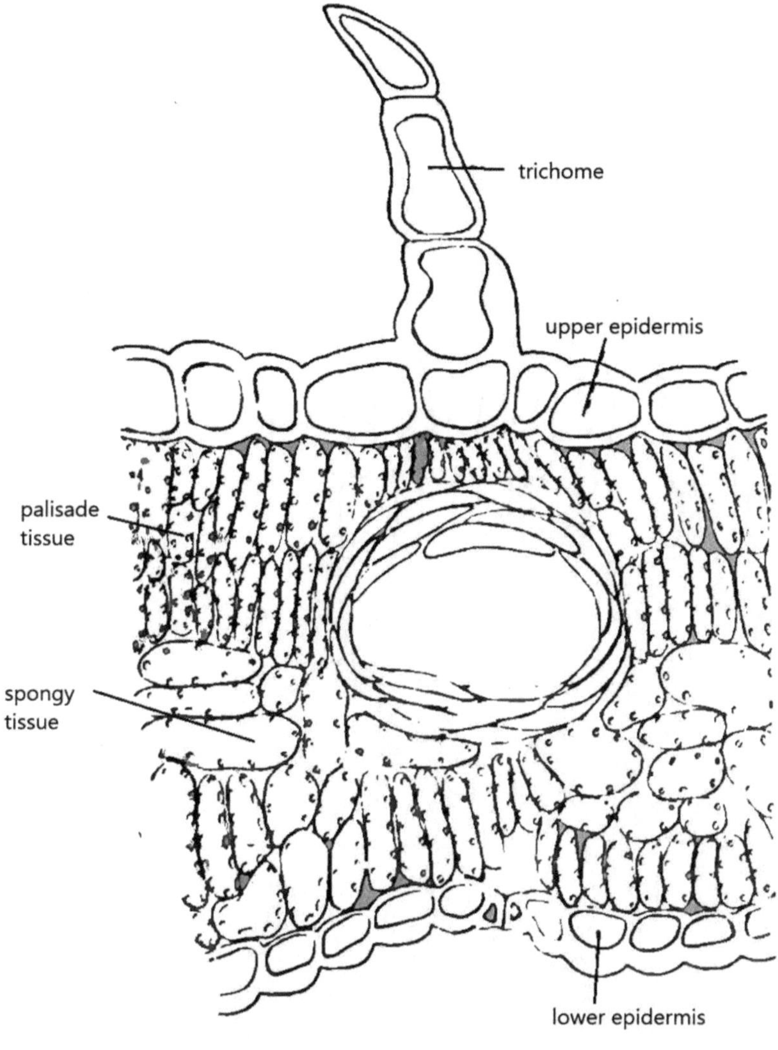

Fig. 145 Cross section through the leaf of *Solidago canadensis* (Hayden 1919)

8 Ecological Anatomy of the Leaf in Psychrophytes (Arctic, Alpine Plants) (General Approach)

The term *psychrophytes* is used by some plant ecologists (Warming 1909a) to designate plants that grow in cold habitats (arctic and high mountain areas—alpine plants) (Gr. *psuchros*—cold, frozen). Although the term may be more appropriate, etymologically, for the plants of arctic areas, it can also be correctly applied to alpine plants (considered here, generally, without much nuance, as those plants that grow in the high areas of the mountains); the vegetation in the polar areas (below the

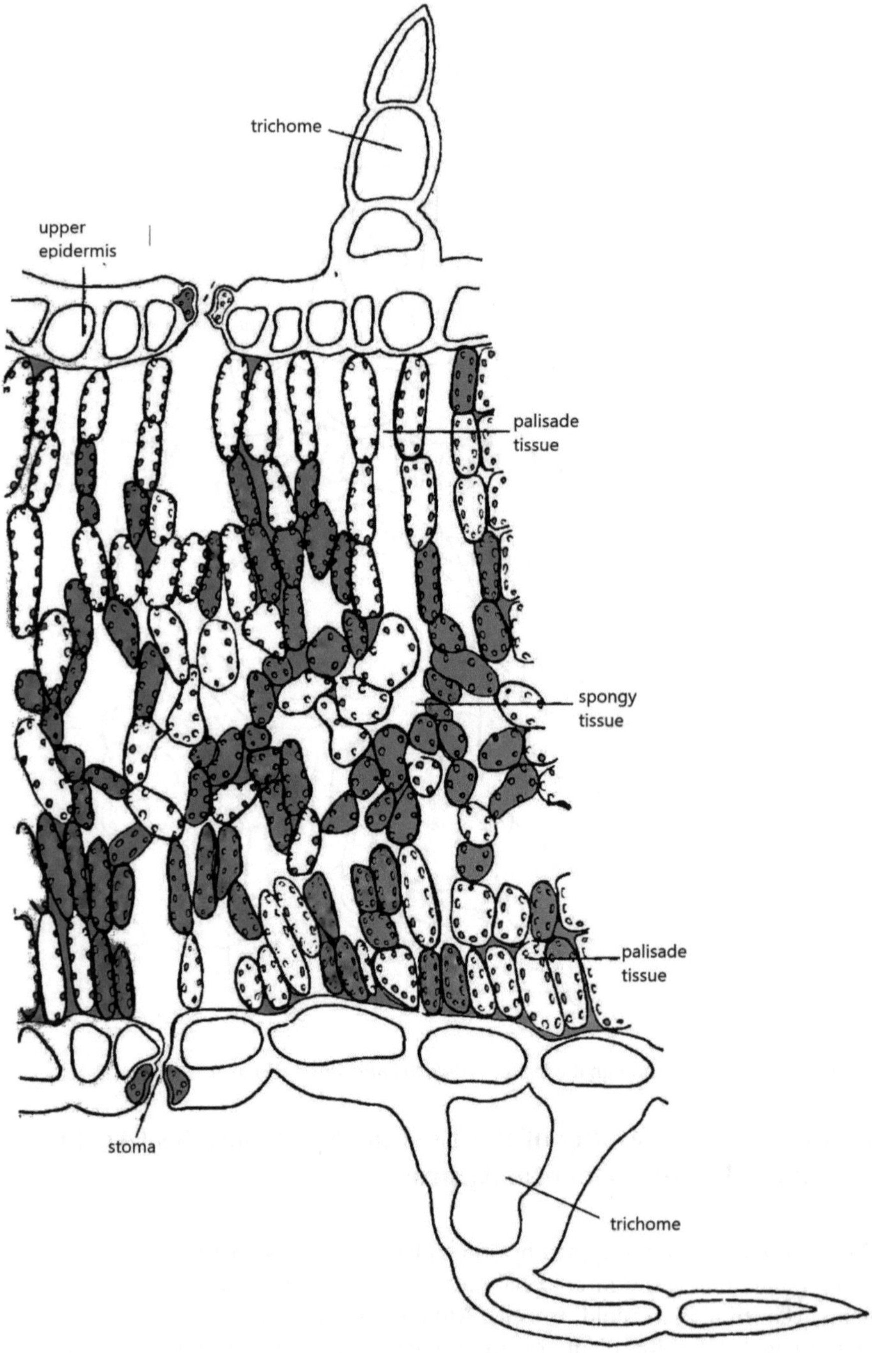

Fig. 146 Cross section through the leaf of *Solidago rigida* (Hayden 1919)

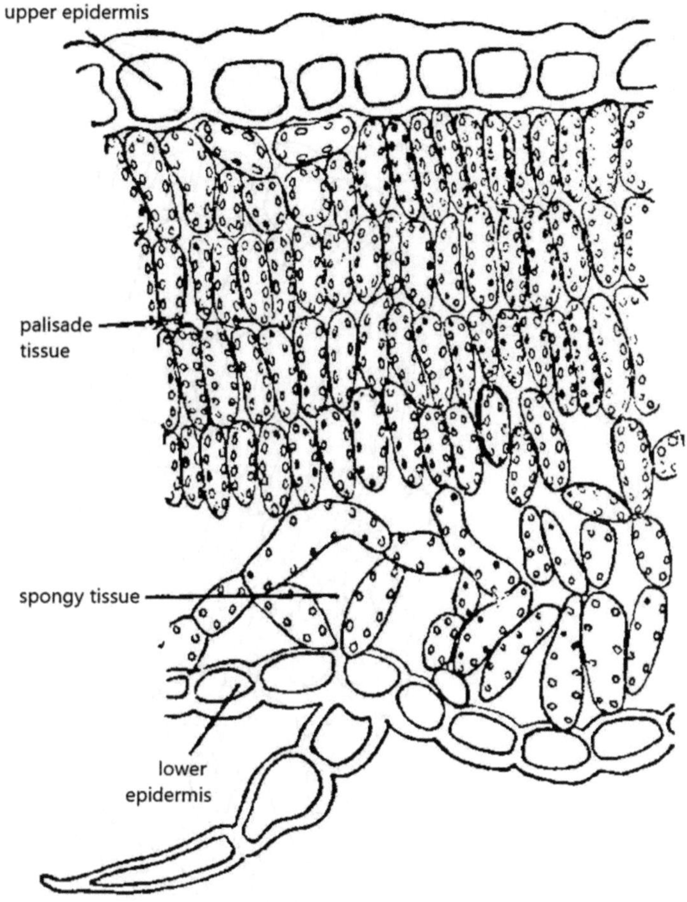

Fig. 147 Cross section through the leaf of *Helianthus grosseserratus* (Hayden 1919)

snow line) and in the high mountain areas, form, according to Warming (1909a), "subglacial communities," which, although they differ in terms of particular conditions—polar and high mountain—still have some common features imposed by the convergence of environmental factors (intense lighting during the day, low air and soil temperature during summer, insolation, strong and dry winds, shortening of the growing season—just to mention a few, very briefly) (Warming 1909a; Schimper 1903; Treshow 1970; Seddon 1971; Crawford 1989; Rundel et al. 1994; Archibold 1995; Collinson 1988; Dawson and Lucas 2005; Ingrouille and Eddie 2006; Lütz 2012; Lambers and Oliveira 2019; Schulze et al. 2019; Pfadenhauer and Klötzli 2020; Körner 2022), which also imposes common morphological (Grigore 2022) and anatomical characters, of xeromorphosis.

It must be said that the literature regarding the biology of psychrophytes (in general) and their anatomical adaptations (in particular) is very rich, with consistent

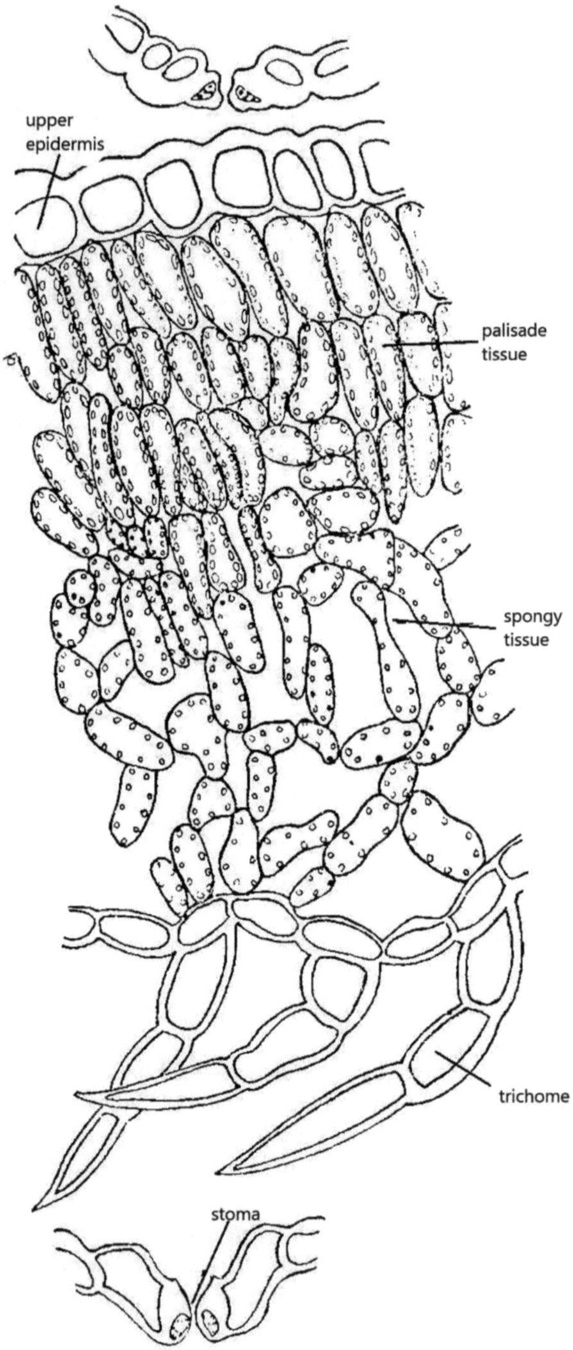

Fig. 148 Cross section through the leaf of *Helianthus tuberosus* (Hayden 1919)

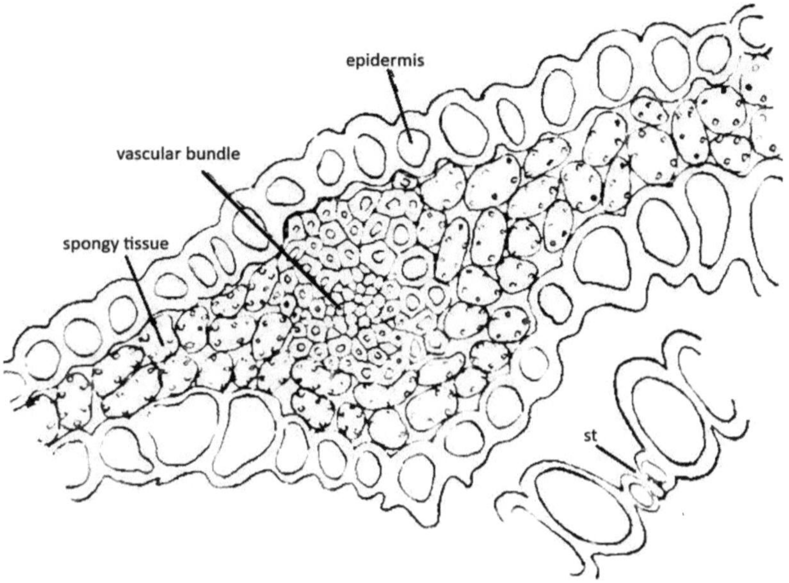

Fig. 149 Cross section through the leaf of *Leersia oryzoides* (Hayden 1919)

results dating back to the second half of the nineteenth century, through classic studies (Wagner 1892; Bonnier 1894a, b; Lazniewski 1896; Leist 1890), continuing, of course, in the following period (Warming 1908, 1909b; Petersen 1908a, b; Mentz 1909; Galløe 1910; Seidelin 1912; Jessen 1911; Heide 1912; Forsaith 1920; Carlquist 1994); at the beginning of the twentieth century, there was an increased interest in alpine plants—especially horticultural—but also aesthetic; some of the works of this kind have beautiful color illustrations, in addition to the strictly biological monographic character, but done with enough care (Robinson 1903; Brown 1907; Farrer 1908; Arber 1910; Wright 1911; Thompson 1912—to mention only a few).

Returning to the part of the structure of psychrophilic plants, we should mention the classic studies of Bonnier (1887, 1888, 1890, 1894a, b), who was particularly concerned with the influence of environmental factors (the influence of the alpine environment) on morphological and anatomical changes in alpine plants, his experimental concerns being characterized, for the time, by a great finesse of the observations, hypotheses, and interpretations made. Under experimental conditions, he worked a lot with ecotypes, i.e., different forms of the same species, cultivated in different environmental conditions, such as in the plains (lowlands) and in the alpine zone, but also with "sub-psychrophytes," i.e., arctic plants and plants from mountain areas (see introductory discussion above). As expected, he observed that plants grown in the alpine environment show some characteristic morphological and anatomical changes compared to lowland forms; as can be seen, we use the term

Fig. 150 Cross section through the leaf of *Polygonum muhlenbergii* (Hayden 1919)

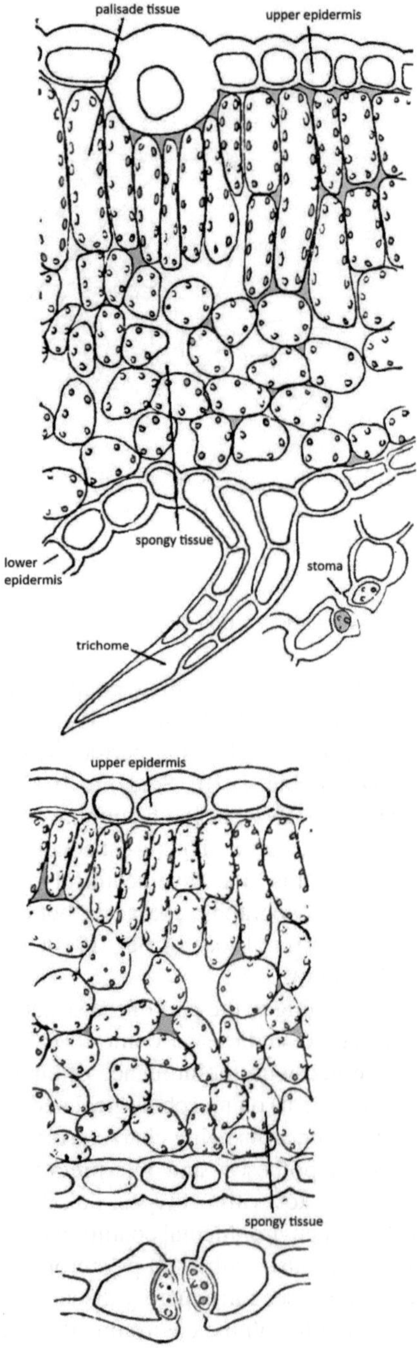

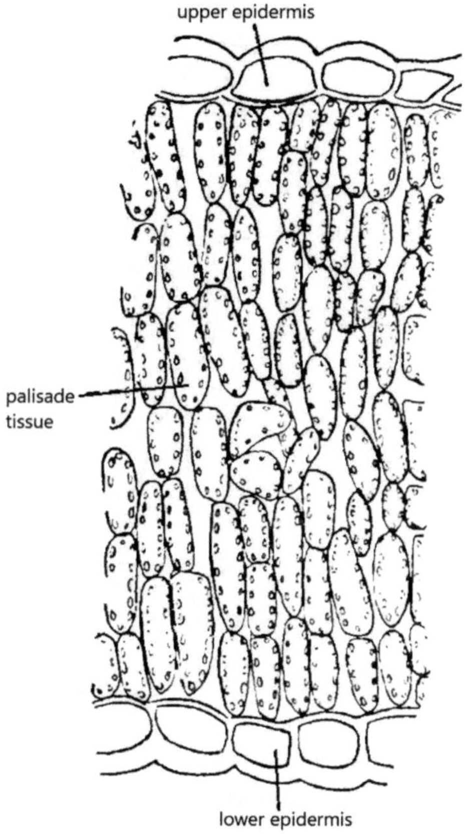

Fig. 151 Cross section through the leaf of *Baptisia leucantha* (Hayden 1919)

"modifications," and not that of adaptations, since, indeed, we cannot speak in this case of adaptation, but of acclimatization, considering the intention of the author.

In a study by Bonnier (1894a), he compared the anatomical structure of arctic plants (collected from the islands of Jan Mayen and Spitzbergen—Arctic Ocean) with that of alpine plants (from the highlands of the Alps and the Pyrenees).

Following his analysis, considering arctic plants and comparing them with alpine plants, he found that these structural differences (Figs. 169, 170, 171, 172, and 173) are mainly as follows:

1. In different organs of the plant, the lignified elements have a tendency to decrease numerically, the walls are less thick, and the vessels have a narrower diameter.
2. The leaves are thicker, but with a less differentiated structure; the palisade tissue is much less individualized (Figs. 169, 173, and 175); intercellular lacunae or meatus are much more developed.
3. The epidermis of leaves (and stems) is less individualized and presents a thinner cuticle (Figs. 169, 173, 174, and 175).
4. All the tissues of the leaf (of the stem and of the root), if we were to compare them with those corresponding to alpine plants, have a more or less pronounced

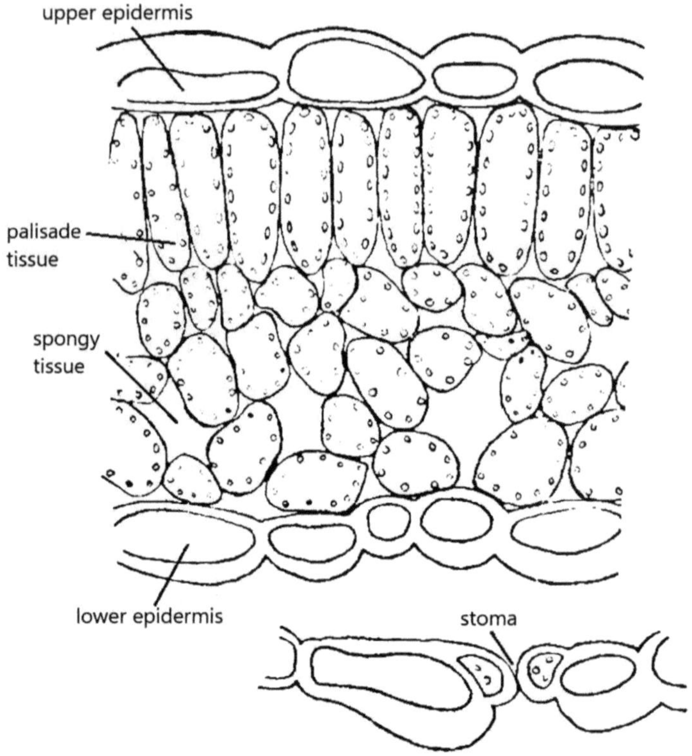

Fig. 152 Cross section through the leaf of *Steironema lanceolata* (Hayden 1919)

tendency to round off their cells, which often even form trabeculae that separate large intercellular spaces, filled with air.

Following these considerations, the anatomical characters of alpine plants can be also delimited.

These structural differences are attributed by Bonnier to the influence of environmental factors, the most important of which are temperature, degree of illumination, air and soil humidity, the chemical nature of the soil, and the composition of soluble substances that are found dissolved in the water absorbed by the roots.

In another study, Bonnier (1894b) experimentally followed the anatomical changes of some plant species, with individuals cultivated in the plain (low altitude) and in the alpine zone (high altitude) or transplanted from the alpine zone to the plain. The results are very interesting, as they certify, once again, the influence of environmental factors on the anatomical structure of the leaf (the discussion here is about the leaf, but Bonnier also investigated other plant organs).

Thus, Bonnier found that lowland plants (ecoforms), cultivated at a higher altitude, acquire, under the influence of the "alpine climate," some characteristic anatomical changes, some of which are accentuated indefinitely with altitude, and

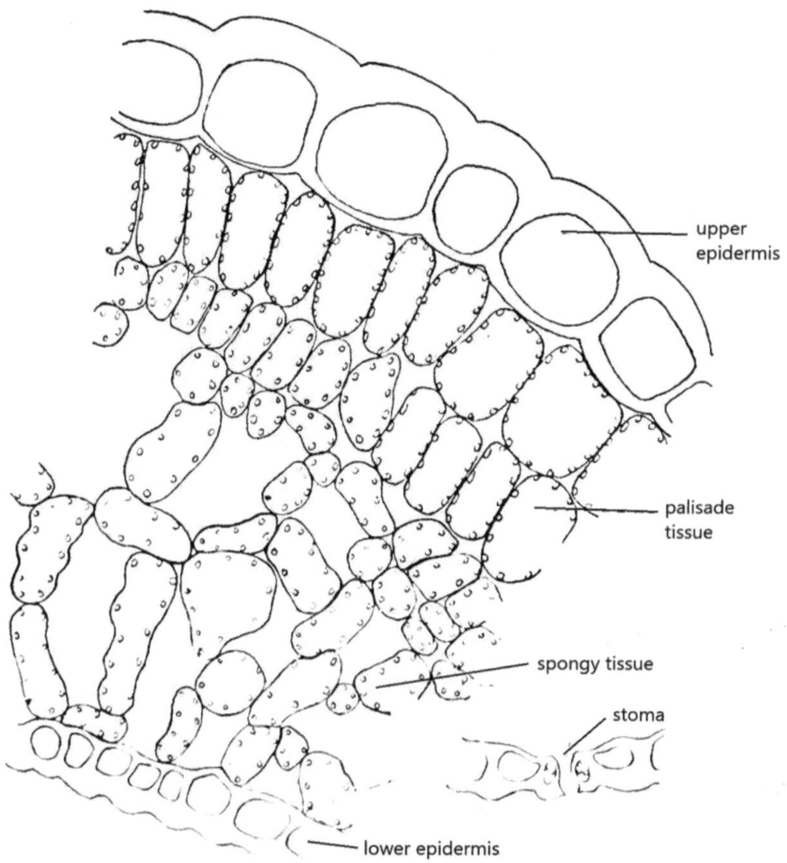

Fig. 153 Cross section through the leaf of *Asclepias verticillata* (Hayden 1919)

others (e.g., chlorenchyma tissues of the leaf) can reach an optimal value below the last altitude limit that the species can support.

These changes are as follows, compared to the reference individuals, corresponding to the lowland area (Figs. 176, 177, 178, 179, 180, 181, 182, 183, and 184):

1. The leaves are generally smaller—with exceptions, sometimes in the subalpine region—hairier, thicker in relation to their surface and often even thicker in absolute value, and darker green in color, by reflection or transparency.
2. Chlorenchyma tissues better arranged to carry out the photosynthetic function; the palisade tissue is more developed, either because its cells are longer and narrower or because the number of rows of palisade cells is greater. This anatomical reality has been also confirmed by other classic research, such as that of Wagner (1892—see the Figs. 185, 186, 187, 188, 189, 190, 191, 192, and 193). In general, there was the assumption that with high location, the formation of the pali-

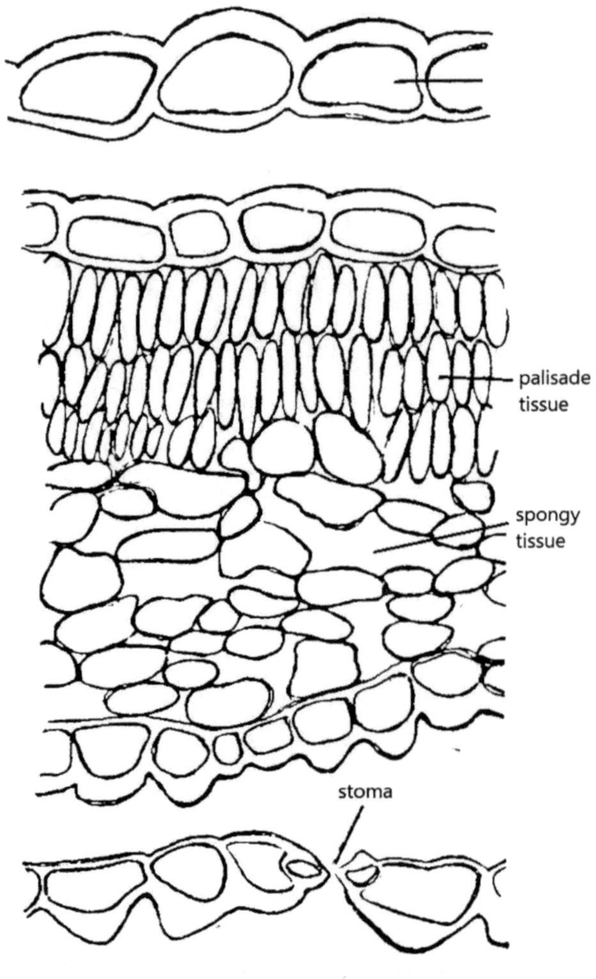

Fig. 154 Cross section through the leaf of *Apocynum cannabinum* (Hayden 1919)

sade is based on the increasing the individual cells, or by the occurrence of new layers of cells, or by combination of both cases.

In addition, the cells contain a greater number of chloroplasts, and each chloroplast often has a greener hue; when there are secretory channels, they have, relatively or absolutely, a larger diameter; the epidermis shows fewer differences than that of the stem, but is still generally better individualized, especially in the evergreen leaves that also show better developed protective subepidermal cells; epidermal cells are usually smaller and often the number of stomata, per unit area, is greater, especially on the upper face of the lamina.

Probably, the most extensive studies related to the morphology, and especially to the anatomy of some psychrophytes (arctic plants in this case), are those published between 1908 and 1912, by a group of Danish botanists (Warming 1908, 1909b; Petersen 1908a, b; Mentz 1909; Galløe 1910; Seidelin 1912; Jessen 1911; Heide

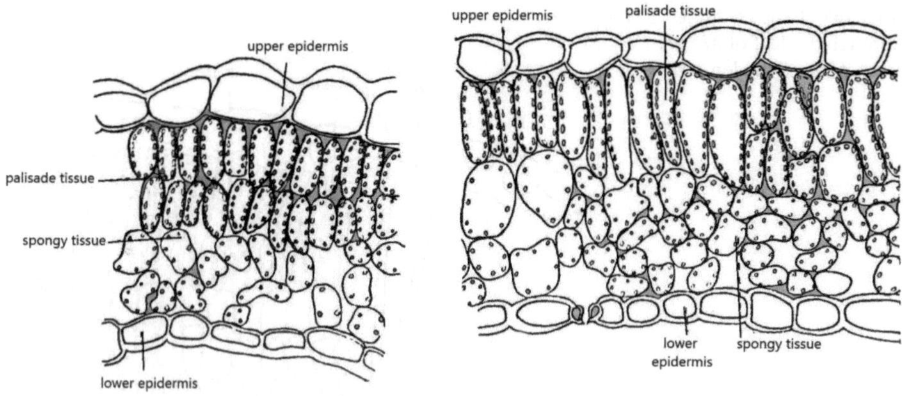

Fig. 155 Cross section through the leaf of *Mimulus ringens* (Hayden 1919)

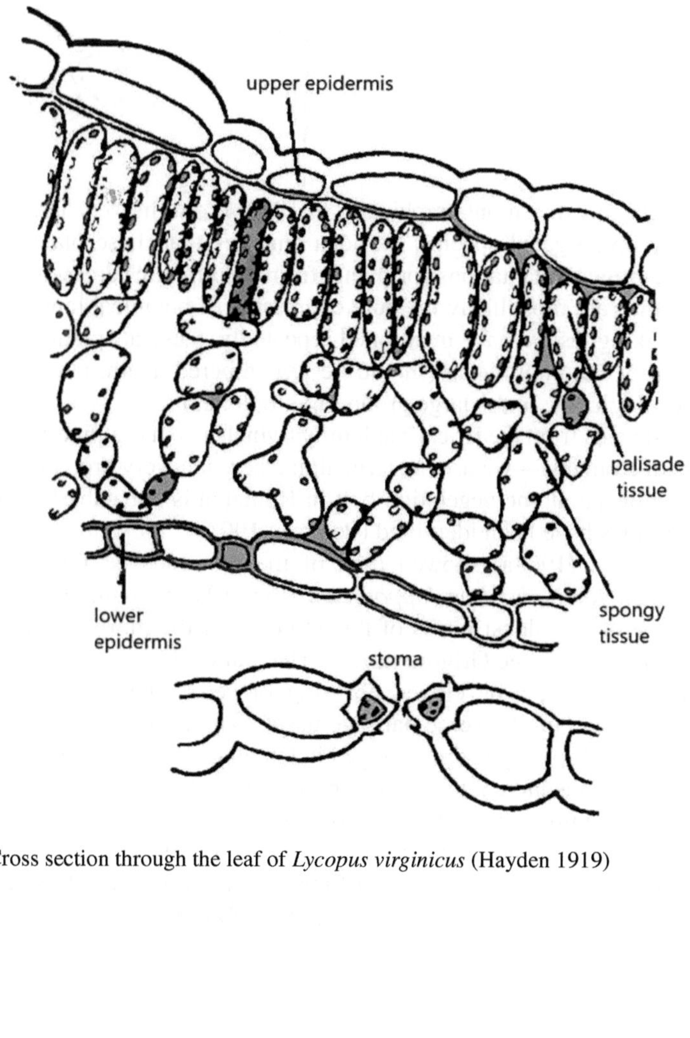

Fig. 156 Cross section through the leaf of *Lycopus virginicus* (Hayden 1919)

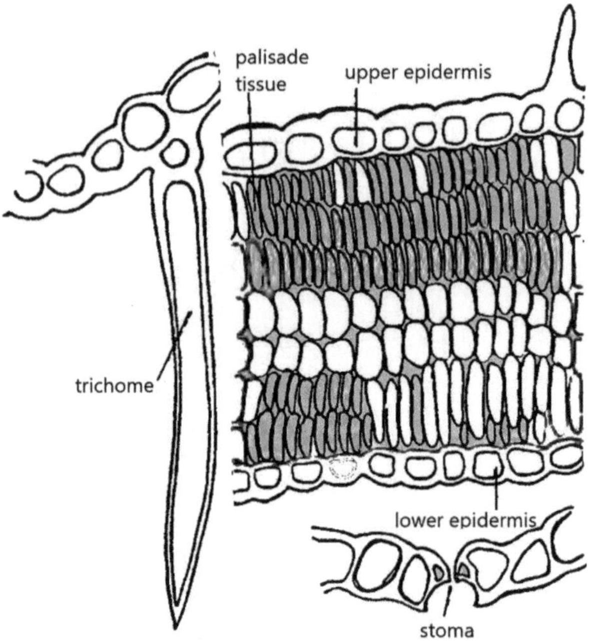

Fig. 157 Cross section through the leaf of *Mentha canadensis* (Hayden 1919)

1912), brought together monographically in 1912. From the very large number of anatomically investigated species, the idea emerges that these plants have some common anatomical adaptations for plants from cold (arctic) areas, and other anatomical features are most likely the pure expression of botanical family characters. Broadly speaking, as we have mentioned repeatedly, these adaptations are built in the direction of xeromorphosis (should be remembered again, psychrophytes are included, ecologically, in the large group of xerophytes).

Thus, in species from the Ericaceae family (which also has alpine representatives in the flora of Romania—we use the term alpine here in a very broad sense, because the presence of the alpine vegetation belt in Romania is debatable), several structural peculiarities have been identified (Petersen 1908a).

Thus, Petersen (1908a) showed that, of the broad-leaved forms that persist through winter, *Rhododendron lapponicum* (Fig. 194) is undoubtedly especially arctic and one of the oldest forms of this region. For the morphological forms of leaves of alpine plants, see Grigore (2022). The leaf of this species has a thick epidermis on both sides; a developed covering of peltate hairs; particularly well-protected stomata; a multilayered palisade tissue, but consisting of small cells; and a slightly branched spongy tissue arranged in lamellae around the large intercellular spaces. Related to the large intercellular spaces, they could be correlated with arctic conditions, for example, the high humidity of the air in these regions. It is probably an adaptation that does not necessarily depend on a particular climate but should rather be connected with the thickness of the epidermis, which offers good protection against the danger of excessive transpiration. On the other hand, it seems that

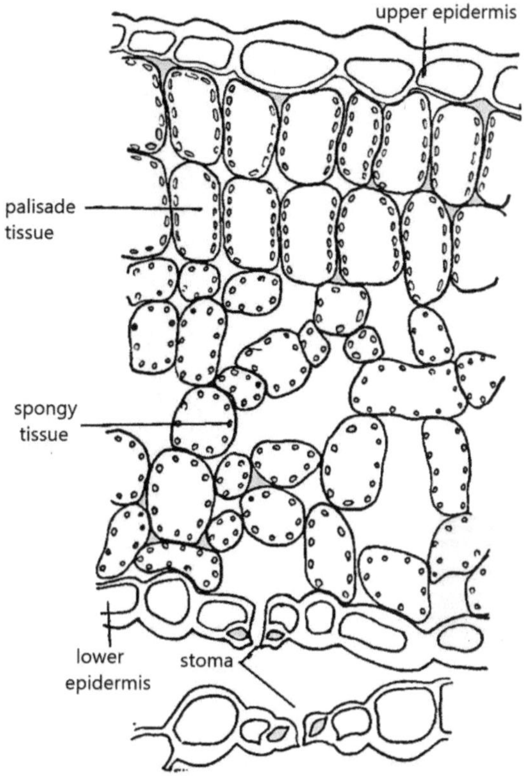

Fig. 158 Cross section through the leaf of *Physostegia virginiana* (Hayden 1919)

the intercellular spaces may become necessary in the present case, because the stiff, coriaceous leaves cannot be moved as freely by the wind as the broad, deciduous leaves.

Another species that is as well protected as *Rhododendron* is *Ledum palustre* var. *decumbens* (*Ledum decumbens*) (Fig. 195). This is a very widespread species in the arctic regions, but it also has a much more southern distribution, in areas with a continental climate and, consequently, a severe winter. The stomata are not as well protected as those of *Rhododendron*, but on the other hand, there is a more defined water-storage tissue in this species. The palisade cells are somewhat longer than those of *Rhododendron*.

Many species of psychrophytes investigated by Petersen (1908a) (*Arctostaphylos uva-ursi*, *Andromeda*, *Lyonia calyculata*, *Vaccinium vitis-idaea*, and *Oxycoccos*) do not have their main distribution only in arctic regions.

Among them, *Lyonia calyculata* (Fig. 196) and *Arctostaphylos uva-ursi* (Fig. 197) resemble each other the most from foliar anatomical point of view, being the best protected against the danger of excessive transpiration: *Lyonia* by the presence of trichomes and *Arctostaphylos* by the thick outer walls of the epidermal cells. These two species have the palisade tissue with small cells characteristic of the *Rhododendron* leaf.

Fig. 159 Cross section through the leaf of *Solidago graminifolia* (Hayden 1919)

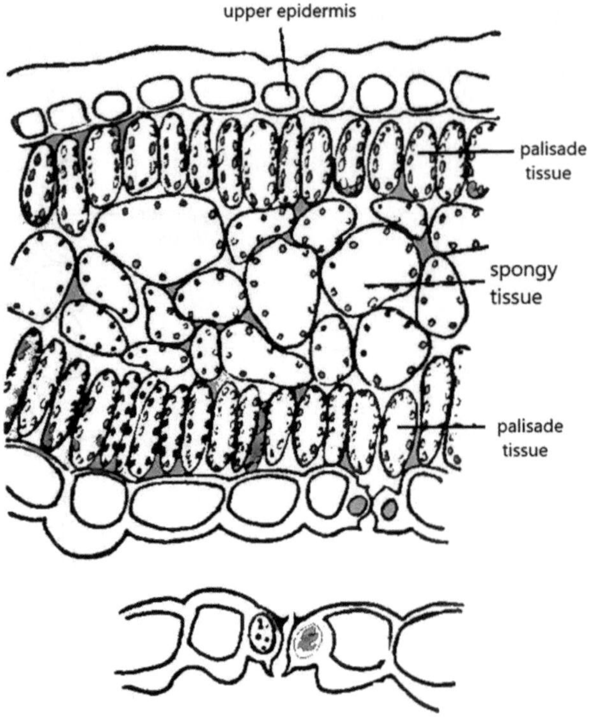

Vaccinium vitis-idaea is among the least protected species, where palisade tissue is found (Fig. 198); this species has a degree of protection approximately equal to that found in *Vaccinium oxycoccos* (Fig. 199) and *Andromeda* (Fig. 200), which have wax on the lower surface of the leaves. In terms of protection against the danger of excessive transpiration, all five species mentioned above have a low protection compared to *Rhododendron* and *Ledum*, probably because their distribution is not so strictly related to the arctic environment. The palisade cells of *Vaccinium oxycoccos* and *Andromeda* are long and occur in fewer and thicker layers than in *Ledum*.

At *Vaccinium vitis-idaea* and *Andromeda*, the lack of enhanced protection seems to be structurally translated by cells of the mesophyll that have rather thickened cell walls. The development of cuticle on the inner walls of the stomatal cells and adjacent parts is common to all forms in which the protecting trichomes are not developed.

All the furrowed and needle-leaved forms have, by their outline, and also by the position of the stomata, a distinct advantage over the broad-leaved forms.

Loiseleuria is the least protected in terms of leaf shape. But, nevertheless, the leaves of this form may be said to be well protected by a thick epidermis, by mucilaginous walls, and by dense pilosity, at the level of the epidermis (Fig. 201); palisade cells are medium in length.

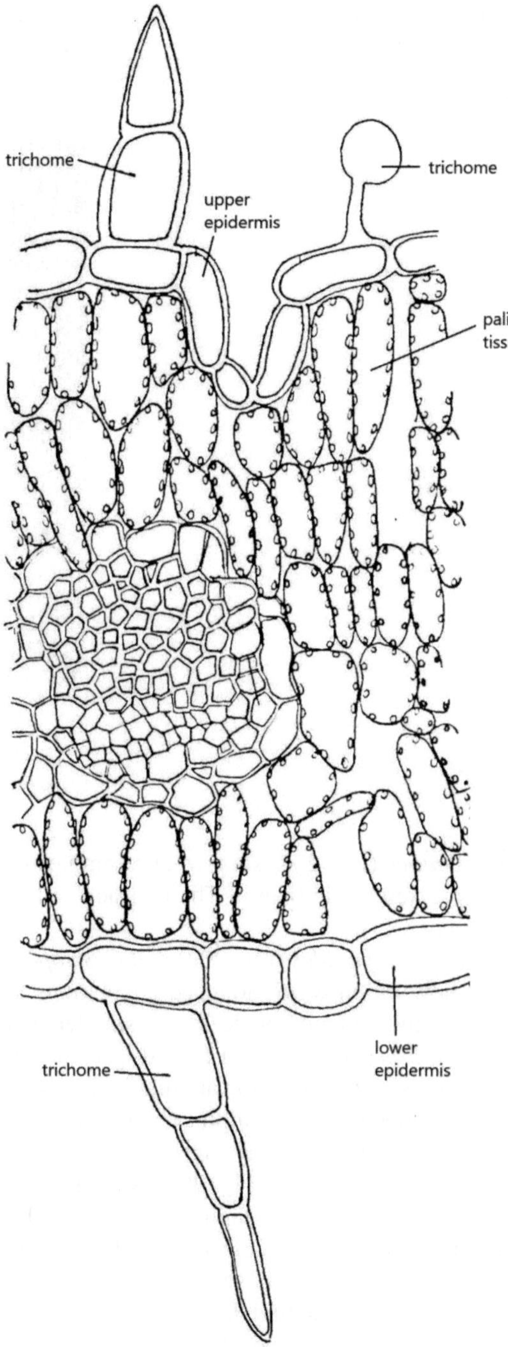

Fig. 160 Cross section through the leaf of *Veronica noveboracensis* (Hayden 1919)

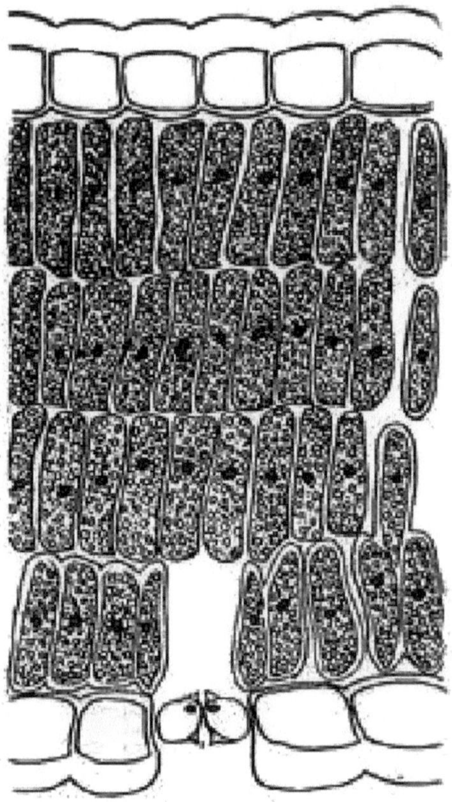

Fig. 161 Cross section through the leaf of *Dalea alopecuroides* (Shimek 1915)

Phyllodoce (Fig. 202) somewhat resembles *Loiseleuria* in structure, but is probably more xerophilic than it. The palisade cells are of medium length, as are those of *Ledum*.

Cassiope hypnoides (Fig. 203) and *Cassiope tetragona* (Fig. 204), especially the latter, show the most pronounced adaptations to the arctic environment; palisade tissue is multilayered, made up of smaller cells. As in the case of *Ledum* and *Rhododendron* species, broad-leaved species, the protective devices against the danger of excessive transpiration are highly developed. These two forms probably originally developed in arctic (alpine) conditions.

Another particular characteristic for some arctic species (*Pirola secunda*, Fig. 205; *Pirola uniflora*, Fig. 206) is the presence of chloroplasts in ordinary epidermal cells, a rare reality in the plant world. Their presence at this level, especially their function, is still unclear. In a few instances, small, atypical chlorophyll-containing chloroplasts have been observed in epidermal cells from *Arabidopsis thaliana*, although the thylakoids, which are responsible for light-dependent photosynthetic reactions, are poorly developed. It is assumed that, in some plants, these chloroplasts could be involved in their immune response, through their accumulation at the interface with the pathogen (Irieda 2022). In psychrophytes, the role of

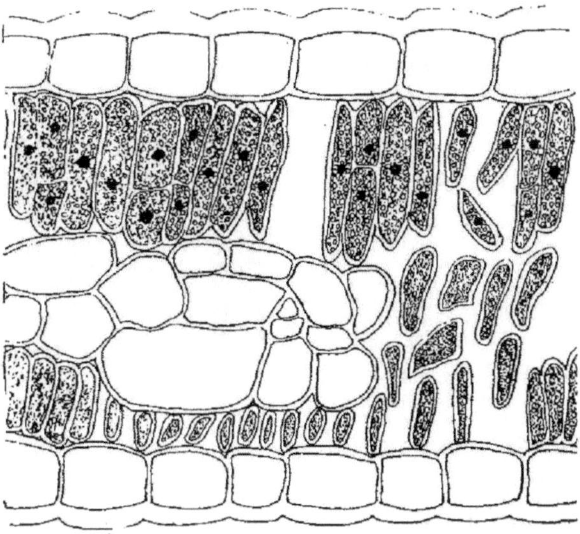

Fig. 162 Cross section through the leaf of *Oxytropis lambertii* (Shimek 1915)

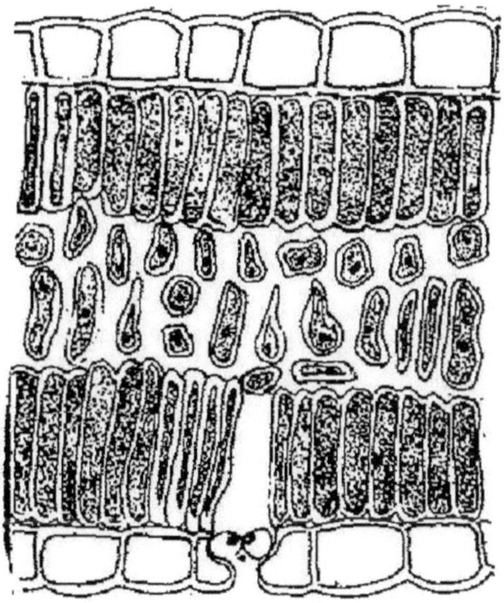

Fig. 163 Cross section through the leaf of *Amorpha canescens* (Shimek 1915)

chloroplasts in epidermal cells remains to be clarified, probably related to environmental factors (pronounced lighting), functioning as a protective screen, or intervene in the efficiency of photosynthesis.

Some psychrophytes show leaves of ericoid type, which were morphologically characterized in a previous paper (Grigore 2022). It shall be reminded that they belong to the functional physiognomic type of plant—small and dwarf shrubs, with green leaves during the summer, and evergreens—being evergreen or with deciduous leaves (chamaephytes), being characterized by winter dormancy, frost tolerant,

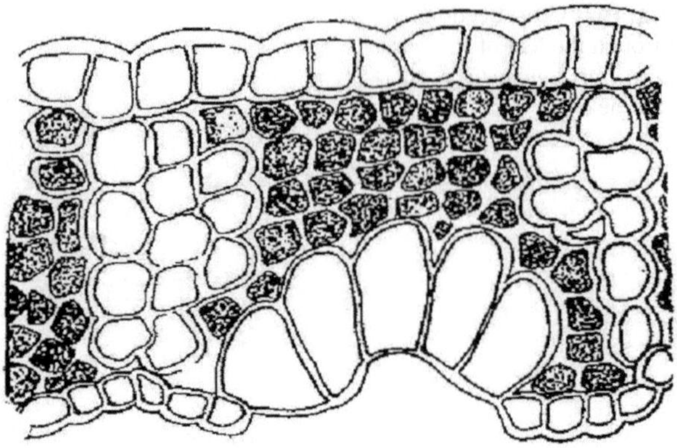

Fig. 164 Cross section through the leaf of *Panicum scribnerianum* (Shimek 1915)

Fig. 165 Cross section through the leaf of *Solidago nemoralis* (Shimek 1915)

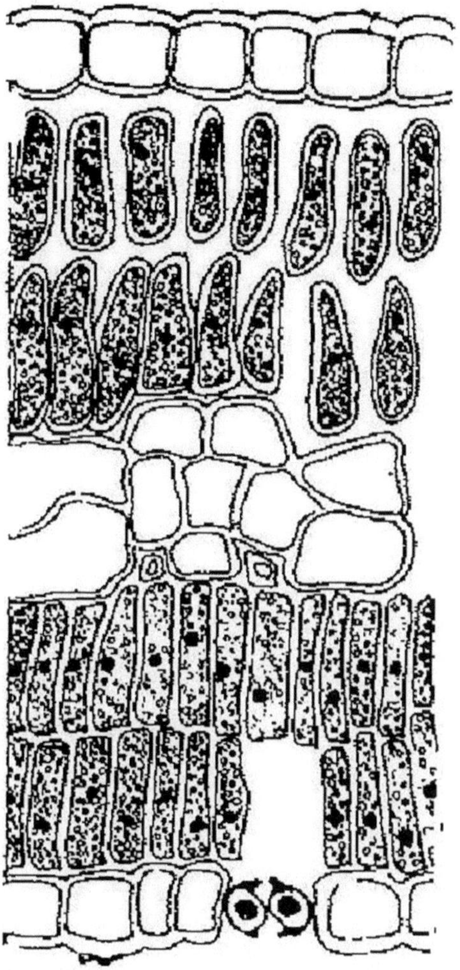

Fig. 166 Cross section through the leaf of *Solidago speciosa* (Shimek 1915)

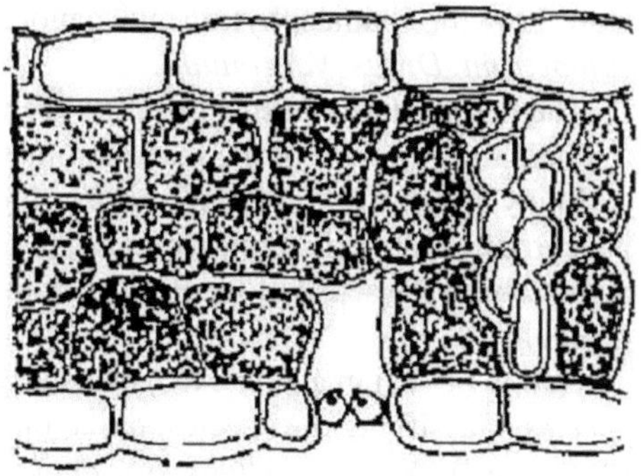

Fig. 167 Cross section through the leaf of *Lactuca ludoviciana* (Shimek 1915)

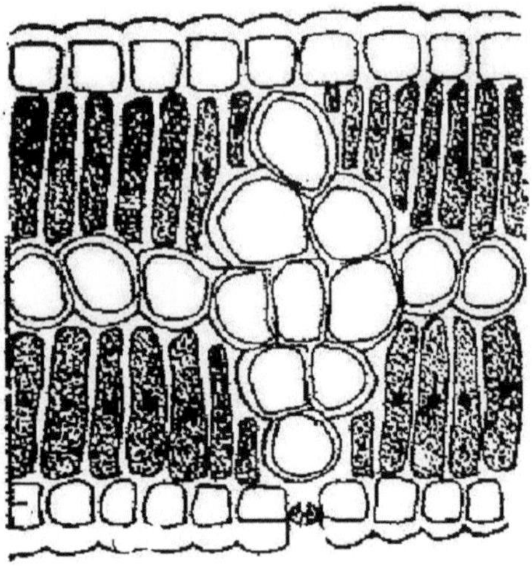

Fig. 168 Cross section through the leaf of *Grindelia squarrosa* (Shimek 1915)

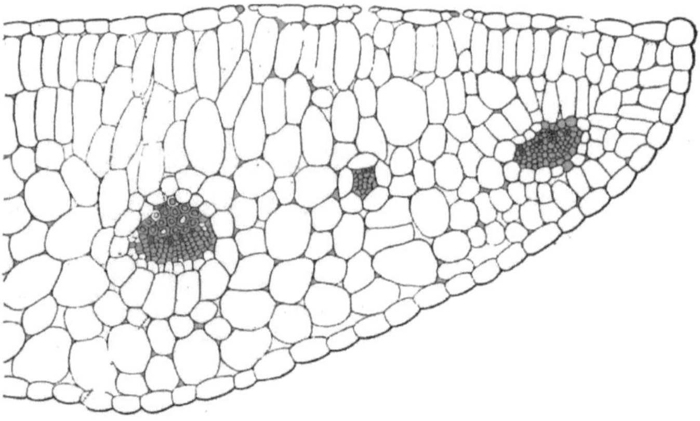

Fig. 169 *Silene acaulis*, arctic zone: cross section through the leaf (Bonnier 1894a)

slow growth, chamaephytes, and nanophanerophytes, and leaf life of 3–4 years (*Empetrum*, *Dryas*, *Vaccinium*).

The leaf anatomy of this type has been shown above (see also Fig. 62). Below is a schematic structure of the leaf anatomy of *Empetrum nigrum* (Fig. 207).

Many species of the Saxifragaceae family can be classified in the sub-group of psychrophytes (arctic, alpine plants); their number is so large that their anatomical approach, in the sense of identifying common adaptive structural elements, is quite difficult.

Galløe (1910) made an extensive study on the anatomical structure of *Saxifraga species*, classifying the species, based on anatomical characters, into several groups ("sections"): *Nephrophyllum*, *Boraphila*, *Dactyloides*, *Trachyphyllum*, *Euaizonia*, and *Porphyrion*, each comprising therefore several species. Therefore, the species included in each "section" show greater anatomical affinities (relationships) with

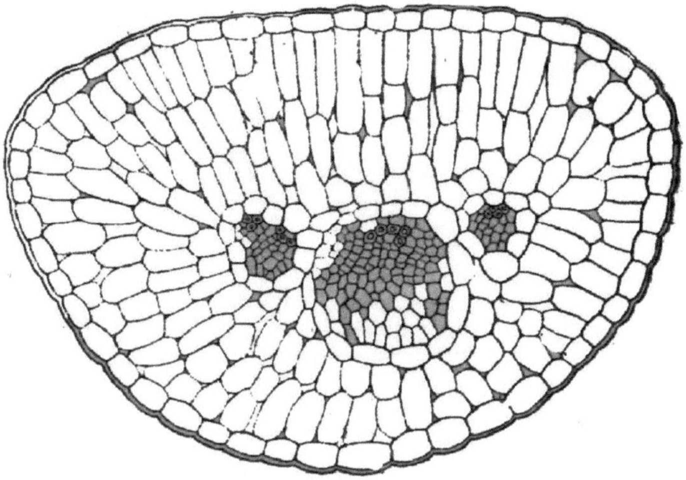

Fig. 170 *Silene acaulis*, alpine zone (Mont Blanc): cross section through the leaf (Bonnier 1894a); individualized palisade tissue, thick cuticle

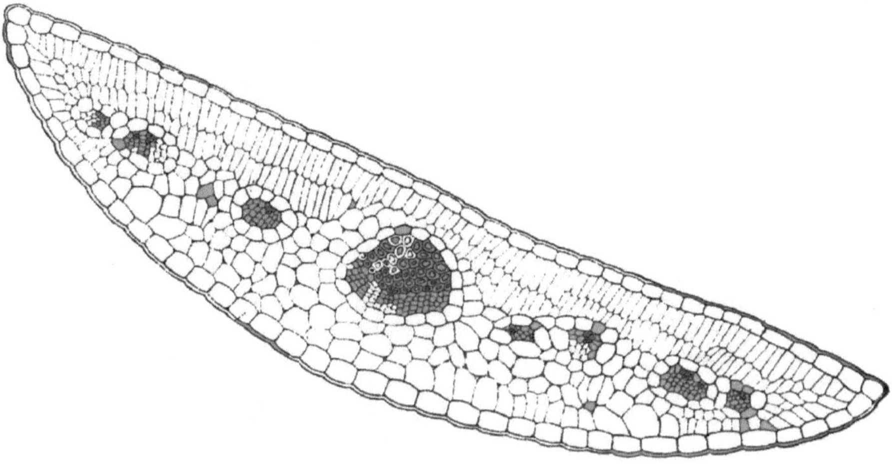

Fig. 171 *Silene acaulis*, alpine zone (Alps Mountains): cross section through the leaf (Bonnier 1894a); individualized palisade tissue, thick cuticle

each other than would species belonging to different "sections." Attention should be paid to the fact that these names do not represent taxonomical units, despite their linguistic resonance.

Of all the anatomical characters, the structure of the hydathodes and the appearance of the radial walls of the epidermis seem, in general, to be the characters that are least influenced by external conditions, and which, with a certain degree of constancy, remain uniform within the same section. It should be reminded that hydathodes are complex structures (considered by some authors to be modified

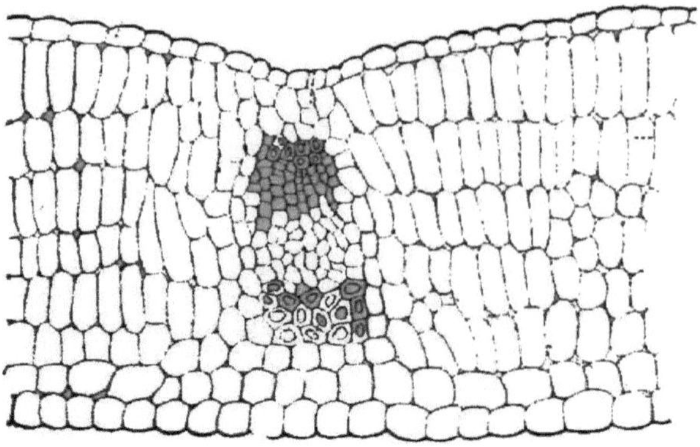

Fig. 172 *Salix reticulata*, arctic zone: cross section through the leaf (Bonnier 1894a)

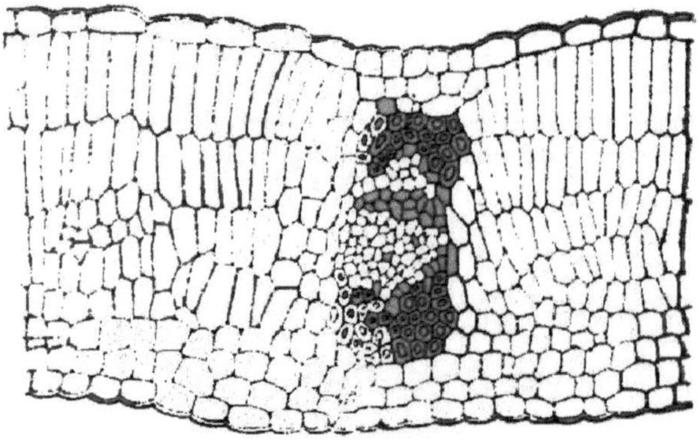

Fig. 173 *Salix reticulata*, alpine zone: cross section through the leaf (Bonnier 1894a); individualized palisade tissue

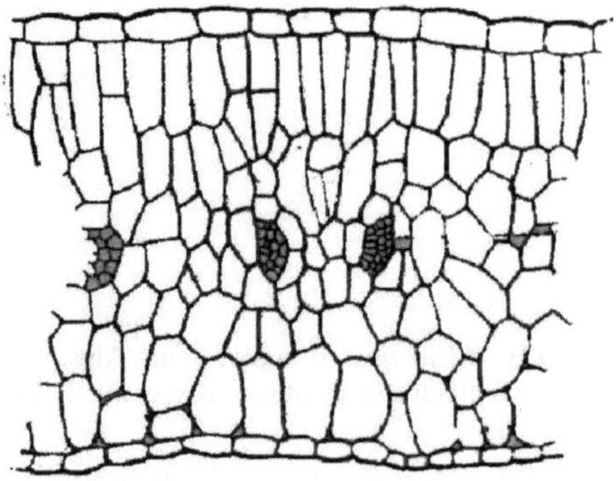

Fig. 174 *Oxyria digyna*, alpine zone: cross section through the leaf (Bonnier 1894a); individualized palisade tissue, compact structure

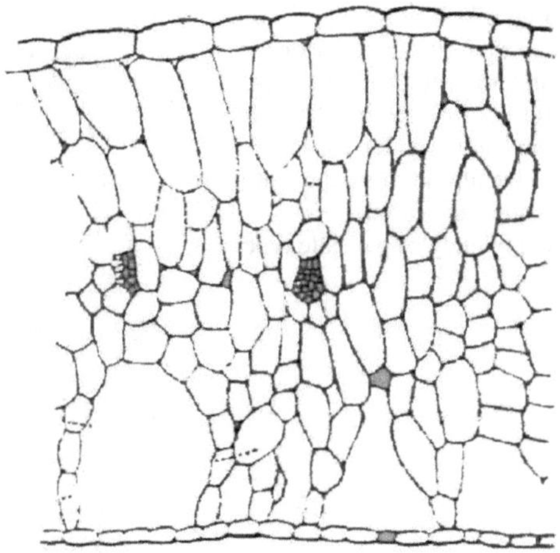

Fig. 175 *Oxyria digyna*, arctic zone: cross section through the leaf; looser structure (Bonnier 1894a)

stomata) involved in the elimination of water in liquid form (through the process of guttation). In plants such as many species of *Saxifraga*, guttation gives the leaves a crusty appearance, because the secretion product of guttation is not pure water, but a mixture with mineral salts, most often calcium carbonate. This genus contains about 400 species, with multiple taxonomic revisions. It is a holoarctic genus that occupies habitats ranging from the arctic tundra to the shaded cliffs of the mountains of southern Europe (Wightman et al. 2017).

Thus, Galløe (1910) described three different types of hydathodes:

1. Marginal hydathodes, actually located at the edge of the leaf, without secretion of calcium carbonate, found in species of the "sections" of *Boraphila* (*Saxifraga nivalis*, Fig. 208; *S. hieraciifolia* and *S. stellaris*, Fig. 209) and *Nephrophyllum* (*Saxifraga cernua* and *S. rivularis*, Fig. 210)
2. Hydathodes located on the upper face of the leaf margin, without cavity, without calcium carbonate secretion, found in species from the *Dactyloides* "sections" (*Saxifraga hypnoides*, Fig. 211; *S. groenlandica*, Fig. 212) and *Trachyphyllum* (*Saxifraga aizoides*; Fig. 213; *S. flagellaris*)
3. Hydathodes located on the upper face of the leaf margin, cavity present, secretion of calcium carbonate, found in species of the "sections" *Euaizofiia* (*Saxifraga cotyledon* and *S. aizoon*, Fig. 214) and *Porphyrion* (*Saxifraga oppositifolia*, Fig. 215)

The family Ranunculaceae contains several genera with psychrophilic species, many of which were investigated by Jessen (1911). Thus, regarding the ecological anatomy of the leaf of some arctic ranunculaceous species, Jessen described two large groups of leaves: submerged leaves and aerial leaves (see the morphology of some aquatic leaves—Grigore 2022, as well as the anatomy of the leaf in aquatic plants—in present work).

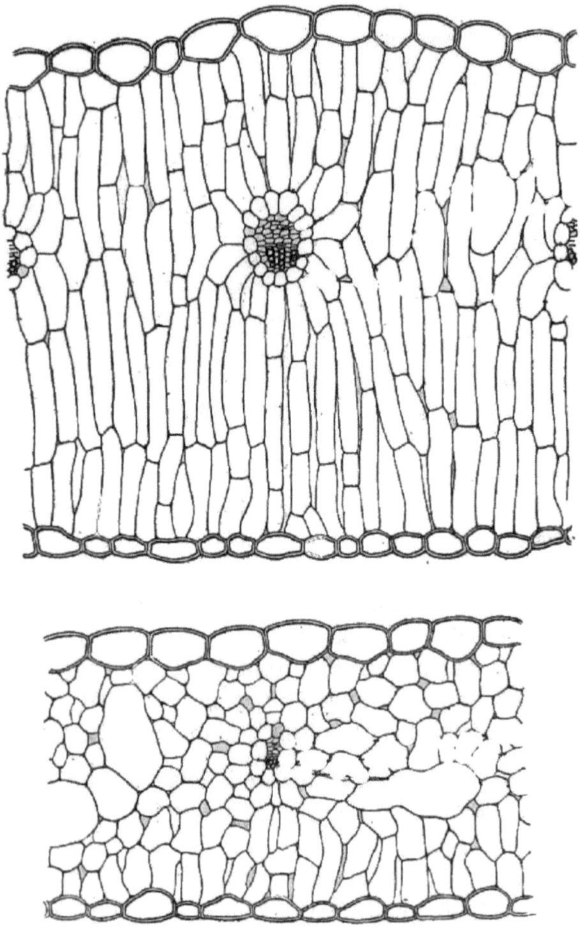

Fig. 176 *Lotus corniculatus*, alpine (top) and lowland (bottom) form: cross section through the leaf (Bonnier 1894b); in the alpine form, the leaf blade has become completely of palisade type

A. Submerged leaves typical of *Batrachium confervoides* and *Ranunculus reptans* are cylindrical and more or less radial; the *Batrachium* leaf has one leading bundle in each lobe, and *R. reptans* invariably has three vascular bundles in each leaf. When these plants grow in dry habitats, the leaves of both species react in a similar way and approach the bifacial type that characterizes the aerial leaves of *R. reptans*. The sunken leaves of *Batrachium* have no stomata, while those of *Ranunculus* can present few stomata (review the general characteristics of the leaves of aquatic plants—in present work); in bifacial leaves, stomata are more numerous on the upper surface.
B. Aerial leaves. All species of *Ranunculus*, except *R. reptans* and *R. glacialis*, can be included, as regards the structure of their leaves, in the same sub-group. The leaves are glabrous or slightly hairy (except *R. acer*), the epidermis has thin-walled cells, and the palisade tissue and spongy parenchyma are homogeneously developed in all species (Figs. 216, 217, 218, 219, 220, 221, 222, 223, 224, and 225); the palisade tissue consists of one to two layers of cells that make up about half the thickness of the mesophyll, and the palisade cells are barrel-shaped,

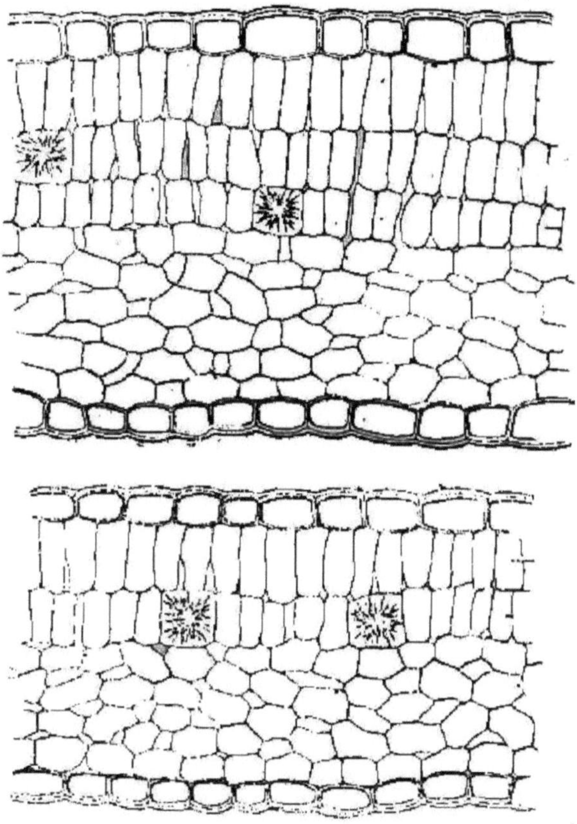

Fig. 177 *Potentilla tormentilla*, alpine (top) and lowland (bottom) form: cross section through the leaf (Bonnier 1894b); in the alpine form, the palisade tissue is more developed (tri-layered) and the cuticle is thicker

somewhat irregularly arranged; the spongy tissue contains numerous intercellular spaces and the cells are star-shaped branches.

At *Ranunculus hyperboreus*, the stomata are more numerous on the upper side, and in *R. pallasii*, their number is equal on both sides, but in other species, they are more numerous on the lower side. In arctic regions, the habitat of these species is among moist mosses. *Ranunculus glacialis* differs from the species mentioned above by the more developed palisade tissue and the looser structure of the spongy tissue. Stomata are more numerous on the upper surface. It grows in localities that are poorer in humus than those where the other *Ranunculus* species grow.

In Greenland, *Anemone richardsoni* (Fig. 220) is considered especially a shade plant; by the structure of the leaf, it is very similar to *Anemone nemorosa*. The stomata are located almost exclusively on the underside of the leaf.

Of the species that were examined by Jessen (1911), *Thalictrum alpinum* (Fig. 221) and *Coptis trifolia* (Fig. 222) are considered to be the most xerophilous. The leaves are coriaceous and the epidermis is thick, especially in *Thalictrum*. The palisade cells are arranged in two to three compact layers; spongy tissue consists of

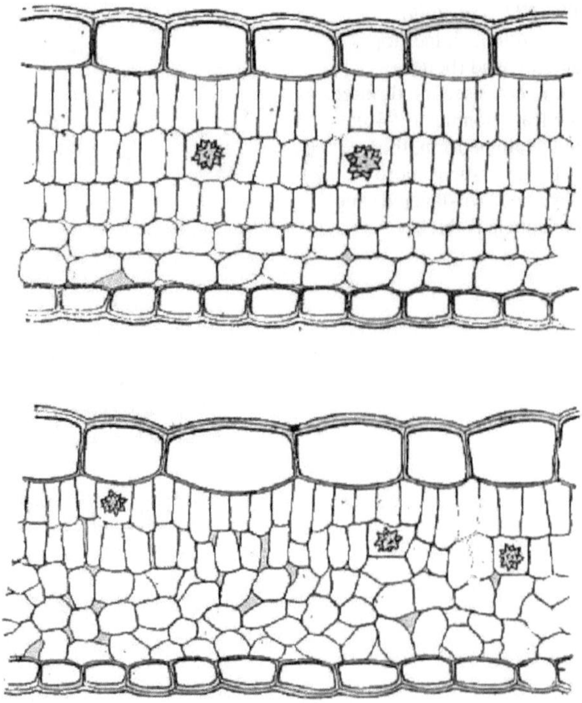

Fig. 178 *Fragaria vesca*, alpine (top) and lowland (bottom) form: cross section through the leaf (Bonnier 1894b); in the alpine form, the palisade tissue is more developed (tri-layered) and the lamina is thicker

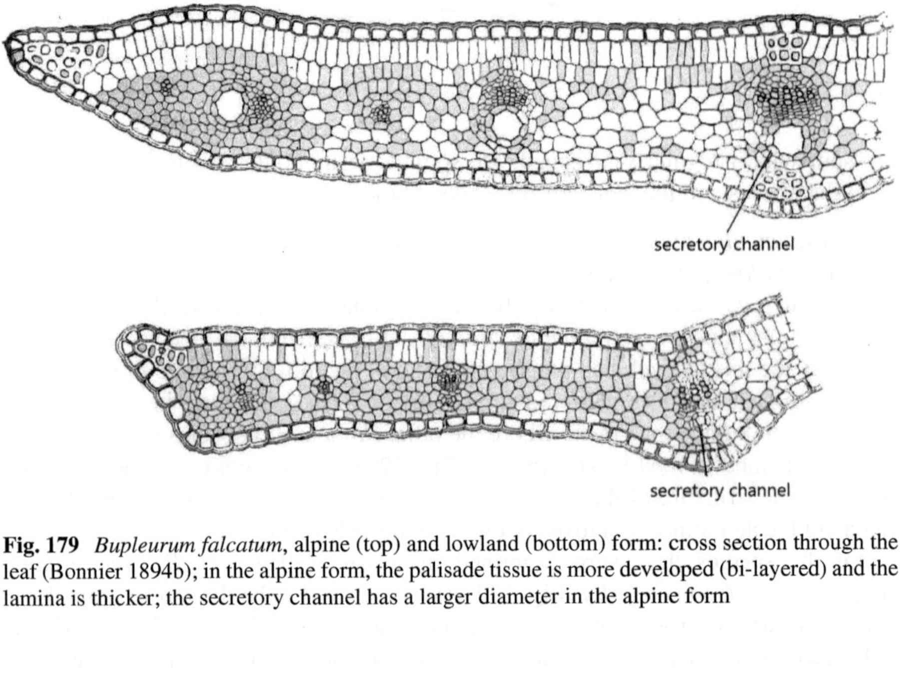

Fig. 179 *Bupleurum falcatum*, alpine (top) and lowland (bottom) form: cross section through the leaf (Bonnier 1894b); in the alpine form, the palisade tissue is more developed (bi-layered) and the lamina is thicker; the secretory channel has a larger diameter in the alpine form

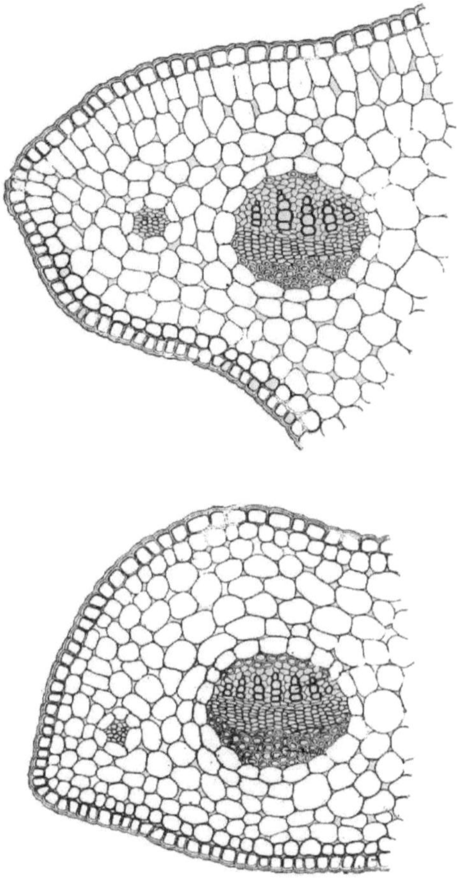

Fig. 180 *Achillea millefolium*, alpine (top) and lowland (bottom) form: cross section through the leaf (Bonnier 1894b); in the alpine form, the palisade tissue is individualized (one- to two-layered), in the lowland form, not; the hypodermis has cells with thicker walls

short, branched cells. In both species, stomata are almost absent on the upper surface, but numerous on the lower surface. *Thalictrum alpinum* grows in very different localities. The fact that the leaves remain green during the winter may suggest that the species *Coptis trifolia* is xerophilous (Jessen 1911).

A particular characteristic, common to all arctic species of the Ranunculaceae family, is the presence of chloroplasts in the epidermal cells, a situation that has been already mentioned for the *Pirola* species (see Figs. 205 and 206). This characteristic is also found in temperate regions in plants that grow in low light, for example, in many aquatic plants (as we have already discussed) and in forests. It has been reported that many northern European plants (full list—Table 1—Lidforss 1907) have chloroplasts in the epidermis, especially that on the lower surface of the leaf; the epidermis is often separated from the mesophyll by large lacunae and then, due to the presence of chloroplasts, it will be more independent in carrying out the photosynthetic function.

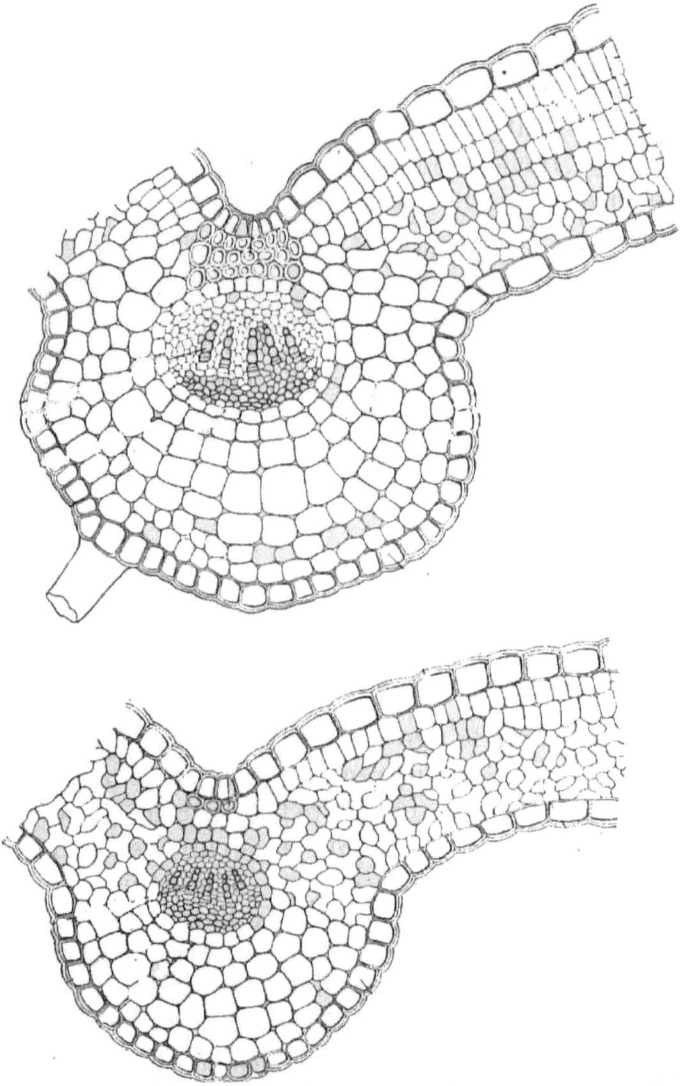

Fig. 181 *Brunella vulgaris*, alpine (top) and lowland (bottom) form: cross section through the leaf (Bonnier 1894b); in the alpine form, the palisade tissue is more developed (three- to four-layered) than in the plain form; also here, the mechanical tissue is more developed, on the upper face, hypodermically, next to the leading bundle

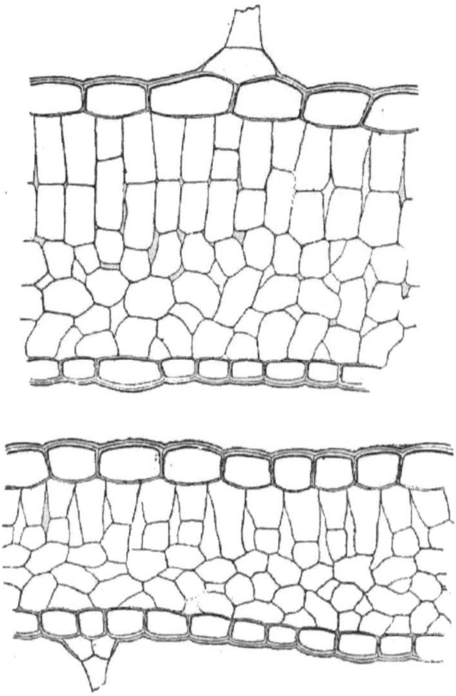

Fig. 182 *Galeopsis tetrahit*, alpine (top) and lowland (bottom) form: cross section through the leaf (Bonnier 1894b); in the alpine form, the palisade tissue is more developed (bi-layered) than in the plain form; the leaf lamina is thicker

Returning to the species of Ranunculaceae, of the 14 species that were investigated (Jessen 1911), the stomata, in most cases, were more numerous on the lower side of the leaf, in the species *R. affinis*, *R. acer*, *R. sulphureus*, *R. nivalis*, *R. pygmaeus*, *R. lapponicus*, *Anemone richardsoni*, *Thalictrum alpinum*, and *Coptis trifolia*. Of these, it seems that in *R. lapponicus*, *Thalictrum alpinum*, and *Coptis trifolia*, *the* stomata are usually absent on the upper side of the lamina; finally, *R. pallasii* has an almost equal number of stomata on both sides of the leaf, a fact that may be related to the position of the leaf.

The following species have a greater number of stomata on the upper side of the leaf: *R. glacialis*, *R. hyperboreus* (its leaves are sometimes floating leaves), *R. reptans*, and probably *Batrachium confervoides*, the latter two in their aerial leaves.

All species of *Ranunculus*, *Batrachium confervoides*, and *Anemone richardsoni* show hydathodes (with epithem) at the apex of the leaves. Typically, hydathode pores occur marginally on a usually upward-facing surface, and in all these species, the epithem is essentially similar in structure; the tracheids spread and terminate at the level of the intercellular spaces, and the mesophyll cells are shortly branched and have strongly undulating walls.

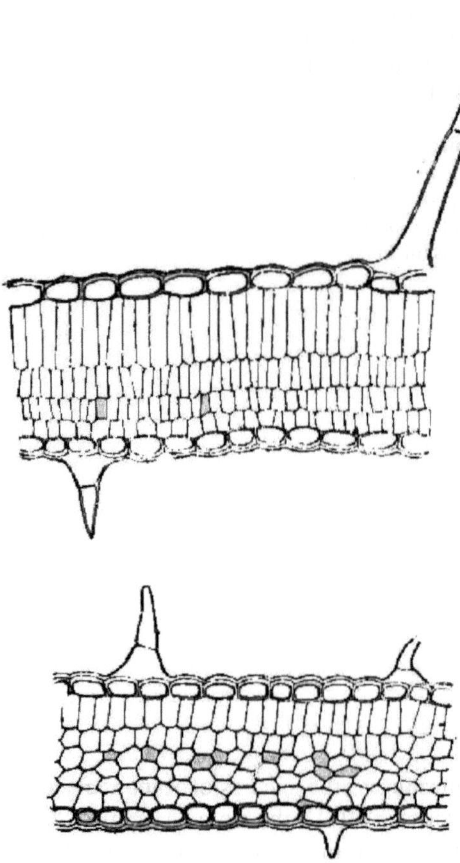

Fig. 183 *Teucrium scorodonia*, alpine (top) and lowland (bottom) form: cross section through the leaf (Bonnier 1894b); in the alpine form, the palisade tissue is more developed (occupying practically the entire thickness of the lamina) than in the plain form; the leaf lamina is thicker

9 Ecological Anatomy of the Leaf in Tropical Alpine Plants

In the previous sub-chapter, we treated the ecological anatomy of the leaf in psychrophytes (arctic, alpine plants) in a general way, taking into account the adaptations of arctic plants that grow in areas under the influence of polar climates, as well as of plants generically called alpine ("mountainous")—if we use the term only to refer to high altitude –, referring above rather to those that grow in habitats under the influence of the continental climate group. We recall that discussions related to climates, vegetation zones, biomes, ecosystems, and habitats—and other concepts associated with plants and vegetation in general—are very complicated and raise numerous problems of interpretation and homogenization of language (Polunin 1960; Riley and Young 1966; Seddon 1971; Savile 1972; Billings 1978; Tivy 1993; Rundel et al. 1994; Archibold 1995; Brown and Lomolino 1998; Collinson 1988; Gibson and Gibson 2006; Lütz 2012; Lambers and Oliveira 2019; Schulze et al. 2019; Pfadenhauer and Klötzli 2020; Körner 2022), which we do not want to deal with and debate in the present paper.

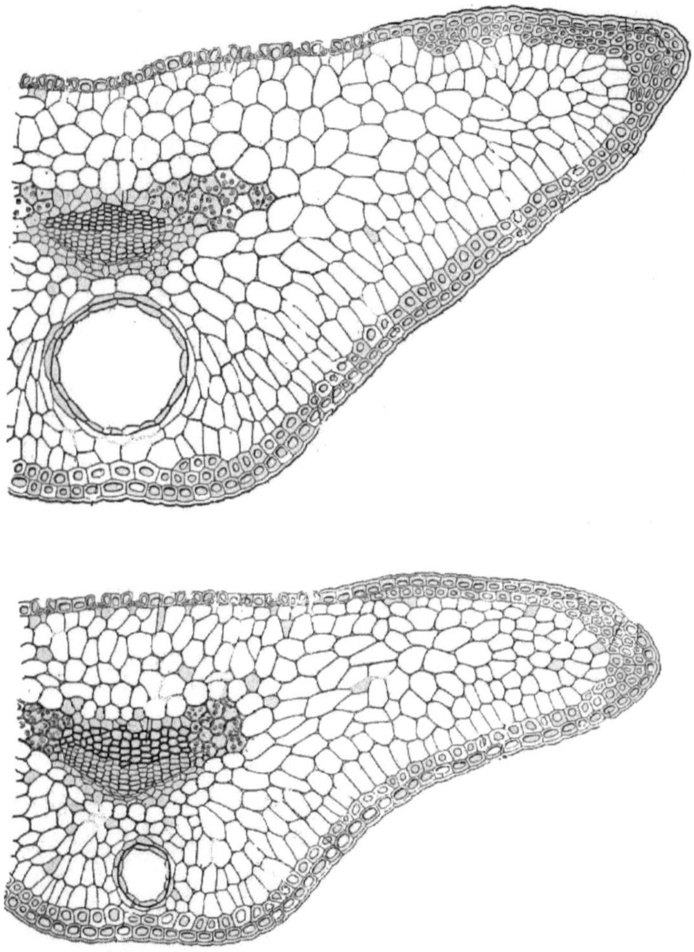

Fig. 184 *Juniperus communis*, alpine (top) and lowland (bottom) form: cross section through the leaf (Bonnier 1894b); in the alpine form, the palisade tissue on the lower face is more developed than in the plain form; the leaf lamina is thicker; stomata on the upper face are more numerous; secretory channels or larger diameter

In addition to the alpine plants whose anatomical adaptations we have previously discussed, this sub-chapter is dedicated to an ecological sub-group of plants, that of tropical alpine plants, i.e., plants that grow at very high altitudes, in mountainous areas under the influence of the tropical climatic group.

The term "alpine tropical" is very versatile; this usually refers to the regions in the tropics that occur between the upper limit of continuous, deciduous forest (often around 3500–3900 m) and the upper limit that supports plant life (often around 4600–4900 m) (Rundel et al. 1994; Smith 1994) (Fig. 226); the term is used generically, but may also include local, regional terms, such as that of "páramo" (Fig. 227)

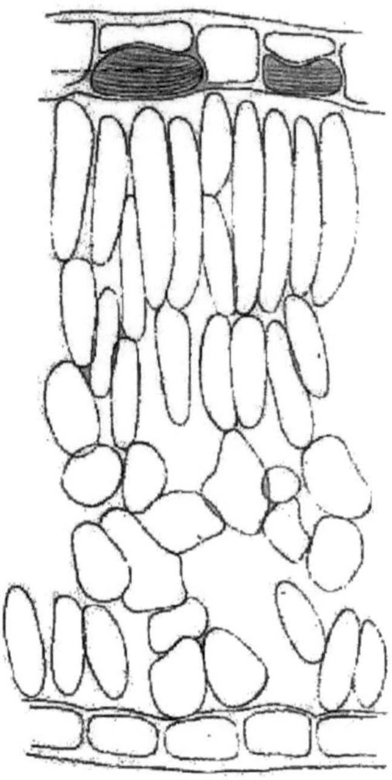

Fig. 185 *Daphne striata*; vegetating on 2200 m—cross section through the leaf (Wagner 1892); the hatched cells from the epidermis indicate mucilage cells

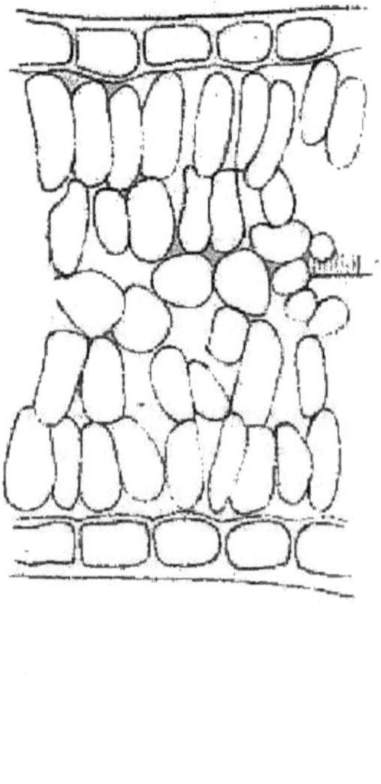

Fig. 186 *Erigeron uniflorus*; vegetating on 2000 m—cross section through the leaf (Wagner 1892)

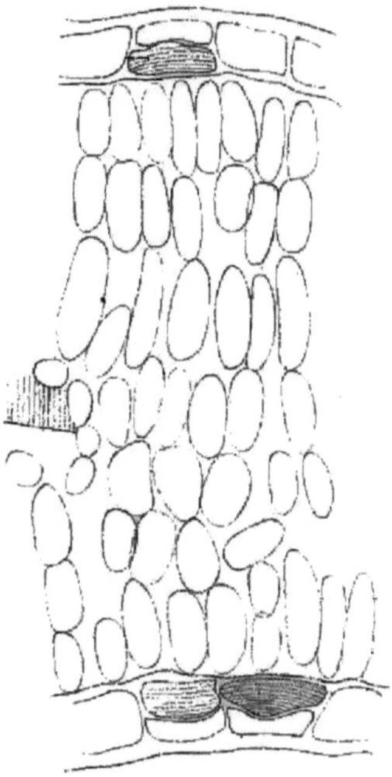

Fig. 187 *Helianthemum alpestre*; cross section through the leaf (Wagner 1892); the hatched cells from the epidermis indicate mucilage cells

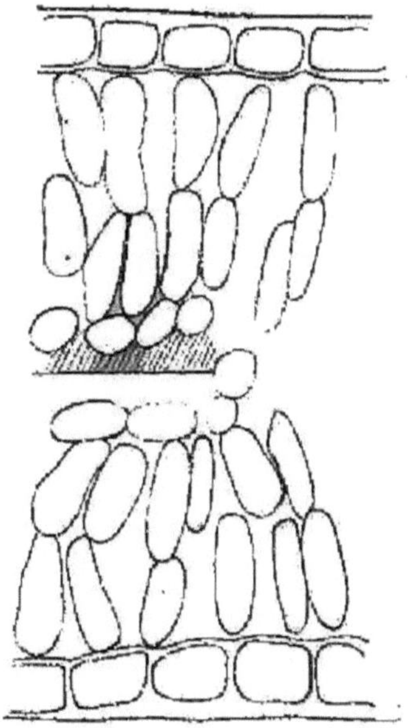

Fig. 188 *Oxytropis halleri*; vegetating on 2200 m cross section through the leaf (Wagner 1892); the hatched cells from the epidermis indicate mucilage cells

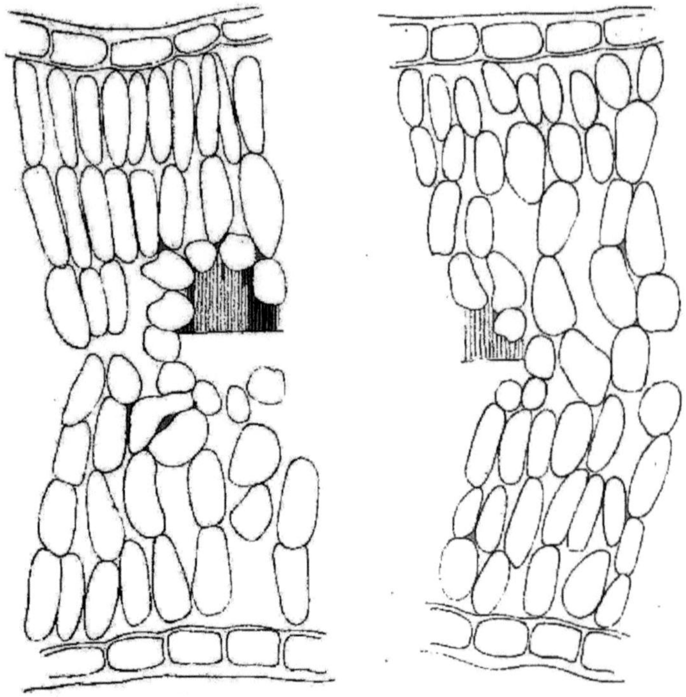

Fig. 189 *Armeria alpina*; left—vegetating on 2200 m and cultivated on botanical garden (right) cross section through the leaf (Wagner 1892); in the alpine form, the palisade tissue is more developed (occupying practically the entire thickness of the lamina)

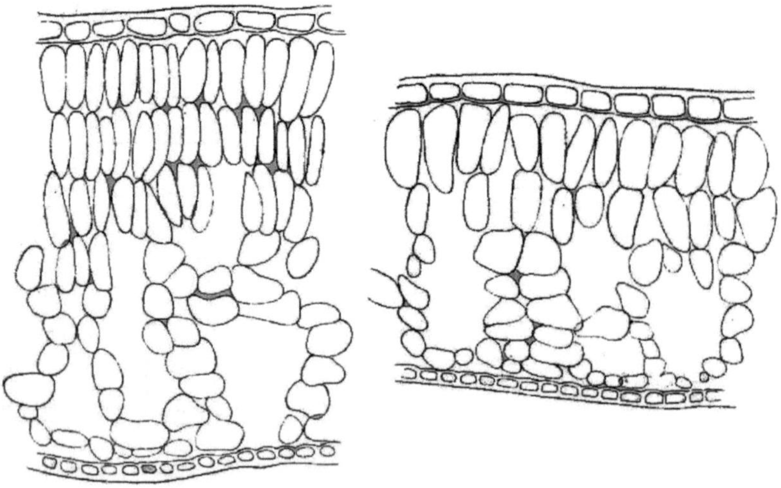

Fig. 190 *Homogyne alpina*; left—vegetating on 2200 m and on 1500 m (right); cross section through the leaf (Wagner 1892); in the form from higher altitude, the palisade tissue is more developed

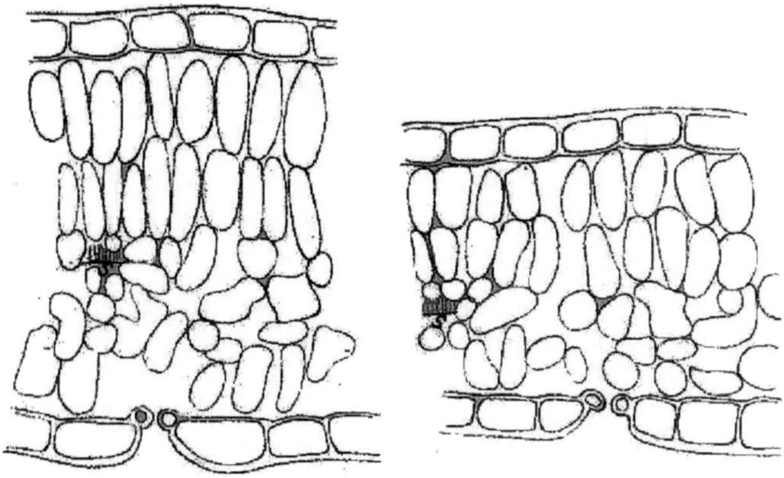

Fig. 191 *Papaver pyrenaicum*; left—alpine form and cultivated in botanical garden (right); cross section through the leaf (Wagner 1892); in the alpine form, the palisade tissue is more developed

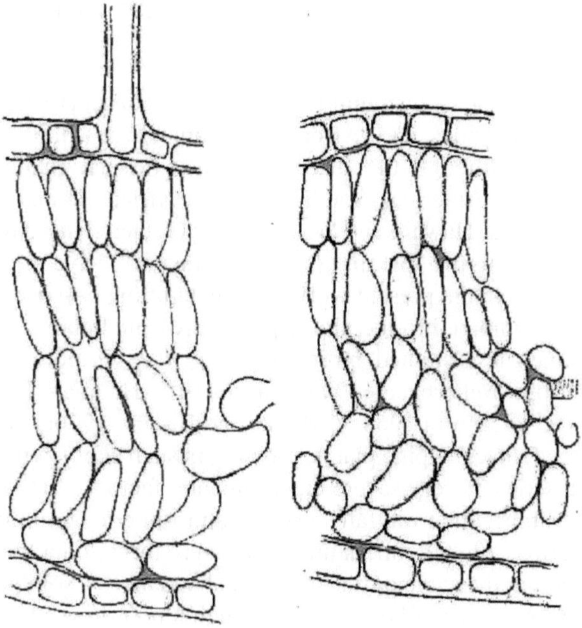

Fig. 192 *Potentilla nitida*; left—alpine form from 2500 m and cultivated in botanical garden (right); cross section through the leaf (Wagner 1892); in the alpine form, the palisade tissue is more developed

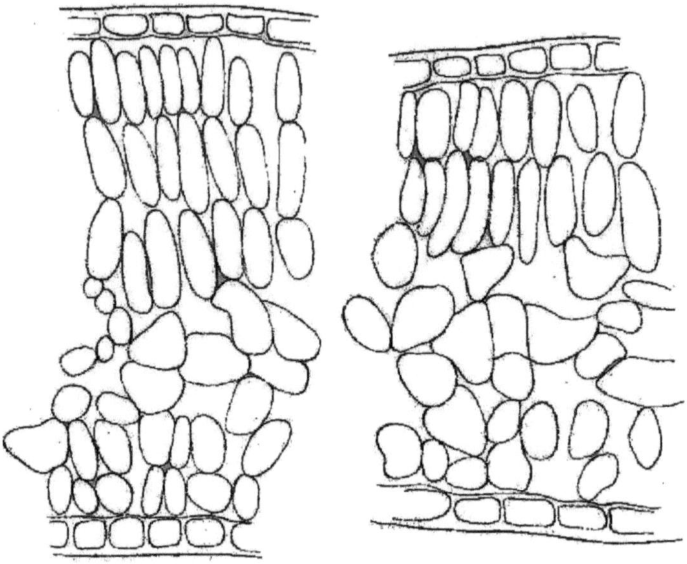

Fig. 193 *Senecio incanus*; left—alpine form from 2500 m and cultivated in botanical garden (right); cross section through the leaf (Wagner 1892); in the alpine form, the palisade tissue is more developed

and “jalca” in the humid Andes from Venezuela to northern Peru, “puna” in the drier central Andes, and “Afroalpine” from Africa (Smith 1994). As a geographical location—although the discussion is somewhat relative—these habitats appear in Central America, South America, Africa, Papua New Guinea, Borneo, Java, Sumatra, Hawaii, and Western New Guinea (Irian Jaya, West Papua).

The terms “páramo (s)” and “puna (s)” have already been introduced and presented in the present work (see, in addition, Grigore 2022).

Compared to the alpine plants presented above, tropical alpine ones have been much less studied morphologically (see Grigore 2022) and anatomically (Fries 1905; Weberbauer 1905; Hauri 1917; Hauman 1918; Espinosa 1932; Carlquist 1957; Cabrera 1968; Carlquist and Bissing 1976; Ancibor 1980).

Tropical alpine plants, growing in such ecosystems (e.g., punas), have to adapt to a series of very severe environmental factors, such as water deficit, in most of the year, irregular rainfall, very low atmospheric humidity, strong solar radiation (insolation), night temperatures below zero throughout the year, large temperature variation in the same day, salty soils, abundance of soluble salts in the soil solution, and the presence of herbivores in a region poor in vegetation anyway.

Therefore, it is obvious that these environmental factors will induce morphological (Grigore 2022) and anatomical adaptations of xeromorphosis. Thus, the adaptive characteristics must undoubtedly be related to water supply, its conservation (storage) by accumulation (in the water storage tissue) or to the reduction of transpiration and protection against cold.

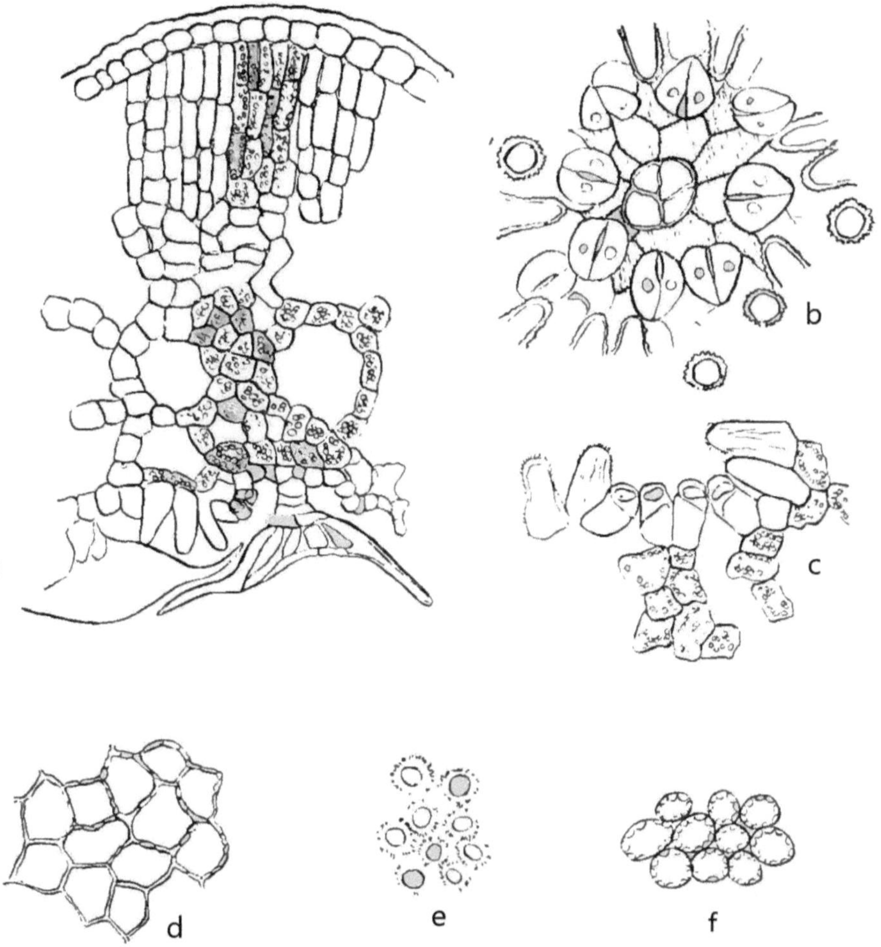

Fig. 194 *Rhododendron lapponicum*: cross section through the leaf (**a**); (**b**) epidermis, with stomata and a peltate hair, front view; (**c**) stomata in cross section; (**d**) upper epidermis, front view; (**e**) lower epidermis, front view; (**f**) palisade cells front view (Petersen 1908a); the development of palisade tissue and thick cuticle at the level of the upper epidermis is observed

Among these morpho-anatomical features at the level of the leaf, only few of them should be mentioned.

Thus, the leaves can be reduced and caducous, without special structures, as in *Fabiana denudata*. Small and leathery leaves (coriaceous—case of sclerophily—i.e., thick, hard, thickened leaves; see Grigore 2022) common in shrub species: *Psila boliviensis*, *Baccharis incarum*, *Fabiana densa*, *Junellia seriphioides*, and so on. The leaves can be fleshy, provided with water-storage tissues, as in *Senecio viridis*, *Werneria poposa*, *Portulaca perennis*, *Portulaca rotundifolia*, *Calycera crenata*, *Echeveria peruviana*, etc.

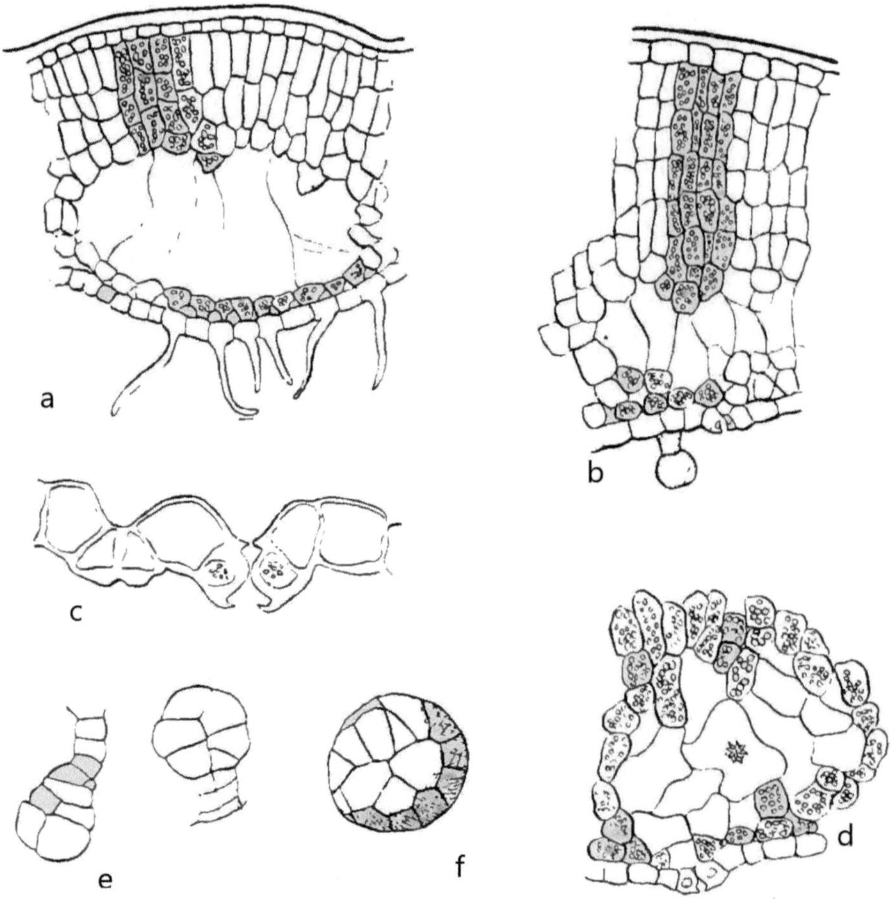

Fig. 195 *Ledum palustre* var. *decumbens*: cross section through the leaf (**a, b**); (**c**) stomata in cross section; (**d**) spongy tissue; (**e, f**) glandular hairs, seen from the side and above (Petersen 1908a); the development of palisade tissue and thick cuticle at the level of the upper epidermis is observed

The leaves may have wavy edges to protect the upper side (in grasses)—or the lower side—from excessive transpiration: *Chiliotrichiopsis kaidelii*, *Baccharis polifolia*, *Baccharis grisebachii*, etc. The wavy leaf forms a large crypt (depression) on each side of the vein where humidity is higher; these depressions are usually protected with protective bristles. Leaves can also be spiniform and rigid, thus reducing transpiration and serving as a defense against herbivores: *Chuquiraga atacamensis*, *Chuquiraga acanthophylla*, *Nassauvia axillaris*, and *Mulinum* sp. In *Tetraglochin cristatum*, the rachis of the compound leaves becomes spiniform; such spiniform leaves are also found in species of Cactaceae.

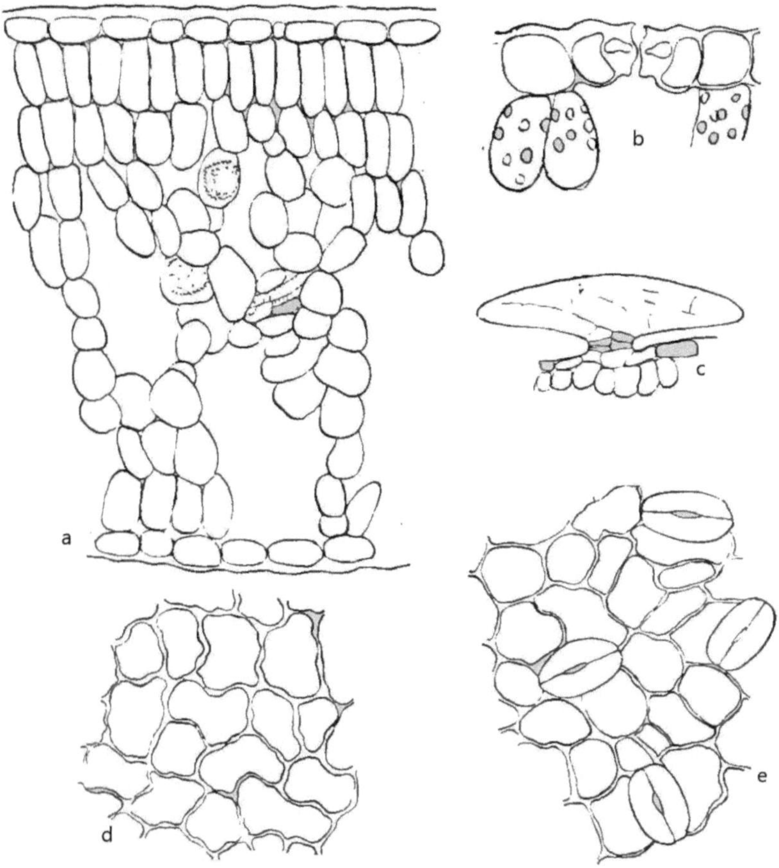

Fig. 196 *Lyonia calyculata*: cross section through the leaf (**a**); (**b**) stomata; (**c**) peltate hair, lower epidermis; (**d**) upper epidermis, front view; (**e**) lower epidermis, front view (Petersen 1908a); the development of the palisade tissue and thick cuticle at the level of the two epidermises is observed

In some species, the leaves are scaly, dense, and appressed (pressed) to the stem; these are often covered with hairs that give the organ a woolly appearance, so that a chamber is formed between the leaf and the stem where the humidity is kept high, and therefore temperature changes are less sudden. This type of leaf is found in *Parastrephia lepidophylla*, *Parastrephia quadrangularis*, *Parastrephia teretiuscula*, *Anthobryum triandrum*, *Pycnophyllum* sp., *Azorella compacta*, and other species. Many species here form cushions (Grigore 2022) and have very reduced leaves, but they are not always attached to the stem.

From an anatomical point of view, the most important anatomical characters of the leaf are the following: very thick cuticle; resin or wax deposits on the epidermis; the outer walls of the epidermal cells are very thick; and hairs developed (giving the leaf a woolly appearance) on the underside of the leaf, as in *Trichocline auriculata*,

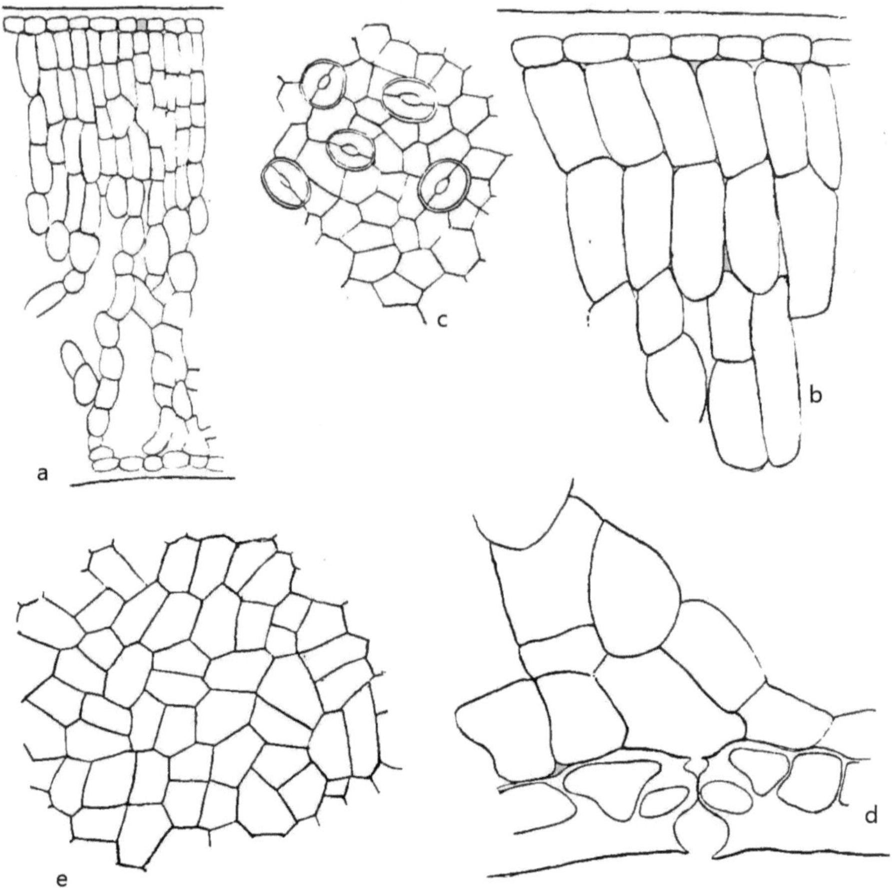

Fig. 197 *Arctostaphylos uva-ursi*: cross section through the leaf (**a**, **b**); (**c**) lower epidermis, front view; (**d**) stomata; (**e**) upper epidermis, front view (Petersen 1908a); the development of the palisade tissue and thick cuticle at the level of the two epidermises is observed

Mutisia ledifolia, *Baccharis polifolia*, or on both sides, as in *Gomphrena meyeniana*, *Senecio spegazzini*, etc. As we have seen repeatedly, this coat of woolly hairs is believed to provide protection from intense solar radiation while also reducing foliar transpiration; at the same time, it is considered that such a coating can also provide protection against sudden changes in temperature.

In other species, the protecting hairs are closely attached to the lower epidermis or both epidermises, as in *Lupinus paniculatus* and various species of *Astragalus*, *Dichondra serum*, etc. They protect the leaves from intense solar radiation and reduce transpiration (see the general part about these trichomes—in present work—sub-chapter dedicated to ecological anatomy of the leaf in xerophytes).

There are also resin-secreting glandular hairs in numerous species growing in puna-type ecosystems, such as *Fabiana densa* (Fig. 228a and A'), *Baccharis*

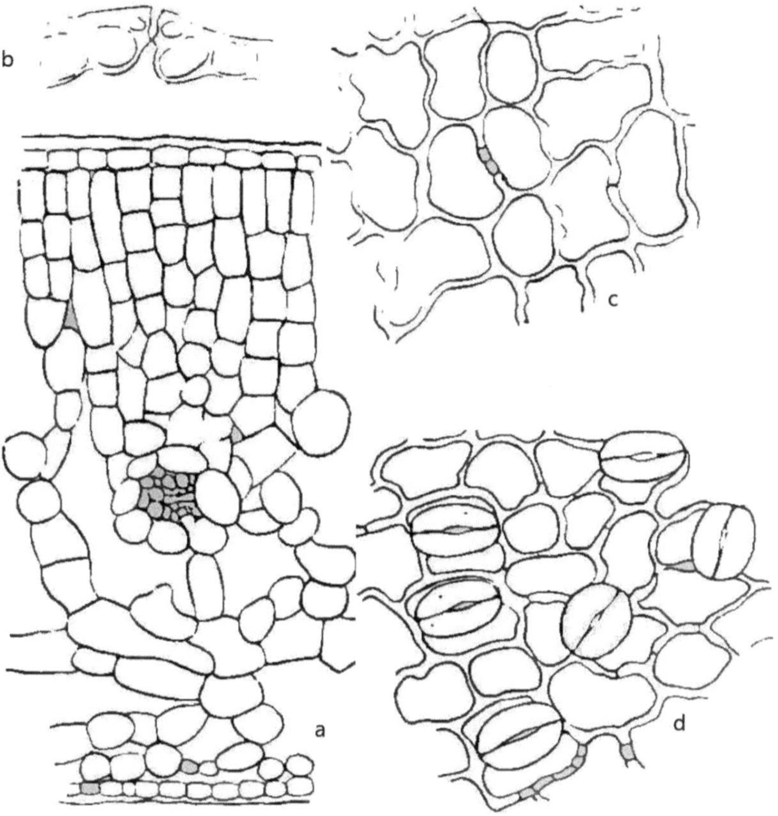

Fig. 198 *Vaccinium vitis-idaea*: cross section through the leaf (**a**); (**b**) stomata; (**c**) upper epidermis, front view; (**d**) lower epidermis, front view (Petersen 1908a); the development of the palisade tissue and thick cuticle at the level of the two epidermises is observed

incarum, and *Calceolaria santolinoides* (Fig. 228c); it is assumed that they too can protect the epidermis against excessive transpiration (Cabrera 1968).

The stomata may be provided with external articles, as in *Fabiana densa* (Fig. 228a) and *Baccharis incarum* or can be located in crypts formed by the revolute edge of the leaf: *Chiliotrichiopsis kaidelii*, *Baccharis polifolia*, etc.

The palisade tissue can be bi- or tri-layered, a common situation in many species; as a rule, it is located under both epidermises. In some species, the palisade tissue is found located only on the abaxial face, as in *Parastrephia lepidophylla* and *Werneria poposa*, due to the position of the leaf, which is pressed ("stuck") to the stem. The spongy tissue is compact. The development of chlorenchyma (palisade tissue) is however a common anatomical feature in tropical alpine plants (from "puna" and "paramos") (Figs. 229 and 231) and in alpine plants in general.

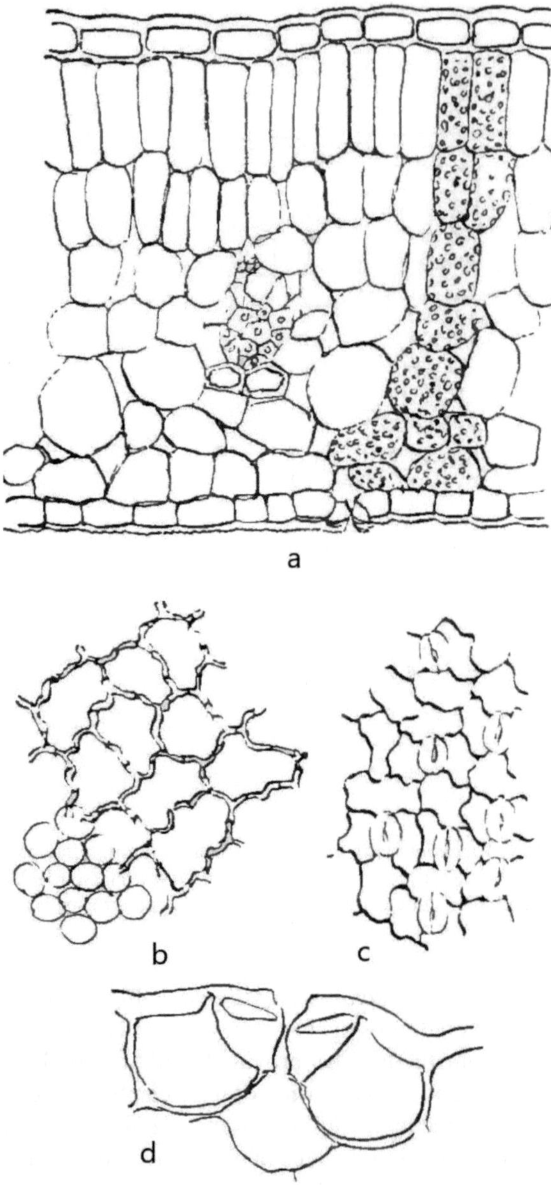

Fig. 199 *Vaccinium oxyoccos*: cross section through the leaf (**a**); (**b**) upper epidermis, front view; (**c**) lower epidermis, front view; (**d**) stomata (Petersen 1908a); the development of the palisade tissue and thick cuticle at the level of the two epidermises is observed

The important development of the sclerenchyma is also noticeable (Fig. 230), as in many grasses (the situation is also valid for other species): *Festuca orthophylla* and *Deyeuxia cabrerae*, but also at *Chuquiraga atacamensis* (family Asteraceae).

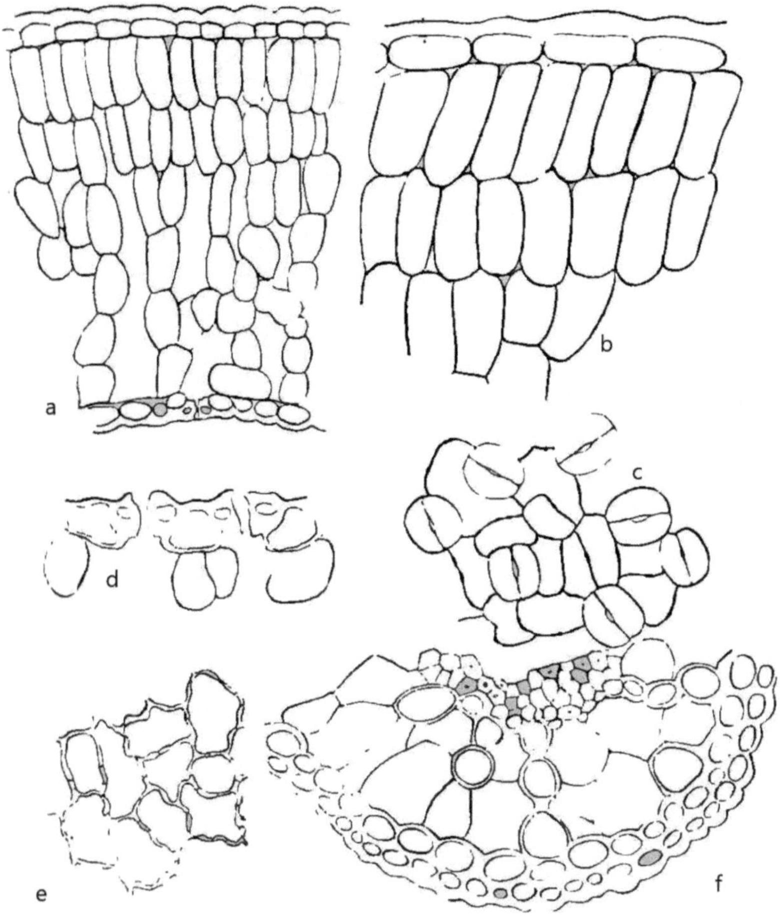

Fig. 200 *Andromeda polifolia*: cross section through the leaf (**a**, **b**); (**c**) lower epidermis, front view; (**d**) stomata; (**e**) upper epidermis, front view; (**f**) leaf fragment—water-storage tissue (Petersen 1908a); the development of the palisade tissue and thick cuticle at the level of the two epidermises is observed

As we have already pointed out for other sub-groups of xerophytes, the leaves can be succulent, due to the development of the water-storage tissue, as in *Calycera crenata* and *Senecio amphibolus*.

Of course, in addition to general considerations regarding the ecological anatomy of the leaf of "puna" plants, there are many variations and nuances regarding some genera or species of plants (Carlquist 1994—and references cited), such as those of *Espeletia*, *Lobelia*, *Senecio*, and *Geranium*.

To mention just one example (Carlquist and Bissing 1976), an anatomical study of alpine species of *Geranium* from the Hawaiian Islands reflects the nature of the xeromorphous adaptations of the species in relation to the nature of environmental

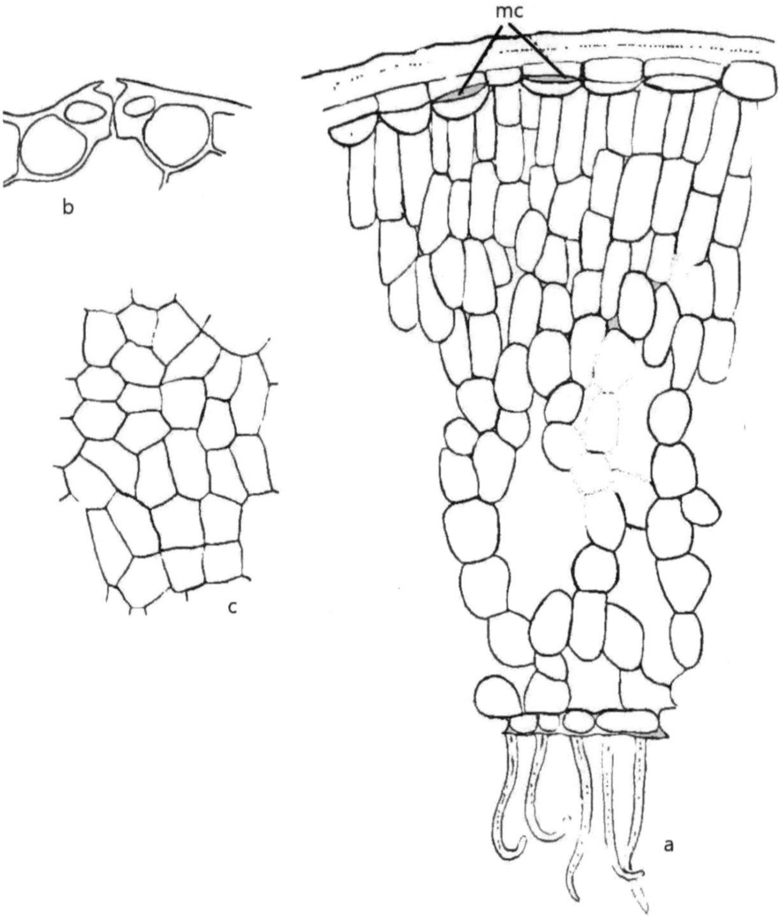

Fig. 201 *Loiseleuria procumbens*: cross section through leaf (**a**); (**b**) stomata; (**c**) upper epidermis, front view; mc, mucilaginous cells (Petersen 1908a); palisade tissue development is observed, thick cuticle at the level of the upper epidermis, stomata (not shown here) are located exclusively at the level of the lower epidermis, pronounced pilosity

factors (discussed, in the general way, in the introductory part of this sub-chapter, with some particularities for these islands: strong insolation with intense UV radiation, radiation coming from the volcanic lava or from the sparks coming off from it).

Thus, among these xeromorphic adaptations can be mentioned: the reduction of the size of the leaf; the tendency of the leaves towards an isolateral structure (to the detriment of the bifacial one); the thick palisade tissue, formed by several rows of cells; a high number of glandular hairs that form a covering, on one face or on both sides of the epidermis; and the development of a hypodermis on one or both sides of the epidermis.

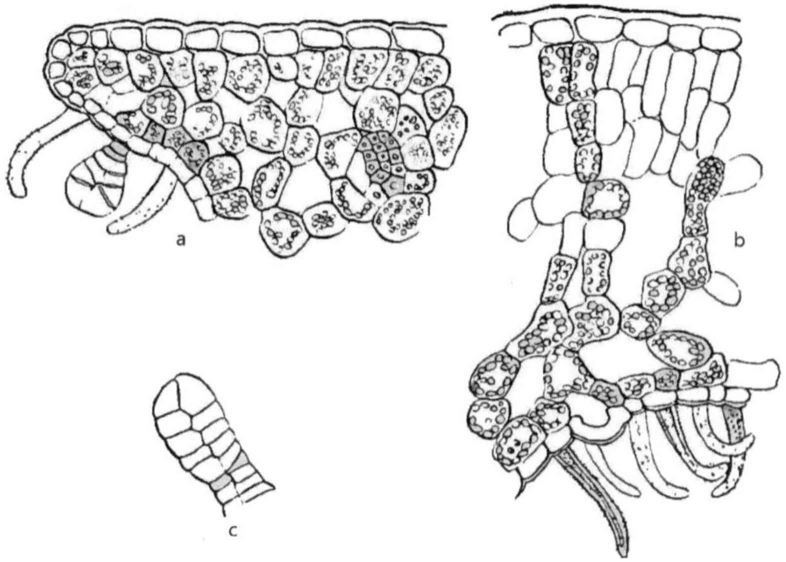

Fig. 202 *Phyllodoce caerulea*: cross section through the leaf (**a**, **b**); (**c**) multicellular hair (Petersen 1908a); thick cuticle is observed at the level of the upper epidermis, pronounced hairiness

In *Geranium cuneatum cuneatum*, the leaves are approximately isolateral (so with palisade tissue under both epidermises), the stomata are twice as numerous on the abaxial face as on the adaxial one, and the cells located in the center of the mesophyll resemble the hypodermic ones.

The ecological anatomy of the leaf from *Geranium cuneatum tridens* reveals the following xeromorphosis adaptations: isolateral leaves, with two rows of palisade cells under both epidermises; a well-defined layer of hypodermal cells on the abaxial side, while on the adaxial side, the subepidermal cells are typically palisade or transitional to the hypodermis, in that their walls are thick and devoid of chloroplasts; two to three layers of cells, similar to the hypodermic ones, located in the central mesophyll; stomata only slightly more numerous at the level of the abaxial epidermis; and profuse protective hairs on both sides of the leaf.

In *Geranium cuneatum hypoleucum*, leaf anatomical features, xeromorphic in nature, include complete isolateral structure, with approximately two rows of palisade cells on the abaxial face and three layers of palisade cells on the adaxial face; a row (sometimes two) of large hypodermal cells below the adaxial epidermis; a layer of small hypodermic cells (some have the form of spongy tissue cells) at the level of the abaxial epidermis; stomata present only on the abaxial face, which corresponds to the absence of a covering of glandular hairs; the presence of protecting hairs; hydathodes of medium size; and the cuticle covering these trichomes that is thick.

Fig. 203 *Cassiope hypnoides*: cross section through the leaf (**a**, **b**); (**c**) stomata; (**d**) epidermis with stomata (Petersen 1908a); thick cuticle is observed at the level of the epidermis

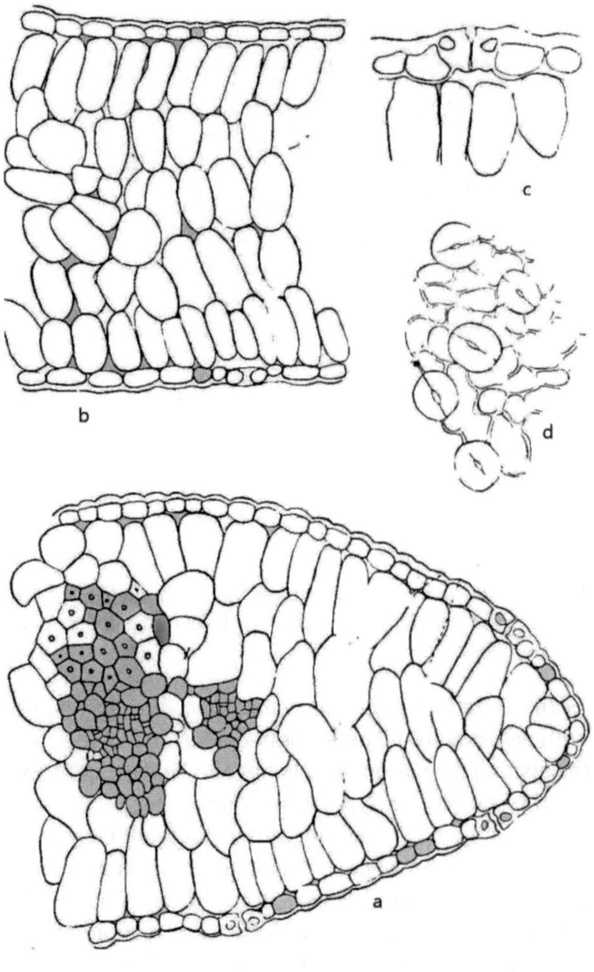

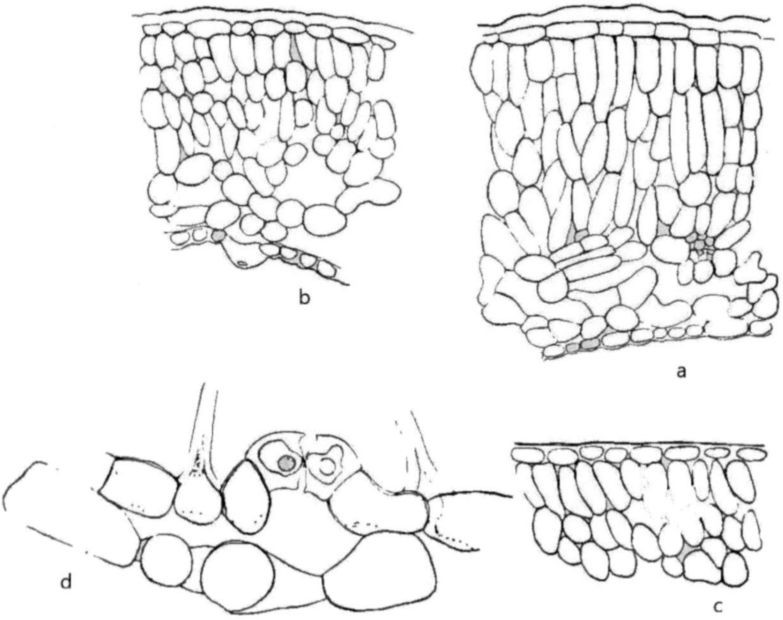

Fig. 204 *Cassiope tetragona*: cross-section through the leaf, different sectors (**a**–**c**); (**d**) stomata (Petersen 1908a); thick cuticle is observed at the level of the epidermis

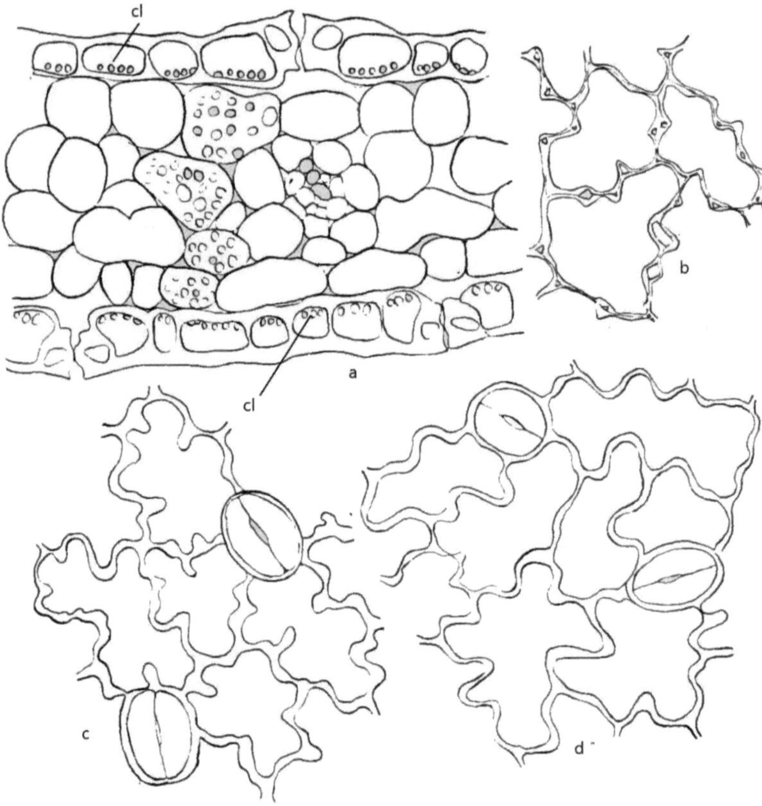

Fig. 205 *Pyrola secunda*: cross section through the leaf (**a**); (**b**) cells of the mesophyll, front view; (**c**) lower epidermis, front view; (**d**) upper epidermis, front view (Petersen 1908a); thick cuticle at the level of the epidermis and the presence of chloroplasts (cl) in the epidermal cells are observed

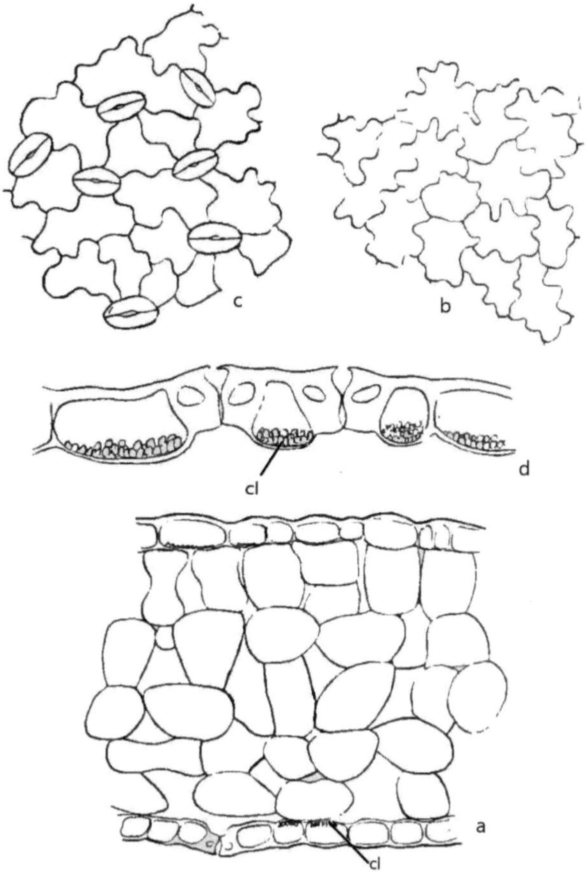

Fig. 206 *Pyrola uniflora*: cross section through the leaf (**a**); (**b**) upper epidermis, front view; (**c**) lower epidermis, front view; (**d**) epidermis detail, with stomata and chloroplasts (cl) in epidermal cells (Petersen 1908a); thick cuticle at the level of the epidermis and the presence of chloroplasts (cl) in the epidermal cells are observed

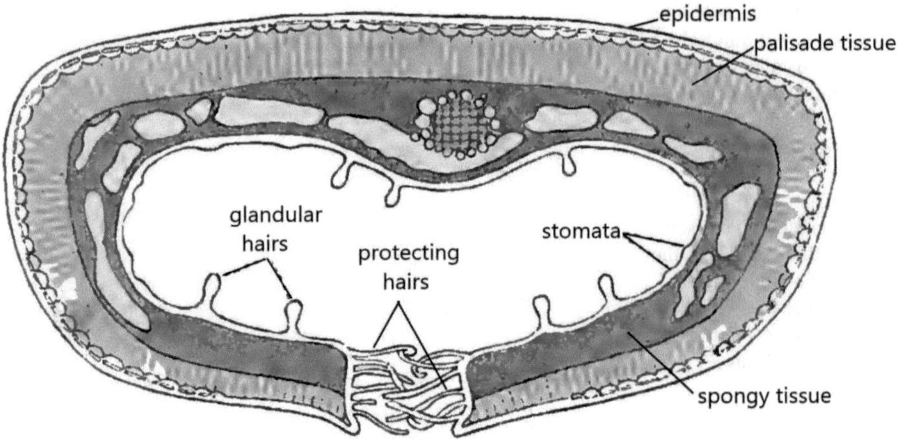

Fig. 207 *Empetrum nigrum*: cross-section through the leaf (Mentz 1909)

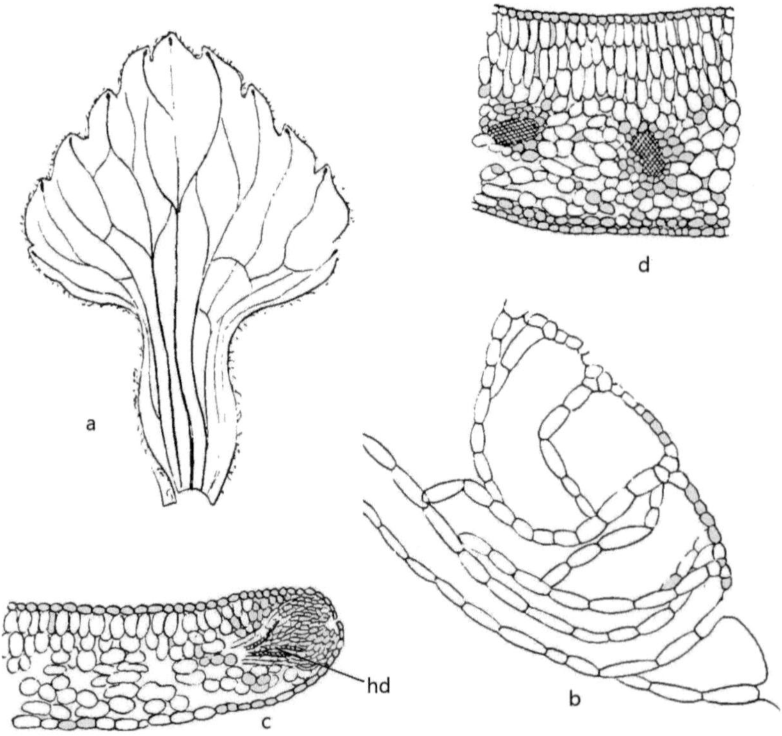

Fig. 208 *Saxifraga nivalis*: (**a**) leaf shape; (**b**) trichomes, the edge of the leaf; (**c**) hydathode (hd), longitudinal section through the leaf; (**d**) leaf, cross section (Galløe 1910)

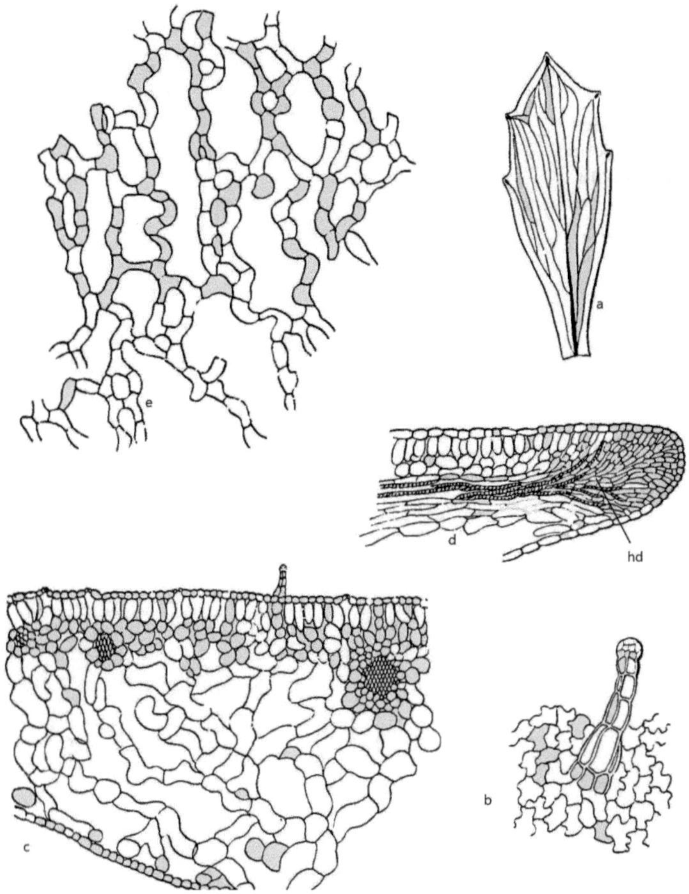

Fig. 209 *Saxifraga stellaris*: (**a**) leaf shape; (**b**) upper epidermis, front view; (**c**) leaf, cross section; (**d**) hydathode (hd), longitudinal section through the leaf; (**e**) spongy tissue (Galløe 1910)

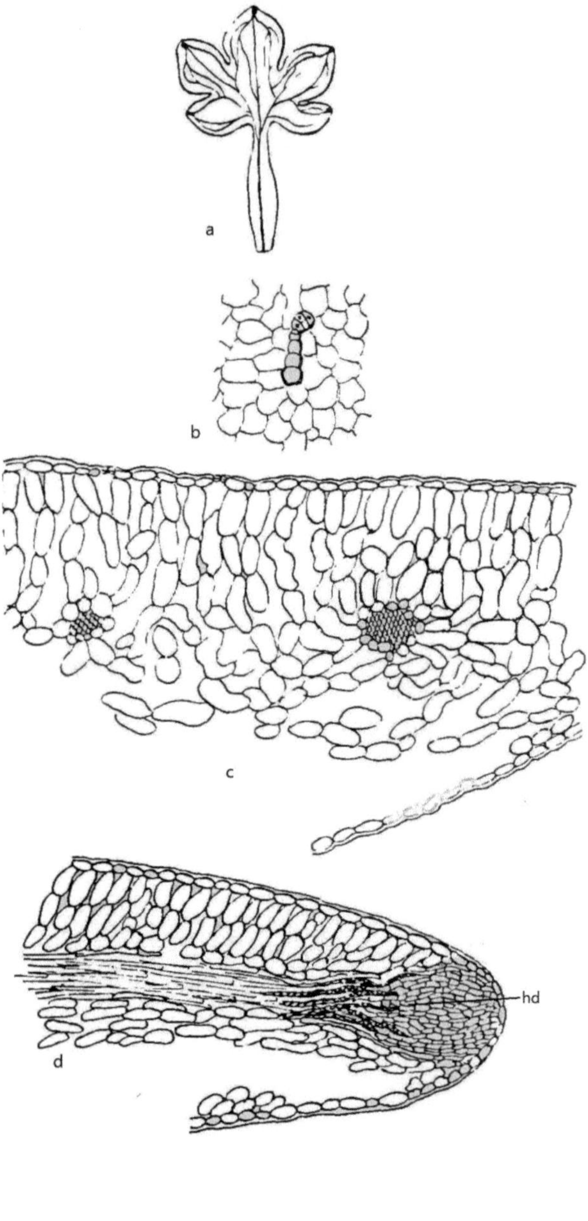

Fig. 210 *Saxifraga rivularis*: (**a**) leaf shape; (**b**) upper epidermis, front view; (**c**) leaf, cross section; (**d**) hydathodes (hd), longitudinal section through the leaf (Galløe 1910)

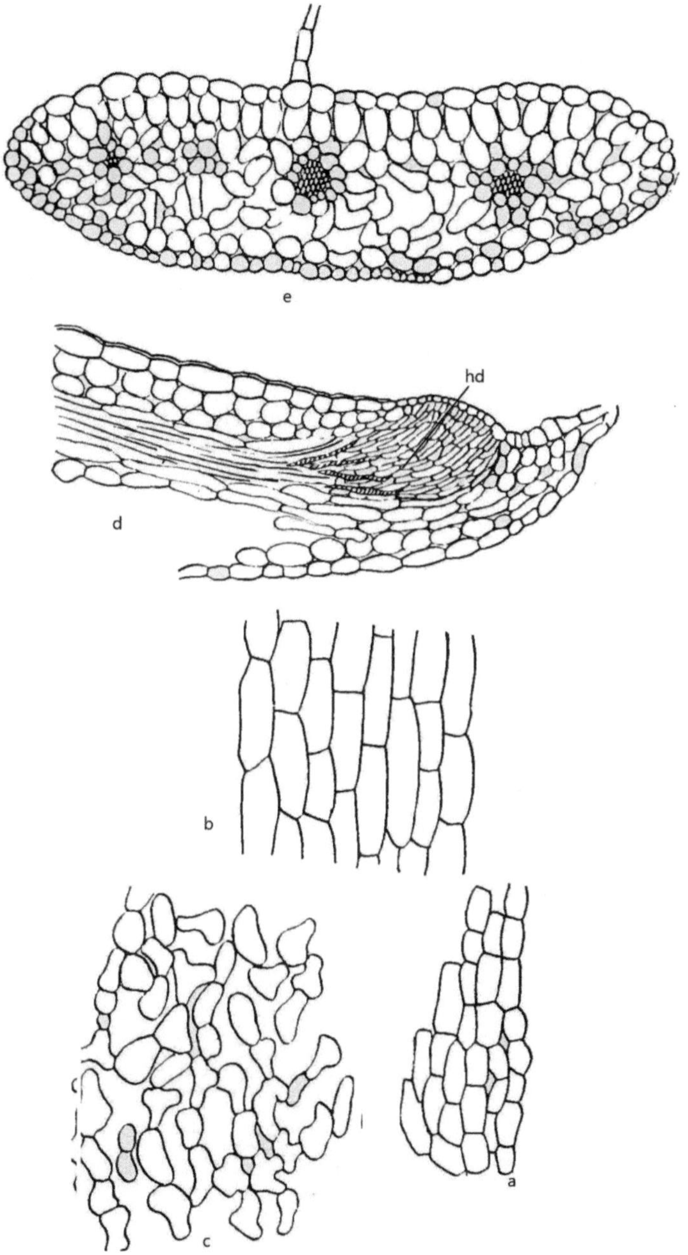

Fig. 211 *Saxifraga hypnoides*: (**a**) spongy tissue, leaf blade, front view; (**b**) spongy tissue, leaf petiole, front view; (**c**) spongy tissue, between the epidermis and the midrib, front view; (**d**) hydathode (hd), longitudinal section through the leaf; (**e**) leaf, cross section (Galløe 1910)

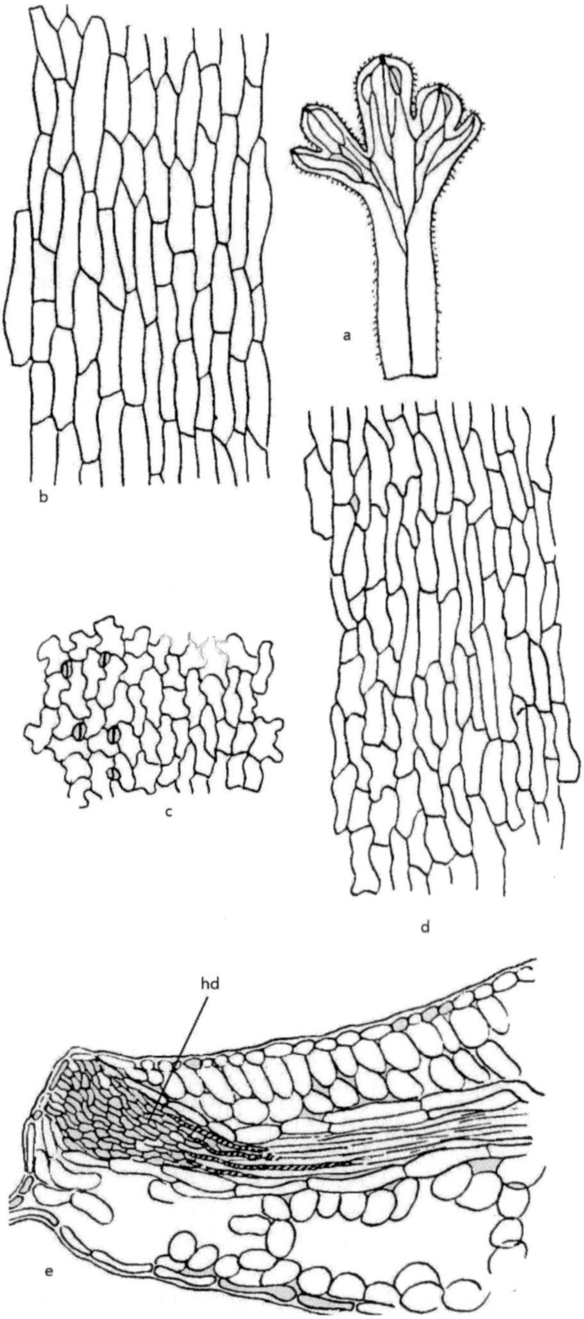

Fig. 212 *Saxifraga groenlandica*: (**a**) leaf shape; (**b**) upper epidermis, front view; (**c**) lower epidermis, front view; (**d**) lower epidermis, front view, midrib; (**e**) hydathodes (hd), longitudinal section through the leaf (Galløe 1910)

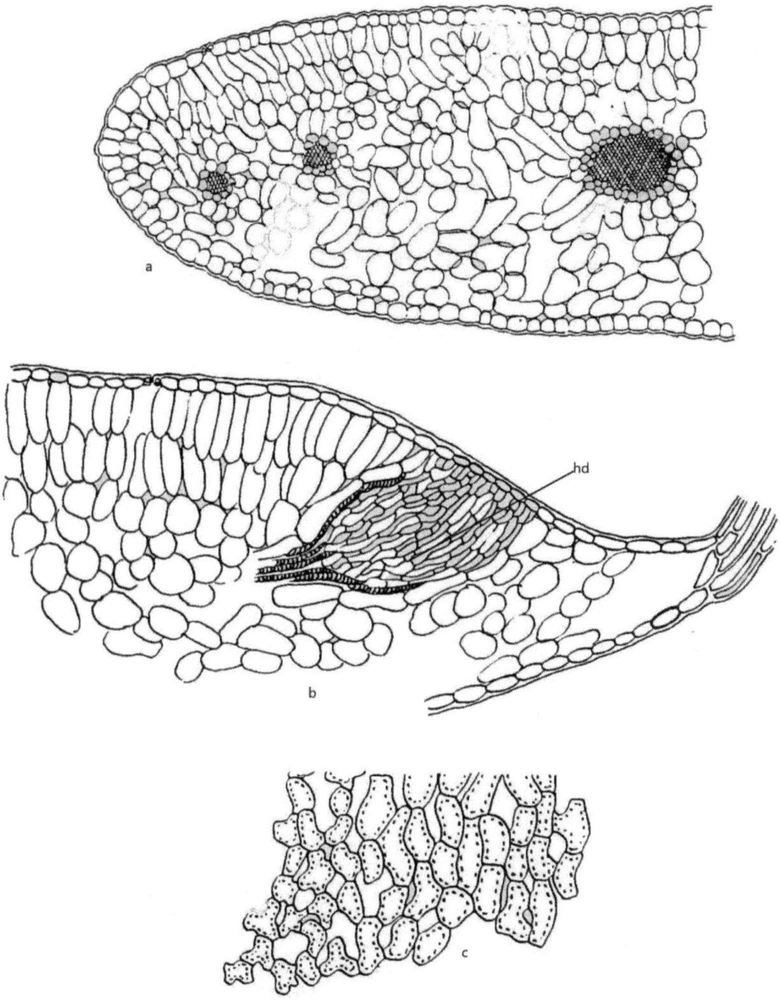

Fig. 213 *Saxifraga aizoides*: (**a**) leaf, cross section; (**b**) hydathode (hd), longitudinal section through the leaf; (**c**) spongy tissue (Galløe 1910)

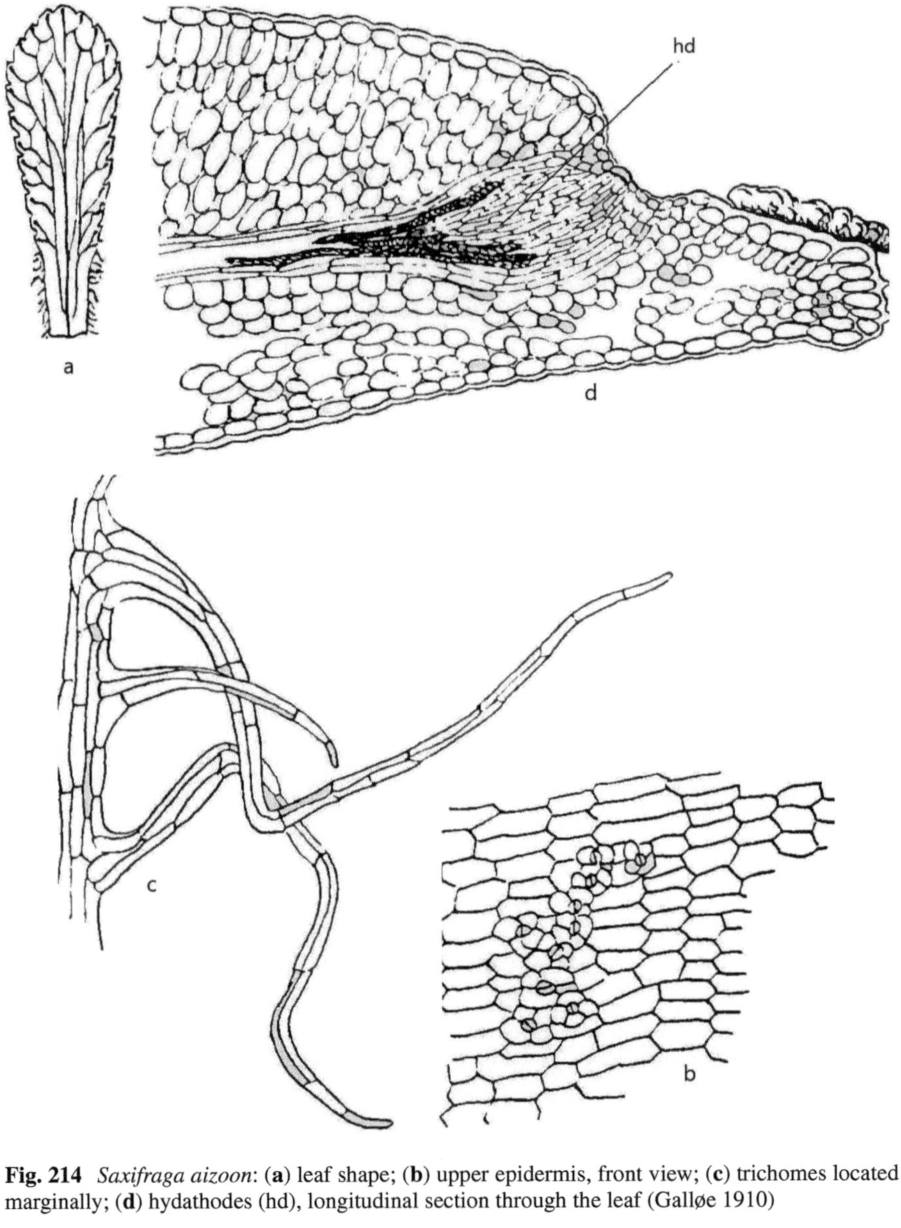

Fig. 214 *Saxifraga aizoon*: (**a**) leaf shape; (**b**) upper epidermis, front view; (**c**) trichomes located marginally; (**d**) hydathodes (hd), longitudinal section through the leaf (Galløe 1910)

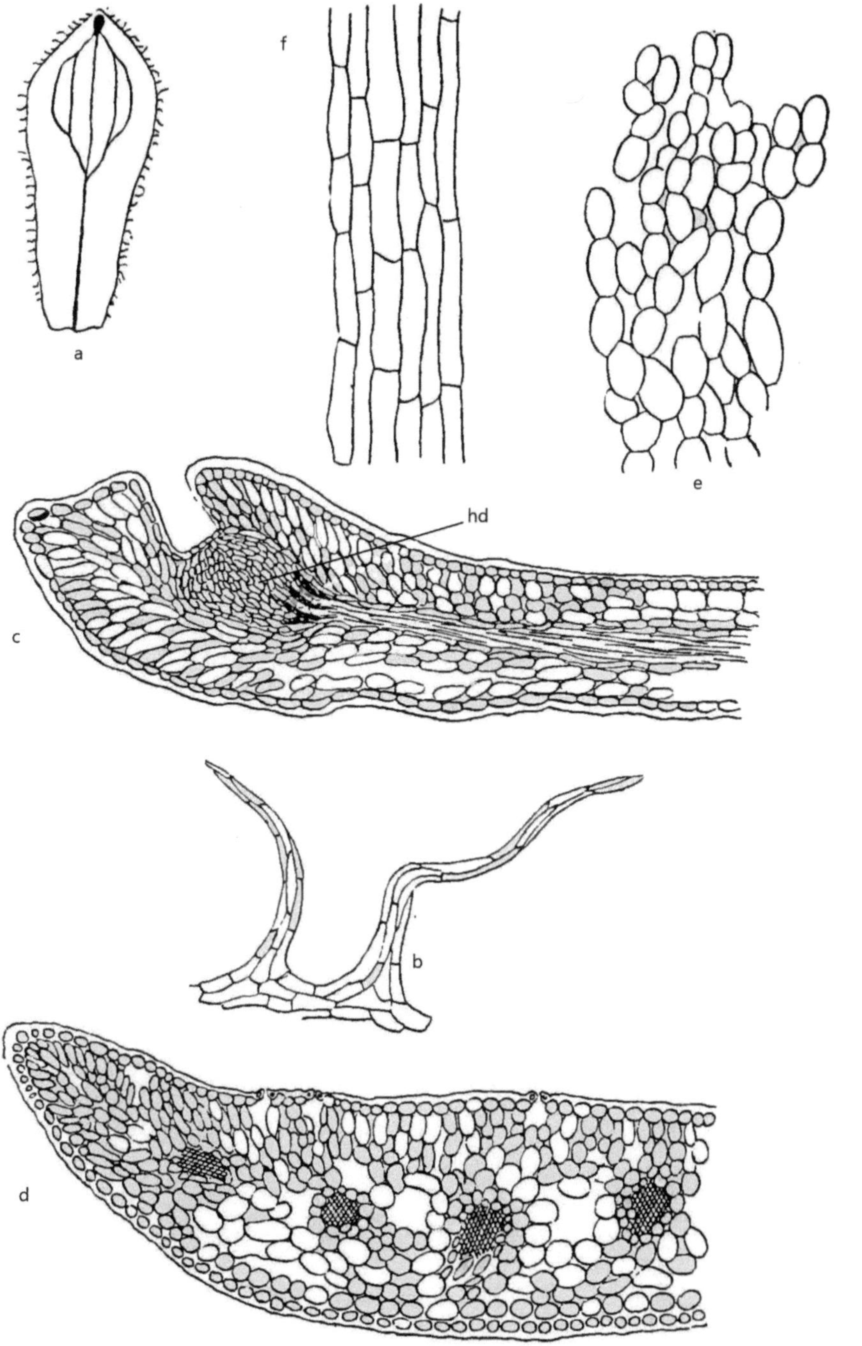

Fig. 215 *Saxifraga oppositifolia*: (**a**) leaf shape; (**b**) trichomes located marginally; (**c**) hydathode (hd), longitudinal section through the leaf; (**d**), leaf, cross section; (**e**) spongy tissue; (**f**) spongy tissue, leaf petiole (Galløe 1910)

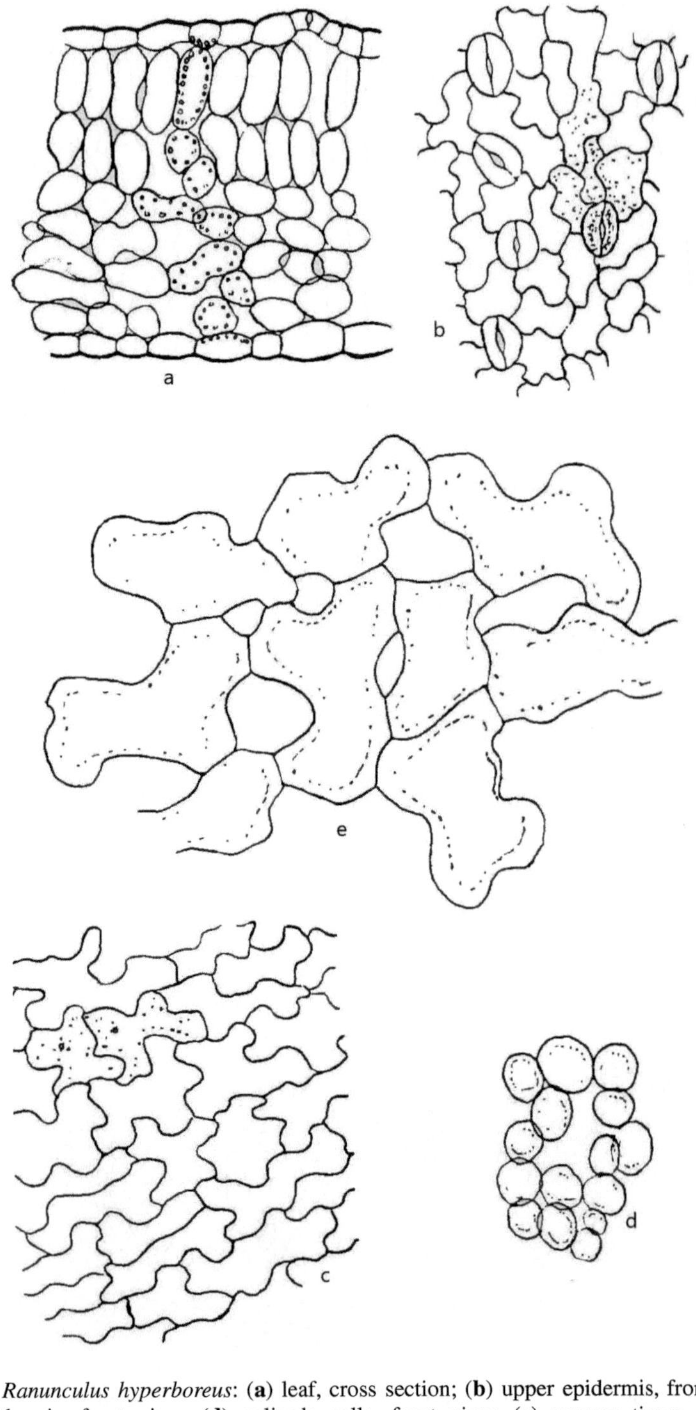

Fig. 216 *Ranunculus hyperboreus*: (**a**) leaf, cross section; (**b**) upper epidermis, front view; (**c**) lower epidermis, front view; (**d**) palisade cells, front view; (**e**) spongy tissue, front view (Jessen 1911)

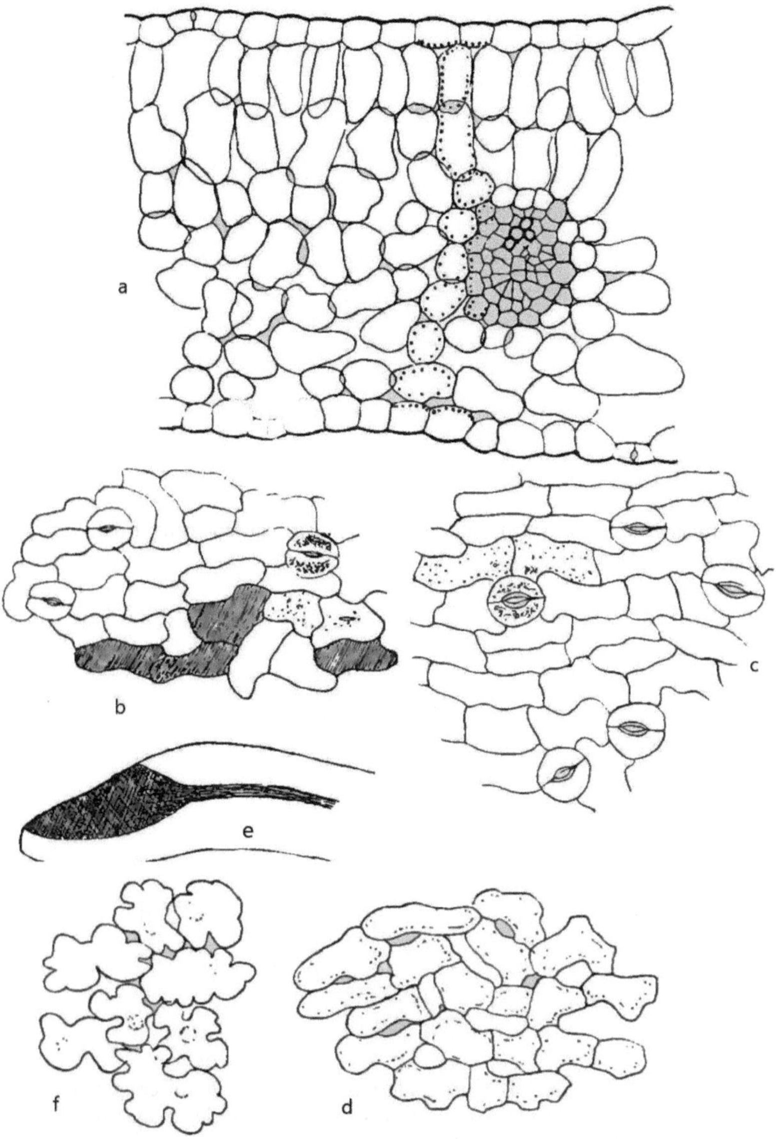

Fig. 217 *Ranunculus pallasii*: (**a**) leaf, cross section; (**b**) upper epidermis, front view (hatched cells are probably tanniniferous); (**c**) lower epidermis, front view; (**d**) spongy tissue, front view; (**e**) longitudinal section through the tip of the leaf, showing the position of the epithema of hydathode; (**f**) cells of the hydathode epithema (Jessen 1911)

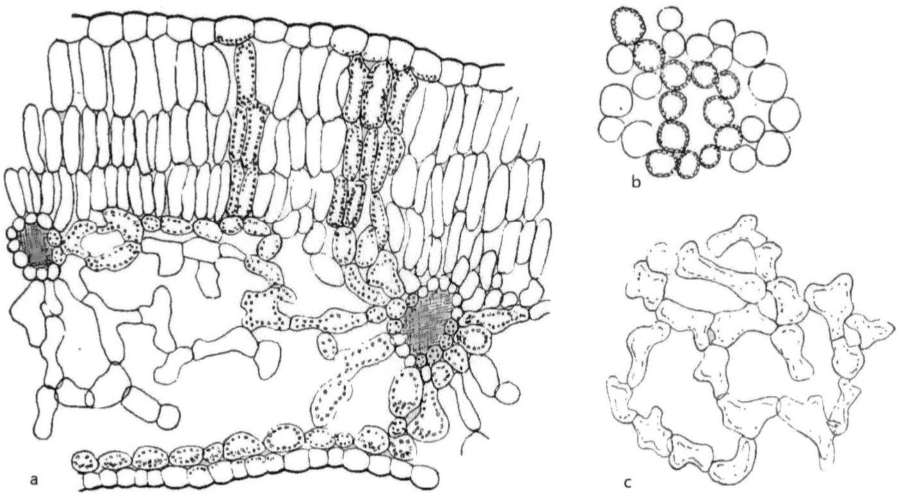

Fig. 218 *Ranunculus glacialis*: (**a**) leaf, cross section; (**b**) palisade cells, front view; (**c**) spongy tissue, front view (Jessen 1911)

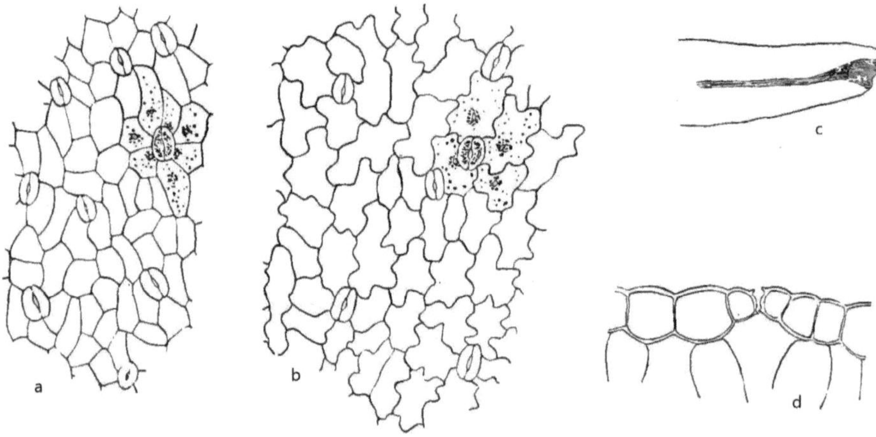

Fig. 219 *Ranunculus glacialis*: (**a**) upper epidermis, front view; (**b**) lower epidermis, front view; (**c**) longitudinal section through the tip of the leaf, showing the position of the hydathode epithema; (**d**) stomata, lower epidermis (Jessen 1911)

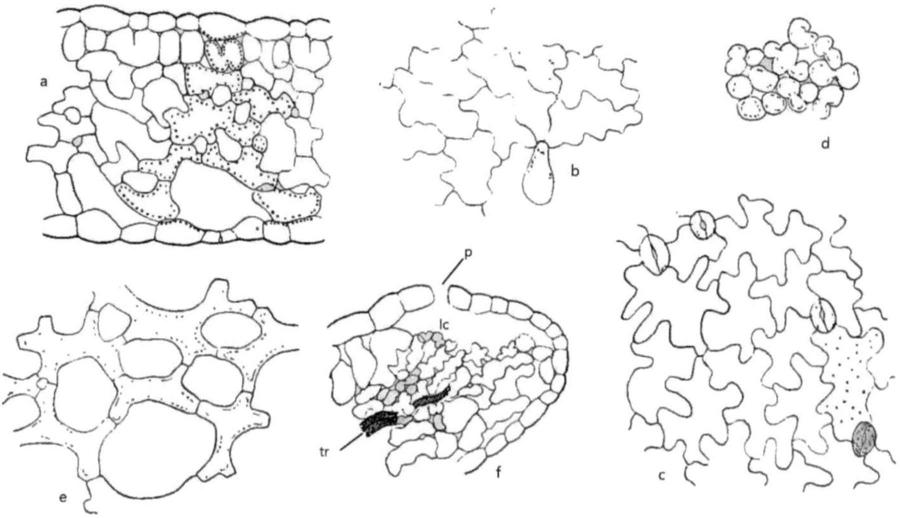

Fig. 220 *Anemone richardsoni*: (**a**) leaf, cross section; (**b**) upper epidermis, front view; (**c**) lower epidermis, front view; (**d**) palisade tissue, front view; (**e**) spongy tissue, front view; (**f**) longitudinal section through the tip of the leaf—hydathode; lc, gap; p, pore; tr, tracheids (Jessen 1911)

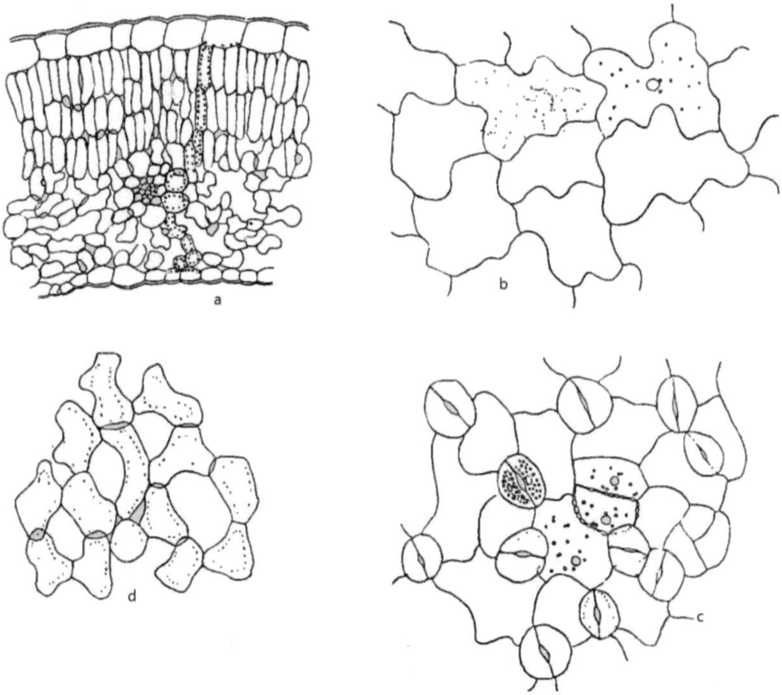

Fig. 221 *Thalictrum alpinum*: (**a**) leaf, cross section; (**b**) upper epidermis, front view; (**c**) lower epidermis, front view; (**d**) spongy tissue, front view (Jessen 1911)

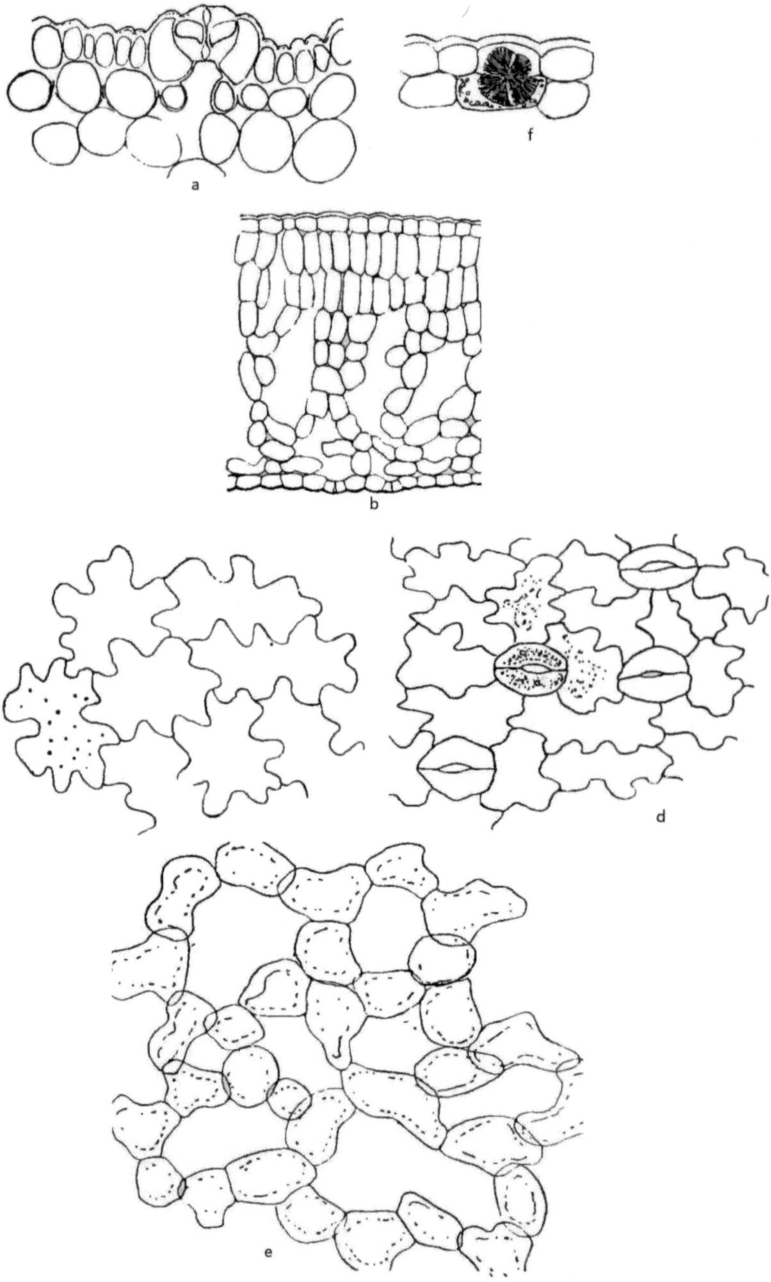

Fig. 222 *Coptis trifolia*: (**a**) stomata, cross section through the petiole of the leaf; (**b**) leaf, cross section; (**c**) upper epidermis, front view; (**d**) lower epidermis, front view; (**e**) spongy tissue, front view; (**f**) spherocrystal from epidermal cell, lower epidermis (Jessen 1911)

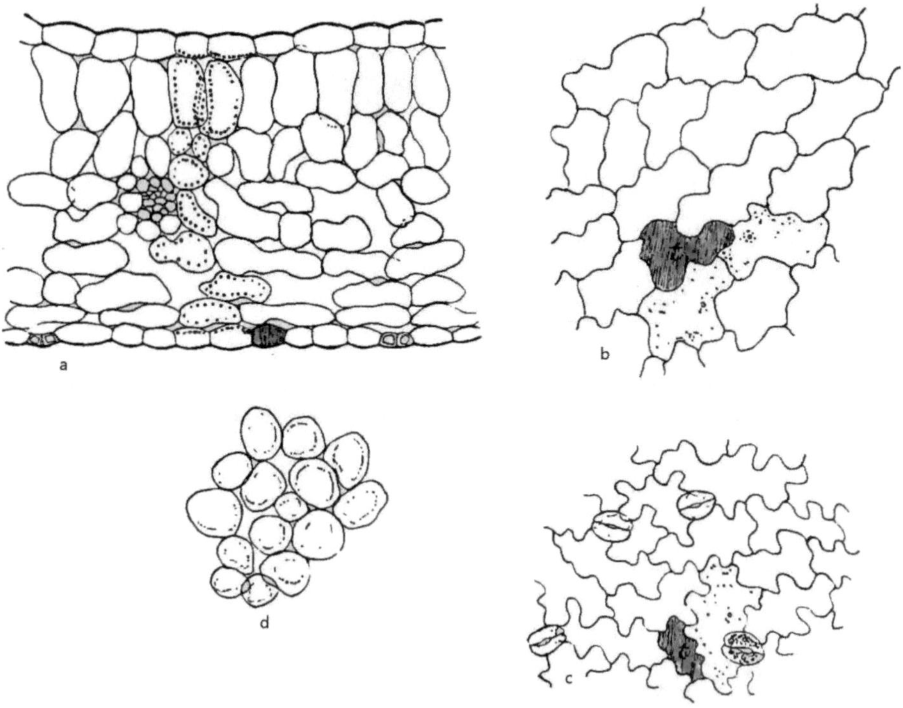

Fig. 223 *Ranunculus lapponicus*: (**a**) leaf, cross section; (**b**) upper epidermis, front view; (**c**) lower epidermis, front view; (**d**) palisade tissue, front view (Jessen 1911)

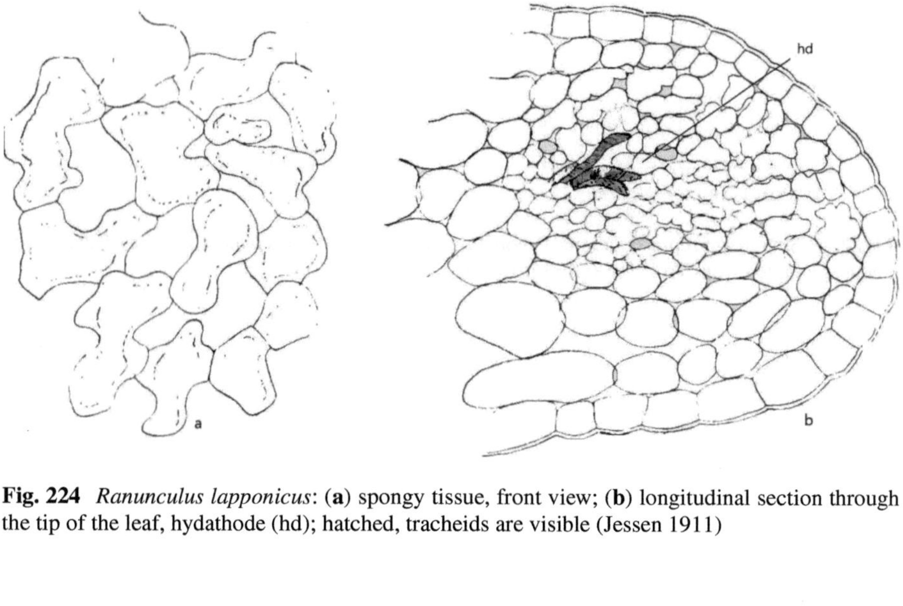

Fig. 224 *Ranunculus lapponicus*: (**a**) spongy tissue, front view; (**b**) longitudinal section through the tip of the leaf, hydathode (hd); hatched, tracheids are visible (Jessen 1911)

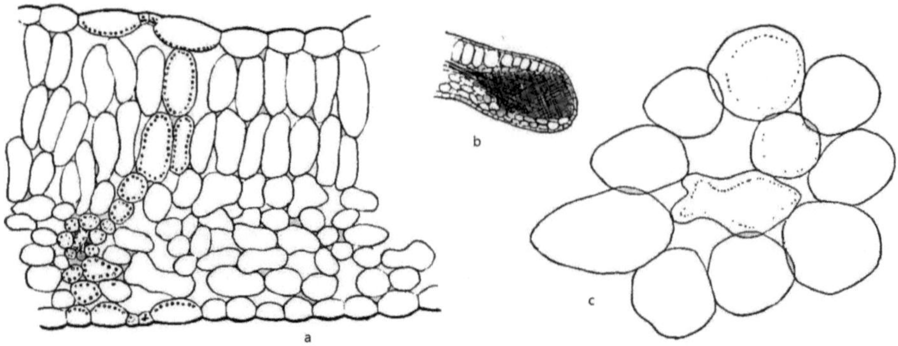

Fig. 225 *Ranunculus affinis*: (**a**) leaf, cross section; (**b**) longitudinal section through the tip of the leaf—hydathode; hatched, tracheids are observed in the epithema; (**c**) palisade tissue, front view (Jessen 1911)

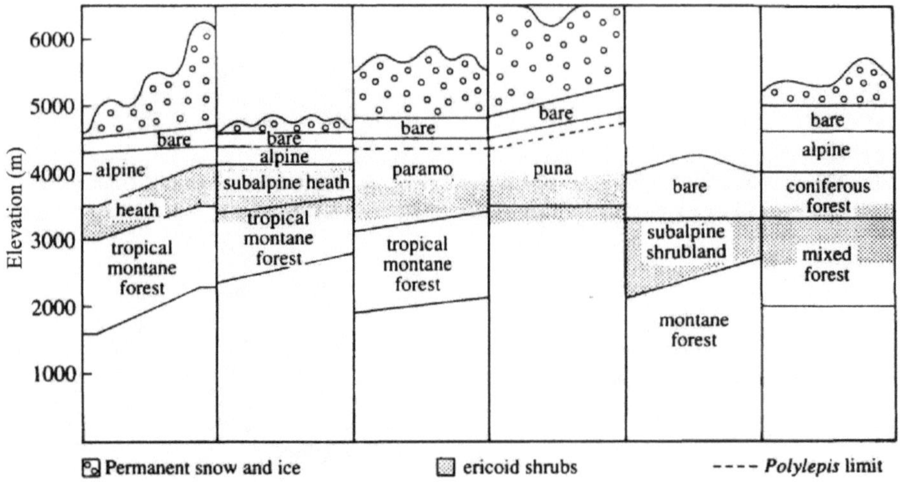

Fig. 226 Schematic representation of mountain vegetation—South America, East Africa (modified and adapted from Troll 1968; Smith 1994)

Table 1 "Winter" species (original term) showing chloroplasts in epidermal cells (Lidforss 1907)

Chloroplasts present in cells of both epidermises	*Oxycoccus palustris, Pyrola uniflora, P. secunda, P. minor, Diapensia lapponica, Arctostaphylos uva-ursi, A. alpina, Silene acaulis, Saxifraga stellaris, Polypodium vulgare*
Chloroplasts present predominantly in the upper epidermis	*Ledum palustre, Cassandra calyculata, Andromeda polifolia, Azalea procumbens, Empetrum nigrum*
Chloroplasts present predominantly in the lower epidermis	*Bellis perennis, Hieracium pilosella, Crepis* sp., *Vinca minor, V. angustifolia, V. latifolia, Gentiana acaulis, Ajuga reptans, Thymus chamedrys, T. corsicus, Galeobdolon luteum, Lamium purpureum, L. amplexicaule, Pentstemon* sp., *Veronica hederifolia, V. chamedrys, V. serpyllifolia, Pachysandra procumbens, Geum urbanum, G. rivale, Saxifraga cordifolia, S. crassifolia, Viola tricolor* var. *arvensis, V. odorata, Geranium molle, G. pusillum, Asarum europaeum, Reseda luteola, Aubrietia graeca, Capsella bursa pastoris, Erysimum* sp., *Barbarea vulgaris, Epimedium chrysanthemum, Ranunculus gramineus, R. abortivus, Anemone hepatica, Papaver argemone, Chelidomum majus, Asplenium trichomanes, A. septentrionale, Scolopendrium officinale*

Fig. 227 Map showing the current distribution of areas covered by páramos (red color)

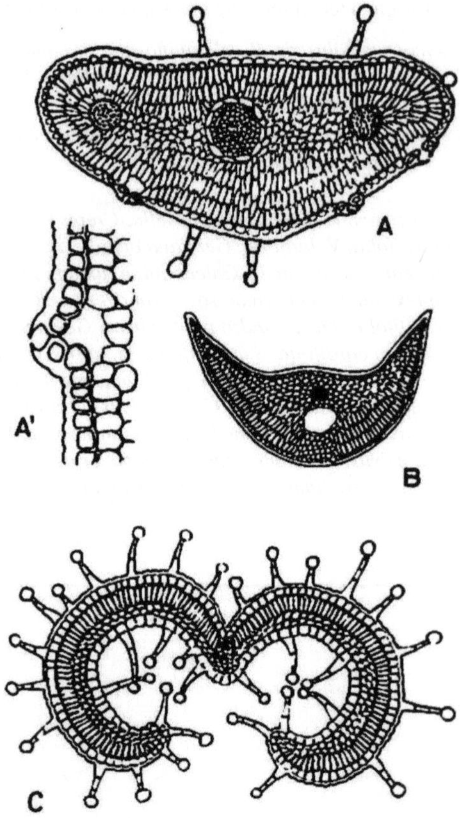

Fig. 228 Cross sections through the leaf—"puna" plants*:* (**a**) *Fabiana densa*; A', stomata, stem epidermis; (**b**) *Werneria incisa*; (**c**) *Calceolaria santolinoides* (Cabrera 1968)

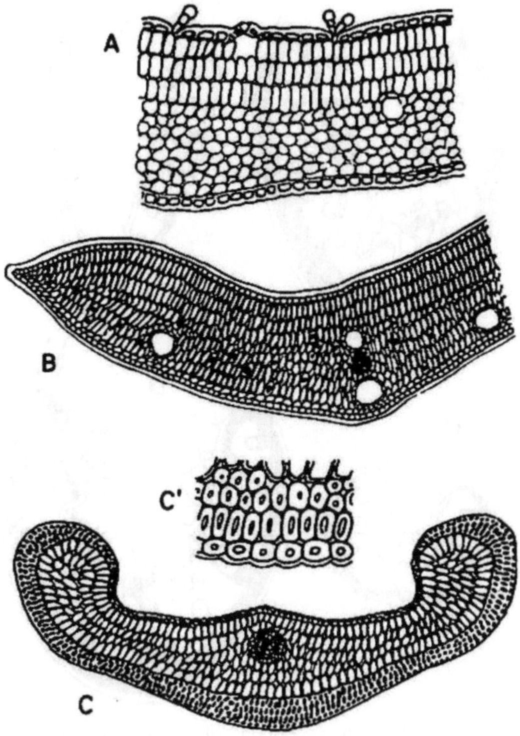

Fig. 229 Cross sections through the leaf—"puna" plants*:* (**a**) *Baccharis incarum*; (**b**) *Azorella compacta*; (**c**) *Chuquiraga atacamensis*; C ', the same species, sclerenchyma detail (Cabrera 1968)

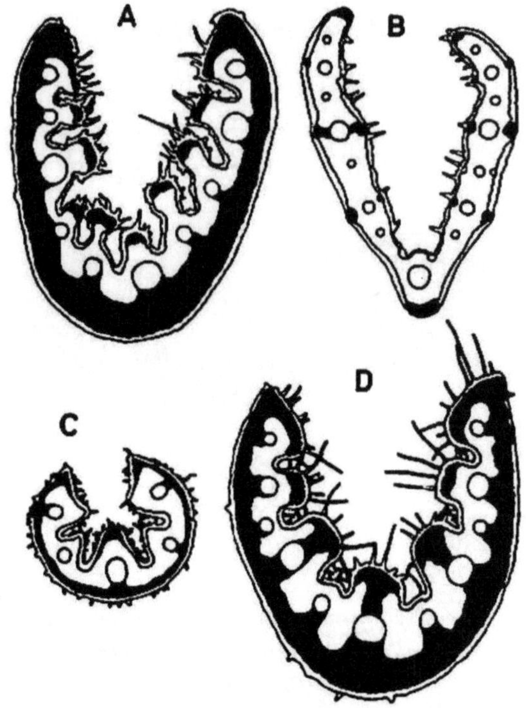

Fig. 230 Cross sections through the leaf—"puna" plants: (**a**) *Deyeuxia cabrerae*; (**b**) *Poa gymnantha*; (**c**) *Stipa henrardiana*; (**d**) *Festuca orthophylla* (Cabrera 1968)

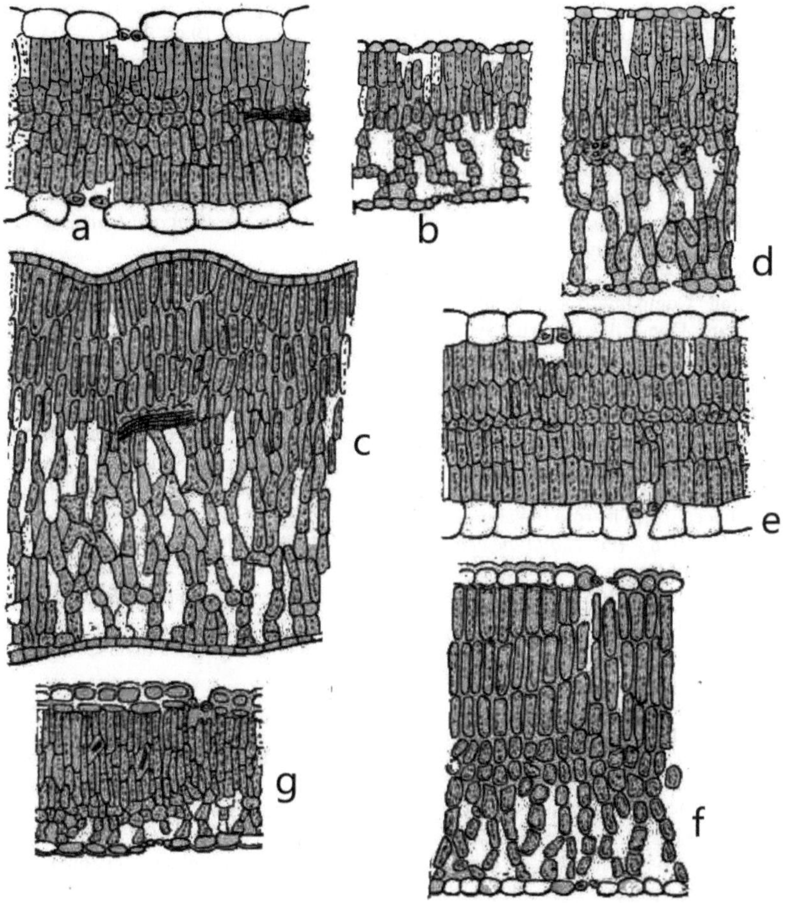

Fig. 231 Cross sections through the leaf—plants of American Cordillera; (**a**) *Oxalis adenophylla*; (**b**) *Caltha andicola*; (**c**) *Calandrinia picta*; (**d**) *Calandrinia paffinis*; (**e**) *Tropaeolum polyphyllum*; (**f**) *Calycera* ssp.; (**g**) *Dioscorea volckmanni* (Reiche 1907)

Bibliography

Albernethy G. A., Fountain D. W., McManus M. T., 1998 – Observations on the leaf anatomy of *Festuca novae-zelandiae* and biochemical responses to a water deficit. *New Zeal. J. Bot.*, 36(1): 113–123

Ali Z., Merrium S., Habib-Ur-Rahman M., Hakeem S., Saddique M. A. B., Sher M. A., 2022 – Wetting mechanism and morphological adaptation; leaf rolling enhancing atmospheric water acquisition in wheat crop – a review. *Environ. Sci. Pollut. Res. Int.*, 29(21): 30967–30985

Alvarez J. M., Rocha J. F., Machado S. R., 2003 – Ultrastructural aspects of bulliform cells in two Cerrado Grass species. *Acta Microscopia*, 8, 57–58

Alvarez J. M., Rocha J. F., Machado S. R., 2008 – Bulliform cells in *Loudetiopsis chrysothrix* (Ness) Conert and *Tristachya leiostachya* Nees (Poaceae): Structure in relation to function. *Braz. Arch. Biol. Technol.*, 51(1): 113–119

Ancibor E., 1980 – Estudio anatomico de la vegetacion de la puna de Jujuy. II. Anatomia de las plantas en Cojin. *Boletin de la Sociedad Argentina de Botánica*. 19 (1–2): 157–202

Arber E. A. N., 1910 – *Plant life in alpine Switzerland*, London, John Murray, Albemarle Street, W.

Arber A., 1920 – *Water plants. A study of aquatic angiosperms*, Cambridge at the University Press

Archibold O. W., 1995 – *Ecology of world vegetation*, Springer-Science+Business Media, B.V.

Ballego-Campos I., Forzza R. C., Paiva E. A. S., 2020 – More than scales: Evidence for the production and exudation of mucilage by the peltate trichomes of *Tillandsia cyanea* (Bromeliaceae: Tillandsioideae). *Plants*, 9: 763

Barnes C. R., 1898 – *Plant life considered with special reference to form and function*, New York, Henry Colt & Company

Barbour M. G., Burk J. H., Pitts W. D., 1980 – *Terrestrial plant ecology*, The Benjamin/Cummings Publishing Company, Inc., Menlo Park, California; Reading, Massachusetts; London; Amsterdam; Don Mills, Ontario; Sydney

Baron W. M. M., 1979 – *Organization in plants*, third ed., John Wiley & Sons, New York

Baylis GT., (1940–1941) – Leaf anatomy of the New Zealand mangroves. *Trans. Proc. Royal Soc. New Zeal.*, 70: 164–170

Benzing D. H., 1990 – *Vascular epiphytes. General biology and related biota*, Cambridge University Press, Cambridge, New York, Port Chester, Melbourne, Sydney

Benzing D. H., Seemann J., Renfrow A., 1978 – The foliar epidermis in Tillandsioideae (Bromeliaceae) and its role in habitat selection. *Am. J. Bot.*, 65: 359–365

Billings W. D., 1978 – *Plants and the ecosystem*, third ed., Wadsworth Publishing Company, Inc., Belmont, California

Blanc M., 1912 – Sur les diaphragmes des canaux aérifères des plantes. *Revue gén. de Bot.*, 24: 233–243

Bonnier G., 1887 – Note sur les cultures comparées des mêmes espèces à diverses altitudes. *Bull. Soc. Bot. de France*, 34, 2 ème sr. (T IX): 467–475

Bonnier G., 1888 – Étude expérimentale sur l'influence du climat alpin sur la végétation et les fonctions des plantes. *Bull. Soc. Bot. de France*, 35, 2 ème sr. (T X): 436–439

Bonnier G., 1890 – Influence des hautes altitudes sur les fonctions des végétaux. *Comptes rendus hebd. des sances de l'Acad. des Sci.*, 111: 377-380

Bonnier G., 1894a – Les plantes arctiques comparées aux mêmes espèces des Alpes et des Pyrénées. *Rev. Gén. Bot.*, 6: 505–527

Bonnier G., 1894b – Recherches expérimentales sur l'adaptation des plantes au climat alpin. *Ann. Sci. Nat., septième sér., Bot.*, 20(4–6): 217–358

Brown S., 1907 – *Alpine flora of the Canadian Rocky Mountains*, G. P. Putnam's Sons, New York and London

Brown J. H., Lomolino M. V., 1998 – *Biogeography*, second ed., Sinauer Associates, Inc. Publishers, Sunderland, Massachusetts

Cabrera A. L, 1968 – Ecologia vegetal de la Puna. *Colloq. Geographicum*, Band 9: Geo-ecology of the mountainous regions of the tropical Americas / Geo-ecologia de las regiones montanosas de las Americas tropicales: 91–116

Cannon W. A., 1921 – *Plant habits and habitats in the arid portions of South Australia*, Washington, D. C., published by the Carnegie Institution of Washington

Carlquist S., 1957 – Leaf anatomy and ontogeny in *Argyroxiphium* and *Wilkesia* (Compositae). *Am. J. of Bot.*, 44: 696–705

Carlquist S., 1994 – Anatomy of tropical alpine plants. In *Tropical alpine environments. Plant form and function* Rundel P. W., Smith A. P., Meinzer F. C. (eds.), Cambridge University Press, pp. 111–128

Carlquist S., Bissing D. R., 1976 – Leaf anatomy of Hawaiian geraniums in relation to ecology. *Biotropica*, 8: 248–259

Chermezon H., 1910 – Recherches anatomiques sur les plantes littorales. *Ann. Sci. Nat., sér. 9, Bot.*, 20 :117–129, 270–274, 299–307

Clayton W. D., Renvoize S. A., 1986 – *Genera graminum – grasses of the world*, Kew Bulletin, Additional Series XIII. London: Her Majestic Stationary Office
Clements F. E., 1907 – *Plant physiology and ecology*, New York, Henry Holt and Company
Collinson A. S., 1988 – *Introduction to world vegetation*, second ed., London, Unwin Hyman, Boston, Sydney, Wellington
Coulter J. M., Barnes C. R., Cowles H. C., 1911 – *Textbook of botany for colleges and universities, vol. II. Ecology*, New York, Cincinnati, Chicago, American Book Company
Crawford R. M. M., 1989 – *Studies in plant survival. Ecological case histories of plant adaptation to adversity*, Blackwell Scientific Publications, Oxford, London, Edinburgh, Boston, Palo Alto, Melbourne
Daubenmire R. F., 1974 – *Plants and environment. A textbook of plant autecology*, third ed., John Wiley & Sons, New York, London, Sydney, Toronto
Dawson J., Lucas R., 2005 – *The nature of plants. Habitats, challenges, and adaptations*, Timber Press, Portland, Cambridge
de Bary A., 1884 – *Comparative anatomy of the vegetative organs of the phanerogams and ferns*, Oxford: Clarendon
Denis M., 1919 – Recherches anatomiques sur quelques plantes littorales de Madagascar. *Rev. gén Bot.*, 31: 33–140
Dolzmann P., 1964 – Elektronenmikroskopische Untersuchungen an den Saughaaren von *Tillandsia usneoides* (Bromeliaceae). I. Feinstruktur der Kuppelzelle. *Planta*, 60: 461–472
Dolzmann P., 1965 – Electron microscopic investigations on the absorptive hairs of *Tillandsia usneoides* II. Observations on the fine structure of plasmodesmids. *Planta*, 64: 76–80
Dring M. J., 1982 – *The biology of marine plants*, Edward Arnold (Publishers) Limited, London
Esau K., 1960 – *Anatomy of seed plants*, John Wiley & Sons, Inc., New York, London
Esau K., 1965 – *Plant anatomy*, second edition, John Wiley and Sons, New York, London, Sydney
Espinosa R. 1932 – Oekologische Studien über Kordillerenpflanzen (morphologisch und anatomisch dargestellt). *Bot. Jahrb.*, 65: 120–211
Etherington J. R., 1975 – *Environment and plant ecology*, Hohn Wiley & Sons, London, New York, Sydney, Toronto
Fahn A., Cutler D. F., 1992 – *Xerophytes*, vol. 13, *Handbuch der Pflanzenanatomie*, Gebrüder Borntraeger, Berlin
Farrer R., 1908 – *Alpines and bog-plants*, London, Edward Arnold
Font-Quer P., 1970 – *Diccionario de Botánica*, Labor: Barcelona
Forsaith C. C., 1920 – Anatomical reduction in some alpine plants. *Ecology*, 1(2): 124–135
Fries R. E., 1905 – Zur Kenntnis der alpinen Flora im nördlichen Argentinien. *Nova Acta Reg. Societatis Scientiarum Upsaliensis*, Ser. 4, 1, 1: 1–205
Galløe O., 1910 – The structure and biology of arctic flowering plants. *Saxifragaceae.* 2. The biological leaf-anatomy of the arctic species of *Saxifraga. Medd. Grønl.*, 36: 237–294
Gibson J. P., Gibson T. R., 2006 – *Plant ecology*, Chelsea House Publishers, New York
Goebel K., 1891–1893 – *Pflanzenbiologische Schilderungen*, zweiter teil, erste lieferung, zweiter teil, zweite lieferung, Marburg N. G. Elwert'sche Verlagsbuchhandlung
Graham L. E., Graham J. M., Wilcox L. W., 2014 – *Plant biology*, second ed., Pearson Education Limited, Essex, England
Grigore M.-N., 2008 – *Introducere în Halofitologie. Elemente de anatomie integrativă*, Editura PIM, Iași
Grigore M.-N., 2021 – Definition and classification of halophytes as an ecological group of plants. In: Grigore M.-N. (Ed.), *Handbook of halophytes. From molecules to ecosystems towards biosaline agriculture*, Springer, pp. 3–50
Grigore M.-N., 2022 – *Morfologia plantelor. Aspecte ecologice*, Editura Universității „Ștefan cel Mare" Suceava
Grigore M.-N., Toma C., 2010a – *Halofitele. Aspecte de anatomie ecologică.* Editura Universității „Alexandru Ioan Cuza", Iași

Grigore M.-N., Toma C., 2010b – *Structuri secretoare de săruri la halofite. O abordare integrativă.* Editura Academiei Române, Bucureşti

Grigore M.-N., Toma C., 2010c – A proposal for a new halophytes classification, based on integrative anatomy observations. *Muz. Olteniei Craiova. Studii şi Comunicări, Ştiinţele Naturii*, 26(1): 45–50

Grigore M.-N., Toma C., 2011a – Halofitele, o categorie ecologică polimorfă. Între seceta fiziologică a solului şi stresul salin. *Revista Botanică (Chişinău)*, 2(3): 38–46

Grigore M.-N., Toma C., 2011b – Ecological implications of bulliform cells on halophytes, in salt and water stress natural conditions. *Studia Universitatis `Vasile Goldiş`, Ser. Şt. Vieţii*, 21(4): 785–792

Grigore M.-N., Toma C., 2017 – *Anatomical adaptations of halophytes. A review of classic literature and recent findings*, Springer International Publishing

Grigore M.-N., Toma C., 2021a – Morphological and anatomical adaptations of halophytes. A review. In: Grigore M.-N. (Ed.), *Handbook of halophytes. From molecules to ecosystems towards biosaline agriculture*, Springer, pp. 1079–1222

Grigore M.-N., Toma C., 2021b – Integrative anatomy of halophytes from Mediterranean climate. In: Grigore M.-N. (Ed.), *Handbook of halophytes. From molecules to ecosystems towards biosaline agriculture*, Springer, pp. 1293–1328

Grigore M.-N., Ivănescu L., Toma C., 2014 – *Halophytes. An integrative anatomical study.* Springer, Cham, Heidelberg, New York, Dordrecht, London

Grigore M.-N., Toma C., Boşcaiu M., Zamfirache M.-M., Ivănescu L., 2012 – Anatomical and ecological observations on psammo-halophytes species (Eastern part of Spain). *Lucr. Şt. (Horticultură), USAMV „Ion Ionescu de la Brad", Iaşi*, 55(2): 19–24

Haberlandt G., 1914 – *Physiological plant anatomy*, Macmillan and Co, London

Hauman L., 1918 – La vegetation des Hautes Cordilleres de Mendoza. *Anal. Soc. Cient. Argentina*, 86: 121–188, 225–348

Hauri H., 1917 – Anatomische Untersuchungen an Polsterpflanzen nebst morphologischen und ökologischen Notizen. *Beih. Bot. Centralbl.*, 33(1): 275–293

Hayden A., 1919 – The ecologic foliar anatomy of some plants of a prairie province in central Iowa. *Am. J. Bot.*, 6: 69–85

Heide F., 1912 – The structure and biology of arctic flowering plants. *Lentibulariaceae (Pinguicula). Medd. Grønl.*, 36: 441–481

Henslow G., 1895 – *The origin of plant-structures by self-adaptation to the environment*, London, Kegan Paul, Trench, Trübner, & Co, Ltd, Paternoster House, Charing Cross Road

Hopkins W. G., Hüner N. P. A., 2009 – *Introduction to plant physiology*, fourth edition, Wiley, John Wiley & Sons, Inc., Hoboken

Ingrouille M., 1992 – *Diversity and evolution of land plants*, Chapman & Hall, London, Glasgow, New York, Tokyo, Melbourne, Madras

Ingrouille M., Eddie B., 2006 – *Plants: Evolution and diversity*, Cambridge University Press

Irieda H., 2022 – Emerging roles of motile epidermal chloroplasts in plant immunity. *Int. J. of Molec. Sci.*, 23(7): 4043. https://doi.org/10.3390/ijms23074043

Jessen K., 1911 – The structure and biology of arctic flowering plants. *Ranunculaceae. Medd. Grønl.*, 36: 3330–440

Körner C., 2022 – *Alpine plant life. Functional plant ecology of high mountain ecosystems*, third ed., Springer Nature

Lambers H., Oliveira R. S., 2019 – *Plant physiological ecology*, third ed., Springer, Cham, Switzerland

Larcher W., 1975 – *Physiological plant ecology*, Springer-Verlag, Berlin, Heidelberg, New York

Lazniewski W., 1896 – Beiträge zur Biologie der Alpenpflanzen. *Flora*, 82: 224–267

Leist K., 1890 – Ueber den Einfluss des alpinen Standortes auf die Ausbildung der Laubblätter. *Mitt. der Natur. Ges.*, 1215–1243: 159–201

Lésage P., 1890 – Recherches expérimentales sur les modifications des feuilles chez les plantes maritimes. *Rev. gén. Bot.*, 2: 55–65;106–121; 163–173

Lidforss B., 1907 – Die wintergrüne Flora. *Lunds Universitets Årsskrift*, 2, 2(13): 1–83

Luo Y., Che-Ling H., Helliker Brent R., Katifori E., 2021 – Leaf Water Storage and Robustness to Intermittent Drought: A Spatially Explicit Capacitive Model for Leaf Hydraulics. *Front. Plant Sci.*, 12: 725995

Lütz C. (ed.), 2012 – *Plants in alpine regions. Cell physiology of adaptation and survival strategies*, Springer, Wien, New York

Mader A., Langer M., Knippers J., Speck O., 2020 – Learning from plant movements triggered by bulliform cells: the biomimetic cellular actuator. *J. R. Soc. Interface*: 17:20200358

Mauseth J. D., 1988 – *Plant anatomy*, Menlo Park, Calif. : Benjamin/Cummings Pub. Co.

Mentz A., 1909 – The structure and biology of arctic flowering plants. *Empetraceae. Empetrum nigrum* L. *Medd. Grønl.*, 36: 155–168

Mer E., 1880 – Des modifications de forme et de structure que subissent les plantes, suivant qu'elles végètent à l'air ou sous l'eau. *Bull. Soc. Bot. France*, 2 ème sér. (T II), 28: 50–55

Merrium S., Ali Z., Tahir M. H. N., Habib-ur-Rahman M., Hakeem S., 2022 – Leaf rolling dynamics for atmospheric moisture harvesting in wheat plant as an adaptation to arid environments. *Environ. Sci. Pollut. Res.*, 29: 48995–49006

Metcalfe C. R., 1960, 1971 – *Anatomy of the Monocotyledons. I. Gramineae, V. Cyperaceae.* Clarendon Press, Oxford

Mez C. 1904 – Physiologische Bromeliaceen-Studien I. Die Wasser-Ökonomie der extrem atmosphärischen Tillandsien. *Jb. wiss. Bot.* 40: 157–229

Minden von, M., 1899 – Beitrage zur anatomischen und physiologischen Kenntnis Wasser-secernierender Organe. *Bibliotheca Botanica*, 9(46): 1–76

Nawazish S., Hameed M., Naurin S., 2006 – Leaf anatomical adaptations of *Cenchrus ciliaris* L. from the Salt Range, Pakistan against drought stress. *Pakistan J. Bot.*, 38(5): 1723–1730

Nedukha O. M., 2022 – Effects of moderate drought on leaf bulliform cells of aquatic and coastal population of *Phragmites australis. Turk. J. Bot.*, 46: 459–472

Ogden E. C., 1974 – *Anatomical patterns of some aquatic vascular plants of New York*, BULLETIN 424, New York State Museum and Science Service

Oppenheimer H. R., 1959 – Adaptation to drought: xerophytism. In *Plant-water relationships in arid and semi-arid conditions. Reviews of research*, UNESCO, Printed by C. J. Bucher, Lucerne, Swilzerland, pp. 105–138

Öpik H., Rolfe S. A., Villis A. J., 2005 – *The physiology of flowering plants*, fourth ed., Cambridge University Press, Cambridge, New York, Melbourne, Madrid, Cape Town, Singapore, São Paulo

Papini A., Tani G., di Falco P., Brighigna L., 2010 – The ultrastructure of the development of *Tillandsia* (Bromeliaceae) trichome. *Flora*, 205: 94–100

Petersen H. E., 1908a – The structure and biology of arctic flowering plants. *Ericineae (Ericaceae, Pirolaceae)*. 2. The biological anatomy of the leaves and of the stem *Medd. Grønl.*, 36: 73–138

Petersen H. E., 1908b – The structure and biology of arctic flowering plants. *Diapensiaceae. Diapensia lapponica* L. *Medd. Grønl.*, 36: 139–154

Pfadenhauer J. S., Klötzli F. A., 2020 – *Global vegetation. Fundamentals, ecology and distribution*, Springer Nature Switzerland

Polunin N., 1960 – *Introduction to plant geography and some related sciences*, McGraw-Hill Book Company, Inc., New York, Toronto, London

Porsch O., 1903 – Zur Kenntnis des Spaltöffnungsapparates submerser Pflanzenteile. *Sitzungsber. d. Math.-naturwiss. Klasse d. k. Akad. d. Wissens. Wien*, 112: 97–138

Porsch O., 1905 – *Der Spaltoffnungsapparat im Lichte der Phylogenie*, Jena, Verlag von Gustav Fischer

Prodan I., 1939 – *Flora pentru detrminarea şi descrierea plantelor ce cresc în România*. II. (ediţia a II-a). Edit. Cartea Românească, Cluj-Napoca

Ranwell D. S., 1972 – *Ecology of salt marshes and sand dunes*, London, Chapman and Hall

Raux P. S., Gravelle S., Dumais J., 2020 – Design of a unidirectional water valve in *Tillandsia. Nature Commun.*, 11, 396

Reiche K., 1907 – *Grundzüge der Pflanzenverbreitung in Chile*. In *Die vegetation der Erde*, vol. 8, Leipzig, Verlag von Wilhelm Engelmann

Riley D., Young A., 1966 – *World vegetation*, Cambridge at the University Press

Ristic Z., Cass D. D., 1991 – Leaf anatomy of *Zea mays* L. in response to water shortage and high temperature: a comparison of drought-resistant and drought-sensitive lines. *Bot. Gaz.*, 152(2): 173–185

Robinson W., 1903 – *Alpine flowers for gardens*, third ed., London, John Murray, Albermarle Street

Rundel P. W., Smith A. P., Meinzer F. C. (eds.), 1994 – *Tropical alpine environments. Plant form and function*, Cambridge University Press

Safiallah S., Mahdi Hamdi S.M, Grigore M.-N., Jalili S., 2017 – Micromorphology and leaf ecological anatomy of *Bassia* halophyte species (Amaranthaceae) from Iran. *Acta Biol. Szeged.*, 61(1):85–93

Sam O., Jerez E., Dell'Amico J., Ruiz-Sanchez M. C., 2000 – Water stress induced changes in anatomy of tomato leaf epidermis. *Biol. Plant.*, 43(2): 275–277

Sauvageau C., 1891 – Sur les feuilles de quelques monocotylédones aquatiques. *Ann. Sci. Nat., 7ème sér. Bot.*, 13: 103–296

Savile D. B. O., 1972 – *Arctic adaptations in plants*, Research Branch, Canada Department of Agriculture, Otawa

Schenck H., 1886 – Vergleichende Anatomie der submersen Gewächse. *Bibliotheca Bot.*, 1: 1–67

Schimper A. F. W., 1884 – Ueber bau und Lebensweise der Epiphyten Westindiens. *Botanisches Zentralblatt*, 17: 192–195; 223–227; 253–258; 284–294; 319–326; 350–359; 381–389

Schimper A. F. W., 1891 – Die indo-malaysche Strandflora. *Bot. Mit. Trop.*, 3: 1–204

Schimper A. F. W., 1903 – *Plant geography upon a physiological basis*, Oxford: Clarendon Press

Schulze E.-D., Beck E., Buchmann N., Clemens S., Müller-Hohenstein K., Scherer-Lorenzen M., 2019 – *Plant ecology*, 2nd ed., Berlin: Springer Nature

Sculthorpe, C. D., 1967 – *The biology of aquatic vascular plants*, London, Edward Arnold Publishers

Seddon B., 1971 – *Introduction to biogeography*, Gerald Duckworth and Co. Ltd., London

Seidelin A., 1912 – The structure and biology of arctic flowering plants. *Hippuridaceae, Halorrhagidaceae* and *Callitrichaceae*. *Medd. Grønl.*, 36: 295–332

Shimek E., 1915 – The ecological histology of prairie plants. *Proc. of the Iowa Acad. of Sci.*, 22(1), 121–126

Silva S., Soares A. M., Oliveira L. E. M., Magalhaes P. C., 2001 – Respostas fisiologicas de gramineas promissoras para revegetacao ciliar de reservatorios hidreletricos, submetidas a deficiencia hidrica. *Ciencia Agrotecnica*, 25: 124–133

Silva H., Martinez J. P., Baginsky C., Pinto M., 1999 – Effect of water deficit on the leaf anatomy of six cultivars of the common bean, *Phaseolus vulgaris*. *Rev. Chil. Hist. Natural*, **72**, 2: 219–235

Smith A. P., 1994 – Introduction to tropical alpine vegetation. In *Tropical alpine environments. Plant form and function*, Rundel P. W., Smith A. P., Meinzer F. C. (Eds.), Cambridge University Press, pp. 1–20

Solereder H., 1908 – *Systematic anatomy of the Dicotyledons. A handbook for laboratories of pure and applied Botany*, vol. 1, 2. Oxford: Clarendon.

Stover E. L., 1951 – *An introduction to the anatomy of seed plants*, D. C. Heath and Company Boston

Şerbănescu-Jitariu G., Toma C., 1980 – *Morfologia şi anatomia plantelor*, Editura Didactică şi Pedagogică, Bucureşti

Thompson H. S., 1912 – *Sub-alpine plants or flowers of the Swiss xylems and meadows*, London, George Routledge & Sons, Limited New York: E. O. Dutton & CO

Tivy J., 1993 – *Biogeography. A study of plants in the ecosphere*, third ed., Longman Scientific & Technical, Copublished in the United States with John Wiley & Sons, Inc., New York

Toma C., Dumitru E., 1973 – Contribuţii la studiul histo-anatomic al organelor vegetative de la *Carex ligerica* J. Gay. *Stud şi Com. Muz. Şt. Nat. Suceava*, 3: 165–180

Tomlinson P. B., 2016 – *The botany of mangroves*, second ed., Cambridge University Press

Treshow M., 1970 – *Environment & Plant responses*, McGraw-Hill Book Company, New York, St. Louis, San Francisco, Dusseldorf, London, Mexico, Panama, Sydney, Toronto

Troll C., 1968 – The Cordilleras of the tropical Americas. Aspects of climatic, phytogeographical and agrarian ecology. *Colloq. Geographicum*, 9: 15–56

Turner I. M., 2001 – *The ecology of trees in the tropical rain forest*, Cambridge University Press

Urcelay J. C., 1923 – Contribución al estudio de la adaptación de las plantas para diminuir la transpiración. Estudio anatómico de algunas especies de las estepas españolas. *Trab. Museo Nacional de Ciencias Nat. de Madrid, Ser. Bot.*, 18: 5–75

Vesque J., 1882 – L'espèce végétale considérée au point de vue de l'anatomie compare. *Ann. Sci. Nat., sixième sér., Bot.*, 13: 5–135

von Marilaun A. K., 1902 – *The natural history of plants*, vol. 1, 2, London Blackie & Son, Limited, 50 Old Bailey, E.G., Glasgow and Dublin

Volkens G., 1887 – *Die Flora der Aegyptisch-Arabischen Wüste auf Grundlage anatomisch-physiologischer Forschungen*, Berlin: Gebrüder, Borntraeger; 1887

Wagner A., 1892 – Zur Kenntniss des Blattbaues der Alpenpflanzen und dessen biologischer Bedeutung. *Sitz. der Kaiser. Akad. Der Wissenschaft. Matem-Naturwissenschaft. classe*, 101 (6): 487–548

Walter H., Steiner M., 1936 – Die ökologie der Ost-Afrikanischen mangroven. *Zeit für Bot.*, 30(3/4): 65–193

Wang X., Huang J., Peng S., Xiong D., 2023 – Leaf rolling precedes stomatal closure in rice (*Oryza sativa*) under drought conditions. *J. Exp. Bot.*, 74 (21): 6650–6661

Warming E., 1908 – The structure and biology of arctic flowering plants. *Ericineae (Ericaceae, Pirolaceae)*. 1. Morphology and biology. *Medd. Grønl.*, 36: 1–72

Warming E., 1909a – *Ecology of plants. An introduction to the study of plant communities* (English edition prepared by P. Groom, I. B. Balfour), Oxford: Clarendon Press

Warming E., 1909b – The structure and biology of arctic flowering plants. *Saxigragaceae*. 1. Morphology and biology. *Medd. Grønl.*, 36: 169–236

Warming E., 1918 – *Eug. Warming's Lehrbuch der ökologischen Pflanzengeographie*, Berlin Verlag von Gebrüder Borntraeger

Weberbauer A., 1905 – Anatomische und biologische Studien über die Vegetation der Hochanden Perus. *Bot. Jhb.*, 37: 60–94

Westermaier M., 1884 – Uber Bau und Funktion des pflanzlichen Hautgewebesystems. *Jahrb. für Wissenschaft. Bot.*, 14: 43–81

Wightman R., Wallis S., Aston P., 2017 – Hydathode pit development in the alpine plant *Saxifraga cochlearis*. *Flora*, 233: 99–108

Wright W. P., 1911 – *Alpine flowers and rock gardens*, London: Headley Brothers, Bishopsgate

Yang X., Wang J., Mao X., Li C., Li L., Xue Y., He L., Jing R., 2022 – A locus controlling leaf rolling degree in wheat under drought stress identified by Bulked Segregant Analysis. *Plants*, 11(16): 2076

MIX
Papier aus verantwortungsvollen Quellen
Paper from responsible sources
FSC® C105338

If you have any concerns about our products,
you can contact us on
ProductSafety@springernature.com

In case Publisher is established outside the EU,
the EU authorized representative is:
Springer Nature Customer Service Center GmbH
Europaplatz 3, 69115 Heidelberg, Germany

Printed by Libri Plureos GmbH
in Hamburg, Germany